T0215601

The Consultative Group on International Agricultural Research (CGIAR) is an association of agricultural research centres which together represent an important force in genetic conservation of crops and their wild relatives. Under the CGIAR umbrella, the centres are collectively custodians of international genetic resource collections for crops that provide 75% of the world's food energy. This volume considers the status of the key collections, in each case providing details of the botany, distribution and agronomy of the species concerned in addition to extensive information on germplasm conservation and use. The book presents a unique synthesis of knowledge drawn from the CGIAR centres, providing an invaluable source of reference for all those concerned with monitoring, maintaining and utilising the biodiversity of our staple crop species.

Biodiversity in Trust

Biodiversity in Trust

Conservation and Use of Plant Genetic Resources in CGIAR Centres

Edited by

DOMINIC FUCCILLO
International Agricultural Programs, University of Arkansas, Fayetteville, USA

LINDA SEARS
International Plant Genetic Resources Institute, Rome, Italy

PAUL STAPLETON
International Plant Genetic Resources Institute, Rome, Italy

CAMBRIDGE
UNIVERSITY PRESS

CAMBRIDGE UNIVERSITY PRESS
Cambridge, New York, Melbourne, Madrid, Cape Town, Singapore, São Paulo

Cambridge University Press
The Edinburgh Building, Cambridge CB2 8RU, UK

Published in the United States of America by Cambridge University Press, New York

www.cambridge.org
Information on this title: www.cambridge.org/9780521593656

First published 1997
This digitally printed version 2007

A catalogue record for this publication is available from the British Library

ISBN 978-0-521-59365-6 hardback
ISBN 978-0-521-59653-4 paperback

Contents

Preface

The Earth's natural resources are finite and vulnerable. Realization of this fact underlay the drafting of the Convention on Biological Diversity at the Earth Summit in Rio in 1992. The signature of the Convention by countries, both developed and developing, formalized their pledge to stem the rapid loss of biodiversity and sustain this vital resource for present and future generations.

Arguably, the most important component of biodiversity is the genetic diversity of plant species involved in food and agriculture – crops and forage species for livestock feed. This diversity created in farmers' fields over the millennia and by scientific research institutions over the last century is complemented by diversity present in wild relatives of the crop and forage species. Together, these genetic resources provide the raw material for further selection and improvement to meet the food security needs of the world's rapidly rising population.

Although technological advances in plant breeding and agricultural methods have led to dramatic increases in the amount and quality of food available today, it is estimated that more than 800 million people throughout the world do not have enough food to meet basic nutritional needs. The majority live in regions of the developing world where environmental and economic constraints impede their benefiting from technological advances.

To achieve global food security will require efforts on many fronts. The 16 Centres of the CGIAR[1] share a mission to make their contribution through research on sustainable agriculture in developing countries. They work in close collaboration with partners in national research systems, to develop resource-efficient technologies for sustainable improvement in the productivity of agriculture, forestry and fisheries. They also conduct research into agriculture-related policy and assist in capacity-building. The twin pillars of their research are productivity and natural resources management, underpinned by a commitment to the conservation and sustainable use of biodiversity.

Together, the Centres of the CGIAR represent the largest international effort on the conservation and sustainable use of crop, forage, livestock, agroforestry, forestry and aquatic genetic resources. The CGIAR's initial focus was on the genetic resources of major food crops and their conservation *ex situ* for use by breeders in the Centres and in national crop improvement programmes. This involvement broadened in the early

---

[1] The CGIAR is an informal association of 52 public and private sector members from countries worldwide and is cosponsored by the Food and Agriculture Organization of the United Nations, the United Nations Development Programme, the United Nations Environment Programme and the World Bank. It supports a network of 16 international agricultural research Centres located in 12 developing and 3 developed countries.

1990s to include livestock, aquatic and forestry species, and attention to *in situ* conservation. In 1994, recognising the benefits of greater collaboration, the CGIAR established the System-wide Genetic Resources Programme (SGRP) to draw together the activities and strengths of the individual Centres into a system-wide effort and to enhance the efficiency and effectiveness of the CGIAR's contribution to the implementation of the Convention on Biological Diversity.

To date, the CGIAR Centres, in collaboration with partner national agricultural research systems, have made their major contribution to the well-being of the poor and commitment to safeguarding the earth's biodiversity through the *ex situ* conservation and use of plant genetic resources. Today, more than half a million samples of crop and forage species are maintained in the CGIAR's 11 genebanks. They include landraces, nondomesticated species, advanced and old cultivars and breeding lines. The collections represent both insurance against genetic erosion and a source of tolerances to diseases and pests, climatic and other environmental stresses, improved quality and yield traits for crop improvement.

The collections held by CGIAR Centres have been assembled over the past two decades through donations from other genebanks, through collaborative collecting expeditions with national programmes, from breeders around the world and from within the Centres themselves. In recognition of the multiple origins of the materials and the fact that they were donated for the benefit of international agricultural research, the Centres signed agreements in 1994 to place the collections in trust for the world community under the intergovernmental authority of the FAO. The terms of these agreements stipulate that the materials will be maintained by the CGIAR to international technical standards, and will continue to be made available to all, with the understanding that no intellectual property protection is to be applied to the material.

In the interests of transparency with regard to the origins and location of the genetic resources in the trust collections, and recognising the importance of information as a key factor in facilitating access to and use of genetic resources, the SGRP has been proactive in providing information on the CGIAR collections. The System-wide Information Network for Genetic Resources (SINGER) and International Crop Information System (ICIS) are being developed through system-wide and inter-Centre efforts with this aim in mind.

This book is a further means of providing additional, complementary information on the collections. The chapters describe the efforts of the CGIAR Centres and their national programme partners to conserve and improve crops and forage species that are critical to the food security of the world's most disadvantaged people. They cover staples such as rice, wheat, maize and potatoes that are common the world over. Rice and wheat alone are estimated to provide 50% of the world's food energy supply. Others may be less commonly known, such as cassava, yam, sorghum and millet, but are critical to the nutrition of many millions of people who subsist in the harsher environments of the developing world.

The original idea to produce a CGIAR publication on the global status of the conservation and use of CGIAR mandate crops was proposed by Drs Masa Iwanaga and Jan Engels of IPGRI. This idea was endorsed by the CGIAR Inter-Centre Working Group on Genetic Resources. The efforts of the many CGIAR scientists who have subsequently contributed to bringing this project to fruition are gratefully acknowledged.

Contributors

Walter Amoros	CIP, Apartado 5969, Lima, Peru
Carlos Arbizu	CIP, Apartado 5969, Lima, Peru
Elisabeth Arnaud	INIBAP, Parc Scientifique Agropolis II, 34980 Montferrier-sur-Lez, France
Robert Asiedu	IITA, PMB 5320, Oyo Road, Ibadan, Oyo State, Nigeria
Marc Barré	Cooperative Tripsacum Project, ORSTOM and CIMMYT, 06600 Mexico, DF, Mexico
Steve Beebe	CIAT, Apartado Aereo 6713, Cali, Colombia
Julien Berthaud	Cooperative Tripsacum Project, ORSTOM and CIMMYT, 06600 Mexico, DF, Mexico
Merideth Bonierbale	CIAT, Apartado Aereo 6713, Cali, Colombia
Salvatore Ceccarelli	ICARDA, PO Box 5466, Aleppo, Syria
Jonathan H. Crouch	IITA, PMB 5320, Oyo Road, Ibadan, Oyo State, Nigeria
Kenton Dashiell	IITA, PMB 5320, Oyo Road, Ibadan, Oyo State, Nigeria
J.M.J. de Wet	45, Tidewater Farm Road, Greenland, NH 03840, USA
A.G.O. Dixon	CIAT, Apartado Aereo 6713, Cali, Colombia
H. Jesse Dubin	CIMMYT, Lisboa 27, Apdo. Postal 6-641, 06600 Mexico, D.F., Mexico
Ali M. Abd El Moneim	ICARDA, PO Box 5466, Aleppo, Syria
William Erskine	ICARDA, PO Box 5466, Aleppo, Syria
Christian Fatokun	IITA, PMB 5320, Oyo Road, Ibadan, Oyo State, Nigeria
R. Shaun B. Ferris	IITA, PMB 5320, Oyo Road, Ibadan, Oyo State, Nigeria
R.A. Fischer	GPO 1571, Canberra ACT 2601, Australia
Ali Golmirzaie	CIP, Apartado 5969, Lima, Peru
Claudia Guevara	CIAT, Apartado Aereo 6713, Cali, Colombia
Elcio P. Guimaraes	CIAT, Apartado Aereo 6713, Cali, Colombia. Present address: EMBRAPA/CNPAF, CP 179, Goiânia, Goiás 74001-970, Brazil
Jean Hanson	ILRI, PO Box 5689, Addis Ababa, Ethiopia
Rigoberto Hidalgo	CIAT, Apartado Aereo 6713, Cali, Colombia
Jean-Pierre Horry	INIBAP, Parc Scientifique Agropolis II, 34980 Montferrier-sur-Lez, France
Zosimo Huamán	CIP, Apartado 5969, Lima, Peru
Michael T. Jackson	IRRI, PO Box 933, 1099 Manila, Philippines
Chris Johansen	ICRISAT, Patancheru, A.P. 502 324, India
David R. Jones	INIBAP, Parc Scientifique Agropolis II, 34980 Montferrier-sur-Lez, France

Monty Jones	WARDA, 01 BP 2551, Bouaké 01, Côte d'Ivoire
T.J. Kelley	ICRISAT, Patancheru, A.P. 502 324, India
Peter C. Kerridge	CIAT, Apartado Aereo 6713, Cali, Colombia
Jan Konopka	ICARDA, PO Box 5466, Aleppo, Syria
Olivier Leblanc	Cooperative Tripsacum Project, ORSTOM and CIMMYT, 06600 Mexico, DF, Mexico
Genoveva C. Loresto	IRRI, PO Box 933, 1099 Manila, Philippines
Brigitte Maas	CIAT, Apartado Aereo 6713, Cali, Colombia
Nicolas Mateo	INIBAP, Parc Scientifique Agropolis II, 34980 Montferrier-sur-Lez, France
A. Mujeeb-Kazi	CIMMYT, Lisboa 27, Apdo. Postal 6-641, 06600 Mexico, D.F., Mexico
N. Quat Ng	IITA, PMB 5320, Oyo Road, Ibadan, Oyo State, Nigeria
Shou Yong Choy Ng	IITA, PMB 5320, Oyo Road, Ibadan, Oyo State, Nigeria
S.N. Nigam	ICRISAT, Patancheru, A.P. 502 324, India
Rodomiro Ortiz	IITA, PMB 5320, Oyo Road, Ibadan, Oyo State, Nigeria
R. Jorge Peña	CIMMYT, Lisboa 27, Apdo. Postal 6-641, 06600 Mexico, D.F., Mexico
Claudine Picq	INIBAP, Parc Scientifique Agropolis II, 34980 Montferrier-sur-Lez, France
P. Parthasarathy Rao	ICRISAT, Patancheru, A.P. 502 324, India
K.E. Prasada Rao	ICRISAT Asia Center, Patancheru 502 324, Andhra Pradesh, India
R.P.S. Pundir	ICRISAT, Patancheru, A.P. 502 324, India
K.N. Rai	ICRISAT, Patancheru, A.P. 502 324, India
Seepana Appa Rao	Lao-IRRI Project, PO Box 4195, Vientiane, Laos
K.N. Reddy	ICRISAT, Patancheru, A.P. 502 324, India
V. Gopal Reddy	ICRISAT, Patancheru, A.P. 502 324, India
P. Remanandan	ICRISAT, Patancheru, A.P. 502 324, India
Larry D. Robertson	ICARDA, PO Box 5466, Aleppo, Syria
Yves Savidan	Cooperative Tripsacum Project, ORSTOM and CIMMYT, 06600 Mexico, DF, Mexico
N.P. Saxena	ICRISAT, Patancheru, A.P. 502 324, India
Ken D. Sayre	CIMMYT, Lisboa 27, Apdo. Postal 6-641, 06600 Mexico, D.F., Mexico
A.K. Singh	ICRISAT, Patancheru, A.P. 502 324, India
B.B. Singh	IITA, PMB 5320, Oyo Road, Ibadan, Oyo State, Nigeria
K.B. Singh	ICRISAT-ICARDA Chickpea Project, ICARDA, PO Box 5466, Aleppo, Syria
Laxman Singh	ICRISAT, Patancheru, A.P. 502 324, India
U. Singh	ICRISAT, Patancheru, A.P. 502 324, India
Bent Skovmand	CIMMYT, Lisboa 27, Apdo. Postal 6-641, 06600 Mexico, D.F., Mexico
John W. Stenhouse	ICRISAT, Patancheru, A.P. 502 324, India
Suketoshi Taba	CIMMYT, Lisboa 27, Apdo. Postal 6-641, 06600 Mexico, D.F., Mexico
Jan Valkoun	ICARDA, PO Box 5466, Aleppo, Syria
H.A. van Rheenen	ICRISAT, Patancheru, A.P. 502 324, India
Dirk Vuylsteke	IITA, PMB 5320, Oyo Road, Ibadan, Oyo State, Nigeria
Nelson M. Wanyera	IITA, PMB 5320, Oyo Road, Ibadan, Oyo State, Nigeria
D.P. Zhang	CIP, Apartado 5969, Lima, Peru

Acronyms

AFRNET	African Feed Resources Network, coordinated by ILRI
ALAD	Arid Lands Agricultural Development Program, Beirut, Lebanon (precursor institution of ICARDA)
ARI	Agricultural Research Institute, Yezin, Myanman
ATFGRC	Australian Tropical Forages Genetic Resources Centre (of CSIRO), Brisbane, Australia
AVRDC	Asian Vegetable Research and Development Center, Taiwan
BPI	Bureau of Plant Industry, Philippines
CARI/PGRC	Central Agricultural Research Institute. Gannoruwa, Peradeniya, Sri Lanka/Plant Genetic Resources Center. Peradeniya, Sri Lanka
CATIE	(Spanish acronym) Centro Agronómico Tropical de Investigación y Enseñanza. Turrialba, Costa Rica
CENARGEN	(Portuguese acronym) Centro Nacional de Recursos Genéticos e Biotecnologia (of EMBRAPA), Brasília, DF, Brazil
CGIAR	Consultative Group on Agricultural Research, Washington, DC, USA
CIAT	(Spanish acronym) Centro Internacional de Agricultura Tropical. Cali, Colombia
CIMMYT	Centro Internacional de Mejoramiento de Maíz y Trigo, Mexico
CIP	Centro Internacional de la Papa, Peru
CIRAD	Centre de Coopération Internationale en Recherche Agronomique pour le Développement, France
CNPAF	Centro Nacional de Pesquisa de Arroz e Feijao
CNPGC	Centro Nacional de Pesquisa de Gado de Corte (of EMBRAPA), Campo Grande, MS, Brazil
CNPGL	Centro Nacional de Pesquisa de Gado de Leite (of EMBRAPA), Juiz de Fora, MG, Brazil
CNPMF	(Portuguese acronym) Centro Nacional de Pesquisa de Mandioca e Fruticultura Tropical
CRBP	Centre Régional Bananiers et Plantains, Cameroon
CRIFC/MARIF	Central Research Institute for Food Crops, Bogor, Indonesia/Malang Research Institute for Food Crops, Malang, Indonesia
CSIRO	Commonwealth Scientific and Industrial Research Organisation, Australia
CTCRI	Central Tuber Crops Research Institute. Trivadrum, Kerala, India
DSIR	Department of Scientific and Industrial Research, Palmerston North, New Zealand
EMBRAPA	(Portuguese acronym) Empresa Brasileira de Pesquisa Agropecuária
EMVT	Département d'élevage et de Médicine Vétérinaire of CIRAD (formerly, IEMVT)

EPAMIG	Empresa de Pesquisa Agropecuária de Minas Gerais, Brazil
ESARRN	Eastern and Southern Africa Root Crops Research Network
FAO	Food and Agriculture Organization of the United Nations, Rome, Italy
FHIA	Fundación Hondureña de Investigación Agrícola, Honduras
FLHOR	Department of Fruits and Horticultural Products, CIRAD, France
FONAIAP	Fondo Nacional de Investigaciones Agropecuarias, Venezuela
GTZ	Deutsche Gesellschaft Für Technische Zusammenarbeit, Germany
IAEA	International Atomic Energy Agency, Austria
IAN	(Spanish acronym) Instituto Agronómico Nacional, Cacupé, Paraguay
IAR	Institute of Agriculture Research, Treetown, Sierra Leone
IARC	International Agricultural Research Centre
IBP	Instituto biotechnología de las Plantas, Cuba
IBPGR	International Board for Plant Genetic Resources, Rome, Italy (now IPGRI)
IBTA-PROINPA	Instituto Boliviano de Tecnología Agropecuaria - Programa de Investigacion en Papa, Bolivia
ICARDA	International Center for Agricultural Research in the Dry Areas, Aleppo, Syria
ICRAF	International Centre for Research in Agroforestry, Nairobi, Kenya
ICRISAT	International Crops Research Institute for the Semi-Arid Tropics, Patancheru, India
IDEA	International Institute of Advances Studies, Caracas, Venezuela
IDESSA	Institut des Savanes, Bouaké, Côte d'Ivoire
IDRC	International Development Research Centre, Canada
IGER	Institute of Grassland and Environmental Research, Aberystwyth, UK (formerly, Welsh Plant Breeding Station)
IIA	(Spanish acronym) Instituto de Investigación Agrícola - El Vallecito-Santa Cruz, Bolivia
IIA	(Spanish acronym) Instituto de Investigaciones Agropecuarias, Panamá
IICA	Interamerican Institute for Cooperation in Agriculture
IITA	International Institute of Tropical Agriculture, Ibadan, Nigeria
ILCA	International Livestock Centre for Africa, Addis Ababa, Ethiopia (now ILRI)
ILRAD	International Laboratory for Research on Animal Diseases (now ILRI, Kenya)
ILRI	International Livestock Research Institute (created from merger of ILCA and ILRAD, Ethiopia and Kenya)
INIA	Instituto Nacional de Investigacao Agronomica, Maputo, Mozambique
INIA	Instituto Nacional de Investigación Agraria, Peru
INIAP	(Spanish acronym) Instituto Nacional de Investigación Agropecuarias, Quito, Ecuador
INIBAP	International Network for the Improvement of Banana and Plantain, France
INIFAP	(Spanish acronym) Instituto Nacional de Investigaciones Forestales y Agropecuarias. Veracruz, México
INIVIT	(Spanish acronym) Instituto Nacional de Investigación de Viandas Tropicales. Santo Domingo, Villa Clara, Cuba
INRA	Institut National de Recherche Agronomique, Rabat, Morocco
INTA	(Spanish acronym) Instituto Nacional de Tecnología Agropecuaria. Buenos Aires, Argentina
IPGRI	International Plant Genetic Resources Institute, Rome, Italy (formerly, IBPGR)
IPK	Institut für Planzengenetik und Kulturpflanzenforschung,

	Gatersleben, Germany
IRAT	Institut de Recherches Agronomiques Tropicales et de Cultures Vivrières, Nogent-sur-Marne, France
IRI	Instituto de Pesquisas, Brazil
IRRI	International Rice Research Institute, Philippines
ISRA/CDH	Institut Senegalais de Recherches Agricole/Centro pour le Development de l'horticulture, Dakar, Senegal
KUL	Katholieke Universiteit Leuven, Belgium
MARDI	Malaysian Agricultural Research and Development Institute, Serdang, Selangor, Malaysia
NARI	National Agricultural Research Institute
NARS	National Agricultural Research System
NBPGR	National Bureau of Plant Genetic Resources, New Delhi
NFTA	formerly Nitrogen Fixing Tree Association, Hawaii, USA; now Forest, Farm, and Community Tree Network (FACT Net)
NRCRI	National Root Crops Research Institute. Umudike, Umuahia, Nigeria
OFI	Oxford Forestry Institute, Oxford, UK
ORSTOM	formerly Office de la Recherche Scientifique et Technique d'Outre-mer; now Institut Français de Recherche Scientifique pour le Développement en Coopération, France
PGRU/CRI	Plant Genetic Resources Centre/Crops Research Institute, Bunso, Ghana
PRCRTC/IPB	Philippine Root Crop Research and Training Center. Baybay, Leyte, Philippines/Institute of Plant Breeding, Los Baños, Laguna, Philippines
QDPI	Queensland Department of Primary Industries, Australia
RABAOC	Réseau de Recherche en Alimentation du Bétail en Afrique Occidentale et Centrale (West and Central African Feed Research Network), coordinated by CIAT, EMVT, and AFRNET
RFCRC	Rayong Crops Research Centre, Huay Pong, Rayong, Vietnam
RIEPT	Red Internacional de Evaluación de Pastos Tropicales (International Tropical Pastures Evaluation Network), coordinated by CIAT
SADAF	South Australian Department of Agriculture and Fisheries, Adelaide, Australia
SADC	Southern African Development Community
SCATC/UCRI /GAAS	South China Academy of Tropical Crops, Baudap Xomcin, Hainan China/Upland Crops Research Institute/Guang Dong Academy of Agricultural Science, Wushan, Guangzhou
SEAFRAD	Southeast Asian Forage and Feed Research and Development Network, coordinated by CIAT and CSIRO
SINGER	System-wide Information Network on Genetic Resources
SPII	Seed and Plant Improvement Institute, Karaj, Iran
SRCV	Statique de Recherge sur les cultures vivieres, Niaouli, Attogr, Benin
TBRI	Taiwan Banana Research Institute, Taiwan
TROPIGEN	Network for Amazonan Genetic Resources
UNA	(Spanish acronym) Universidad Nacional Agraria, Facultad de Agronomía, Managua, Nicaragua
UNC	Universidad Nacional de Cajamarca, Cajamarca, Peru
UNSAAC	Universidad Nacional de San Antonio Abad del Cusco, Cusco, Peru
UNSCH	Universidad Nacional de San Cristobal de Huamanga, Ayacucho, Peru
USAID	United States Agency for International Development
USDA	United States Department of Agriculture, USA
WADA	West Australian Department of Agriculture, Perth, Australia
WANA	West Asia and North Africa
WANANET	West Asia and North Africa Network for Genetic Resources
WARDA	West Africa Rice Development Association, Côte d'Ivoire

Chapter 1

Cassava

M. Bonierbale, C. Guevara, A.G.O. Dixon, N.Q. Ng, R. Asiedu and S.Y.C. Ng

Cassava (*Manihot esculenta* Crantz) is a major food source for more than 500 million people in Africa, Latin America and Asia. Cultivated for its starchy roots, this New World native species claims a world production of 152 million t (FAO 1995), ranking seventh in yield production among all crops and fourth as an important source of calories in the tropics (Cock 1985; Balogapalan *et al.* 1988). The storage roots form the basic carbohydrate component of the diet and the leaves are consumed as a preferred green vegetable in many parts of Africa, providing protein, mineral and vitamins (Hahn 1989). Two CGIAR Centres, CIAT and IITA, share the mandate for cassava. Genetic diversity of the *Manihot* genus is eroding in the face of expansion of agriculture in the American tropical lowlands, while in important cassava-growing regions of Africa, hundreds of traditional varieties have been abandoned during the 20th century (Nweke and Polson 1990; Nweke *et al.* 1994), and newer production regions in Asia tend to rely on single-cultivar plantations.

BOTANY AND DISTRIBUTION

Cassava is a member of the Euphorbiaceae, subfamily Crotonoideae and tribe Manihotae. The genus *Manihot* contains nearly 100 species of herbs, shrubs and trees among which the production of latex and cyanogenic glucosides is common (Rogers and Fleming 1973; Bailey 1976) and these are grouped into 19 taxonomic sections (Rogers and Appan 1973). Previous species names – *M. utilissima*, *M. dulcis*, *M. aipi* and *M. palmata* – probably have been attributed to cassava because of existing intraspecific morphological diversity and the variable content of cyanogenic compounds in the roots (Rogers 1965). Among Euphorbiaceae, *Manihot* and *Hevea* (the rubber tree) have the same somatic chromosome number ($2n=36$), but the family is highly variable in karyotype, with basic numbers of $x=6, 7, 8, 9, 10$ and 11 (Perry 1943). Despite evidence that the basic chromosome number of *Manihot* is 9 and that the genome may have a polyploid constitution (Magoon *et al.* 1969; Umanah and Hartman 1973), there have been no reports of somatic counts of 18 chromosomes in the genus (Bai *et al.* 1993). It is expected that current research toward the development of a molecular genetic map will provide a better definition of the structure of the cassava genome (Gomez *et al.* 1995; Fregene *et al.* 1996).

Common names for the crop include *yuca* (Spanish), *manioc* (French) and *mandioca* (Portuguese). Native American names reflect the importance and antiquity of cassava in the Neotropics (Table 1.1). *Maniyua/maniva* (bitter) and *aipi/aipim* (sweet) refer to the different tastes of cassava. Cassava production is limited to 30°N and 30°S. The total area harvested is about 16 million hectares with 60, 24 and 16% in Africa, Asia and Latin America, respectively. Cassava production statistics for 1994 are presented in Table 1.2.

Table 1.1. Some native names for cassava in the Americas.

Native name	Dialect or tribe	Region or estate/country
Yuca	Taíno	Antillas/Haiti
Entaha	Malibú	Bajo Magdalena/Colombia
Enbutac	Tamalameque	Bajo Magdalena/Colombia
Aro	Muzo	Eastern/Colombia
Rumu	Quechua	Peru
Mandiva	Tupinamba	East Coast/Brazil
Mandioc, typiaca	Tupi, Guarani	Southern Brazil/Brazil
Maniyua, maniva	Nheêgatú	North Coast Brazil/Brazil
Aipi, aipim, macaxera	Nheêgatú	North Coast Brazil/Brazil
Mandió	Guaraní	Paraguay
Tsanimba, mama	Maynas	Marañón estuary/Peru
Yujumka	Aguaruna,Huambisa	Andean amazonian/Peru
Yucuta	Antillano	Antillas/
Kasabi	Arawak	Haiti, Guyanas
Guahcamote	Sinaloa	Mexico
Camvirí	Puinave	Orinoco estuary/Colombia
Pebeyanajaba	Guahibo	Orinoco estuary/Colombia
Tubirike	Curripaco	Guaviare estuary/Colombia
Aratöinesajue	Piaroa	Orinoco estuary/Colombia
Yejédëkë	Tucano	Inírida river/Colombia
Pejek	Ayoreo	Paraguayan Chaco

Sources: Patiño 1964; Barnes 1975; Boster 1984; Toro *et al.* 1985; Mejia 1991; Schmeda-Hirschmann 1994.

Origin, Domestication and Diffusion

Cassava has been studied since as early as 1886, when Alphonse de Candolle placed its geographic origin in the lowland tropical Americas (Smith 1968). It shares the Brazilian-Paraguayan centre of origin with peanuts, cacao, rubber and other crops (Vavilov 1992). Rogers (1963) identified two geographic centres of speciation: (1) the drier areas of western and southern Mexico and portions of Guatemala, and (2) the dry northeastern portions of Brazil. Nassar (1978a, 1978b) identified four areas of diversity of the wild species: (1) central Brazil, (2) northeastern Brazil, (3) southwestern Mexico, and (4) western Mato Grosso (Brazil) and Bolivia.

The crop may have been cultivated in Colombia and Venezuela from 3000 to 7000 years ago (Reichel-Domatoff 1957, 1965; Rouse and Cruxent 1963, cited in Hershey 1987). Ugent and coworkers (1986) cite evidence for domestication on the Peruvian coast earlier than 4000 BC. According to Vavilov (1939) the areas of distribution of cassava were more accessible to primitive peoples than the humid tropical forests. Sauer (1952 quoted by Smith 1968) proposed the heart of domestication as northwestern South America. The Mesoamerican region extending from the northwestern coast of Mexico and covering parts of Guatemala, El Salvador and Nicaragua is also a potential area for early domestication (Rogers 1965), where the wild species *M. aesculifolia*, *M. pringlei* and *M. isoloba* may have contributed to the cultigen through extensive hybridization, a theory that has been questioned (Brucher 1989). Rogers and Fleming (1973) concluded that the cultigen is a complex species with multiple sites of initial cultivation. Allem (1987, 1994) proposed that *M. esculenta* is not a cultigen, but derived from two primitive forms, and described three subspecies, the domesticated *M. esculenta* subsp. *esculenta* (which includes all known cultivars), and the two wild forms, *M. esculenta* subsp. *peruviana* and *M. esculenta* subsp. *flabellifolia*. Wild relatives of cassava are distributed in a wide range of ecologies from Arizona (in the USA) to Argentina (Rogers and Appan 1973).

Early selection and distribution of sweet or low-cyanide cassava roots may be correlated with cultural preferences (Dufour 1994). Cultivars with high cyanogenic

Table 1.2. Major cassava-producing countries, 1994.

Location	Area ('000 ha)	Yield (t/ha)	Production (million t)
World	**15819**	**9.6**	**152.5**
Africa	**9481**	**7.7**	**72.8**
Zaire	2430	8.1	19.6
Nigeria	2000	10.5	21.0
Mozambique	908	3.6	3.3
Tanzania	693	10.4	7.2
Ghana	607	7.2	4.4
Angola	400	2.5	1.0
Uganda	380	9.0	3.4
Madagascar	336	6.7	2.3
Asia	**3745**	**12.9**	**48.5**
Thailand	1383	13.8	19.1
Indonesia	1295	11.6	15.0
India	235	22.8	5.3
Vietnam	285	9.2	2.6
China	231	15.2	3.5
Philippines	212	8.9	1.9
South America	**2375**	**12.7**	**30.1**
Brazil	1838	13.1	24.0
Paraguay	174	14.6	2.6
Colombia	196	10.2	2.0
Central America and Caribbean	**200**	**4.9**	**1.0**
Cuba	70	4.1	0.3
Haiti	85	4.0	0.3
Dominican Republic	21	5.8	0.1

Source: FAO 1995.

potential are preferred by some cultures, apparently for the quality of certain processed cassava products, and especially where predators are a problem (Boster 1984; Mejia 1991). Renvoize (1972) refers to climatic factors as influencing the distribution of the two types, and although both types are now grown in Mexico, the earliest accounts indicate that only sweet cassava was cultivated there. The Orinoco estuary (Colombia) may have been an area of concentration of bitter varieties and the Peruvian Andean Amazonian of sweet varieties (Mejia 1988).

Vavilov (1939) suggested early cultivation of cassava in equinoctial America, as recounted in chronicles documented by Patiño (1964). Cassava was carried by the Arawak tribes of Central Brazil to the Caribbean Islands and Central America in the 11th century (Brucher 1989), by the Portuguese to the west coast of Africa, via the Gulf of Benin and the Congo River at the end of the 16th century (Jones 1959), and to the east coast via the islands of Reunion, Madagascar and Zanzibar at the end of the 18th century (Barnes 1975; Jennings 1976). The crop arrived in India about 1800. The Spaniards took it into the Pacific, but it was not widely used as a food crop there until the 1960s (Jennings 1976). Gulick *et al.* (1983) have defined primary, secondary and tertiary levels of diversity for *Manihot esculenta* in modern times. Important secondary diversity lies in Africa, outside the crop's centre of origin.

Reproductive Biology

Cassava is monoecious and predominantly outcrossing mediated by protogyny, which leads to a high degree of heterozygosity in plants and among populations produced from botanical seed (Byrne 1984; Hershey and Jennings 1992). Mature seed

Table 1.3. Useful characters[†] reported in *Manihot* species.

Manihot species	Swollen roots	HCN[‡]	Comments	Reference
aesculifolia	+	H	Close relative of *M. esculenta*; high range adaptation	Rogers and Appan 1973; Nassar and Cardenas 1986
alutacea	–	H	Cold tolerance; resistance to soil toxicity, mite resistance	Rogers and Appan 1973; Nassar 1986
angustiloba	+	H	Drought tolerance; high starch content	Rogers and Appan 1973; Nassar and Cardenas 1986; Chavez 1990
anisophylla	NA	NA	Cold tolerance	Rogers and Appan 1973
anomala	+	L-H	High photosynthetic rate; adapted to humid conditions; variable genepool	Rogers and Appan 1973; Byrne 1984; Nassar 1986
attenuata	NA	NA	Cold tolerance	Rogers and Appan 1973; Chavez 1990
brachyandra	+	NA	Grown in NE Brazil with considerable tuber production	Nassar and Fitchner 1978
brachyloba	–	NA	Resistant to mealybug	Byrne 1984
caerulescens	+	M	Drought tolerance; rubber producer	Rogers and Appan 1973; Nassar 1986
carthaginensis	+	NA	Drought tolerance; salt tolerance; xerophytic zones; high protein content	Lopez and Herrera 1970; Rogers and Appan 1973; Chavez 1990
catingae	–	–	Several virus resistance	Brucher 1989
chlorosticta	NA	NA	High salinity resistance	Nassar and Cardenas 1986; Chavez 1990
davisiae	+	H	Drought tolerance; salinity tolerance	Rogers and Appan 1973; Nassar and Cardenas 1986
dichotoma	–	NA	Resistant to mealybug; high photosynthetic rate; CMD and BSV resistance; Paraiba 10 cultivar from cross with *M. esculenta*; resistance to drought and several viruses	Jennings 1976; Brucher 1989; Chavez 1990
epruinosa	+	NA	Grown with considerable tuber production; adaptation to limestone soil	Rogers and Appan 1973; Nassar and Fitchner 1978; Nassar 1986
esculenta subsp. *flabellifolia*	–	–	High protein content	CIAT 1993
falcata	–	M	Potential for breeding small-stature plants	Byrne 1984
filamentosa	–	–	Potential as forage, low levels of deterioration	CIAT 1993
foetida	–	NA	High photosynthetic rate	Rogers and Appan 1973; Byrne 1984
glaziovii	–	L	Potential rubber crop, resistance to CMD, BSD, CBB and mealy bug; drought tolerance; used in several breeding programmes	Rogers and Appan 1973; Jennings 1976; Hahn *et al.* 1980; Brucher 1989
gracilis	–	M	High root protein; reduced growth	Rogers and Appan 1973; Nassar 1978c; Brucher 1989
grahami	–	H	Resistant to *Coelosternus manihoti*; cold tolerance	Rogers and Appan 1973; Chavez 1990
guaranitica	+	H	Adaptation to sandy soils	Brucher 1989
irwinii	–	–	Adaptation to acid, low-P soil	Chavez 1990
longepetiolata	–	M	Dwarf cultivars	Byrne 1984
melanobasis	–	–	Virus resistant	Jennings 1959
nana	–	L-H	Dwarf cultivars	Nassar 1978c

Manihot species	Swollen roots	HCN[‡]	Comments	Reference
neusana	−	−	Resistance to Coelosternus manihoti	Chavez 1990
oligantha subsp. nesteli	+	−	High protein content	Nassar 1978c, 1986; Brucher 1989
orbicularis	−	−	Adaptation to acid, low-P soils	Chavez 1990
paviaefolia	+	M	Adaptation to poor soils	Byrne 1984
peltata	NA	H	Resistant to soil toxicity	Byrne 1984
pentaphylla	+	M	Adaptation to limestone soils; highly variable species	Nassar and Fitchner 1978
pohlii	NA	NA	Resistant to Coelosternus manihoti	Rogers and Appan 1973; Chavez 1990
pringlei	+	L	Resistant to CMD	Rogers and Appan 1973; Nassar and Cardenas 1986; Chavez 1990
procumbens	−	H	Tolerant to poor soil, acid and high Al content	Nassar and Fitchner 1978; Nassar 1986
pruinosa	+	L	Adaptation to limestone soils	Nassar and Fitchner 1978
pseudoglaziovii	NA	NA	Resistant to CBB and drought tolerant	Chavez 1990
pusilla	NA	M	Dwarf habit	Nassar and Fitchner 1978
quinquepartita	+	NA	Mealybug resistance	Mahon et al. 1977; Rogers and Appan 1973
reptans	−	H	Coelosternus manihoti and CBB resistance; high soil range adaptation	Chavez 1990
rhomboidea	+	H	−	Rogers and Appan 1973
rubricaulis	NA	NA	Drought and cold tolerance; adaptation to high altitude	Rogers and Appan 1973; Nassar 1986; Nassar and Cardenas 1986
subspicata	−	−	Drought resistance	Nassar and Cardenas 1986
stipularis	−	M	Resistant to soil toxicity and cold; dwarf horticultural use; adaptation to high altitude	Rogers and Appan 1973; Nassar 1986
tomentosa	−	H	Drought tolerance	Nassar 1978c
tripartita	+	M	Adaptation to acid, low-P soils; high protein content; adapted to long dry season	Nassar 1978c, 1986; Brucher 1989
tristis subsp. saxicola	+	H	Adaptation for soils of granitic origin, high albumin content	Bolhuis 1953; Rogers and Appan 1973; Brucher 1989
walkerae	+	H	−	Rogers and Appan 1973
zehntneri	+	L	−	Rogers and Appan 1973; Nassar 1978c; Nassar and Cardenas 1986

[†] L=Low, M=Medium; H=High; NA=Not available; +=Present; − =Absent.

[‡] HCN: refers to reports of cyanogenic potential of storage roots.

lustrous testa and a caruncle with the embryo and cotyledons imbibed in a large endosperm element (Rogers and Appan 1973; Leon 1987). Cassava varieties are heterozygous individuals, which are propagated vegetatively to reproduce the genotype (Kawano 1980). The cultivated germplasm has erratic flowering habits and apical dominance, generally producing a single woody stem with 2-3 levels of lateral branching. Although some branching is useful for increasing the leaf area index, too much branching conflicts with the need for uniform vegetative stakes for propagation and may lead to a lower root yield as a consequence of enhanced interplant competition (Cook *et al.* 1979). The application of a group of 53 morphological descriptors proposed by IPGRI (Gulick *et al.* 1983) has resulted in a non-anatomical model for characterization of cassava genotypes.

GERMPLASM CONSERVATION AND USE

In the late 1960s and early 1970s, two CGIAR Centres (CIAT and IITA) and NARs started with the continuous assembly, characterization and integration into breeding programmes of components of *ex situ* collections of cassava germplasm. The genetic resources of cassava consist of local or introduced landraces, improved cultivars and related wild species (Gulick *et al.* 1983; Hershey 1987). Exchange of landrace germplasm from *ex situ* collections seeks to match characteristics to the ecological, production and use requirements of the target area utilizing genotypes that are predicted to perform well (Hershey 1984). Genes from closely related wild species are also accessible through hybridization (Asiedu *et al.* 1994). Desirable characteristics of *Manihot* species are compiled in Table 1.3.

Conservation Methods

Most countries that have active selection or breeding programmes and maintain local germplasm collections were catalogued recently (IPGRI 1994). Table 1.4 records the number of accessions and conservation methods used in these programmes. In 1979 CIAT established a tissue culture laboratory in its Genetic Resources Unit to handle the *in vitro* conservation of cassava germplasm and facilitate international exchange (CIAT 1991). Procedures for meristem culture and virus indexing were established at IITA in 1980, leading to an agreement with Nigeria's Plant Quarantine Authority and Inter Africa Phytosanitary Council to permit movement of *in vitro*, virus-tested cassava germplasm in Africa (IITA 1992). Recommendations are available for conservation methods and characterization procedures through a network of collaborators with Secretariat at CIAT (IPGRI 1994). Clones are maintained at CIAT and IITA under the following conditions: constant temperature of 23-25°C; 12-hour day/night photoperiod provided with 1000-1500 lux illumination; slightly modified basal Murashige Skoog (MS) medium; 25 x 150 mm test tubes capped with aluminium foil and firmly sealed with saran wrap (Roca *et al.* 1989; CIAT 1991; Ng 1991). Five tubes are maintained per clone.

The field genebank at CIAT holds 4695 cassava clones from 23 countries of Latin America and Asia, and 334 representatives of 26 *Manihot* species (Table 1.5). By June 1995, the CIAT *in vitro* bank held a total of 5632 cassava clones and 29 *Manihot* species. About 87% of the cassava clones held in the CIAT base collection are landraces and the remainder improved varieties. Specific gaps in representation of Latin American germplasm are recognized for Mesoamerica in Honduras and Nicaragua; for the Caribbean region, in the Dominican Republic and Haiti; and for South America, the Guyanas, the highland ecozones of Ecuador, Peru and Bolivia, and the acid savanna Chaco region of Paraguay, Bolivia and Argentina. A core collection of 630 accessions from 23 countries has been assembled to represent the genetic diversity of the base collection held at CIAT in a manageable size (Iglesias *et al.* 1993; Hershey *et al.* 1994). Evaluation of the core collection facilitates assessment of available variability for characters that are costly to measure and helps to orient the evaluation of the base collection.

Table 1.4. Cassava germplasm in some national and international centres[†].

Region/ Country	Natl. collection		Institute/programme responsible	Observations (estimated composition)
	Field	*In vitro*		
South America				
Colombia	4695	5632	CIAT[‡]	Incl. 2001 (Colombia), 22 (other ctry)
Brazil	4132	988	CNPMF/CENARGEN	Held at 12 institutions
Paraguay	360	120	IAN	–
Ecuador	101	–	INIAP	–
Argentina	177	120	INTA	–
Bolivia	18	–	IIA	–
Central America				
Costa Rica	154	–	CATIE	71 landraces
Mexico	225	–	INIFAP	105 landraces
Panama	50	–	IIA	44 landraces
Nicaragua	37	–	UNA	16 landraces
Caribbean				
Cuba	495	–	INIVIT	385 landraces
Dominican Rep.	46	–	–	30 landraces
Southern Africa				
Tanzania	410	–	RTCP Zanzibar/Tanz.	215 landraces
Malawi	520	–	RTCP Lilongwe	200 landraces
Uganda	413	–	RTCP Kampala	200 landraces
Kenya	250	–	RTCP Katumani	213 landraces
Mozambique	92	–	INIA	19 landraces
West and Central Africa				
Benin	228	–	SRCV	113 landraces
Cameroon	203	–	–	–
Côte d'Ivoire	300	–	–	–
Gabon	42	–	–	–
Ghana	161	1	PGRC/CRI	161 landraces
Liberia	50	–	–	–
Nigeria	435	5	NRCRI	417 landraces
Senegal	57	–	ISRA/CDH	11 landraces
Sierra Leone	89	–	IAR	43 landraces
Togo	734	–	–	–
Zaire	250	–	–	–
IITA	2161	320	IITA	Incl. germpl. from Nigeria, Benin Rep., Ghana, Congo, Kenya, Brazil, CIAT
Asia - Oceania				
China	86	–	SCATC/UCRI/GAAS	38 landraces
India	1507	30	CTRCI	1397 landraces
Indonesia	251	–	CRIFC/MARIF	158 landraces
Israel	5	–	Isr. Genebank for Agric. Crops, Volcani Center	Introduced breeding lines
Malaysia	92	–	MARDI	69 landraces
Myanmar	21	–	ARI	10 landraces
Pakistan	3	–	Plant Introd. Centre, NARCentre, Islamabad	Introduced landraces
Philippines	384	–	PRCRTC/IPB	176 landraces
Sri Lanka	112	56	CARI/PGRC	56 landraces
Thailand	250	–	RFCRC	41 landraces
Vietnam	36	10	Hung Loc Agric. Centre, Dong Nai Prov.	20 landraces

[†] Source: IPGRI 1994; [‡] Source: GRU files 1995.

Table 1.5. Number of accessions maintained as germplasm at CIAT, Colombia, June 1995.

Source	CIAT code	No. of accessions	
		In vitro bank	Field bank
Cultivated			
Argentina	MARG	72	16
Bolivia	MBOL	3	3
Brazil	MBRA	1334	824
Colombia	MCOL	2001	1911
China	MCHN	2	2
Costa Rica	MCR	148	144
Cuba	MCUB	77	74
Dominican Republic	MDOM	5	5
Ecuador	MECU	117	107
Fiji	MFJI	6	6
Guatemala	MGUA	91	90
Indonesia	MIND	51	51
Malaysia	MMAL	67	67
Mexico	MMEX	102	97
Nigeria	MNGA	19	18
Panama	MPAN	43	40
Paraguay	MPAR	231	181
Peru	MPER	405	402
Philippines	MPHI	6	6
Puerto Rico	MPTR	15	15
Thailand	PTAI	31	8
United States	MUSA	10	9
Venezuela	MVEN	249	230
CIAT/ICA Hybrids		400	242
Genetic stock		**147**	**147**
Subtotal		**5632**	**4695**
Wild species			
29 spp. *in vitro*, 26 spp. field,		349	334
3 undefined species		4	–
Total		**5985**	**5029**

Quarantine restrictions hinder intercontinental transfer of vegetative cassava germplasm (Thurston 1973; Frison and Feliu 1991). IITA holds the CGIAR mandate for conservation and evaluation of African germplasm. The cassava genebank at IITA consists of 2161 accessions comprised of hybrid clones from its breeding programme (55%) and local varieties from five countries (45%) (Table 1.6). IITA maintains a field genebank of 64 accessions of wild *Manihot* species from 12 of the 19 sections in the genus (Table 1.7). Both CIAT and IITA use the software ORACLE to document information about cassava genetic resources and this system has recently been discussed with Latin American NARS for eventual uniform information management. IITA has also developed a database (FoxPro software) for operation on personal computers. These systems will be fully integrated into the System-wide Information Network for Genetic Resources for public access through Internet.

Seed of several *Manihot* species was collected in the early 1980s during missions in Guatemala, Costa Rica, Mexico, Nicaragua, Brazil and Panama (IPGRI 1994) and during 1992-95, the collection and taxonomic work of EMBRAPA Brazil was facilitated through a grant supported by IPGRI, with contribution and participation from IITA and CIAT

Table 1.6. Existing cassava germplasm accessions at the Genetic Resources Unit of IITA and their sources of origin.

Country or region	Advanced breeding line	Local cultivar	Wild species	Total
Nigeria	1193	197	–	1390
Republic of Benin	–	300	–	300
Congo	–	60	–	60
Ghana	–	90	–	90
Kenya	–	11	–	11
CIAT	–	–	27	27[†]
Brazil	–	21	259[‡]	280
Unknown	3	–	–	3
Total	1196	679	286	2161

[†] Recently introduced local cultivars and wild *Manihot* species are not included.
[‡] Majority of the samples are yet to be multiplied.

Table 1.7. Accessions of wild *Manihot* species held by IITA, in glasshouse (G) and field (F).

Section	*Manihot* species	Access. held	Place held	Hybridized	Useful cassava traits
Brevipetiolatae	*stricta*	1	G	+	–
Caerulescentes	*caerulescens*	2	F, G		Drought tolerance
Caerulescentes	*heptaphylla*	1	G		–
Carthaginensis	*carthaginensis*	2	G		Drought and salt tolerance
Crotalariaeformes	*reptans*	1	G		CBB resistance
Glaziovianae	*brachyandra*	3	F, G	+	–
Glaziovianae	*catingae*		G	+	–
Glaziovianae	*dichotoma*	2	G		CM and ACMD resistance; high photosynthetic rate
Glaziovianne	*epruinosa*	15	F, G	+	–
Glaziovianae	*glziovii*	2	F, G		CM, CBB and ACMD resistance; drought tolerance
Graciles	*fruticulosa*	1	G		
Graciles	*gracilis*	1	F, G	+	High root protein; low cyanide
Heterophyllae	*grahami*	3	G		Cold tolerance
Heterophyllae	*pohlii*	5	F, G	+	–
Heterophyllae	*tristis*	8	F, G	+	High root protein
Parvibracteatae	*aesculifolia*	1	G		–
Parvibrateatae	*chlorosticta*	2	F		–
Peruvianae	*leptophylla*	3	F, G	+	–
Peruvianae	*quinquepartita*	1	G	+	CM resistance
Quinquelobae	*falcata*	1	F, G		–
Quinquelobae	*divergens*	1	G		–
Sinuarae	*anomala*	3	F, G	+	High photosynthetic rate
Tripartitae	*tripartita*	2	F, G	+	Drought tolerance
?	*neusana*	2	F, G		–

Table 1.8. Distribution of *Manihot* germplasm (*in vitro*) from CIAT in the last three years.

Sector	Number of accessions (and samples)		
	1992	1993	1994
Centre staff in host country	163[†] (163)[‡]	29 (29)	81 (81)
Centre staff in other countries	–	–	–
Other IARCs	–	–	–
NARS in developing countries	175 (182)	243 (290)	185 (279)
NARS in developed countries	21 (47)	200 (262)	131 (191)
Private sector in developing countries	7 (7)	6 (6)	–
Private sector in developed countries	–	3 (3)	–
Others	–	–	–
Total of different accessions sent by year	366	481	397
Total of material sent by year	(399)	(590)	(551)

[†] Number of different accessions sent to each sector, with a mean of 5 plantlets/accession.
[‡] Number of samples sent (could include repeated accessions).

Table 1.9. Distribution of cassava virus-tested plantlets during 1989 to 1995 from IITA.

	No. of genotypes				
Year	Improved	Local	No. of packages	No. of countries	No. of plantlets
1989	29	0	27	17	1100
1990	39	1	23	16	1045
1991	39	1	24	21	1330
1992	49	2	33	19	3365
1993	104	2	26	18	3007
1994	128	2	33	19	2385
1995	235	11	32	24	8291

(Allem 1995). A documentation system has been implemented recently to record passport data of the *Manihot* species accessions introduced to CIAT. Two networks of Latin American national institutions for plant genetic resources, REMEFI and TROPIGEN, have included *Manihot* as priority species for conservation.

The total annual cost of conservation per cassava clone at CIAT in Colombia has recently been estimated at $US 24.00 *in vitro* and $US 30.00 in the field (Epperson *et al.* 1995). Cryoconservation is an alternative method under development and is expected to provide an efficient and economical alternative for conservation of meristems of cassava clones (CIAT 1994). Experiments also have been conducted to determine conditions for storing cassava pollen (Laton-Cortez 1993).

In addition to improving cassava genepools with *ex situ* collections as a genetic base, during 1979-94 more than 3891 *in vitro* accessions, including 1531 different clones from the 'in trust' collections and samples of material, have been distributed from CIAT (Table 1.8), mainly for use in crop improvement programmes of developing countries. Over the past 7 years, IITA has distributed approximately 620 improved clones and 19 local varieties to its national counterparts (Table 1.9). The collections held in trust through the IARCs also ensure the preservation of national collections. Several countries in Latin America (Paraguay, Peru and Argentina) have received their germplasm back from CIAT after disease indexing, cleaning and further characterization.

The use of cassava descriptors has assisted curators in organizing activities in their germplasm collections (Gulick *et al.* 1983; IPGRI 1994). A revised version of IPGRI

Table 1.10. Cassava minimum descriptors list.

Minimum descriptors		
1.	Shoot colour	3 – Light green, 5 – Dark green, 7 – Green-purple, 9 – Purple
2.	Shoot pubescence	0 – Absent, 1 – Present
3.	Shape of central lobe	1 – Ovoid; 2 – Elliptic-lanceolate; 3 – Obovate-lanceolate; 4 – Oblanceolate; 5 – Lanceolate; 6 – Linear; 7 – Pandurate; 8 – Linear pyramidal; 9 – Linear-pandurate; 10 – Linear-hostatilobada
4.	Petiole colour	1 – Green-yellow; 2 – Green; 3 – Green with little red; 5 – Green with much red; 7 – Red; 9 – Purple
5.	Stem cortex colour (collenchyma)	1 – Yellow, 2 – Light green, 3 – Dark green
6.	Stem external colour	3 – Orange, 4 – Green-yellow; 5 – Gold; 6 – Dark brown; 7 – Silver; 8 – Grey; 9 – Dark brown
7.	Phyllotaxia length	3 – Short (<8 cm); 5 – Medium (8-15 cm); 7 – Large (>15 cm)
8.	Root peduncle presence	0 – Sessile; 3 – Pedunculate; 5 – Both
9.	Root external colour	1 – White or cream; 2 – Yellow; 3 – Light brown; 4 – Dark brown
10.	Root cortex colour	1 – White or cream, 2 – Yellow, 3 – Pink, 4 – Purple
11.	Root pulp colour	1 – White, 2 – Cream, 3 – Yellow, 4 – Pink
12.	Root epidermis texture	3 – Smooth; 7 – Rugose
13.	Flowering	0 – Absent; 1 – Present
Supplementary descriptors		
1.	HCN	
2.	Allozyme	

Sources: IPGRI 1994 and Workshop on Cassava Genetic Resource Management, Cruz das Almas, CNPMF Brasil. Oct 22-27 1995 (In press).

(1994) minimal descriptors elaborated during the Latin American Workshop for Cassava Genetic Resources held at CNPMF, Cruz das Almas, 22-27 October 1995, is presented in Table 1.10.

Table 1.11 lists the estimated number of landrace varieties in Asia and the Americas, for an appreciation of the potential genetic base of production. Characterization of the base collection at CIAT for allelic diversity at two esterase (isozyme) loci has been used as a method of quantifying cultivar diversity. When the Nei index of diversity was calculated for accessions originating from 12 countries in Latin America, countries such as Panama, Ecuador, Paraguay and Mexico showed relatively high diversity among accessions compared with the Caribbean region (Cuba and Dominican Republic) (Iwanaga *et al.* 1993; Hershey 1994). In Africa, genetic diversity of cassava is considered to be highest in the humid and subhumid regions of West and Central Africa, moderate in the coastal regions of East Africa and the islands of Madagascar and lowest in the dry Savanna and high-altitude regions of East and Central Africa (Gulick *et al.* 1983). Patterns of diversity for useful traits are also beginning to emerge through the evaluation of accessions held at CIAT for characters of agronomic interest. For example, the majority of cassava accessions identified as having high photosynthetic potential under natural field conditions (measured as net leaf photosynthetic rates in μmol CO_2 m^{-2} s^{-1}) originate as landraces in the subtropical regions of Brazil and Argentina (CIAT 1992, 1993). Wild *Manihot* germplasm from the same area also presents outstanding photosynthetic efficiency (CIAT 1994). Evaluation of the extensive collection of Colombian accessions held at CIAT shows a predominance of landraces with high dry matter in the North Coast and Andean regions and of varieties with low levels of cyanogens in the highlands (Hershey 1987).

Recent progress has been made in determining relationships among *Manihot* species by applying conserved chloroplast sequences from other plants as molecular

Table 1.11. Estimated number of landrace varieties of cassava in some countries of Asia and the Americas.

Region/country	Estimated total landrace varieties[†]	Estimated total landrace varieties at CIAT[‡]
South America		
Argentina	24	72
Bolivia	60	3
Brazil	3110	1273
Colombia	1932	1969
Ecuador	176	117
Paraguay	192	225
Peru	513	405
Venezuela	303	249
Region Total	6310	4313
Meso-America and Carribean		
Costa Rica	59	148
Cuba	66	75
Dominican Rep.	50	5
Guatemala	57	91
Haiti	–	–
Mexico	89	97
Nicaragua	–	–
Panama	45	43
Puerto Rico	8	15
Region Total	374	474
Asia and Oceania		
China	8	2
Fiji	6	6
India	–	–
Indonesia	120	38
Laos	–	–
Malaysia	80	61
Philippines	40	4
Sri Lanka	–	–
Thailand	4	7
Vietnam	–	–
Region Total	258	128 (including 10 clones from USA but probably of Asian origin)
Total	**6942**	**4915**

[†] Source: Hershey 1994 (Africa was not included in the analysis for lack of information. Includes estimates of total representation in CIAT collection, level of duplication and proportion of accessions which are not landrace varieties. A landrace is loosely defined here as a farmer-selected variety, even if introduced from another region or country.)

[‡] Based on current available passport information, GRU files 1995; does not include level of duplication.

phylogenetic markers (Bertram 1993; Fregene *et al.* 1994). In efforts to develop a genus-wide conservation strategy, CIAT has begun to consolidate available information on the geographic distribution of *Manihot* species (Figs. 1.1, 1.2). Characterization for allelic diversity of isozymes has provided quantitative estimates of relatedness among cassava accessions and with wild species (Lefevre and Charrier 1993; Wanyera *et al.* 1994) and the utility of nuclear molecular markers has been shown (Beeching *et al.* 1993; Bonierbale *et al.* 1995; Roa *et al.* 1996; Tonukari *et al.* 1996).

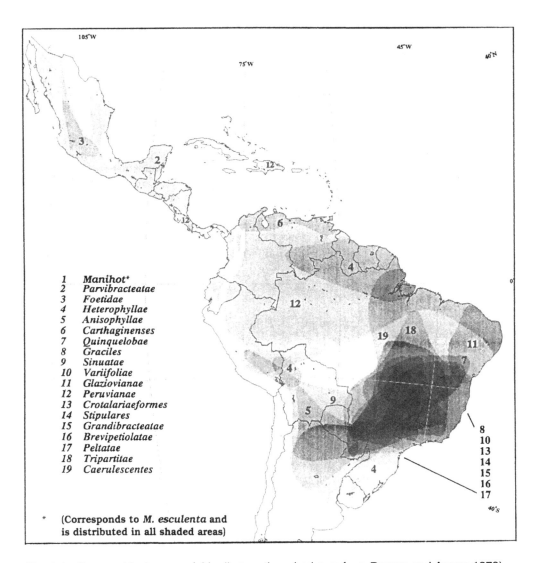

Fig. 1.1. Geographical range of *Manihot* sections (redrawn from Rogers and Appan 1973).

Properties and Uses

The main nutritional component of cassava is carbohydrate energy, or calories, most of which is derived from starch accumulated in the roots. Reflecting this, productivity is measured in terms of dry matter, ranging from 30 to 40% under normal production conditions (Cock 1985). About 85% of the storage root dry matter content is starch.

Cassava roots contain only small amounts of protein, ranging from 1 to 2% on a fresh-weight basis, but are rich in vitamin C and calcium, with acceptable levels of vitamins A and B, and other minerals. Investigations of the quality of protein in cassava germplasm are generally lacking, although Cock (1985) reports that it is good, but deficient in sulphur-containing amino acids. As animal feed, fresh roots are a good source of carbohydrates (Buitrago 1990). Dried cassava is used in rations for poultry, swine and cattle, comprising an international commodity of which the European

Community purchases over 5 million tonnes annually (CIAT 1993). Native and modified starches are important raw materials for many industrial uses such as food processing, paper, textile and adhesive manufacturing and in oil-drilling operations. Derived sugar products include glucose, fructose, maltodextrins and mannitol (Balogapalan *et al.* 1988). A less well-known source of nutrition for both man and animal is the cassava foliage (Ravindram 1993; Bokanga 1994).

A wide range of products known to aboriginal, rural or urban communities has been developed, varying from human and animal foodstuffs and medicinal to fuel potential in industrial countries. Various cassava-based breads and porridge-like products (some fermented and others dried and baked) are important staples in parts of Africa and Latin America (O'Brien *et al.* 1994). Beer-making properties are associated with the yellow varieties as reported by Aguaruna Indians from the Amazon basin in Peru, resulting in the popular refreshments, *mazatto* and *beshu* (Rogers 1965; Boster 1984). Cassava has additional medical uses, being used to control chills and fever and cutaneous eruption, as well as diarrhea. It serves also as a fishing tool, poisoning the prey for easy catching (Duke and Vasquez 1994). Approximately 71% of world cassava production is used for human consumption, while the rest goes to animal feed and industrial uses (Sarma and Kunchai 1991).

Breeding

Domestication of cassava probably began with selection for large roots, more erect plant type with less branched growth and the ability to establish easily from stem

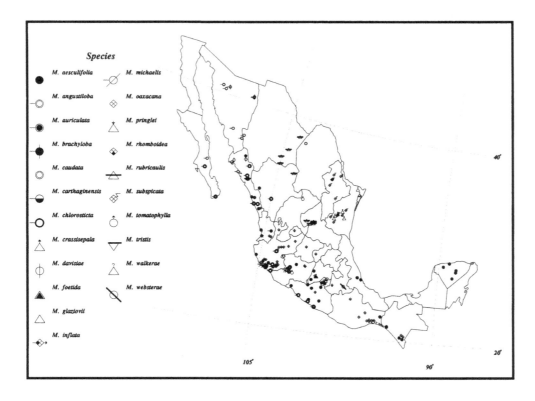

Fig. 1.2. Distribution of *Manihot* species in Mexico.

cuttings. Mixed cropping systems in which several genotypes are grown in the same field probably provided opportunity for new variability to occur in sexual seed, upon which selection was practised (Iglesias 1990). Selection worked toward virtual discontinuation of seed propagation and tended to encourage sterility and/or dormancy in sexual propagation (Jennings 1976). Breeding objectives for cassava have distinct regional differences that are largely determined by biotic or abiotic constraints. Resistance to cassava bacterial blight (*Xanthomonas campestris* pv. *manihoti*), root rots (*Phytopthora* spp., *Diplodia* spp. and others), white flies, mites and thrips are under selection in breeding programmes in Latin America, as are root dry matter content, cyanogen content and several features of root and plant appearance (Hershey and Jennings 1992). The recognition of variability in vitamin A content, associated with yellow root parenchyma, has led to efforts to develop high-vitamin genepools (Moorthy *et al.* 1990). Post-harvest shelf-life and the functional properties of starch in processing are also in early stages of investigation at both CIAT and IITA. Useful variation in tolerance to water- and nutrient-limited conditions, including acid, low-phosphorus and high-aluminium soils, has been found in cultivated germplasm and is being incorporated into improved germplasm (El-Sharkawy and Cock 1987; CIAT 1992; El-Sharkawy *et al.* 1992; Pellet and El-Sharkawy 1993a, 1993b; De Tafur *et al.* 1997).

Since the 1970s, the genetic base for cassava improvement at IITA has consisted of local cultivars from within Nigeria, selected clones from crosses originally made in East Africa between cassava (*M. esculenta*) and its related wild species *M. glaziovii* and segregating progenies from East Africa, Latin America and Asia (Hahn *et al.* 1990; Asiedu *et al.* 1994; IITA 1994; Ng *et al.* 1994). The most pressing early objectives of IITA's programme were to breed high-yielding varieties with resistance to African cassava mosaic disease (ACMD) and cassava bacterial blight (CBB). Other breeding objectives include high-yielding clones with yellow root flesh that have high carotene content and low cyanogenic potential (CNP); good *gari* (grated, fermented and roasted fresh cassava) quality; resistance to cassava mealybug (CM), cassava green mites (CGM) and cassava anthracnose disease (CAD). African national breeding programmes are also concerned with resistance to the devastating ACMD, which is presently threatening survival of the crop and of genetic diversity in severe epidemics (Jennings 1994). A systematic and agro-ecologically based germplasm introduction scheme to broaden the genetic base of cassava in Africa was initiated by IITA and CIAT in 1990 (IITA 1993a). Crosses are made at CIAT, Colombia, between parents of Latin American origin that are adapted to four different agro-ecologies of South America with homologues in Africa and improved ACMD-resistant clones from IITA. By 1994, more than 200 000 botanical hybrid seeds comprising over 1000 families were introduced for evaluation in Nigeria through IITA, observing strict quarantine regulations for intercontinental exchange of germplasm.

Sources of germplasm for cassava improvement in Asia have been local landraces for direct release, selections from open-pollinated seed originating in germplasm collections and controlled intervarietal hybrids, as well as crosses between local and introduced (CIAT) clones (Kawano and Hershey 1985). Indian Cassava Mosaic Virus, related to African Cassava Mosaic Virus, is a threat to production which does not at present affect cassava in Latin America (Mathew 1991). As disease and pest pressure on cassava is low in Asia, primary objectives are to increase productivity and starch content.

Additional useful genetic variability has been detected in the wild relatives of cassava, although it has not been used extensively (Jennings 1959; Nassar 1986). Recent work at IITA and CIAT has confirmed previous reports of higher protein content and insect resistance in *Manihot* species and identified properties that may be associated with their xeromorphic adaptation and efficient photosynthesis (Asiedu *et al.* 1992; CIAT 1993, 1994). IITA has used several *Manihot* species in interspecific hybridization with cassava to transfer their desirable genes into cassava (Hahn *et al.* 1980, 1990). For instance, *M. glaziovii* was introduced into West, East and Central Africa as a source of rubber in the period 1890-1900 but is currently used only as a shade plant in young cash crop plantations to provide resistance to ACMD (Nichols 1947; Hahn *et al.* 1977, 1979) and CBB (Hahn *et al.* 1980). Accessions of *Manihot tristis*, unlike cultivated cassava, are known to contain high protein in the storage roots (Asiedu *et al.* 1992). Crosses between cultivated cassava and *M. tristis* yielded hybrid progenies with higher storage root protein and total amino acid content (IITA 1993b, 1993c). The national programme of Colombia also recognized the relatively high protein content of *M. carthagenensis*, native to the northern littoral regions of South America (Lopez and Herrera 1970).

On average, improved clones from cassava breeding programmes are no more than three generations advanced from local or introduced materials, resulting from the adaptive selection and recombination among superior genotypes with a broad range of favourable characteristics (Jennings and Hershey 1985). Once agronomically acceptable genepools with adequate genetic base are available for a target production area, additional desirable traits may be introduced by a modified backcrossing scheme, using different members of the adapted genepool as recurrent parents to avoid problems of inbreeding depression (Martin 1976; Bueno 1987). The innovative strategy of triploid breeding has been reported as successful via unreduced gametes and by colchicine induction (Hahn *et al.* 1990; CGPRT Newsletter 1995).

Adaptive selection and cassava breeding, facilitated by the organization and evaluation of national and international germplasm collections, have had a significant impact on varietal deployment. An extreme example of the deployment of new varieties may be seen in Asia, where a recent survey of Asian countries (Kawano 1995) estimates that 0.38 million hectares are planted with cultivars released after 1970 and, of those, 87% are with CIAT-related varieties.

Prospects

Cassava is well known for its ability to tolerate drought, maintaining high productivity under conditions not suitable to most crops (Cock 1985; El-Sharkawy 1993). Drought-prone regions, such as Sub-Saharan Africa and Northeast Brazil, have increased cassava production rapidly in the last two decades (El-Sharkawy 1993). Owing to the diversity of its utilization, adaptation and low input requirements, cassava often provides a valuable link for rural farmers to the market economy (Henry and Best 1994). The most dynamic cassava sectors, those of processed pellets for animal feed and starch or starch products, depend on improved production, processing and market technologies; their development is sensitive to both domestic and foreign trade policies and demand issues with regard to alternative raw materials such as grains and sugarcane (Leihner 1992; Henry and Gottret 1995). Cassava's recognition as an orphan crop (Persley 1990), the mandates taken on by the CGIAR and the operation of the Cassava Biotechnology Network have brought attention and basic resources to the research and development needs of cassava. In terms of global efforts toward the conservation and use of cassava's genetic resources, IITA, CIAT and NARS are interested in developing and adopting efficient methods for assessing diversity that can be used for the definition of representative subsets of germplasm for conservation and utilization. It is hoped that methods such as those used to identify duplicates in the collection held at CIAT (Ocampo *et al.* 1995) and to quantify similarity among accessions (Roa *et al.* 1996; Tonukari *et al.* 1996) may be adapted to quantify the diversity in existing collections and assess potential gaps in representation to orient the conservation process. Further efforts in the study and use of genetic diversity are needed to confirm possible new sources of resistance to important pests for incorporation into breeding programmes, such as may have been identified in ACMD-resistant local African cultivars collected recently by IITA (Ng and Rossel 1993).

Acknowledgements
The authors acknowledge the valuable inputs of Drs M. El-Sharkawy, Cassava Physiologist and C. Iglesias, Cassava Breeder of CIAT.

REFERENCES
Allem, A. 1995. Collection of wild strains of cassava in Brazil. IPGRI and CENARGEN/EMBRAPA.

Allem, A.C. 1987. *Manihot esculenta* is a native of the Neotropics. FAO/IBPGR Plant Genet. Resour. Newsl. 71:22-24.

Allem, A.C. 1994. The origin of *Manihot esculenta* Crantz (Euphorbiaceae). Genet. Resour. & Crop Evolution 41 (3):133-150.

Asiedu, R., S.K. Hahn, K.V. Bai and A.G.O. Dixon. 1992. Introgression of genes from wild relatives into cassava. Pp. 89-91 *in* Proceedings of the Fourth Triennial Symposium of the International Society of Tropical Root Crops - Africa Branch held in Kinshasa, Zaire, 1989 (M.O. Akoroda and O.B. Arene, eds.). ISTRC-AB/IDRC/CTA/IITA.

Asiedu, R., S.K. Hahn, K.V. Bai and A.G.O. Dixon. 1994. Interspecific hybridization in the genus Manihot: progress and prospects. Pp. 110-113 *in* Proceedings of the Ninth Symposium of the International Society of Tropical Root Crops held in Accra, Ghana, 1991 (F. Ofori and S.K. Hahn, eds.). ISTRC/Government of Ghana/IITA.

Bai, K.V., R. Asiedu and A.G.O. Dixon. 1993. Cytogenetics of *Manihot* species and interspecific hybrids. Proceedings of the First Int. Meeting of the Cassava Biotechnology Network. Cartagena, Colombia.

Bailey, H. 1976. Hortus Third. A Concise Dictionary of Plants Cultivated in the United States and Canada. McMillan Publishing Co., New York.

Balogapalan, C., G. Padmaja, S.K. Nanda and S.N. Moothy. 1988. Cassava in Food, Feed and Industry. CRC Press, Boca Raton, FL.

Barnes, H. 1975. The diffusion of the manioc plant from South America to Africa: An essay in ethnobotanical culture history. Dissertation, Columbia University.

Beeching, J.P.M., M. Gavalda, H. Noirot, M. Haysom, A. Hughes and A. Charrier. 1993. An assessment of diversity within a collection of cassava (*Manihot esculenta* Crantz) germplasm using molecular markers. Am. J. Botany 72:515-520.

Bertram, R. 1993. Application of molecular techniques to genetic resources of cassava (*Manihot esculenta* Crantz - Euphorbiaceae). Interspecific evolutionary relationships and intraspecific characterization. Dissertation, University of Maryland.

Bokanga, M. 1994. Processing of cassava leaves for human consumption. *In* Proceedings from an International Workshop on Cassava Safety. Wageningen, The Netherlands.

Bolhuis, G. 1953. A survey of some attempts to breed cassava varieties with high content of protein in the roots. Euphytica 20:107-112.

Bonierbale, M.W., M.M. Maya, J.L. Claros and C. Iglesias. 1995. Application of molecular markers to describing the genetic structure of cassava gene pools. Proceedings for the 2nd Int. Sci. Meeting Cassava Biotechnology Network (CBN), 22-26 August 1994, Bogor, Indonesia.

Boster, J. 1984. Classification, cultivation, and selection of Aguaruna cultivars of *Manihot esculenta* (Euphorbiaceae). Adv. Econ. Bot. 1:34-47.

Brucher, H. 1989. Useful Plants of Neotropical Origin and their Wild Relatives. Springer-Verlag, Berlin, Heidelberg, Germany.

Bueno, A. 1987. Hybridization and breeding methodologies appropriate to cassava, pp. 51-66. CIAT, Cali, Colombia.

Buitrago, J. 1990. La yuca en la alimentacion animal. Cali, Colombia.

Byrne, D. 1984. Breeding cassava. Pp. 73-134 *In* Plant Breeding Reviews (J. Janick, ed.). Westpoint.

CGPRT Newsletter. 1995. Palawija News. CPCRT Center, Bogor, Indonesia. 12:1:16.

Chavez, R. 1990. Especies silvestres de *Manihot* un recurso valioso. Yuca Boletin Informativo 14 (1):2-5.

CIAT. 1991. Genetic resources unit progress report 1987-1991. CIAT. Cali, Colombia.

CIAT. 1992. Cassava program annual report 1987-1991. CIAT, Cali, Colombia.

CIAT. 1993. Cassava program annual report 1993. CIAT, Cali, Colombia.

CIAT. 1994. Cassava program annual report 1994. CIAT. Cali, Colombia.

Cock, J. 1985. Cassava: New Potential for a Neglected Crop. Westview Press Inc., Boulder, Colorado.

Cook, J., D. Franklin, G. Sandoval and P. Juri. 1979. The ideal cassava plant for maximum yield. Crop Sci. 19:271-279.

De Tafur, S.M., M.A. El-Sharkawy and F. Calle. 1997. Photosynthesis and yield performance of cassava in seasonally dry and semiarid environments. Photosynthesis (in press).

Dufour, D. 1994. Cassava in Amazonia: lessons in utilization and safety from native peoples. *in* Proceedings from an International Workshop on Cassava Safety. Wageningen, The Netherlands.

Duke, J.A. and R. Vasquez. 1994. Amazonian Ethnobotanical Dictionary. CRC Press, Inc., Boca Raton, FL.

El-Sharkawy, M. 1993. Drought-tolerant cassava for Africa, Asia, and Latin America. Bioscience 43 (7):441-452.

El-Sharkawy, M.A. and J.A. Cock. 1987. Response of cassava to water stress. Plant & Soil 100:345-360.

El-Sharkawy, M.A., A.P. Hernández and C. Hershey. 1992. Yield stability of cassava during prolonged mid-season water stress. Exp. Agric. 28:165-174.

Epperson, J., D. Pachico and C. Guevara. 1995. The cost of maintaining genetic resources of cassava, *Manihot esculenta* Crantz. Acta Hortic. 429:409-413.

FAO. 1995. FAO production year book for 1994. Vol. 48: 93-94.

Fregene, M., J. Vargas, J. Ikea, F. Angel, J. Tohme, R. Asiedu, M. Akoroda and W. Roca. 1994. Variability of chloroplast DNA and nuclear ribosomal DNA in cassava (*Manihot esculenta* Crantz). Theor. Appl. Genet. (89):717-727.

Fregene, M.A., F. Angel, R. Gómez, F. Rodríguez, W.M. Roca, J. Tohme and M. Bonierbale. 1996. A genetic linkage map of cassava (*Manihot esculenta* Crantz). Theor. Appl. Genet. (in press).

Frison, E. and E. Feliu (eds.). 1991. FAO/IBPGR Technical Guidelines for the Safe Movement of Cassava Germplasm. FAO/IBPGR, Rome.

Gomez, R., F. Angel, M. Bonierbale, F. Rodriguez, J. Tohme and W. Roca. 1995. Selecting heterozygous parents and single dose markers for genetic mapping in cassava. Proceedings for the 2nd Int. Sci. Meeting Cassava Biotechnology Network (CBN), 22-26 August 1994, Bogor, Indonesia.

Gulick, P., C. Hershey and J. Esquinas-Alcazar. 1983. Genetic Resources of Cassava and Wild Relatives. International Board for Plant Genetic Resources, Rome, Italy.

Hahn, S.K. 1989. An overview of African traditional cassava processing and utilization. Outlook on Agric. 18:110-118.

Hahn, S.K., K.V. Bai and R. Asiedu. 1990. Tetraploids, triploids and 2n pollen from diploid interspecific crosses with cassava. Theor. Appl. Genet. 79:433-439.

Hahn, S.K., A.K. Howland and E.R. Terry. 1977. Cassava breeding at IITA. Pp. 4-10 *in* Proceedings of the 3rd International Tropical Root Crops Symposium held at Ibadan, Nigeria, December 2-9, 1973 (C.L.A. Leakey, ed.). ISTRC/IITA.

Hahn, S.K., E.R. Terry, K. Leuschner, I.O. Akobundu, C. Okali and R. Lal. 1979. Cassava improvement in Africa. Field Crops Res. 2:193-336.

Hahn, S., E. Terry and R. Leuschner. 1980. Cassava breeding for resistance to Cassava Mosaic Disease. Euphytica 29:673-683.

Henry, G. and R. Best. 1994. Impact of integrated cassava projects among small-scale farmers in selected Latin American countries. *in* Proceedings from an International Workshop on Cassava Safety. Wageningen, The Netherlands.

Henry, G. and V. Gottret. 1995. Global cassava sector trends: reassessing the crop's future. Inter Center Root and Tuber Review.

Hershey, C. 1984. Breeding cassava for adaptation to stress conditions: development of a methodology. Proceedings 6th Int. Soc. Trop. Root Crops. CIP, Lima, Peru.

Hershey, C. 1987. Cassava germplasm resources. Proc. workshop on cassava breeding: a multidisciplinary review. Philippines, 4-7 March 1985.

Hershey, C. and D. Jennings. 1992. Progress in breeding cassava for adaptation to stress. Plant Breed. Abstr. 62:823-831.

Hershey, C., C. Iglesias, M. Iwanaga and J. Tohme. 1994. Definition of a core collection for cassava. *In* Report of the first meeting of the International Network for Cassava Genetic Resources, Aug 18-23 1992. IPGRI, CIAT.

Hershey, C.H. 1994. *Manihot* genetic diversity. Pp. 111-134 *in* Report of the first meeting of the International Network for Cassava Genetic Resources. IPGRI, Rome, Italy.

Iglesias, C. 1990. Introduccion de diversidad genetica mejorada a nivel de campo. Memorias de la tercera reunion panamericana de fitomejoradores de yuca. CIAT, Cali, Colombia.

Iglesias, C., M. Bonierbale, M. El-Sharkawy, C. Lozano, A. Belloti and C. Wheatley. 1993. Focusing basis research for cassava improvement. *In* Proceedings of the International Symposium on Tropical Tuber Crops. CTCRI, Trivandrum, India.

IITA. 1992. Sustainable food production in sub-Saharan Africa. 1: IITA's contribution. IITA, Ibadan, Nigeria.

IITA. 1993a. Summary of review papers for the internal review of Crop Improvement Division, 21-27 November, 1993. IITA, Ibadan, Nigeria.

IITA. 1993b. Annual Report for 1989-93 of the Root and Tuber Improvement Program, Crop Improvement Division, International Institute of Tropical Agriculture, Part II: Cassava: Tissue culture, crop physiology and food technology. IITA, Ibadan, Nigeria.

IITA. 1993c. Annual Report for 1992. IITA, Ibadan, Nigeria.

IITA. 1994. Annual Report for 1993. IITA, Ibadan, Nigeria.

IPGRI. 1994. International Network for Cassava Genetic Resources. Report of the first meeting of the International Network for Cassava Genetic Resources. CIAT, Cali, Colombia. August 18-23 1992. IPGRI, Rome, Italy.

Iwanaga, M., M.E. Ayala, C.H. Ocampo and C. Hershey. 1993. Caracterizacion de la coleccion de Colombia del germoplasma de yuca (*Manihot esculenta* Crantz) por electroforesis PAGE utilizando la isoenzyma aß esterasa. Recursos geneticos horticolas. Actas del II simposio latinoamericano sobre recursos geneticos de especies horticolas. INTA, Mar del Plata, Argentina.

Jennings, D.L. 1959. *Manihot melanobasis* Müll. Arg. - A useful parent for cassava breeding. Euphytica 8:157-162.

Jennings, D.L. 1976. Cassava, *Manihot esculenta* (Euphorbiaceae). Pp. 81-84 *In* Evolution of Crop Plants (N. Simmonds, ed.). Longman, London, UK.

Jennings, D. 1994. Breeding for resistance to African Cassava Mosaic - Geminivirus in East Africa. Trop. Sci. 34:110-122.

Jones, W.O. 1959. Manioc in Africa. Stanford University Press, Stanford, CA, USA.

Kawano, K. 1980. Cassava. Pp. 225-233 *in* Hybridization of Crop Plants. American Society of Agronomy and Crop Science Society of America, Madison, Wisconsin.

Kawano, K. 1995. Green Revolution and Cassava Breeding. Cassava Breeding, Agronomy Research and Technology Transfer in Asia. CIAT, Bangkok, Thailand.

Kawano, K. and C. Hershey. 1985. CIAT cassava germplasm and its use in Asian national breeding programmes. *In* Proceedings of the International Symposium on South East Asian Plant Genetic Resources. Lembaya Biology Nasional. LIPI, Bogor, Indonesia.

Laton-Cortez, M. 1993. Crioconservacion de polen de yuca. Universidad del Valle.

Lefevre, F. and A. Charrier. 1993. Isozyme diversity within African *Manihot* germplasm. Euphytica (66):73-80.

Leihner, D.E. 1992. The contribution of tropical root and tuber crops to peasant farms and the national economy. Plant Res. and Development 36:81-88.

Leon, J. 1987. Botanica de los cultivos tropicales. Servicio editorial IICA, San Jose, Costa Rica, 304-320.

Lopez, L. and H. Herrera. 1970. *Manihot carthaginensis* una yuca silvestre con alto contenido proteico. Contribucion del ICA a la VIII Reunion de Fitotecnica. Bogota, Colombia.

Magoon, M., R. Krishnan and K. Bai. 1969. Morphology of the pachytene chromosomes and meiosis in *Manihot esculenta* Crantz. Cytologia 34:612-626.

Mahon, J.D., S.B. Lowe and L.A. Hunt. 1977. Variation in the rate of photosynthetic uptake in cassava cultivars and related species. Photosynthetica 11:131-138.

Martin, F. 1976. Cytogenetics and plant breeding of cassava: a review. Plant Breed. Abstr. 46:909-916.

Mathew, A. 1991. Cassava Mosaic Virus. Int. J. Trop. Plant Disease 9:131-151.

Mejia, M. 1988. The Orinoco estuary (Colombia): an area of high diversity of bitter cassava (*Manihot esculenta* Crantz). FAO/IBPGR Plant Genetic Resources Newsl. 75/76:27-28.

Mejia, M. 1991. Diversidad de Yuca *Manihot esculenta* Krantz en Colombia. Vision geografico cultural. COA, Corporacion Colombiana para la Amazonia, Araracuara, Bogota, Colombia.

Moorthy, S., J. Jos, R. Nair and M. Shreecumari. 1990. Variability of B-carotene in cassava germplasm. Food Chem. 36:233-236.

Nassar, N. 1978a. Conservation of the genetic resources of cassava (*Manihot esculenta*) - Determination of wild species localities with emphasis on probable origin. Econ. Bot. 32 (3):311-320.

Nassar, N. 1978b. Microcenters of wild cassava, *Manihot* spp. diversity in central Brazil. Turrialba 28 (4):345-347.

Nassar, N. 1978c. Wild *Manihot* species of central Brazil for cassava breeding. Can. J. Plant Sci. 257-261.

Nassar, N. 1986. Genetic variation of wild *Manihot* species native to Brazil and its potential for cassava improvement. Field Crop Res. 13:177-184.

Nassar, N. and F. Cardenas. 1986. Collecting wild cassava in Northern Mexico. Plant Genet. Resour. Newsl. 65:29-30.

Nassar, N.A. and S. Fitchner. 1978. Hydrocyanic acid content in some wild *Manihot* (cassava) species. Can. J. Plant Sci. 58:577-578.

Ng, N.Q. and H. Rossel. 1993. Characterization and evaluation of cassava germplasm. Pp. 13-134 *in* Annual Report for 1992 of the Genetic Resources Unit, Crop Improvement Division, International Institute of Tropical Agriculture. IITA, Ibadan, Nigeria.

Ng, N.Q., R. Asiedu and S.Y.C. Ng. 1994. Cassava genetic resources programme at IITA, Ibadan. Pp. 71-76 *in* International Network for Cassava Genetic Rsources. Report of the First Scientific Meeting of the International Network for Cassava Genetic Resources, CIAT, Colombia, 18-23 August 1992. International Crop Network Series No. 10. IPGRI, Rome, Italy.

Ng, S.Y.C. 1991. *In vitro* conservation and distribution of root and tuber crop germplasm. Pp. 95-106 *in* Proceedings of an International Workshop on Crop Genetic Resources of Africa, Volume II (N.Q.

Ng, P. Perrino, F. Attere and H. Zedan, eds.), 17-20 October, 1988, International Institute of Tropical Agriculture, Ibadan, Nigeria. IITA/IBPGR/UNEP/CNR.

Nichols, R.F.W. 1947. Breeding cassava for virus resistance. East Africa Agric. J. 12:184-194.

Nweke, F.I. and R. Polson. 1990. The dynamics of cassava production in Africa: Preliminary analysis of indicators. A report from Phase 1 of the Collaborative Study of Cassava in Africa (COSCA) for discussion at the 3rd meeting of the steering committee of COSCA in London, August 13-15, 1990.

Nweke, F.I., A.G.O. Dixon, R. Asiedu and S.A. Folayan. 1994. Cassava varietal needs of farmers and the potential for growth in Africa. Collaborative Study of Cassava in Africa (COSCA) Working Paper No. 10. Resource and Crop Management Division. IITA, Ibadan, Nigeria.

O'Brien, G., D. Jones, C. Wheatley and T. Sanchez. 1994. Processing approaches to optimising raw materials and end product quality in the production of cassava flours. *In* Proceedings from an International Workshop on Cassava Safety. Wageningen, The Netherlands.

Ocampo, C., F. Angel, A. Jimenez, G. Laramillo, C. Hershey and C. Iglesias. 1995. DNA fingerprinting to confirm possible genetic duplicates in cassava germplasm. *In* Proceedings of the 2nd Intnl. Sci. Meeting Cassava Biotechnology Network (CBN), 22-26 August 1994, Bogor, Indonesia.

Patiño, V.M. 1964. Plantas cultivadas y animales domesticos en América equinoccial. Tomo 2. Plantas alimenticias. Imprenta Departamental, Cali, Colombia.

Pellet, D. and M.A. El-Sharkawy. 1993a. Cassava varietal response to phosphorus fertilization. I. Yield, biomass and gas exchange. Field Crops Res. 35:1-11.

Pellet, D. and M.A. El-Sharkawy. 1993b. Cassava varietal response to phosphorus fertilization. II. Phosphorus uptake and use efficiency. Field Crops. Res. 35:13-20.

Perry, B. 1943. Chromosome number and phylogenetic relationships in the Euphorbiaceae. Am. J. Botany 30:527-543.

Persley, G. 1990. Beyond Mendel's garden: biotechnology in the service of world agriculture. CAB International, Wallingford, UK.

Ravindram, V. 1993. Cassava leaves as animal feed: potential and limitations. J. Sci. Food Agric. 61:141-150.

Renvoize, B. 1972. The area of origin of *Manihot esculenta* as a crop plant - A review of the evidence. Econ. Bot. 26:352-360.

Roa, AC., M.M. Maya, M. Duque, J. Tohme and M.W. Bonierbale. 1996. AFLP Analysis of relationships among cassava and other *Manihot* species. (in press).

Roca, W., R. Chavez, M. Marin, D. Arias, G. Mafla and R. Reyes. 1989. *In vitro* methods of germplasm conservation. Genome 813-817.

Rogers, D. 1963. Studies of *Manihot esculenta* Crantz and related species. Bull. Torrey Bot. Club 90 (1):43-54.

Rogers, D.J. 1965. Some botanical and ethnological considerations of *Manihot esculenta*. Econ. Bot. 19 (4):369-377.

Rogers, D.J. and S.G. Appan. 1973. Flora Neotropica Monograph No. 13 *Manihot* Manihotoides (Euphorbiaceae), pp. 1-272. Hafner Press, New York.

Rogers, D.J. and H.S. Fleming. 1973. A monograph of *Manihot esculenta* with an explanation of the taximetrics methods used. Econ. Bot. 27:1-113.

Sarma, J. and D. Kunchai. 1991. Trends and prospects for cassava in the developing world. International Food Policy Research Institute, Washington, DC.

Schmeda-Hirschmann, G. 1994. Plant resources used by the Ayoreo for the Paraguayan Chaco. Econ. Bot. 48:252-258.

Smith, C.E. 1968. The New World centers of origin of cultivated plants and the archaeological evidence. Econ. Bot. 22 (3):253-266.

Thurston, H. 1973. Threatening plant diseases. Ann. Rev. Phytopathol. II:27-52.

Tonukari, N.J., G. Thottappilly, N.Q. Ng and H.D. Mignouna. 1996. Genetic Diversity in Benin Republic cassava. (in press).

Toro, B., E. Ibarra and B. Jborja. 1985. Historia, origen y distribucion geografica. Yuca o mandioca (*Manihot esculenta* Euphorbiaceae). Cultivo e industrializacion de la yuca en el Ecuador. Quito, Ecuador.

Ugent, D., S. Pozorski and T. Pozorki. 1986. Archeological manioc (*Manihot*) from coastal Peru. Econ. Bot. 40:78-102.

Umanah, E. and R. Hartman. 1973. Chromosome numbers and karyotypes of some *Manihot* species. J. Am. Soc. Hortic. Sci. 98 (272-274) :

Vavilov, N. 1992. Origin and geography of cultivated plants: [Collected works 1920-1940]. *In* V. Dorfeyev, ed.. Cambridge University Press.

Vavilov, N.I. 1939. The important agricultural crops of pre-Columbian America and their mutual relationship. Izd. Gos. Geogr. O-va (Public. Nat. Dept. Geogr. USSR) 71 (10):1-25.

Wanyera, N., S. Hahn and M. Aken'Ova. 1994. Root crops for food security in Africa. Introgression of ceara rubber (*Manihot glaziovii* Muell-Arg) into cassava (*Manihot esculenta* Crantz): a morphological and electrophoretic evidence. Pp. 125-130 *in* Proceedings Fifth Triennial Symposium of the International Society for Tropical Root Crops (M. Akoroda, ed.). Uganda.

Chapter 2

The Potato

Z. Huamán, A. Golmirzaie and W. Amoros

The potato that is grown worldwide as one of the major food crops belongs to a single species, *Solanum tuberosum* L. In the English language, another common name is Irish potato. The word potato derives from the Spanish word *patata*, which is also the name for this crop in Portuguese and Italian. The name in French is *pomme de terre*. Its most common name in the Spanish spoken in Latin America is *papa*, a word derived from Quechua, the language of the Incas.

Global potato production is around 260 million t on an area of about 18 million ha. In recent years, potato production has spread from its traditional mountainous environment into warmer, drier areas such as Peru's coastal valleys, the plains of India, Bangladesh and Pakistan, and the irrigated oases of North Africa. It is also spreading into warm, humid zones (Horton 1988).

BOTANY AND DISTRIBUTION

Cultivated species and all their wild relatives are classified in the family Solanaceae, genus *Solanum*, subgenus *Potatoe*, section *Petota* (formerly *Tuberarium*), and subsection *Potatoe* (formerly *Hyperbasarthrum*) (Hawkes 1990). In addition to *S. tuberosum*, Latin America houses seven other cultivated species with ploidy levels from diploid ($2n=2x=24$) to pentaploid ($2n=5x=60$) (see Table 2.1).

Origin, Distribution and Diffusion

The potato undoubtedly originated in South America and was domesticated in the highlands of the central Andes, probably between central Peru and central Bolivia. It was cultivated as far back as 7000 years before present (BP), based on archaeological evidence mainly from coastal Peruvian pre-Inca cultures. Potato-like ceramics found in Peru also indicate that the crop was widely known from the beginning of the Christian era (Hawkes 1990). Some controversy surrounds a possible second independent area of domestication in the Chiloé Archipelago of Chile. Ugent *et al.* (1987) reported wild potato remains from southern Chile dating to 13 000 BP that could have been eaten at that time.

The potato was introduced from South America to Europe in the late 16th century, some years after the discovery and conquest of Peru. Salaman (1954) suggested that the first introductions could have come from Peru or from the port of Cartagena on the north coast of Colombia. Juzepczuk and Bukasov (1929), on the contrary, believed that these early introductions could have come from southern Chile. Salaman (1954) argued that the first direct journey from Chile to Europe via the Straits of Magellan was made in 1579, by which time the potato was already known in Europe. From Europe it spread to North America, India, China, Japan and parts of Africa in the 17th century (Hawkes 1967). At present, the potato is grown at latitudes from 55°N to 50°S and at altitudes from sea level up to 4000 m.

Some cultivars of diploid *S. stenotomum* that are the most similar to wild potatoes could have been the first to be domesticated. *Solanum leptophyes* has been postulated as the wild ancestor and the area of domestication might have been in the region of the Lake Titicaca basin. Other diploid species could have been derived from *S. stenotomum* by human selection: such as *S. phureja* for rapid maturity and lack of tuber dormancy, and *S. goniocalyx* for good palatability and flavour (Hawkes 1967, 1990). *Solanum* x *ajanhuiri* was derived from natural crosses between *S. stenotomum* and the wild, frost-resistant diploid *S. megistacrolobum* (Huamán *et al.* 1980). The most important event in potato evolution was the origin of the tetraploid *S. tuberosum* subsp. *andigena*. Cadman (1942) postulated an autotetraploid origin by crosses of unreduced gametes of *S. stenotomum*. Hawkes (1956) proposed a hybrid origin involving unreduced gametes of *S. stenotomum* and the weed diploid species *S. sparsipilum*. Cribb and Hawkes (1986) provided strong evidence to support this allotetraploid origin. *Solanum tuberosum* subsp. *tuberosum* is considered to derive from subsp. *andigena* through selection for adaptation to long photoperiods (Hawkes 1967). Researchers have shown that this adaptation process was accompanied by morphological changes such as shorter plants, longer leaves with broader leaflets, and reduced flowering and fruit setting (Simmonds 1964; Plaisted 1971). The triploid *S.* x *chaucha* derived from crosses between the tetraploid *S. tuberosum* subsp. *andigena* and the diploid *S. stenotomum* (Hawkes 1967; Jackson *et al.* 1977). Some cultivars of this species could be autotriploids totally derived from *S. stenotomum* by crosses between unreduced and reduced gametes (Bukasov 1964). The cross between the wild, frost-resistant tetraploid *S. acaule* and *S. stenotomum* was the origin of the sterile triploid *S.* x *juzepczukii* (Bukasov 1939; Hawkes 1962; Schmiediche *et al.* 1982). Further crosses involving unreduced gametes of *S.* x *juzepczukii* and diploid gametes of *S. tuberosum* subsp. *andigena* resulted in the pentaploid *S.* x *curtilobum* (Bukasov 1939; Hawkes 1962).

Reproductive Biology

All diploid cultivated potato species are self-incompatible and fertile, triploids are mostly sterile, tetraploids are fertile and highly self-compatible, and pentaploids are fertile in crosses with tetraploids. Diploid and tetraploid species in general show very regular meiotic pairing, but tetraploid potatoes function as cytological autotetraploids (Hawkes 1990).

GERMPLASM CONSERVATION AND USE

The potato possesses more related wild species than any other crop plant and 235 wild species are recognized as composing a polyploid series from diploids to hexaploids. These wild potato species are widely distributed from the southwestern United States to most countries in Central and South America. Two clear centres of diversity are recognized: one in central Mexico and the second in the high Andes from Peru through Bolivia to northwest Argentina. No single wild potato species extends throughout the total geographic area and only a few are widespread. Many species are restricted to limited areas and ecological zones. Their altitudinal range of distribution is from sea level to 4500 m. The countries with the largest number of wild potato species are Mexico, with 36, Bolivia (39) and Peru (100). In addition, five species are found in both Peru and Bolivia (Hawkes 1990).

Table 2.1. Geographic coverage of cultivated *Solanum* species conserved in the CIP genebank.

Cultivated species	No. of samples	Geographic coverage
2*n*=2*x*=24		
Solanum stenotomum	268	ARG (5); BOL (83); COL (1); ECU (6); PER (173)
S. phureja	170	BOL (4); COL (111); ECU (35); MEX (1); PER (19)
S. goniocalyx	48	BOL (2); CRI (1); PER (45)
S. x ajanhuiri	10	BOL (9); PER (1)
Solanum 2x hybrids	56	BOL (6); PER (50)
2*n*=3*x*=36		
S. x chaucha	97	ARG (1); BOL (25); ECU (1); PER (70)
S. x juzepczukii	31	ARG (1); BOL (20); PER (10)
2*n*=4*x*=48		
S. tuberosum subsp. andigena	2644	ARG (64); BOL (303); COL (138); ECU (208); GTM (22); MEX (25); PER (1845); VEN (39)
S. tuberosum subsp. tuberosum	144	ARG (6); BOL (2); CHL (127); ECU (1); GTM (1); MEX (3); PER (2); VEN (1)
Solanum 4x hybrids	48	ARG (1); BOL (6); ECU (1); GTM (7); PER (33)
2*n*=5*x*=60		
S. x curtilobum	11	ARG (2); BOL (3); PER (5); VEN (1)
Total	**3527**	

CIP maintains a total of 1567 wild potato accessions, including 112 different tuber-bearing *Solanum* species collected in 12 countries in the Americas. Biosystematic studies on the wild and cultivated potatoes of Bolivia from this collection have been published in a comprehensive monograph (Ochoa 1990) and another volume on the potatoes of Peru is in preparation. Ochoa's studies have also helped to add new sources of variability with the discovery of 50 new wild species among the materials from collecting expeditions made after the creation of CIP in 1971 (Ochoa, pers. comm.).

Despite the numerous collecting expeditions carried out in recent decades to obtain living materials to be conserved in genebanks, 35 wild potato species have not been found where their type specimens were discovered. However, explorations of new areas or more intensive searches within some known areas made by collecting expeditions from several genebanks have resulted in the discovery of new potato species.

Many cultivated potato collections conserved by several institutions in South America were donated to the genebank maintained at CIP and many new accessions were obtained and shared by joint CIP-NARS collecting expeditions. As a result, CIP assembled a potato collection of more than 15 000 accessions of native potato cultivars from nine countries in Latin America. CIP identified duplicate accessions of the same cultivar and selected about 3500 (see Table 2.1) different cultivars among them (Huamán 1986; Huamán and Stegemann 1989).

Native potato cultivars are maintained in a 3-ha field genebank at Huancayo, Peru (3200 m asl), and back-up duplicate sets of tubers of each accession in the collections are placed in cold storage at 4°C at La Molina, near Lima. A duplicate set of the whole cultivated potato collection is also maintained in *in vitro* culture at La Molina. The storage period, using a culture medium for minimum growth, room temperature of 6-7°C and 1000 lux for 16 hours a day, is now extended to 2 years (Lizárraga *et al.* 1991). This *in vitro* collection is totally duplicated at the Instituto Nacional de Investigaciones Agropecuarias (INIAP) in Quito, Ecuador, for security reasons. There is also an ongoing effort to clean this collection of viruses in order to return accessions to interested NARS in the countries where these genetic resources originated. Seed lots obtained from non-sterile accessions are conserved according to international standards. At CIP, dried seeds are stored in laminated aluminium foil packets in cold rooms at 0°C for immediate use or distribution and at –15°C for long-term storage. Again, for security reasons, duplicate seed samples of selected accessions in this collection are kept in the National Seed Storage Laboratory at Fort Collins, Colorado, USA.

Current *ex situ* conservation efforts for wild potato species could also be complemented by protecting geographic areas where they are found in nature. A number of potential areas in Peru, Bolivia and Ecuador have already been identified (Huamán 1994).

The primary centre of diversity of cultivated potatoes is located in the highlands of Andean South America where thousands of native cultivars are still maintained under traditional farming systems. Peru and Bolivia have the largest number of native cultivars with diverse flower and plant characters and a wide range of tuber shapes and colours. The Chiloé Archipelago of Chile is a secondary centre of diversity where hundreds of tetraploid potatoes with adaptation to long photoperiods still grow under cool environments at or near sea level in some of these islands.

Traditional potato cultivars of *S. tuberosum* subsp. *andigena* are most widespread in the highlands of Venezuela, Colombia, Ecuador, Peru, Bolivia and northern Argentina. They are also cultivated in some areas of Mexico and Guatemala. Altitude ranges from about 2000 to 4000 m asl. All other cultivated species have a more restricted geographic distribution. Thus, ancient cultivars of *S. tuberosum* subsp. *tuberosum* are grown mainly on Chiloé Island in southern Chile at about sea level. *Solanum stenotomum, S. goniocalyx* and *S.* x *chaucha* are mostly cultivated from northern Peru to central Bolivia between 3000 and 3900 m asl. *Solanum phureja* is the only species that is cultivated in the warmer Andean valleys, mostly on the eastern side of the mountain range, of Venezuela, Colombia, Ecuador, Peru and Bolivia, from 2000 to 3700 m asl. Some samples of the latter four species have been found in other countries (see Table 2.1), but these might be recent introductions. Frost-tolerant *S.* x *curtilobum, S.* x *juzepczukii* and *S.* x *ajanhuiri* grow between 3800 and 4200 m asl. The first two are cultivated from central Peru to northern Argentina and the third one is generally confined to the high Andean plateau between southern Peru and northern Bolivia.

Evaluation and Use

For many years, CIP has tried to develop closer collaboration among potato genebanks around the world for the conservation and use of potato genetic resources. Significant progress has been made since 1990 and now joint inter-genebank activities involve the potato collections maintained in the United States, Germany, the Netherlands, the United Kingdom, Russia, Argentina and CIP (Bamberg *et al.* 1995). An Inter-genebank Potato Database (IPD) has been developed. It contains 11 590 wild potato accessions maintained in seven potato genebanks, which includes 8700 different accessions of 195 wild potato species. This database shows that the number of accessions of wild potato species maintained in *ex situ* genebanks does not adequately represent the total genepool. Only a few wild species are adequately sampled throughout their geographic area of distribution.

Researchers at CIP have conducted more than 46 000 evaluations (Table 2.2) of reactions of native Andean potato cultivars maintained in the genebank to biotic and abiotic stresses as well as of other desirable traits (Huamán 1994). The potato collection is also available to NARS interested in screening against pathogens that might have a specific importance for a particular country. The evaluation database at CIP shows the presence of a number of Andean cultivars that have accumulated multiple resistances to pests and diseases or a combination of both. Frequently, some of these cultivars also show resistance to several pathotypes or races of the same organism. The breeding potential of Andean cultivars for the improvement of nutritive attributes is also remarkable. A wide range of variation is found for yield potential, time to maturity, length of tuber dormancy, storage qualities and culinary attributes.

The potato genetic resources available in genebanks are still underutilized. Ross (1986) lists 97 European potato cultivars with 13 potato species in their pedigrees. Some of these species are also in the pedigrees of some North American cultivars. Hawkes (1990) provides a list of potato species reported as the most important sources for resistance to biotic and abiotic stresses. The Inter-genebank Potato Database also includes data on 28 593 evaluations of the reactions to the most important diseases, pests and other desirable traits of wild potato accessions conserved in the most important genebanks. About 20% of the crosses made in CIP's breeding programme between 1972 and 1986 included wild species or native cultivars as one or both progenitors (Huamán 1986). So far, a total of 173 new potato cultivars have been named and released by NARS in 10 countries in South America, 15 in Asia, 13 in Africa and one in the Middle East.

Although new and more diverse genetic resources have been incorporated into breeding programmes in the last decade, the diversity used is just a small fraction of the total variation available in genebanks.

Properties and Uses

In rural areas, potato farmers generally eat more potatoes than non-producers. Where potatoes are relatively cheap, such as in the Andes, potato consumption is highest among poor households. In the more common situation where potatoes are relatively expensive, as in most of Africa and Asia, potato consumption is highest among the wealthy (Horton and Anderson 1992).

Fresh potatoes for human consumption is the most important type of utilization in most countries, both developed and developing, and accounts for from 60 to over 80% of total national output. Industrial manufacturing of processed potatoes seems to be only in its infancy in most developing countries with the exception of China (12%), Korea DPR (6%) and Mexico (28%). There is an increasing trend for the use of pre-cooked and processed potatoes to meet the rising demand of the fast food, snack and convenience food industries (FAO/CIP 1995). Over 98% of the potatoes grown in developing countries are consumed domestically, rather than exported. In some regions, like North Africa and the Middle East, potatoes are an important export crop that generates significant returns in foreign exchange (Horton and Anderson 1992). Exports of processed potato products, including frozen French fries, potato chips and starch, are estimated to be over 2 million tonnes more on a fresh weight basis (FAO/CIP 1995).

Breeding

The genetic base of potato breeding needs a permanent broadening and the major constraints in potato breeding are: a complicated tetraploid genetics, the involvement of quantitative traits with large environmental effects, and a difficult monitoring of recombination and introgression by phenotypic evaluation (Ross 1986). Transmission of important traits such as disease and insect pest resistance has been successfully achieved using $2n$ gametes, especially first division restitution $2n$ pollen which occurs widely and frequently in diploid *Solanum* species (Watanabe and Peloquin 1989).

Table 2.2.   Evaluations conducted at CIP using materials from the cultivated potato collection.

Biotic/abiotic stresses and traits evaluated		Evaluated	With useful genes[a]
		Number of accessions	
Fungi	Phytophthora blight in leaves	636	123
	Phytophthora blight in tubers	338	84
	Phytophthora pink rot	1355	2
	Streptomyces scab	520	0
	Spongospora powdery scab	548	33
	Synchytrium wart	869	56
	Angiosorus smut	224	2
	Fusarium dry rot	539	39
	Rhizoctonia black scurf	1062	0
	Phoma gangrene	490	79
	Macrophomina charcoal rot	2000	28
Bacteria	Erwinia soft rot	408	26
	Pseudomonas wilt	206	41
Viruses	Potato virus X	2058	24[bc]
	Potato virus Y	2016	6[c]
	Potato leafroll virus	2392	4[c]
	Potato virus S	2266	−[c]
	Andean potato latent virus	1831	−[c]
	Andean potato moptop virus	2267	−[c]
	Potato virus T	285	−[c]
Viroids	Spindle tuber viroid	3184	−[c]
Nematodes	*Globodera pallida*		
	Race pa2	1287	138
	Race pa3	1285	174
	Cusco population	672	65
	Globodera rostochiensis	580	95
	Meloidogyne incognita acrita	1332	70
	Nacobbus	11	4
Insects	*Phthorimaea* tuber moth	1479	152
	Andean potato weevil		
	Premnotrypes suturicallus	681	59
	Premnotrypes vorax	2000	17
	Premnotrypes latithorax	47	20
Environmental stress	Frost	565	79[d]
	Hail	3184	819
Other desirable traits		806	23
	Diploids with 2*n* pollen		
	Dry matter content	1788	400[e]
	Protein content (total N)	840	45[e]
	Palatability (flavour)	936	29[e]
	Culinary quality (texture)	936	128[e]
	Dormany period in the tubers	2201	115[e]
Total		46 124	2979

[a] Includes resistant and moderately resistant cultivars; [b] Includes hypersensitive cultivars; [c] cultivars with negative serological reactions but no further testing has been made; [d] with two layers of palisade parenchyma on leaf cross-sections; [e] dry matter >24%, protein >10% dry weight, tasty flavour, starchy texture, tuber dormancy about 6 months.

Research conducted at CIP has shown that heritability for yield varied from 0 to 0.65 according to the population, being more frequently from 0.1 to 0.25; for number and weight of tubers, heritability was 0.5 and 0.7 respectively; for tuber initiation, a component for earliness, 0.65; for dry matter content, 0.7-0.85; and for reducing sugar, 0.4. Heritability for immunities to PVY and PVX was high and monogenic dominant; for resistance to PLRV, 0.24; for vertical resistance to late blight controlled by major R genes derived from *S. demissum*, 0.7 and for horizontal resistance, polygenic and non race-specific, 0.4; for resistance to early blight, 0.7-0.8, suggesting few genes with additive effects (Mendoza 1980; Mendoza and Martin 1989).

CIP's breeding programme maximizes the utilization of commercial cultivars and breeding lines, Andean cultivated species and a set of selected wild species containing valuable traits of resistance or tolerance to stresses, pests and diseases. The basic goals are the maintenance of a wide genetic diversity to secure a high yield and stability of performance; an increase in the frequency of genes controlling desirable attributes, such as adaptation, yield *per se* and resistance or tolerance to biotic and abiotic stresses; and the stimulation of the recombination of desirable attributes in the same genotypes and populations. To achieve these goals a recurrent selection with progeny testing for the population improvement has been used (Mendoza and Jatala 1985).

By breeding for multiple traits in its advanced populations, CIP provides a broad range of durable protection against biotic and abiotic stresses that often interact to reduce the impact of single-factor resistance and desirable agronomic traits. At present, CIP's potato breeding populations combine resistance to potato leafroll virus (PLRV), potato virus X (PVX) and potato virus Y (PVY) with heat tolerance, for the warm lowlands; late blight with PVX and PVY resistances, for use across all potato-growing agro-ecologies; late blight and cyst nematode resistance, for the temperate and highland areas; and *Alternaria*, PVX and PVY resistance with heat tolerance, for the lowlands (Mendoza 1994).

Prospects

Although in the last few decades considerable effort has been devoted to evaluating the potato genetic resources to determine sources of desirable genes, a large number of accessions in genebanks still remain to be evaluated. There is no doubt that both wild potato species and Andean cultivars contain many useful genes that could solve many problems that affect potato productivity, especially in developing countries. Fortunately, the genetic resources maintained at CIP and in other potato genebanks around the world are beginning to be used with more frequency to produce new potato cultivars. Most of these new cultivars have genetic resistance to some diseases and pests that reduce the use of pesticides that affect our environment and, therefore, significantly reduce production costs to farmers. Thus, almost 200 new potato cultivars have been named and released in the last 20 years in more than 40 countries around the world.

The potato genetic resources that are held in trust at CIP are freely available for distribution in two forms: as clonal materials, either as *in vitro* plantlets or as tubers produced under quarantine conditions from pathogen-tested stocks, and as true seed lots tested for potato spindle tuber viroid (PSTVd). CIP produces annually a pathogen-tested list of potato cultivars for international distribution that contains the most relevant data for use by researchers. CIP's policy emphasizes free access to both genetic resources and data related to each accession in the genebank.

REFERENCES

Bamberg, J.B., Z. Huamán and R. Hoekstra. 1995. International cooperation in potato germplasm. Pp. 177-182 *in* International Germplasm Transfer: Past and Present (R.R. Duncan, D.M. Kral and M.K. Viney, eds.). Crop Sci. Soc. of America, Special Publication No. 23. CSSA, Madison, WI, USA.

Bukasov, S.M. 1939. The origin of potato species. Physis. Buenos Aires 18(2):269-284.

Bukasov, S.M. 1964. The origin of species of the series Andigena, Section *Tuberarium*, genus *Solanum*. International Congress of Botany, Edinburgh. 10:341-342.

Cadman, G.H. 1942. Autotetraploid inheritance in the potato: some new evidence. J. Genet. 44:33-52.

Cribb, P.J. and J.G. Hawkes. 1986. Experimental evidence for the origin of *Solanum tuberosum* subsp. *andigena*. *In* Solanaceae: Biology and Systematics (W.G. D'arcy, ed.). Columbia University Press, New York.

FAO/CIP. 1995. Potatoes in the 1990s. Situation and Prospects of the World Potato Economy. International Potato Center/Food and Agriculture Organization of the United Nations, Rome.

Hawkes, J.G. 1956. Taxonomic studies on the tuber-bearing Solanums. I. *Solanum tuberosum* and the tetraploid species complex. Proc. Linn. Soc. 166:97-144.

Hawkes, J. G. 1962. The origin of *S.* x *juzepczukii* Buk. and *S. curtilobum* Juz. et Buk. Z. für Pflanzenzucht. 47:1-4.

Hawkes, J.G. 1967. The history of potato. Masters Memorial Lecture, 1966. J. Roy. Hort. Soc. 92:207-224; 249-262; 288-302; 364-365.

Hawkes, J.G. 1990. The Potato. Evolution, Biodiversity and Genetic Resources. Belhaven Press, London.

Horton, D. 1988. Underground Crops. Long-term Trends in Production of Roots and Tubers. Winrock International.

Horton, D.E. and J.L. Anderson. 1992. Potato production in the context of the world and farm economy. Pp. 794-815 *in* The Potato Crop (Paul Harris, ed.). Chapman and Hall.

Huamán, Z. 1986. Conservation of potato genetic resources at CIP. CIP Circular 14(2):1-7.

Huamán, Z. 1994. *Ex situ* conservation of potato genetic resources at CIP. CIP Circular 20(3):1-7.

Huamán, Z. and H. Stegemann. 1989. Use of electrophoretic analyses to verify morphologically identical clones in a potato collection. Plant Varieties and Seeds 2:155-161.

Huamán, Z., J.G. Hawkes and P.R. Rowe. 1980. A biosystematic study of the origin of the diploid potato, *Solanum ajanhuiri*. Euphytica 31:665-675.

Jackson, M.T., J.G. Hawkes and P.R. Rowe. 1977. The nature of *Solanum chaucha* Juz. et Buk., a triploid cultivated potato of South American Andes. Euphytica 26:775-783.

Juzepczuk, S.W. and S.M. Bukasov. 1929. [A contribution to the question of the origin of the potato]. Proc. U.S.S.R. Congr. Genet. Plant and Animal Breed. 3:592-611.

Lizárraga, R., A. Panta, U. Jayasinghe and J. Dodds. 1991. Cultivo de tejidos para la eliminación de patógenos. Guía de Investigación CIP 3. CIP, Lima, Peru.

Mendoza, H.A. 1980. Development of lowland tropic populations. Pp. 40-55 *in* Utilization of the Genetic Resources of the Potato III. CIP, Lima, Peru.

Mendoza, H.A. 1994. Development of potatoes with multiple resistance to biotic and abiotic stresses: The International Potato Center Approach. *In* Advances in Potato Pest. Biology and Management (G.W. Zehnder, M.L. Powelson, R.K. Jansson and K.V. Raman, eds.). APS Press.

Mendoza, H.A. and P. Jatala. 1985. Breeding potatoes for resistance to root knot nematodes *Meloidogyne* spp. Pp. 217-224 *in* An Advanced Treatise on *Meloidogyne* (Sasser and Carter, eds.). North Carolina State University, USA.

Mendoza, H.A. and C. Martin. 1989. Breeding for resistance to early blight (*Alternaria solani*). Pp. 119-137 *in* Fungal Diseases of the Potato. CIP, Lima, Peru.

Ochoa, C.M. 1990. The Potatoes of South America: Bolivia. Cambridge University Press, Cambridge, UK.

Plaisted, R.L. 1971. A project to duplicate 400 years of potato evolution. New York's Food and Life Sci. 4:24-26.

Ross, H. 1986. Potato breeding: Problems and perspectives (W. Horn and G. Robbelen, eds.). J. Plant Breed. Suppl. 13:132.

Salaman, R.N. 1954. The origin of the early European potato. J. Linn. Soc. (Bot.) 55:185-190.

Schmiediche, P.E., J.G. Hawkes and C.M. Ochoa. 1982. The breeding of the cultivated potato species *Solanum juzepczukii* Buk. and *S. curtilobum* Juz. et Buk. II. Euphytica 31:695-707.

Simmonds, N.W. 1964. Studies of the tetraploid potatoes. II. Factors in the evolution of the Tuberosum group. J. Linn. Soc. (Bot.) 59:43-56.

Ugent, D., T. Dillehay and C. Ramírez. 1987. Potato remains from a late Pleistocene settlement in south central Chile. Econ. Bot. 4(1):17-27.

Watanabe, K. and S.J. Peloquin. 1989. Occurrence of 2*n* pollen and *ps* gene frequencies in cultivated groups and their related wild species in tuber-bearing Solanums. Theor. Appl. Genet. 78:329-336.

Chapter 3

Sweetpotato

Z. Huamán and D.P. Zhang

The sweetpotato belongs to a single species, *Ipomoea batatas* (L.) Lam. In Spanish, the most common names are *batata, camote* and *boniato*; in French, *patate douce*; in Portuguese, *batata doce*; in Italian, *batata dolce*; in Chinese, *gan shu*, and in Quechua, the Inca language, *kumara* or *apichu*. In the United States, the deep orange, moist-fleshed cultivars are sometimes mistakenly called yam, which belongs to the totally different genus *Dioscorea*.

Sweetpotato is the world's seventh most important food crop after wheat, rice, maize, potato, barley and cassava. It is grown in more developing countries than any other root crop. World sweetpotato production is around 124 million t in an area of about 9.2 million ha.

BOTANY AND DISTRIBUTION
The sweetpotato and the wild species closely related to it are classified in the family Convolvulaceae, genus *Ipomoea*, subgenus *Eriospermum*, section *Eriospermum* (formerly *Batatas*) and series *Batatas* (Austin and Huamán 1996).

Linnaeus described the cultivated sweetpotato in 1753 as *Convolvulus batatas*. In 1791, the botanist Lamarck described it as *Ipomoea batatas*. It is a hexaploid plant with $2n=6x=90$ chromosomes, although some plants morphologically quite similar to *I. batatas* with $2n=4x=60$ have been described and named, but they are considered synonyms of this species (Austin 1977).

Origin, Distribution and Diffusion
Abundant evidence shows that sweetpotato was spread widely through the migration routes of people in the New World tropics before the discovery of America. Two main groups of sweetpotato were known during that period: the *aje* (an Arawakan word) group, which was starchy and had a slightly sweet taste, and the *batata* group, which was also starchy but markedly sweet in taste (Austin 1988). O'Brien (1972) showed linguistic and historical evidence indicating that this crop had reached southern Peru and southern Mexico by about 2000 to 2500 BC.

Linguistic evidence has also shown three lines of dispersal of the sweetpotato from America. The *kumara* line is prehistoric and based on lexical parallels between the Quechua name and the Polynesian word *kumara*. This could explain the transfer of sweetpotato by Peruvian or Polynesian voyagers from northern South America to eastern Polynesia around 400 AD. The *batata* line dates from the first voyage of Columbus in 1492, which resulted in the introduction of West Indian sweetpotatoes to western Europe. Portuguese explorers transferred sweetpotatoes grown in western Mediterranean Europe to Africa, India and the East Indies in the 16th century. By 1594, the plant was recorded in South China and in southern Japan by 1698. The *camote* (name derived from the word *camotli* in the Mayan language Nahuatl) line represents the direct transfer of Mexican sweetpotatoes by Spanish trading galleons between Acapulco and Manila, Philippines, in the 16th century (Yen 1982).

It is generally accepted that the sweetpotato is of American origin. *Ipomoea batatas* is not known in the wild state and plants found growing wild are remnants from abandoned cultivated fields or plants coming from sweetpotato seeds, which continue growing by vegetative propagation. Several wild *Ipomoea* species having some morphological resemblance to *I. batatas* have been considered as potential wild ancestors of the sweetpotato. One of these species is *I. trifida*, which Nishiyama (1963) collected in Mexico and reported as a 6x *I. trifida* (accession K123) and claimed to be morphologically similar to *I. batatas*, except that it produced only slightly swollen storage roots. Jones (1967) demonstrated that K123 could be an *I. batatas* derivative found growing in the wild. He pointed out that other characteristics considered by Nishiyama as typical of wild plants, such as the twining habit, are also observed in sweetpotatoes and that many genotypes derived from sweetpotato seeds do not produce storage roots. Furthermore, he reported that F_1 (K123 x *I. batatas*) hybrids produced abundant seeds and the chromosome pairing in metaphase I of the hybrids was similar to that of crosses between sweetpotatoes.

It is not yet defined whether *I. batatas* is an alloploid or an autopolyploid, but sexual polyploidization through the production of unreduced gametes might have facilitated the evolution of *I. batatas* to the hexaploid level. The formation of unreduced pollen has been reported in diploid *I. trifida* (Orjeda *et al.* 1990) and in some tetraploid and hexaploid *I. batatas* (Bohac *et al.* 1992). Freyre *et al.* (1991) also reported 2n egg production in 3x *I. trifida* that generated 6x genotypes in their progenies. Nishiyama (1971) and Austin (1988) suggested an alloploid origin and that *I. trifida* is one of the species most closely related to the sweetpotato. Nishiyama (1971) proposed that sweetpotato might have originated from 2x *I. leucantha*, which produced 4x *I. littoralis*; 2x x 4x crosses between these two species might have originated 3x *I. trifida*, from which 6x *I. trifida* were derived. Further selection and domestication of these wild plants might have produced 6x *I. batatas*. Nishiyama *et al.* (1975) reported similarities in some plant characters, sexual compatibility and behaviour between sweetpotato and artificial 6x hybrids produced from *I. leucantha* and *I. littoralis*. On the basis of numerical analysis of key morphological characters, Austin (1988) hypothesized that *I. triloba* and *I. trifida* are the species that contributed the sweetpotato genome. He also considered that *I. tiliacea* may have been involved in the origin of sweetpotato. With cytogenetical evidence, Shiotani (1988) proposed that sweetpotato has the genomic structure of an autohexaploid with the B genome that also exists in autotetraploids and diploids of the *I. trifida* complex.

Austin (1988) postulated that the centre of origin of *I. batatas* was somewhere between the Yucatán peninsula of Mexico and the mouth of the Orinoco River in Venezuela, where *I. trifida* and *I. triloba* might have been crossed and might have produced the wild ancestor of *I. batatas*. Native people in the area may have discovered the sweetpotato and brought it into cultivation. By at least 2500 BC, the cultigen had most likely been spread by the Mayas and Incas to almost the limits for cultivation in Central and South America that existed at the time when the Europeans arrived. Carbon-dated sweetpotatoes discovered in the Chilca canyon on the coast of Peru were estimated to be from 8000 to 10 000 years before present, which indicates that sweetpotato may be among the world's earliest domesticates (Engel 1970; Yen 1974).

Reproductive Biology
Ipomoea batatas is a self-incompatible species.

GERMPLASM CONSERVATION AND USE

The genus *Ipomoea* comprises 600 to 700 species. Over half are concentrated in the Americas, where there may be 400 taxa, classified within the subgenera *Eriospermum*, *Quamoclit* and *Ipomoea*. These three subgenera contain 10 sections, seven of which were originally confined to the Americas before the species were dispersed as cultigens, medicinal plants and weeds (Austin and Huamán 1996).

Subgenus *Eriospermum*, section *Eriospermum*, series *Batatas* contains, in addition to *I. batatas*, 13 wild species closely related to the sweetpotato. All of these species except *I. littoralis* are endemic to the Americas. Two are considered to be of hybrid origin. *Ipomoea* x *leucantha* has been determined to be intermediate *I. cordatotriloba* x *I. lacunosa* hybrids, and *I.* x *grandifolia* has been hypothesized to include derivatives of *I. cordatotriloba* x *I. batatas* hybrids (Austin 1978). Two species that used to be considered within this group were *I. peruviana* from Peru and Ecuador, now classified within section *Eriospermum*, series *Setosae*, and *I. gracilis* from Australia, now in section *Erpipomoea*.

Since 1985, CIP has carried out 90 collecting expeditions with active participation of NARS of 16 countries in Latin America and the Caribbean. Sweetpotato genetic resources collected included a total of 1157 wild accessions: 532 accessions of 11 wild species in series *Batatas* (Table 3.1), 419 accessions of 52 other wild species and 206 accessions of other wild or weedy materials.

During those collecting expeditions, abundant samples of sweetpotato cultivars native to Latin America were also obtained (Huamán and De la Puente 1988). The cultivated collection at CIP was further expanded by donations of smaller collections from other countries, the transfer of sweetpotato collections maintained in other international centres such as AVRDC in Taiwan and IITA in Nigeria, and donations of breeding lines or advanced cultivars from several countries. The genebank at CIP now maintains a total of 5526 cultivated accessions (Table 3.2), comprising 4168 accessions of native and advanced cultivars from 57 countries (22 in the Americas, 26 in Asia and 9 in Africa) and 1358 breeding lines.

Altitudes of the sites where sweetpotatoes have been collected range from 0 to 3000 m. In Latin America, sweetpotatoes have been found at 1900-2500 m in Bolivia, Colombia and Venezuela, and up to 3000 m in Ecuador and Peru. In Asia, sweetpotatoes have been found growing from 1900 to 2700 m asl only in New Guinea.

The primary centre of diversity of sweetpotatoes is in northwestern South America (Colombia, Ecuador and Peru) and parts of Central America (such as Guatemala) where a great diversity of native sweetpotatoes, weeds and wild *Ipomoea* exists. Secondary centres of sweetpotato diversity outside of the Americas are in China, Southeast Asia, New Guinea and East Africa (Yen 1982; Austin 1983, 1988). Sweetpotato germplasm found outside the Americas, however, has been reported to contain only a small sample of the Latin American variability (Yen 1974).

Because of the asexual propagation of sweetpotato cultivars, numerous duplicate accessions of the same cultivar have been found in the cultivated collection maintained at CIP. There is an ongoing effort to identify these duplicate accessions. Those that are morphologically identical and produce the same electrophoretic banding patterns or DNA fingerprints are considered duplicates. The number of Peruvian accessions in the collection has so far been reduced from 1939 to 1099. Similar work conducted with 1373 accessions from Saint Vincent, the Dominican Republic, Jamaica, Paraguay, Argentina, Brazil and Mexico showed that 731 of them might comprise only 169 different cultivars.

Studies of genetic diversity based on RAPD markers or DNA amplified fingerprints (DAF) showed that sweetpotato exhibits a very high degree of genetic polymorphism. Several accessions cluster together based on their geographic source; other accessions from South America and New Guinea also cluster together, suggesting an evolutionary relatedness; and other New Guinea sweetpotatoes are dispersed across many clusters,

Table 3.1. Geographic coverage of wild *Ipomoea* species, series *Batatas*, conserved in the genebank maintained at CIP.

Ipomoea species	No. of samples	Geographic coverage
2n=2x=30[†]		
I. cynanchifolia (C)	3	BRA (3)
I. x leucantha [§] (C)	14	ARG (4) ; COL (5); ECU (2); MEX (1); PER (1); VEN (1)
I. ramosissima (C)	31	ARG (3);BOL (6); COL (2); ECU (1); NIC (2); PER (17)
I. triloba (C)	51	COL (13); CUB (7); DOM (4); ECU (4); MEX (6); PER (10); PRY (1); VEN (6)
I. umbraticola [‡] (C)	5	MEX (2); NIC (3)
I. lacunosa (C)	0	-
I. littoralis [‡] (I)	0	-
I. tenuissima (C)	0	-
2n=4x=60		
I. tiliacea (I)	41	CUB (2); DOM (5); GTM (1); JAM (6); MEX (2); NIC (7)
I. tabascana [‡] (C)	1	MEX (1)
2n=2x, 4x		
I. cordatotriloba [‡] (C) (ex *I. trichocarpa*)	90	ARG (18); BOL (3); COL (1); MEX (2); PRY (66)
I. trifida (I)	170	COL (24); CUB (8); ECU (3); GTM (53); MEX (2); NIC (56); VEN (24)
2n=2x?		
I. grandifolia [§] (C)	126	ARG (79); BRA (2); PRY (39); URY (6)
Total	532	

[†] Ploidy by Jones 1974; Nishiyama *et al.* 1975; Austin 1988, and [‡] by Jarret *et al.* 1992; Ozias-Akins and Jarret 1994.

C= Self-compatible, I= self-incompatible (Nishiyama *et al.* 1975; [§] CIP's data).

indicating some genetic divergence, probably caused by adaptation to isolated highland ecological conditions (Jarret and Austin 1994; He *et al.* 1995; Zhang *et al.* 1996).

The conservation of the sweetpotato collections at CIP is carried out in three forms. The first form is in the field genebank at San Ramón (800 m asl) by asexual propagations to facilitate its evaluation and characterization. The sweetpotato field genebank is planted in about 2 ha and contains about 2000 accessions. Most accessions are from Peru but some pathogen-tested accessions come from other countries. About 1000 other accessions from other countries are grown in pots in a quarantine screenhouse. The second form is *in vitro* culture using a combination of low temperature (16-18°C) and osmotic stress that allows maintenance of the cultures an average of one year between transfers (Lizárraga *et al.* 1992). The *in vitro* collection is a back-up of the field genebank, and it is used for the clean-up of pathogens of selected genotypes and for their international distribution. Most accessions donated by AVRDC and IITA are maintained in *in vitro* culture. As a security measure, a back-up duplicate set of the whole sweetpotato collection is stored in the *in vitro* laboratory of the International Institute of Advanced Studies (IDEA) in Caracas, Venezuela. The third form is as true seed to secure its long-term conservation. *Ipomoea* species produce hard-coated seeds of the orthodox type, which means that these seeds can be dried to about 4-5% moisture content and stored at subzero temperatures for several decades.

Properties and Uses
Few plants have the sweetpotato's versatility in uses. These range from consumption of fresh roots or leaves to processing into animal feed, starch, flour, candy

Table 3.2. Geographic coverage of *Ipomoea batatas* conserved in the genebank maintained at CIP.

Area and country	Number of accessions	Country	Number of accessions
America			
Argentina (ARG)	106	Jamaica (JAM)	52
Bolivia (BOL)	78	Mexico (MEX)	22
Brazil (BRA)	149	Nicaragua (NIC)	11
Chile (CHL)	1	Panama (PAN)	47
Colombia (COL)	174	Paraguay (PRY)	73
Costa Rica (CRI)	40	Peru (PER)	1099
Cuba (CUB)	207	Puerto Rico (PRI)	38
Dominican Republic (DOM)	114	Saint Vincent (VCT)	10
Ecuador (ECU)	172	United States (USA)	212
Guatemala (GTM)	100	Uruguay (URY)	2
Honduras (HND)	8	Venezuela (VEN)	86
Asia			
Australia (AUS)	3	New Caledonia (NCL)	2
Bangladesh (BGD)	4	New Hebrides (NHB)	2
Burma (BUR)	3	New Zealand (NZL)	7
China (CHN)	38	Niue (NIU)	5
Cook Islands (COK)	6	Papua New Guinea (PNG)	474
Fiji (FJI)	4	Philippines (PHL)	51
Hong Kong (HKG)	1	Singapore (SGP)	3
Indonesia (IDN)	31	Solomon Islands (SLB)	63
Japan (JPN)	142	Sri Lanka (LKA)	5
Korea (KOR)	10	Taiwan (TWN)	324
Lao Peoples Republic (LAO)	8	Thailand (THA)	94
Malaysia (MYS)	12	Tonga (TON)	18
Morocco (MAR)	1	Vietnam (VNM)	2
Africa			
Burundi (BDI)	5	Nigeria (NGA)	18
Cameroon (CMR)	4	Rwanda (RWA)	4
Egypt (EGY)	2	South Africa (ZAF)	2
Kenya (KEN)	2	Uganda (UGA)	4
Madagascar (MDG)	2		
Others			
AVRDC hybrids	38	RCB (Peru) hybrids	282
IITA hybrids	1038	Unknown country	11
		Total	5526

and alcohol. Per capita consumption of fresh roots, according to FAO, averaged 9 kg in Africa, 19 kg in Asia, 79 kg in Oceania, 3 kg in Latin America, 5 kg in Japan and 2 kg in the United States. As a food crop, sweetpotato has declined in Asia and Latin America. Feed and fodder uses, however, have become increasingly important. Sweetpotato use for animal feed currently totals 42% of production in China, 40% in Brazil, 30% in Madagascar, 11% in North Korea and 10% in Vietnam, Cuba and Peru. Sweetpotato processing into starch, flour and noodles, among other products, is also growing. Recent reports from China indicate that more than 50% of sweetpotato output goes to processors in some provinces. Processing is also increasing in Vietnam and there is an increasing interest in it in several countries of Africa and Latin America. Sweetpotato is also being incorporated as a cover crop to minimize soil erosion and to control weeds in others crops such as citruses (CIP 1996).

Breeding
Sweetpotato breeding started in the 1920s. One of the first breeding programmes for sweetpotato was developed by Julian C. Miller of Louisiana State University (Yen 1976). Today, the use of improved cultivars contributes more than any other factor to the increase in sweetpotato yield worldwide. The most used procedure for sweetpotato breeding is by polycrosses in randomized crossing nurseries (Jones 1965). A polycross combined with recurrent mass selection is recommended as an effective way to combine favourable genes and alleles in parental genotypes (Jones *et al.* 1986). This is particularly useful for long-term population improvement when dealing with low-heritability traits. Many sweetpotato traits are highly influenced by genotype by environment interactions (Collins *et al.* 1987). Within-clone and within-plot variations are usually highly significant. Therefore, selection based on individual plants is unreliable. Table 3.3 lists the most important bacterial and fungal diseases, their pathogens and available resistance sources (Clark 1988; Clark and Moyer 1988).

In cultivated sweetpotato and its wild relatives, sources of resistance to most diseases are not yet clearly understood. But many diseases depend almost exclusively on the use of resistant cultivars, such as fusarium wilt and soil rot in the southeastern United States. Variation in reaction to almost all the other important bacterial and fungal diseases of sweetpotato – bacterial soft rot, circular spot, scab, Java black rot, bacterial wilt, scurf, violet root rot, witches' broom, black rot and rhizopus soft rot – has been reported, but resistance to these diseases has not been exploited commercially (Clark 1988). Cultivars used in the United States have resistance to fusarium wilt originally derived from a single source. Frequently, quantitative resistance or tolerance has been the only available form of resistance for many sweetpotato diseases (Jones *et al.* 1979; Clark and LaBonte 1992). So far, little has been done on the quantitative genetics of these diseases, such as heritability estimation (Clark and LaBonte 1992).

Some fourteen different viruses or virus-like agents have been identified as affecting sweetpotato (Moyer and Larsen 1991). Virus diseases can cause substantial damage in both temperate and tropical regions. They are particularly serious in East Africa (Skoglund and Smit 1994). The development of virus-resistant or tolerant cultivars has been the most effective means of reducing sweetpotato losses from virus infection (Hahn *et al.* 1981). Few accessions resistant to sweet potato feathery mottle virus (SPFMV) were found after a preliminary screening in the germplasm bank held at CIP. In addition, resistance to the SPFMV complex has shown strong genotype by environment effects. Accessions identified as resistant in Peru, for instance, were susceptible in Africa. A few African cultivars have been identified as having high-level field resistance to sweetpotato virus complex. These cultivars are now used intensively as resistance sources in breeding programmes (CIP 1995).

Most sweetpotatoes grown around the world are produced on unirrigated land. Therefore, drought is the most important abiotic stress, at both transplanting time and at the start of storage-root formation. Sweetpotato is considered a drought-tolerant crop. Its roots can penetrate to about 2 m in the soil. This deep penetration enables the crop to survive under drought conditions because it can absorb water from the deeper soil layers (Bouwkamp 1985). Although relatively high yields can be achieved under dry conditions, irrigation experiments have clearly demonstrated that yield can be significantly increased by irrigation in areas where rainfall distribution is erratic or insufficient.

Table 3.3. The most important bacterial and fungal diseases, their pathogens and resistance sources in sweetpotato.

Disease	Pathogen	Resistant germplasm
Bacterial soft rot	*Erwinia chrysanthemi*	No useful resistance identified
Bacterial wilt	*Pseudomonas solanacearum*	Resistance available
Soil rot or pox	*Streptomyces ipomoea*	High-level resistance available
Witches' broom	Mycoplasma-like organism	Unknown
Black rot	*Ceratocystis fimbriata*	No high-level resistance
Foot rot	*Plenodomus destruens*	Unknown
Fusarium rot	*Fusarium oxysporum*	No high-level resistance
	Fusarium solani	Has highly resistant cultivar
Fusarium wilt	*Fusarium oxysporum* f.sp. *batatas*	Has highly resistant cultivar
Java black rot	*Diplodia gossypina*	No useful resistance identified
Rhizopus soft rot	*Rhizopus stolonifer*	No useful resistance identified
Scurf	*Monilochaetes infuscans*	No useful resistance identified
Scab	*Elsinoe batatas*	Has resistant cultivar identified
Violet root rot	*Helicobasidium mompa*	No useful resistance identified

Timely controlled irrigation experiments conducted at CIP have demonstrated large cultivar differences in drought tolerance, which is correlated with ability for deep rooting (Pallais 1995, pers. comm.) and extensive development of the root system in the early stage of growth (Yen *et al.* 1964). There is great potential to enhance sweetpotato's ability to cope with drought.

The sweetpotato collection maintained at CIP has been subject to a systematic evaluation to identify sources of desirable genes that could be used in breeding programmes. So far, more than 13 000 evaluations (Table 3.4) have been made and many accessions in the collection have been identified as potential sources of genes for resistance to pests and diseases, good agronomic characters, long storability, high productivity and high nutritive value.

A large amount of genetic variation exists for the most important economic traits in sweetpotato. So far, however, only a small fraction of the germplasm available in nature has been used in sweetpotato breeding. Fortunately, in the past 15 years efforts have been made to broaden the genetic base of modern sweetpotato cultivars and several cultivars have been developed with excellent horticultural quality combined with high levels of resistance to many insects, diseases and nematodes (Jones *et al.* 1986). However, a lot of genetic diversity still remains available in genebanks to produce cultivars with increased pest and disease resistance, which can grow reliably with low inputs under adverse climatic and soil conditions.

Wild *Ipomoea* species have been reported to have a number of desirable traits that can be useful for sweetpotato breeding (Sakamoto 1976; Kobayashi 1978). The US cultivar HiDry is a selection from open-pollinated progenies of the Japanese cultivar Minamiyutaka, which was selected from a cross between a cultivated sweetpotato and a wild *Ipomoea* species (Sakamoto 1976). The main constraints to using wild species in sweetpotato breeding are related to ploidy level and lack of production of large-sized storage roots. Ploidy differences between wild and cultivated species also make progeny testing difficult (Orjeda 1995).

In the past four years, 15 sweetpotato varieties (13 in Peru and 2 in Egypt) have been released by NARS from materials selected from native cultivars or by breeding methods using parents from the collections maintained at CIP.

Table 3.4. Results of preliminary evaluations made at CIP on reactions to biotic and abiotic stresses and other desirable traits in sweetpotato cultivars.

Traits evaluated	Number of accessions	
	Evaluated	With useful genes[†]
Nematodes		
Meloidogyne root knot	2072	329
Pratylenchus root lesion	20	16
Ditylenchus brown ring rot	356	26
Insects		
Euscepes sweetpotato weevil	1354	218
Fungi		
Diplodia Java black rot	351	268
Plenodomus foot rot	147	42
Viruses		
Feathery mottle virus (SPFMV)	1800	15
Sunken vein virus (SPSVV)	255	0[‡]
Environmental stress		
Tolerance to salinity	605	23
Excess soil moisture and aluminium toxicity	463	6
Adaptation to hot environment	463	20
Nutritive quality		
Dry matter content	1506	(>45%) 47
Starch content	902	(>75%) 35[§]
Beta carotene content	779	(>10%) 4[§]
Protein content (Total N)	902	(>10%) 14[§]
Sugar content	902	(< 3%) 4[§]
		(>15) 69[§]
Total	**13269**	**1204**

[†] Includes Resistant, Moderately Resistant and Tolerant cultivars.
[‡] Cultivars with negative serological reactions are being studied.
[§] On a dry weight basis.

Prospects

Although much progress has been made in collecting sweetpotato genetic resources, re-collecting of closely related wild *Ipomoea* species is needed from areas where no living materials have been obtained by previous expeditions or when few living accessions of a given species exist in *ex situ* genebanks. More comprehensive genetic diversity studies are also needed to determine whether the cultivated genepool available in genebanks adequately represents the genetic diversity still in existence in farmers' fields in Latin America, Africa and Asia.

The genetic resources maintained at CIP are freely available for distribution in two forms: first, as clonal materials, either as *in vitro* plantlets or as storage roots produced under quarantine conditions from pathogen-tested stocks; second, as true-seed lots. A pathogen-tested list of sweetpotato cultivars for international distribution is produced annually by CIP and it contains the most relevant data for use by researchers. Our policy emphasizes free access to both genetic resources and data related to each accession in the collections.

Sweetpotato will become more significant in the future, particularly in the developing world. As the human population grows, farmers will have to use more marginal land with diminishing resources. Today, sweetpotato has shifted from a staple food to a much more diverse pattern of uses, such as animal feed and processing in many developing countries. Demand will increase for cultivars with high starch, high yield and a low use of inputs. Sweetpotato is thought to have the largest potential for yield improvement of any major crop in Asia. But resistance to insects, diseases and abiotic stresses needs to be incorporated in order to produce the crop more efficiently under adverse soil and climatic conditions.

REFERENCES

Austin, D.F. 1977. Hybrid polyploids in *Ipomoea* section *Batatas*. J. Hered. 68:259-260.

Austin, D.F. 1978. The *Ipomoea batatas* complex. I. Taxonomy. Bull. Torrey Bot. Club 105:114-129.

Austin, D.F. 1983. Variability in sweetpotato in America. Proc. Am. Soc. Hort. Sci. Tropical Region 27(Part B):15-26.

Austin, D.F. 1988. The taxonomy, evolution and genetic diversity of sweet potatoes and related wild species. Pp. 27-59 *in* Exploration, Maintenance and Utilization of Sweet Potato Genetic Resources. Report of the First Sweet Potato Planning Conference 1987. International Potato Center, Lima, Peru.

Austin, D.F. and Huamán Z. 1996. A synopsis of *Ipomoea* (Convolvulaceae) in the Americas. Taxon 45(1):3-38.

Bohac, J.R., A. Jones and D.F. Austin. 1992. Unreduced pollen: proposed mechanism of polyploidization of sweetpotato (*Ipomoea batatas*). HortScience 27:611.

Bouwkamp, J.C. 1985. Production requirements. Pp. 9-33 *in* Sweet Potato Products: a Natural Resources for the Tropics (J.C. Bouwkamp, ed.). CRC Press, Boca Raton, Florida.

CIP. 1995. 1993-1994 Progress report, International Potato Center, Lima, Peru

CIP. 1996. Sweetpotato facts. International Potato Center, Lima, Peru.

Clark, C.A. 1988. Principal bacterial and fungal diseases of sweetpotato an their control. Pp. 275-290 *in* Exploration, Maintenance and Utilization of Sweetpotato Genetic Resources - Report of the First Sweetpotato Planning Conference 1987.

Clark, C.A. and D.R. LaBonte. 1992. Disease factors in breeding and biotechnology for sweetpotato. Pp. 484-494 *in* Sweetpotato Technology for the 21st Century (W.A. Hill, C.K. Bonsi and P.A. Loretan, eds.). Tuskegee University, Tuskegee, Alabama, USA.

Clark, C.A. and J.W. Moyer. 1988. Compendium of Sweetpotato Diseases. APS Press, St. Paul, MN.

Collins, W.W., L.G. Wilson, S. Arrendell and L.F. Dickey. 1987. Genotype x environment interactions in sweet potato yield and quality factors. J. Am. Soc. Hort. Sci. 112:579-583.

Engel, E. 1970. Exploration of the Chilca Canyon. Curr. Anthropol. 11:55-58.

Freyre, R., M. Iwanaga and G. Orjeda. 1991. Use of *Ipomoea trifida* (HBK.) G. Don germplasm for sweetpotato improvement. 2. Fertility of synthetic hexaploids and triploids with $2n$ gametes of *I. trifida*, and their interspecific crossability with sweet potato. Genome 34:209-214.

Hahn, S.K., E.R. Terry and K. Leuschner. 1981. Resistance of sweet potato to virus complex. HortScience 16:535-537.

He, G., C.S. Prakash and J.L. Jarret. 1995. Analyses of genetic diversity in a sweetpotato (*Ipomoea batatas*) germplasm collection using DNA amplification fingerprinting. Genome 38(5):938-945.

Huamán, Z. and F. De la Puente. 1988. Development of a sweet potato genebank at CIP. CIP Circular 16(2):1-7.

Jarret, R.L. and D.F. Austin. 1994. Genetic diversity and systematic relationships in sweetpotatoes [(*Ipomoea batatas* (L.) Lam.] and related species as revealed by RAPD analyses. Genet. Res. Crop Evol. 41:165-173.

Jarret, R.L., N. Gawel and A. Whittemore. 1992. Phylogenetic relationships of the sweetpotato [*Ipomoea batatas* (L.) Lam.]. J. Am. Soc. Hort. Sci. 117(4):633-637.

Jones, A. 1965. A proposed breeding procedure for sweetpotato. Crop Sci. 5: 191-192.

Jones, A. 1967. Should Nishiyama's K123 (*Ipomoea trifida*) be designated *I. batatas*? Econ. Bot. 21:163-166.

Jones, A. 1974. Chromosome numbers in the genus *Ipomoea*. J. Hered. 55:216-219.

Jones, A., P.D. Dukes and J.M. Schalk. 1986. Sweetpotato breeding. Pp. 1-35 *in* Breeding Vegetable Crops (M.J. Bassett, ed.). AVI Publ. Co., Westport, Connecticut.

Jones, A., J.M. Schalk and P.D. Dukes. 1979. Heritability estimates for resistance in sweet potato to soil insects. J. Am. Soc. Hortic. Sci. 104(3):424-426.

Kobayashi, M. 1978. Sweet potato breeding method using wild relatives in Japan. Proc. Symp. Pp. 1-8 *in* Trop. Agric. Research.

Lizárraga, R., A. Panta, N. Espinoza and J.H. Dodds. 1992. Tissue Culture of *Ipomoea batatas*: micropropagation and maintenance. CIP Research Guide 32. International Potato Center, Lima, Peru.

Moyer, J.W. and R.C. Larsen. 1991. Management of insect vectors of viruses infecting sweetpotato. Pp. 341-358 *in* Sweetpotato Pest Management: a Global Perspective (R.K. Jansson and R.K. Raman K.V., eds.). Westview, Boulder, Colorado, USA.

Nishiyama, I. 1963. The origin of the sweetpotato plant. Pp. 119-128 *in* Plants and the Migrations of Pacific Peoples (J. Barrau, ed.). Bishop Museum Press, Honolulu.

Nishiyama, I. 1971. Evolution and domestication of the sweet potato. Bot. Mag. Tokyo 84:377-387.

Nishiyama, I., T. Miyazaki and S. Sakamoto. 1975. Evolutionary autoploidy in the sweet potato (*Ipomoea batatas* (L.) Lam.) and its progenitors. Euphytica 24:197-208.

O'Brien, P.J. 1972. The sweet potato: its origin and dispersal. Am. Anthropologist 74:343-365.

Orjeda, G. 1995. Ploidy manipulations for sweet potato breeding and genetic studies. PhD. Thesis, Faculty of Science, University of Birmingham, UK.

Orjeda, G., R. Freyre and M. Iwanaga. 1990. Production of 2*n* pollen in diploid *Ipomoea trifida*, a putative wild ancestor of sweetpotato. J. Hered. 81:462-467.

Ozias-Akins, P. and R.L. Jarret. 1994. Nuclear DNA content and ploidy levels in the genus *Ipomoea*. J. Am. Soc. Hort. Sci. 119(1):110-115.

Sakamoto, S. 1976. Breeding of a new sweet potato variety, Minamiyutaka, by the use of wild relatives. J. A.R.Q. 10:183-186.

Shiotani, I. 1988. Genomic structure and the gene flow in sweet potato and related species. Pp. 61-73 *in* Exploration, Maintenance and Utilization of Sweet Potato Genetic Resources. Report of the First Sweet Potato Planning Conference 1987. International Potato Center, Lima, Peru.

Skoglund, L.G. and N.E.J.M. Smit. 1994. Major Diseases and Pests of Sweetpotato in Eastern Africa. International Potato Center (CIP), Lima, Peru.

Yen, D.E. 1974. The sweet potato and Oceania. Bishop Museum Bull., Honolulu 236:1-389.

Yen, D.E. 1976. Sweetpotato *Ipomoea batatas* (Convolvulaceae). Pp. 42-45 *in* Evolution of Crop Plants (N.W. Simmonds, ed.). London.

Yen, D.E. 1982. Sweet potato in historical perspective. Pp. 17-30 *in* Sweet Potato (R.L. Villareal and T.D. Griggs, eds.). Proceedings of First International Symposium, AVRDC Publ. No. 82-172.

Yen, C.T. , C.V. Chu and C.L. Sheng. 1964. Studies on the drought resistance of sweetpotato varieties. Crop Sci. (China) 3: 183-190.

Zhang, D.P., M. Ghislain, Z. Huamán and A. Golmirzaie. 1996. RAPD analyses of genetic diversity in sweetpotato cultivars from South America and Papua New Guinea. International Plant Genome Conference, January 14-18, 1996, San Diego, California. p. 43. (Abstr.).

Chapter 4

Other Andean Roots and Tubers

C. Arbizu, Z. Huamán and A. Golmirzaie

There are nine species of other Andean root and tuber crops (ARTC). These crops are adapted to three different agro-ecological conditions of the Andes. First is the warm Andean valley, where five species that produce edible roots or rhizomes are cultivated: arracacha, achira, yacón, mauka and ahipa. Second are the temperate Andean valleys, where the tuber crops ulluco, oca and mashua are grown associated with Andean potatoes. Third are the Andean highlands, where a root-hypocotyl crop called maca is grown because of its frost tolerance.

The ARTC are classified in different taxonomic families, and are therefore totally different crops (Table 4.1). They also differ in their underground edible part, agro-ecology, storage behaviour, propagation, adaptation, use and economic potential. Andean farmers use these crops for food in different ways. Crops such as yacón and ahipa are eaten raw, whereas others have to be cooked. Most of them store starch, but yacón stores sugar.

The tuber crops oca, ulluco and mashua share the same Andean ecological niches of cultivated potatoes and have a high degree of phenotypic diversity, especially in their tubers. They are grown from 2500 to 4000 m asl from the Andes of Venezuela to northwestern Argentina, with the highest cultivation frequency from 3000 to 3900 m from central Peru to central Bolivia (Arbizu and Robles 1986; King 1988; Franco *et al.* 1989; Tapia *et al.* 1996). Small plots of these crops are grown up to 4200 m in the central highlands of Peru. These crops are also considered to be sensitive to daylength (Bukasov 1930; Palmer 1982). They are grown under the short-day conditions of the Andes (11-12 hours), with rainfall ranging from about 400 to 700 mm distributed during the growing season.

ANDEAN TUBER CROPS

Oca (Oxalis tuberosa *Mol.*)

The name oca is derived from the Quechua word *okka*, *oqa* or *uqa*. Other names are *apilla* in Aymara; oca, oxalis and yam (New Zealand) in English; *truffette acide* in French; *knollen-sauerklee* in German; in Spanish spoken in South America, other names besides *oca* are *ibias*, *cuiba*, *huisisai*, *macachin* and *miquichi*. In Mexico, it receives potato names such as *papa roja*.

Oca is an annual tuberous herb, with succulent green, yellow, pink, red and purplish red stems. Its leaves resemble clover in shape, and the flower structure of heterostily facilitates cross-pollination. The flower sets many viable seeds (Cortez 1978, 1985) in capsules of explosive dehiscence. Tubers vary in shape and skin colour. In the Andes, the crop is propagated only by planting the tubers. Oca tubers are ready for harvesting 7 months after planting.

Table 4.1. Main features of Andean root and tuber crops.

Crop	Botanical name	Family	Edible part	Altitude (m)	Uses Current	Potential
Oca	Oxalis tuberosa	Oxalidaceae	Tuber	3000-4000	Staple (boiled, baked)	Industrial starch for bakery
Ulluco	Ullucus tuberosus	Basellaceae	Tuber	3000-4000	Staple (soups, stews)	Cash crop, medicine
Mashua	Tropaeolum tuberosum	Tropaeolaceae	Tuber	3000-4000	Staple (boiled, baked)	Industrial starch, phar-maceutical
Arracacha	Arracacia xanthorrhiza	Apiaceae	Root	1000-3200	Staple (soups, puddings)	Industry, instant food
Achira	Canna indica	Cannaceae	Rhizome	2000-2700	Baked, industrial starches	Noodles, bakery products
Yacón	Smallanthus sonchifolius	Asteraceae	Root	1300-3200	Snacks	Industry, diet and diabetic food
Maca	Lepidium meyenii	Brassicaceae	Hypocotyl	3900-4500	Juices, cocktail mixes	Pharma-ceutical
Mauka	Mirabilis expansa	Nyctaginaceae	Root	2300-3200	Salty and sweet preparations	Industrial starch
Ahipa	Pachyrhizus ahipa	Leguminosae	Root	1500-3000	Green and fruit salads	Cash crop, insecticide

Knuth (1930) classified oca within the family Oxalidaceae, genus *Oxalis*, section *Tuberosae*, species *Oxalis tuberosa*. There are more than 80 other species of *Oxalis* in the Andes, particularly in Peru (Macbride 1949; Ferreyra 1986; Pool 1993). Some of them set small tubers, but only *O. tuberosa* is cultivated.

Some wild species of *Oxalis* have been found to be diploid ($2n=2x=16$), tetraploid ($2n=4x=32$) and hexaploid ($2n=6x=48$), but *O. tuberosa* is octoploid ($2n=8x=64$) (de Azkue and Martínez 1990; Valladolid et al. 1994; Valladolid 1996).

The wild ancestor of the cultivated oca is unknown. Oca is an ancient crop that might have been domesticated between central Peru (10°S) and central Bolivia (20°S) in pre-Columbian times, probably more than 4000 years ago (Hawkes 1989). The migration of pre-Columbian people extended its cultivation northward and southward from this area. At the time of the Spanish conquest, oca was cultivated from Venezuela (8°N) to northwestern Argentina and Chile (25°S). Spanish chroniclers such as Garcilaso and Cobo gave some detailed information about oca cultivation, its morphological features and its uses in the Andes (Yacovleff and Herrera 1934-35). Although Yacovleff and Herrera (1934-35) indicated that remains of oca tubers were not recovered from sites in the Peruvian west coast, Towle (1961) claimed to have found one specimen of a small, dried tuber of oca in 1948 in the Inca ruins of Pachacamac (1200-1500 AD) near Lima. Oca plants and tubers are also represented in large vessels (about 0.9 m high) of the Wari culture (600-1100 AD) in Ayacucho, Peru.

The greatest genetic diversity and multiple uses occur from the central Andes of Peru to central Bolivia (Rea and Morales 1980; Arbizu and Robles 1986; King 1988). Bitter and sweet cultivars of oca identified more than 50 years ago are still found in this area.

At present, oca is cultivated from the Andes of Venezuela southward to northwestern Argentina and northeastern Chile and from about 2500 to 4000 m asl. In

Mexico, oca is cultivated from 2400 to 3000 m asl (King 1988) and in New Zealand close to sea level (NRC 1989). Although most cultivars are grown under short days in the Andes, cultivars grown in New Zealand at 40-46°S are adapted to long photoperiods and might have originated in central or southern Chile (King 1988).

Andean genebanks maintain 3899 accessions of oca (Table 4.2), but the number of different cultivars in these collections is still unknown. Morphological characterization of 232 Peruvian accessions of oca resulted in the identification of 134 morphotypes (Vivanco and Arbizu 1995). Similarly, morphological characterization of 171 Ecuadorian accessions of oca identified 31 morphotypes (Tapia *et al.* 1996).

Oca is eaten boiled or baked like potato throughout the Andes. It can also be eaten in soups and stews; raw ocas are eaten as snacks. At times of overproduction, oca tubers are dehydrated into *kaya*, which is consumed in desserts (puddings), porridges and omelets. *Kaya* is much preferred by rural Andean women after childbirth, who believe that *kaya* helps them recover quickly. This is likely not because of its nutritional value, but because of four antibiotics (penicillin, streptomycin, ampicillin and nystatin) reported to be fixed, probably during the soaking stage of *kaya* processing. These antibiotics apparently protect rural Andean women from infection after childbirth (Flores 1991).

Ulluco *(Ullucus tuberosus Caldas)*

The name ulluco is derived from the Quechua word *ulluku* (*ullu* means male organ) (Soto 1976). Other names in Aymara are *ulluma* or *illaco*; in Spanish, besides *ulluco*, it is called *melloco, papa lisas, michuri, micuchi, tiguiño, timbos, camarones de tierra, chigua* and *rubas*. The names for wild ullucos in Peru are *atuqpa ulluku, atuq lisas* and *kita lisas*.

Ulluco is an erect, compact, succulent and mucilaginous herb, with glabrous and ridged green to red pinkish stems and alternate leaves. Flowers are bisexual and rarely set seeds. Ulluco is considered an outbreeder and its ability to set seeds depends on the genotype. In addition, environmental factors strongly influence seed production (Pietilá 1995). Tubers are of different shapes and colours and are stable under different environments (Rousi *et al.* 1989; Pietilá and Rousi 1991). The cropping cycle of ulluco is about 7 months.

This crop is classified within the family Basellaceae, genus *Ullucus*, species *Ullucus tuberosus* (Sperling 1987). The most recent taxonomic treatment of this species includes all wild ullucos within *U. tuberosus* subsp. *aborigineus* and all cultivated forms within *U. tuberosus* subsp. *tuberosus* (Sperling 1987).

Both cultivated and wild ullucos were found to be diploid ($2n=2x=24$) by Sperling (1987), but cultivated diploid and triploid ($2n=3x=36$) forms also have been reported (Cárdenas and Hawkes 1948; Gandarillas and Luizaga 1967; Lescano 1985; Larkka 1991; Larkka *et al.* 1992; Méndez *et al.* 1994; Méndez 1995). Méndez (1995) and Méndez *et al.* (1994) also reported a tetraploid ($2n=4x=48$) Peruvian ulluco. All wild ullucos have so far been found to be triploid (Larkka 1991; Larkka *et al.* 1992; Méndez *et al.* 1994; Méndez 1995).

According to Sperling (1987), the wild ancestor of the cultivated ulluco would be *U. tuberosus* subsp. *aborigineus*. Like *Oxalis tuberosa*, ulluco was probably brought into cultivation more than 4000 years ago (Hawkes 1989). Its domestication could have occurred between the central Andes of Peru and Bolivia. Martins (1976) identified ulluco starch from 4050 to 4250 BP in tuber remains from Ancón-Chillón, Peru. Large ceremonial vessels of the Robles Moqo style of Wari culture (600-1100 AD), which existed in Ayacucho, Peru, show representations of ulluco. The first Spanish chroniclers did not mention ulluco, probably because they confused it with potato. Linguistic evidence shows that distribution of ulluco north of Ecuador occurred near the end of colonial times (Sperling 1987).

Ulluco also shows a high degree of genetic diversity and multiple uses from the central Andes of Peru to central Bolivia (Rea and Morales 1980; Arbizu and Robles 1986; King 1988; Rousi *et al.* 1989). The provinces of Cañar, Pichincha, Imbabura and Chimborazo in Ecuador were also found to be important centres of ulluco diversity

Table 4.2. Estimated accessions of ARTC maintained by Andean germplasm banks.

Common name	ECU[†]	PER[‡]	BOL[§]	CIP	Total
Oca	171	2961	303	464	3899
Ulluco	287	1689	63	437	2476
Mashua	89	828	39	76	1032
Arracacha	93	436	–	48	577
Achira	30	297	–	35	362
Yacón	32	378	–	33	443
Maca	–	–	–	33	33
Mauka	11	103	–	5	119
Ahipa	64	–	–	5	69
Total	**777**	**6692**	**405**	**1136**	**9010**

[†] INIAP.
[‡] INIA, UNSAAC, UNSCH, UNMSM, UNC.
[§] IBTA-PROINPA.

(Castillo *et al.* 1988).   This high genetic diversity could be explained by sexual reproduction. But ulluco rarely set seeds in the highlands and therefore most variation probably originated through somatic mutation (Benavides 1976; Rea 1980; Rousi *et al.* 1989) followed by aesthetic selection carried out by Andean farmers (Hawkes 1983).

Andean genebanks maintain 2476 accessions of cultivated ulluco (Table 4.2), but the number of different cultivars is unknown.   These collections likely contain a high number of duplicate accessions of the same cultivar.   Thus, morphological chraracterization of 160 Peruvian accessions indicated 86 morphotypes (Vivanco and Arbizu 1995). Similarly, out of 287 Ecuadorian accessions, only 57 morphotypes were identified (Tapia *et al.* 1996).

The ulluco tubers are consumed by Andean people as staple food usually eaten in soups and stews with meat and vegetables. There are a number of traditional dishes in Peru, Bolivia and Ecuador. Contemporary dishes incorporate ulluco tubers in salads. Ulluco leaves are also eaten in soups and salads. Dehydrated ulluco is usually eaten in soups and stews in the Peruvian highlands.

Ulluco is also used as a medicine in the Peruvian highlands. In some rural areas, some well-known ulluco cultivars are preferred for treating childbirth problems. Frozen slices of ulluco tubers are also used to treat traumatic injuries and mumps, and ground ulluco is used to treat kidney pain.

Mashua (**Tropaeolum tuberosum** *Ruiz and Pavón*)

The name mashua is derived from the Quechua names *maswa* or *mashwa*.   Other names are *añu* in Quechua; *isaño* in Aymara; in Spanish, *mashua* in Peru and Ecuador, and *cubio* or *navo* in Colombia; and mashua in English.

This tuber crop is an annual herb that resembles nasturtium.   Plants are totally glabrous, with green to dark purplish stems, with twining petioles and alternate and peltate leaves. Flowers are orange, with fused sepals forming a spurred calyx. They are also zygomorphic and single and have long peduncles. Mashua sets many viable seeds with high germination rates (Cortez 1985).   Tubers vary in shape and colour and are harvested about 7 months after planting.

Mashua is classified within the family Tropaeolaceae and genus *Tropaeolum*, section *mucronata*, species *Tropaeolum tuberosum* Ruiz and Pavón, which has two subspecies: *tuberosum* and *silvestre*. The first subspecies sets tubers and the second one does not (Sparre 1973; Sparre and Anderson 1991).   Whereas the cultivated subsp. *tuberosum* produces the secondary compound p-methoxybenzyl isothiocyanate, the wild subsp. *silvestre* releases benzyl, 2-propyl and 2-butyl isothiocyanates (Kjaer *et al.* 1978; Johns and Towers 1981).

Preliminary studies on the chromosome number of mashua indicate that the plant appears to have 42 chromosomes. This confirms previous information given by Darlington and Janaki-Ammal (1945).

Although 86 wild species of *Tropaeolum* grow from Mexico to temperate South America (Sparre 1973; Sparre and Anderson 1991), the wild prototype of mashua is unknown. This crop has been under cultivation for thousands of years. Its representations appear on ceremonial vessels of the Robles Moqo style from the Wari culture in Peru (600-1100 AD). The first chroniclers, Garcilaso and Cobo, mentioned añu and isaño as having anti-aphrodisiacal properties. It is said that when the Inca army went to conquer new territories, the soldiers were fed with mashua so as to forget their wives (Yacovleff and Herrera 1934-35).

The greatest diversity of the crop and its uses appears to occur from the central Peruvian Andes to central Bolivia (Arbizu and Robles 1986; King 1988; Rea and Morales 1980). Some cultivars mentioned by Herrera in 1934 are still grown in the Andes of Cusco, Peru. The provinces of Cañar and Carchi in Ecuador were also found to be important centres of mashua diversity (Castillo *et al.* 1988), which can be explained by sexual reproduction, although the crop is traditionally propagated by planting tubers.

Mashua is cultivated from the Andes of Venezuela to northwestern Argentina at altitudes ranging from 2600 to 4000 m. But its greatest concentration is located from central Peru to central Bolivia (Rea and Morales 1980; Arbizu and Robles 1986; King 1988; Franco *et al.* 1989; Tapia *et al.* 1996). Although mashua is a short-day adapted crop, King (1988) mentions production of mashua tubers in New Zealand under long-day photoperiod.

Andean genebanks maintain 1032 accessions of mashua (Table 4.2), but the number of different cultivars is unknown. Preliminary evaluations of genetic diversity involving 64 Peruvian accessions of mashua showed 44 morphotypes (Vivanco and Arbizu 1995), whereas 89 Ecuadorian accessions of mashua showed 23 morphotypes (Tapia *et al.* 1996).

Mashua has a higher protein content than oca. A cultivar with 16% protein on a dry weight basis has been found (Valladolid, pers. comm.). Boiled mashua used to feed weaning pigs has been reported to produce a higher weight gain and feed efficiency than other rations containing raw tubers and the grain control (Bateman 1961). Similarly, yearling calves fed a ration containing boiled mashua had the same performance as those fed the conventional ration containing cereals (Ramos *et al.* 1976). Mashua was found to be high in the amino acids isoleucine, lysine and valine. It also had higher levels of calcium, iron, riboflavin and ascorbic acid than potato (King 1988).

Before mashua is eaten, its tubers have to be exposed to sunlight for 4-6 days to prevent the rather noticeable odour in the tubers caused by isothiocyanates. In some rural areas of the Andes, mashua tubers are popular among women but not with men, who usually refuse to eat them because they believe that mashua causes impotence. Thus, mashua has the reputation of being an anti-aphrodisiac. Experiments with male rats fed with mashua showed a 45% drop in testosterone/dihydrosterone in their blood levels (Johns *et al.* 1982).

The plant has been shown to have insecticidal and nematocidal properties because it contains isothiocyanates (Johns *et al.* 1982). Traditionally, mashua is considered a repellent plant for pests and in some parts of the Andes mashua is planted as a fence around potato fields to protect them from pests. Mashua is also traditionally used as a medicine to cure kidney ailments and as a diuretic for humans. Some liquid mixtures based on mashua can be used to treat rumen problems in cattle.

ANDEAN ROOT CROPS

Arracacha (**Arracacia xanthorrhiza** *Bancroft*)

The name arracacha is derived from the Quechua word *raqacha*. Other names are *lakachu* in Aymara; *pueb* in Amusha, the language of an Amazonian tribe; other names in Spanish in South America besides *arracacha* are *virraka, zanahoria blanca* and *apio criollo*; in Portuguese it is *mandioquinha salsa, batata baroa, batata aipo* and *aipo do Perú*, among others; in English it is arracacha, white carrot, Peruvian carrot and Peruvian parsnip; in French it is *arracacha, pomme de terre céléri* and *panème*.

This root crop is a perennial plant with short and cylindrical stems and large and numerous green to purple compound leaves with long petioles. The upper part of arracacha resembles celery. It flowers and sets seeds rather occasionally in the Andes. The flowers are yellow or purple, small and arranged in an umbel inflorescence. Storage roots are clustered around the stem; their shape ranges from ovoid to fusiform and their colour ranges from white to yellow. Arracacha roots are very perishable. For centuries, however, Andean farmers in Peru have dehydrated slices of arracacha roots into *qawi*. Harvesting of arracacha takes place from 10 to 12 months after planting.

Arracacha is classified within the family Apiaceae (Umbelliferae), genus *Arracacia*, species *Arracacia xanthorrhiza*. The number of species in the genus distributed in the mountainous regions of Mexico, Guatemala, Costa Rica, Panama, Peru and Bolivia ranges from 30 to 36 (Constance 1949; Mathias and Constance 1962, 1976; Hiroe 1979).

Little is known about the reproductive biology of arracacha. According to dos Santos (pers. comm.), the rate of cross-pollination in arracacha is high. Giordano *et al.* (1994) reported the use of open-pollinated seeds harvested from commercial production fields to generate segregating populations for breeding purposes. In Brazil, an ongoing breeding programme uses cultivated arracacha and two wild species (dos Santos, pers. comm.).

Arracacha is the only Apiaceae domesticated in the New World (León 1967). The wild ancestor is unknown. Constance (1949) indicated that *Arracacia equatorialis* and *A. andina* may be the closest relatives of the cultivated arracacha. However, he considers now that *A. andina* is a synonym of *A. xanthorrhiza* and that *A. equatorialis* is probably not related to the cultivated arracacha (Constance 1996, pers. comm.). This crop is among the most ancient cultivated plants of South America (Bukasov 1930). Some drawings on pottery from the Nazca culture in Peru dated about 2000 years ago show storage roots debated to be either cassava (Yacovleff and Herrera 1934-35) or arracacha (Hodge 1954). The first Spanish chroniclers mentioned arracacha as "certain roots as thick as the arm, similar in flavour and odour to carrots but without a hard medulla or stem in the centre as carrots, all this fruit and root is eaten very well" (Yacovleff and Herrera 1934-35). From the Andes, arracacha was introduced into Mexico, Guatemala, Costa Rica, Panama and Puerto Rico, apparently at the end of colonial times. The plant was also introduced into Jamaica by the British (Hodge 1954) and then into India and Sri Lanka at the end of the last century. A number of unsuccessful attempts were made to introduce arracacha into the United States, England, France and Switzerland after the 1820s (Hodge 1954).

Diversity of the crop appears to be evenly distributed in the warm Andean valleys of Colombia, Ecuador, Peru and Bolivia. Arracachas in this area are white, yellow or purple. Foliage of Colombian and Peruvian arracachas also ranges from yellowish green to purplish. Hodge (1954) reported that the Quillacingas from the Putumayo region in Colombia distinguish up to 11 different cultivars. Similarly, more than 12 cultivars are named and grown by farmers in the upper part of La Convención valley in Cusco, Peru (Arbizu and Robles 1986; Meza 1995a).

In warm Andean valleys, where potatoes are difficult to grow, arracacha is a good substitute. It is grown under somewhat wet climates from 1500 to 3200 m asl, with temperatures ranging from 12 to 22°C and rainfall of 600-1200 mm. Low temperatures result in a long cropping cycle and temperatures above 22°C appear to prevent storage-root development. Arracacha appears to be a short-day plant, but the range of variation for daylength is unknown (Bukasov 1930; NRC 1989).

Andean genebanks maintain 577 accessions of arracacha. The number of different cultivars is unknown. Morphological characterization of 93 Ecuadorian accessions showed 17 morphotypes (Tapia *et al.* 1996). Preliminary morphological characterization of Peruvian arracachas showed 16 morphotypes out of 32 accessions studied (Blas and Arbizu 1995).

In the Andes, a number of salty and sweet dishes contain arracacha. The storage roots can be eaten boiled, fried, or baked in soups, stews, omelettes, pudding or *humitas* (Andean cake wrapped with maize shucks). A typical Andean stew called *puchero* in Perú and *sancocho* in Colombia is made using arracacha roots as the main ingredient. *Chicha*, Andean beer, is usually made from maize, but also can be made using arracacha roots. The young stems and leaves are used in salads; the rootstock and leaves are used to feed pigs. In some rural areas, Andean women also use arracacha leaves in infusions like tea to regulate menstruation.

This crop shows great potential. Brazil has a growing demand for arracacha roots from the industry that produces instant soups and baby food. The crop also has potential for use as flour and precooked dehydrated flakes for school meals (Santos and Hermann 1994).

Achira (Canna indica L.= C. edulis Ker-Gawler)

The name of this crop is derived from the Quechua word *achira*. Other names in Spanish are *achira*, *capacho*, *sagú*, *tasca*, *chisqua*, *adura* and *luano*; in Portuguese it is *araruta gigante*; in Vietnamese it is *dong* or *khoai* or *cu* combined with the words *rieng, tay* and *dao*; in English it is achira and Queensland arrowroot.

Achira is a perennial plant with large alternate leaves; it produces long and voluminous rhizomes with segments. The inflorescence is a raceme with two or, less often, one flower that exserts progressively from the central axis; flowers can be yellowish orange or red. The fruit is a capsule and the seeds are hard coated and usually need scarification to germinate. Rhizomes of achira are fully enlarged for harvesting 8-10 months after planting.

The edible achira is classified within the family Cannaceae, genus *Canna,* species *Canna indica* (Maas and Maas 1988). Achira was previously known as *C. edulis* (Kranzlin 1912; Motial 1982; NRC 1989). However in a recent taxonomic revision of the family in the Flora of Ecuador by Maas and Maas (1988), *C. edulis* is considered a synonym of *C. indica*. Furthermore, Brako (1993) also recognized such synonymy. Whereas *C. indica* is grown because of its starchy rhizomes, other species such as *C. glauca, C. iridiflora* and *C. flacida* from the New World and *C. speciosa, C. nepalensis, C. orientalis* and *C. reevesii* from Southeast Asia are grown as ornamentals (Khoshoo and Mukherjee 1970; Motial 1982; Maas and Maas 1988; Brako 1993).

The basic chromosome number of achira is $x=9$ and the crop has diploid ($2n=2x=18$) and triploid ($2n=3x=27$) cultivars (Darlington and Janaki-Ammal 1945; Gonzales and Arbizu 1995). Edible achira cultivars are predominantly inbreeders. The stigma develops at the same level as the anthers and self-fertilization occurs before the flowers open. Diploid achiras set viable seeds. Ornamental achiras, on the other hand, have exserted stigmas that apparently prevent self-fertilization together with accumulating sterility (Mukherjee and Khoshoo 1970).

The genus comprises from 25 to 60 species distributed in tropical and subtropical regions of America and Asia (Kranzlin 1912; Segeren and Maas 1971; Maas and Maas 1988). The highest concentration of species occurs in the Americas. Most of these species produce fleshy, starchy rhizomes at different degrees of success. To a certain extent, tuberous rhizomes are produced by *C. paniculata* which grows in Peru, Brazil and Chile and *C. iridiflora* from the Peruvian Andes (Kranzlin 1912; Khoshoo and Mukherjee 1970; Segeren and Maas 1971; Motial 1982; Maas and Maas 1988; Brako 1993). These species might be related to the cultivated *C. indica* as they are sympatric. Vavilov (1951) postulated that achira was domesticated in the central Andes (Ecuador, Peru and Bolivia). Other authors consider that the plant was most probably domesticated in Peru (Ugent *et al.* 1984), either in the mountainous areas of the south (Herrera 1942a, 1942b) or in the northwest Peruvian coast (Cohen 1978). On the other hand, Sauer (1952) and Gade (1966) postulated the rainforest of Colombia, where achira is intensively grown, as the place where it was first taken into cultivation. Ancient achira cultivation was shown by a number of archaeological remains and pottery found in several sites in Peru which were dated as far back as 2500 BC (Herrera 1934; Towle 1961; Ugent *et al.* 1986; Valdez 1994).

Great diversity of achiras has been found in the upper valley of Apurímac in Peru, where up to eight named cultivars have been reported (Arbizu 1994; Arbizu and Robles 1986; Meza 1995b). In the Patate area of Tungurahua, Ecuador, three named cultivars are grown for starch production (Espinoza *et al.* 1993) and two named cultivars are grown in the departments of Huila and Cundinamarca, Colombia (Morales 1969). Many other areas in Peru, Ecuador and Colombia grow achira on a lower scale and two to three local cultivars are most commonly grown at each site (Arbizu 1994). This relatively low degree of variability in achira may be due to its self-fertilization.

It is cultivated in warm Andean valleys from about 1000 to 2900 m, with better adaptation from 2000 to 2650 m (Arbizu 1994; Arbizu and Robles 1986). It is susceptible to low temperatures. Achira appears to be daylength neutral (NRC 1989). Achira is currently cultivated in the New World from Mexico and the Caribbean to Brazil and Argentina. It is also grown in Hawaii, Australia and Southeast Asia (Kranzlin 1912; Macbride 1936; León 1964; Cárdenas 1969; NRC 1989; Arbizu 1994; Ho and Hao 1995). Statistics on the area cultivated with achira in the Andes are not available, but in Vietnam about 30 000 ha have been reported (Ho and Hao 1995).

Andean genebanks maintain 362 accessions of achira (Table 4.2) but, as in the case of the other ARTC, the number of different cultivars is unknown.

Achira is basically eaten after baking in an earthen oven. When baked, it can last several weeks under ordinary storage conditions. In this form, the starchy rhizomes are marketed as a delicacy in rural areas and some cities such as Cusco and Pauza, Ayacucho, Peru. Achira also can be consumed boiled. A pudding is made from its starch for children and persons convalescing in some parts of Colombia and Peru (García-Barriga 1974; Arbizu 1994). Its roots and rhizomes are used as a diuretic in some parts of Colombia (García-Barriga 1974). The foliage and rhizome residues are used to feed pigs. At present, achira is becoming a cash crop in Huila, Colombia; Patate, Ecuador; Pauza, Peru (Morales 1969; Espinoza *et al.* 1993; Arbizu 1994) and in Vietnam (Ho and Hao 1995).

Yacón (Smallanthus sonchifolius *(Poeppig & Endlicher) Robinson*=Polymnia sonchifolia *Poeppig & Endlicher)*

The name yacón is derived from the Quechua word *yakun*, which makes reference to the watery storage roots. Other names are *aricoma* in Aymara; *jicama* and *arboloco* in Spanish, besides *yacón*; and yacon, yacon strawberry and jiquima in English.

This crop is a perennial compact herb with pilose stems up to 2 m high, with opposite triangular or hastate leaves, with long storage roots that have white to orange yellowish flesh. The plant sets viable seeds in the Andean valleys (Ortega, pers. comm.) but the crop is vegetatively propagated. Expanded roots can be harvested 7-8 months after planting. Yacón appears to be a daylength-neutral plant (NRC 1989).

According to Robinson (1978), yacón is classified within the family Asteraceae (Compositae), genus *Smallanthus*, species *Smallanthus sonchifolius*. Yacón was previously classified as *Polymnia sonchifolia* (Wells 1965). However, Robinson (1978) noted important morphological differences between the two genera. Within the genus, yacón is the most important species because it has storage roots that are rich in inulin. Other important species with some medicinal properties are *S. uvedalius* (*Polymnia uvedalia*=*Osteospermum uvedalia*), *S. glabratus* (=*P. glabrata*) and *S. maculatus* (=*P. maculata*) (Wells 1965; Uphof 1968; Lipp 1971; Robinson 1978). First reports on the ploidy level of yacón have shown that the plant appears to have 60 chromosomes (Nakanishi 1992).

The wild ancestor of *S. sonchifolius* is unknown. However, *S. connatus* (=*P. connata*=*Gymnolomia connata*) from Uruguay, Brazil, Paraguay and Argentina has shown to have some similarities with *S. sonchifolius*. This includes the production of tuberous roots used as human food (Wells 1965; Robinson 1978). Safford (1917) identified molded representations of the root and stems of yacón in Peru, but he did not indicate the culture that produced them. Root remains of yacón were found in Argentina associated with the Candelaria culture (1-1000 AD) (Zardini 1991). Clothes with embroideries of the Nazca culture (100-1000 AD) show elongated and curved roots of yacón (O'Neal and Whitaker 1947). The first written record of yacón was given by the Spanish chronicler Bernabé Cobo, who described yacón roots by the middle of the 17th century (Yacovleff and Herrera 1934-35).

The genus *Smallanthus* comprises 21 species distributed in the New World from 500 to about 4000 m asl. The highest concentration of species occurs in Peru, Colombia and Venezuela (Wells 1965; Robinson 1978).

The plant is grown from the Andes of Venezuela to northwestern Argentina (León 1964; Wells 1965; Robinson 1978; Zardini 1991) at altitudes ranging from 300 to 3300 m (Arbizu and Robles 1986; Tapia *et al.* 1996). Major production areas in Colombia are Cundinamarca, Boyacá and Nariño (2600-3000 m) (Rea 1992); in Ecuador, Loja, Azuay, Cañar, Imbabura, Bolívar, Chimborazo and Pichincha (2000-3000 m) (Tapia *et al.* 1996); in Peru, Cajamarca, Ancash, Junín, Huancavelica, Ayacucho, Cusco and Puno (1300-3500 m) (Arbizu and Robles 1986); in Bolivia, Larecaja, Camacho, Muñecas, Bautista Saavedra, Cochabamba, Chuquisaca and Santa Cruz (2500-3600 m) (Rea 1992); and in Argentina, Jujuy and Salta at about 2000 m (Zardini 1991). In the Andes, yacón is generally grown in the border of maize or potato fields and in home gardens. It was introduced in Japan in 1985 (Asami *et al.* 1989) and at present, the cultivated area is increasing (CIP 1994).

Andean genebanks maintain 443 accessions of yacón, but the number of different cultivars involved is unknown.

The storage roots of yacón have up to 86% water content and a high concentration of calcium, free fructose, glucose and sucrose. The most prominent amino acids are asparagine, glutamine, proline and arginine (Asami *et al.* 1989). Yacón is valued because it produces fructans, which are considered to be low-calorie products. It also favours human intestinal flora and alleviates hyperpilemia (Hata *et al.* 1983; Hidaka *et al.* 1987). It is a natural source of inulin, a polymer composed mainly of fructose. The human body has no enzyme to hydrolize inulin, so it passes unmetabolized through the digestive tract (NRC 1989). Therefore, yacón is most suited for dieters and people suffering from diabetes. Furthermore, the fructosylsucroses do not induce dental caries (Ikeda *et al.* 1982).

Yacón is known as the 'fruit of the poor' in the Andes. Its roots are juicy; they have a very good flavour and are eaten raw like fruit. In some parts of the Peruvian Andes, farmers produce sugar from the storage roots (Arbizu and Robles 1986). Yacón is also used as a flavouring for ice cream. Cobo mentioned that the taste of yacón roots is better if they are exposed to sun. The crop also stores well for more than 20 days (Yacovleff and Herrera 1934-35).

Maca (**Lepidium meyenii** *Walpers*)

The name maca is derived from the Quechua word *maca*; the English name, besides maca, is Peruvian ginseng.

This crop is an annual herb, although it is managed as a biennial crop (Quiros *et al.* 1996). Its leaves are arranged in a rosette form; they are alternate, basal and crowded (12-20). New leaves are continuously produced at the centre of the rosette. The underground hypocotyl, which is the edible part of the plant, can reach 6 cm in diameter and the skin colour can be white, yellow, gray, reddish and purple (Tello *et al.* 1992). In the first year, maca grows vegetatively, forming an underground hypocotyl 6-8 months from planting and in the second year it flowers and sets seeds (Tello *et al.* 1992). Flowers are cleistogamous and are produced for approximately 2 months. Every day 2-3 new flowers open per raceme and remain open for about 3 days. A plant produces more than 30 000 small seeds and 1000 seeds weigh approximately 0.6 g (Aliaga 1995).

Maca is classified within the family Brassicaceae, genus *Lepidium*, species *Lepidium meyenii*. The genus comprises about 150 species of annual, biennial or perennial herbs widely distributed in temperate regions of the world (Bailey and Bailey 1935; Willis 1973). *Lepidium meyenii* from the New World and *L. sativum* from the Old World are cultivated (Bailey and Bailey 1935).

The basic chromosome number of the genus *Lepidium* is $x=8$ (Darlington and Janaki-Ammal 1945) and *L. meyenii* is a disomic polyploid of $2n=8x=64$ chromosomes (Quiros *et al.* 1994, 1996). It is a self-pollinating plant, which explains the morphological uniformity of plant populations under field conditions (Aliaga 1995; Quiros *et al.* 1994, 1996).

The wild ancestor of maca is unknown. Wild *Lepidium* species grow in the Andes from Ecuador to Argentina, but their relationship with the Peruvian cultivated *Lepidium* is unknown (King 1988). There are no accounts on the origin of maca. Rea (1992) postulated that its domestication might have taken place 2000 years ago in the villages around Chinchaycocha Lake, Junín, Peru. The Spaniard Juan Tello de Soto Mayor visited the Junín area in 1549 and received maca hypocotyls as a tribute and used them to improve the fertility of the livestock of Castille. He also mentioned another visit to Huánuco in 1572, where he found that the Chinchaycocha people had used maca for bartering since Inca times as there was no other crop to grow at such altitudes (above 4000 m) (Rea 1992).

The centre of diversity for maca is located in the vicinity of Chinchaycocha Lake in the departments of Pasco and Junín in Peru, such as the villages of Carhuamayo, Huayre, Uco, Junín, Ninacaca, Vico and Ondores. Diversity of the plant is also found in the rural villages of Yanacancha and Achipampa and the highlands of San Juan de Jarpa in the department of Junín (Chacón 1990; Tello *et al.* 1992).

Except for the presence or absence of purple pigmentation, the foliage of maca does not show variation. The hypocotyl skin colour, however, shows up to 13 combinations involving white, yellow, gray, red, purple and black (Tello *et al.* 1992). Low levels of polymorphism have been observed by Quiros *et al.* (1994).

Maca is grown in small plots of a restricted area in Peru's central Andes from 4000 to 4500 m asl, with rainfall ranging from 500 to 700 mm distributed throughout the growing season. It grows well under either short- or long-day conditions (Quiros *et al.* 1994, 1996). With the exception of frost-tolerant bitter potatoes, maca is the only crop that can survive frost, hail and snow and still have consistent yields above 4000 m asl.

Ex situ collections of maca at present appear to exist only at CIP, where 33 accessions are maintained (Table 4.2).

Maca has more protein content (10-14% on a dry weight basis) than any other Andean root and tuber crop (Tello *et al.* 1992) and higher levels of iron and calcium than potato and carrot. The dry hypocotyls contain high levels of the amino acids leucine and isoleucine and the palmitic, linoleic, oleic and stearic fatty acids, as well as high levels of sterols (Dini *et al.* 1994).

Andean farmers eat fresh maca traditionally baked in *watias* or *pachamanca*, whereas dehydrated hypocotyls are used more extensively after being soaked overnight in water and then boiled to prepare juices, puddings, marmelades, drinks and other regional preparations. The flour of maca is also used for other salty and sweet dishes. Boiled mashed hypocotyls are also prepared to eat as *utunkas* (Tello *et al.* 1992). More recently, maca flour has been sold in gelatin capsules or as sweets in many health-food stores and supermarkets in Lima because of its reputation of enhancing female fertility, treating impotency and improving stamina.

Mauka (Mirabilis expansa *Ruiz and Pavón*)

The name of this crop is derived from the Aymara word *mauka*. Other names in Spanish besides *mauka* are *miso, tazo, chago, yuca inca, arracacha de toro* and *pega-pega*. In English, it is called mauka.

Mauka is a compact perennial herb that is prostrate and has branching basal shoots from which a mass of opposite coriaceous leaves arises. Flower colour ranges from white to purple. Storage roots are long and fusiform, with cream to yellow skins and cream flesh (Seminario 1993). Although the plant produces numerous viable seeds, it is propagated by cuttings from the underground or from the upper stems (Rea and León 1965; Seminario 1993). The edible parts of mauka are the storage roots and the expanded underground stems, which can be up to 5 cm wide and 50 cm long (Rea and León 1965; Rea 1992). The storage roots and the underground parts can be harvested 8-9 months after planting.

Mauka is classified within the family Nyctaginaceae, genus *Mirabilis*, species *Mirabilis expansa* (Liesner 1993). It seems that *M. expansa* is the only species within the genus that sets edible tuberous roots. The genus appears to have been introduced into Europe by 1525 from plants sent from Peru (Soukup 1970).

Preliminary observations on the chromosome number of *M. expansa* have indicated that the plant has 58 chromosomes.

The wild ancestor of the cultivated mauka is unknown. There are several wild allies from Central America to Chile (Macbride 1937; Herrera 1941; Weberbauer 1945; Rea 1992). Although *M. expansa* was described in 1794, little is known about its origin and history of cultivation. Archaeological records have not been reported for mauka.

The genus is found in the Americas (Burkill 1966) and has 60 species distributed from Mexico to Chile, from about 100 to 3300 m asl, with apparently the greatest concentration of diversity in the South American Andes (Macbride 1937; Herrera 1941; Ferreyra 1986). Nine species of *Mirabilis* have been reported in Peru by Soukup (1970) and Ferreyra (1986) and 10 by Liesner (1993). In Peru, a great diversity in several plant characters of mauka has been reported in the northern departments of Cajamarca, Amazonas and La Libertad (Seminario 1993). The cultivation of mauka at several sites in the southern department of Puno has been reported recently (Vallenas 1995). Other areas of diversity are the departments of La Paz and Cochabamba in Bolivia (Rea 1992) and the provinces of Cotopaxi and Pichincha in Ecuador (Tapia *et al.* 1996). Although it was reported that the Bolivian and Peruvian maukas produce only purple flowers and those from Ecuador produce white ones (Rea and León 1965; NRC 1989), flower colours for Peruvian maukas are purple, whitish purple and white (Seminario, pers. comm.).

The plant grows under the short-day conditions of the Andes, under average rainfall of 680 mm and average temperature of 14°C (Seminario 1993). In Ecuador, mauka is grown from 2400 to 3000 m asl (Tapia *et al.* 1996); in Peru, from 2200 to 3100 m (Franco *et al.* 1989); and in Bolivia at about 2900 m (Rea and León 1965).

Statistics on area under cultivation with mauka are unknown. It is grown in maize fields and in home gardens at many sites in Peru, Bolivia and Ecuador.

Some 119 accessions of mauka are maintained in Andean genebanks with an unknown number of different cultivars. Systematic collections of the crop were carried out by Franco *et al.* (1989) from 1985 to 1989 in the warm Andean valleys of the northern departments of Cajamarca and Amazonas, Peru.

There is little information on the nutritive value of mauka. It has been reported to have a protein content ranging from 4 to 5% and higher levels of fat than those of other ARTC crops, except for maca (Rea 1992; Seminario 1993).

The storage roots of mauka are a good substitute for cassava, arracacha and sweetpotato. Roots are eaten after being exposed to sunlight to prevent their rather unpleasant bitter taste. They are eaten boiled or baked in soups, stews, puddings and desserts, accompanied by molasses or sugarcane juices. After mauka is boiled, the water can be used as a soft drink. The upper part of the plant is used to feed pigs.

Ahipa (Pachyrhizus ahipa (Weddell) Parodi)

Ahipa is a name derived from the Quechua word *aqipa* or *asipa*. Other names in Aymara are *konori* or *villu*; in Spanish, *ahipa* or *enana*; in English, yam bean.

This crop is an erect, semi-erect, or twining herbaceous plant, with very short inflorescences with 2-6 flowers per lateral axis; the twining petals curve outward following anthesis. Each plant sets only one swollen root, which is thickened at the top end and tapers toward the tips "radish-like" (Ørting *et al.* 1996).

Its taxonomic classification is within the family Leguminosae, genus *Pachyrhizus*, with three cultivated species – *Pachyrhizus ahipa*, *P. erosus* and *P. tuberosus* – and two wild relatives, *P. panamensis* and *P. ferrugineus* (Sørensen 1988).

Pachyrhizus ahipa is cultivated in Bolivia and is perhaps extinct or restricted to a few valleys in northen Peru (Ørting *et al.* 1996). It has also been cultivated in Argentina since early times (Towle 1961) and plants found in Jujuy and Salta might have been derived from seeds introduced by Bolivian workers (Sørensen 1990).

Pachyrhizus erosus is found from sea level to 1700 m asl from southwestern Mexico to northwest Costa Rica. This species appears to have been introduced to Brazil, the Philippines, Indonesia, India, Oceania, the Far East and the west coast of Africa (Sørensen 1988).

Pachyrhizus tuberosus is widely cultivated in the Amazonian region of South America from sea level to 1500 m asl. This species was introduced to the Caribbean and is found in Puerto Rico, Jamaica, Hispaniola and Trinidad. It also was distributed to botanical gardens in Calcutta, Sri Lanka and parts of Australia from Trinidad (Sørensen 1988).

The wild species *P. panamensis* is found in Panama and southwestern Ecuador at altitudes ranging from 0 to 800 m asl and *P. ferrugineus* grows from Mexico to Colombia from sea level to 1600 m (Sørensen 1988).

Preliminary observations on the chromosome number of ahipa indicate that the species has 22 chromosomes. Ahipa is an inbreeder and it is propagated by seeds selected by farmers on the basis of size and shape. Seed is produced by selecting the most vigorous plants and pruning the flowers of the remaining plants to increase storage-root size. Another method is to leave the first pod for seed production and to remove the subsequent flowers (Ørting *et al.* 1996).

The wild ancestor of *P. ahipa* is unknown but could be derived from wild forms that grow on the eastern Andean slopes toward the Amazon basin (Brücher 1989). However, Rea (1995) claimed to have found wild ahipas in Guanay (450 m), La Asunta (700 m) and San Pedro (2500 m) in the department of La Paz, Bolivia. Archaeological evidence shows an ancient cultivation of ahipa in Peru. Storage-root remains were found in the bundled mummies of the Nazca culture. Ceramic and embroideries of Nazca and Mochica cultures show representations of ahipa roots (Yacovleff 1933; Yacovleff and Herrera 1934-35; Brücher 1989). The Spanish chronicler Bernabé Cobo also described the plant as "...a bulky root like a leg .... resembling radish ... very watery and sweet ...used as a fruit" (Yacovleff and Herrera 1934-35).

The greatest diversity and use appear to occur in the Bolivian rural communities located in the mid-elevation tropics of the departments of Cochabamba (Terrazas 1995) and La Paz (Rea 1995).

Ahipa is grown in fertile, warm Andean valleys from 1500 to 3000 m asl in the departments of La Paz and Cochabamba, Bolivia and in Salta and Jujuy, Argentina. At present, *P. ahipa* appears not to be cultivated in Peru (Yacovleff 1933; Yacovleff and Herrera 1934-35), whereas *P. tuberosus* is cultivated in the mid-elevation tropics toward the Amazon basin in the departments of San Martín and Cusco. Ahipa reacts as a day-length-neutral plant (NRC 1989), but cultivation takes place under the short days of the Andes, with rainfall ranging from 400 to 700 mm and temperatures from about 5 to 30°C.

Andean genebanks maintain some 69 accessions of *Pachyrhizus*.

The dry matter content of ahipa roots ranges from 19 to 25%; about half is sugar, 10% is protein and 40% is starch (Ørting *et al.* 1996). Although there is no report on the insecticidal properties of ahipa, compounds such as rotenone, pachyrhizid and erosone found in *P. erosus* need to be investigated in *P. ahipa* (Sørensen 1990).

Ahipa root is eaten raw as a refresher in snacks (Terrazas 1995; Ørting *et al.* 1996). The flesh of ahipa root usually maintains its colour and crunchy features once it is peeled. The cultivated species *P. erosus* is a popular food in Mexico, Central America and some tropical Asian countries (NRC 1989). Ahipa is currently grown in Bolivia as a cash crop and its prices are sometimes comparable to or even higher than those of potato (Ørting *et al.* 1996).

CONSERVATION OF ANDEAN ROOT AND TUBER CROPS

The *ex situ* conservation of ARTC was started in 1958 by the Interamerican Institute for Cooperation in Agriculture (IICA) of the Organization of American States - Andean Zone. Jorge León and Martín Cárdenas established in the village of Candelaria, Cochabamba, Bolivia, a field collection of 148 accessions of oca, 91 of ulluco and 60 of mashua. Ocas were collected in Mexico, Venezuela, Colombia, Ecuador, Peru and Bolivia; ullucos in Venezuela, Colombia, Ecuador, Peru, Bolivia and Argentina; and mashuas in Peru, Bolivia and Argentina. When funding ended in the mid-1960s, these collections were distributed to the universities of Huancayo, Ayacucho and Cusco in Peru and to other Andean institutions (Rea, pers. comm.). In 1965, IICA also established a field genebank of arracacha at San Mateo, Lima, Peru, with 50 accessions from Colombia, Ecuador, Peru and Bolivia. When funding ended in 1967, this collection passed to the University of Cajamarca, Peru; University of Los Andes, Mérida, Venezuela; Colombian Agricultural Institute, Colombia and Agronomic Institute of Campinas, Brazil (Rea 1984).

In the early 1980s, IBPGR provided funds to collect the genetic resources of ARTC in Peru (Arbizu and Robles 1986; Franco *et al.* 1989). USAID also provided funding in 1984 to maintain an *in vitro* collection of ulluco, oca and mashua at the University of San Marcos, Lima, Peru, which by 1995 comprised 456 accessions of ulluco, 231 of oca and 112 of mashua (Estrada, pers. comm.). CIP, in collaboration with GTZ from Germany, began in 1990 an ARTC project to assist Andean NARS in the conservation and utilization of these valuable genetic resources. Swiss funding made available to CIP in 1993 has allowed a comprehensive conservation strategy for ARTC (Biodiversity Project) in close collaboration with NARS of Peru, Bolivia, Ecuador and Brazil. This initiative has resulted in closer cooperation between CIP and NARS of Ecuador, Peru and Bolivia. This project also has considered the development of efficient strategies for the conservation of ARTC biodiversity and systematical use of ARTC germplasm within and outside the Andes. Strategies for maintaining field and *in vitro* collections also have been improved and exchanged (International Potato Center 1995).

Andean genebanks now maintain 9010 cultivated accessions of nine species of ARTC (Table 4.2). With the exception of maca and ahipa, which are seed-propagated, the other crops are maintained vegetatively in field genebanks. The NARS involved are the Instituto Nacional de Investigaciones Agropecuarias (INIAP) at the Santa Catalina Experiment Station near Quito, Ecuador. In Peru, the Instituto Nacional de Investigación Agraria (INIA) at its experiment stations of Los Baños (Cajamarca), Canchán (Huánuco), Santa Ana (Huancayo), Andenes (Cusco) and Illpa (Puno); the Universidad Nacional de San Antonio Abad del Cusco (UNSAAC) at Kayra and Quillabamba, Cusco; the Universidad Nacional de San Cristobal de Huamanga (UNSCH) in Ayacucho and the Universidad Nacional de Cajamarca (UNC) in Cajamarca are involved. In Bolivia, the Instituto Boliviano de Tecnología Agropecuaria (IBTA) and Programa de Investigacion en Papa (PROINPA) at Toralapa, Cochabamba, maintain genetic resourcers. CIP also maintains in trust 1136 accessions of cultivated ARTC and 90 of their wild allies, totaling 1226 accessions from seven South American countries (Table 4.3).

Activities related to on-farm (*in situ*) conservation of ARTC are also included in the Biodiversity Project aiming at developing effective strategies for *in situ* conservation of ARTC. Six research subprojects are under execution with Andean farmers in several sites of Peru and Bolivia. The selected research sites are 16 microcentres of ARTC diversity in Peru and six in Bolivia. Factors that contribute to the management and use of ARTC and their main constraints are also being studied (International Potato Center 1995).

Breeding

Conventional sexual breeding has not been attempted on ARTC aside from arracacha. Apart from the economic and social factors, the main constraint appears

Table 4.3. Number of accessions of Andean root and tuber crops maintained in trust by CIP.

Common name	Country							
	COL	ECU	PER	BOL	ARG	CHL	BRA	Total
Oca	–	–	336	80	41	7	–	464
Ulluco	4	2	320	72	39	–	–	437
Mashua	–	–	72	4	–	–	–	76
Arracacha	1	–	27	–	–	–	20	48
Achira	3	4	20	1	–	7	–	35
Yacón	–	2	26	4	1	–	–	33
Maca	–	–	33	–	–	–	–	33
Mauka	–	–	5	–	–	–	–	5
Ahipa	–	–	3	2	–	–	–	5
Wild allies	–	4	84	2	–	–	–	90
Total	8	12	926	165	81	14	20	1226

COL= Colombia, ECU= Ecuador, PER= Perú, ARG= Argentina, CHL= Chile, BRA= Brazil.

to have been the lack of knowledge of the basic reproductive biology of ARTC. In the case of arracacha, only open-pollinated seeds harvested from commercial production fields have been used to generate segregating populations in order to select the best genotypes (Giordano *et al.* 1994). Although most ARTC are known to be late-maturing crops, this problem can be overcome through breeding for selection of early maturing cultivars. Thus, Santos *et al.* (1994) have reported early maturing arracacha clones (6 months) yielding approximately 1 kg per plant as opposed to the commercial clones with 0.25 kg and late maturity. The material was selected from seeds produced by commercial cultivars of arracacha of more than 10 months from planting to harvesting. It is likely that similar selection efforts could also produce early maturing genotypes in other ARTC that produce viable seeds.

REFERENCES

Aliaga, R. 1995. Biología floral de la maca. Tesis Ing. Agr., Facultad de Agronomía, Universidad Nacional Agraria La Molina, Lima, Peru.

Arbizu, C. 1994. The agroecology of achira in Peru. CIP Circular 20(3):12-13.

Arbizu, C. and E. Robles. 1986. Catálogo de los recursos genéticos de raíces y tubérculos andinos. Universidad Nacional de San Cristóbal de Huamanga, Facultad de Ciencias Agrarias, Prog. de Investigaciones en Cultivos Andinos, Ayacucho, Perú.

Asami, T., M. Kubota, K. Minamisawa and T. Tsukihashi. 1989. Chemical composition of yacón, a new root crop from the Andean highlands. [in Japanese] Jpn. J. Soil Sci. Plant Nutr. 60(2):122-126.

Bailey, L.H. and E.Z. Bailey. 1935. Hortus, a concise dictionary of gardening, general horticulture and cultivated plants in North America. The MacMillan Co., New York.

Bateman, J.V. 1961. Una prueba exploratoria de alimentación usando *"Tropaeolum tuberosum"*. Turrialba 11(3): 98-100.

Benavides, A.S. 1976. Variabilidad clonal en ulluco (*Ullucus tuberosus* Loz.). Fitotecnia Latinoamericana 4(2):91-98

Blas, R. and C. Arbizu. 1995. Estudios preliminares sobre la variación de la arracacha (*Arracacia xanthorrhiza* Bancroft). *In* Resúmenes del Primer Congreso Peruano de Cultivos Andinos "Oscar Blanco Galdós", Universidad Nacional de San Cristóbal de Huamanga, Facultad de Ciencias Agrarias, Programa de Investigación en Cultivos Andinos, Ayacucho, Perú, 11-16 setiembre, 1995, Cultivos Andinos 5(1):17.

Brako, L. 1993. Cannaceae. P. 326 *in* Catalogue of the Flowering Plants and Gymnosperms of Peru (L. Brako and J.L. Zaruchi, eds.). Missouri Botanical Garden.

Brücher, H. 1989. Useful plants of neotropical origin and their wild relatives. Springer-Verlag, Berlin.

Bukasov, S.M. 1930. The cultivated plants of Mexico, Guatemala and Colombia. Bull. Applied Botany, Genetics and Plant Breeding, Suppl. 47, Leningrad.

Burkill, I.H. 1966. A dictionary of the economic products of the Malay Peninsula. Vol. II, Ministry of Agriculture and Co-operatives, Kuala Lumpur, Malaysia.

Cárdenas, M. 1969. Manual de plantas económicas de Bolivia. Imprenta Icthus, Cochabamba, Bolivia.

Cárdenas, M. and J.G. Hawkes. 1948. Número de cromosomas de algunas plantas nativas cultivadas por los indios en los Andes. Revista de Agricultura, Universidad Mayor de San Simón, Cochabamba, Bolivia 5(4):30-32.

Castillo, R., C. Nieto and E. Peralta. 1988. El germoplasma de cultivos andinos en Ecuador. Pp. 323-331 *in* Memorias del VI Congreso Internacional sobre Cultivos Andinos, Quito, Ecuador, 30 mayo-2 junio 1988. Instituto Nacional de Investigaciones Agropecuarias (INIAP).

Chacón, G. 1990. La maca (*Lepidium peruvianum* Chacón sp. nov.) y su hábitat. Revista Peruana de Biología 3(2):171-272.

CIP. 1994. Yacón in Hokkaido, Japan. CIP Circular 20(2):11.

Cohen, M.N. 1978. Archeological plant remains from the central coast of Perú. Nawpa Pacha 16:23-50.

Constance, L. 1949. The South American species of *Arracacia* (Umbelliferae) and some related genera. Bull. Torrey Bot. Club 76(1):39-52.

Cortez, H. 1978. Avances de las investigaciones en oca. Pp. 227-243 *in* Memorias del I Congreso Internacional sobre Cultivos Andinos, Ayacucho, Perú, 25-28 Octubre 1977.

Cortez, H. 1985. Avances en las investigaciones en tres tubérculos andinos: Oca (*Oxalis tuberosa*), olluco (*Ullucus tuberosus*), maswa, isaño o añu (*Tropaeolum tuberosum*). Pp. 62-82 *in* Avances en las investigaciones sobre tubérculos alimenticios de los Andes, 2nd. ed. (M.E. Tapia, ed). Prog. Nac. de Sistemas Andinos de Producción Agropecuaria, INIPA, Perú.

Darlington, C.D. and E.K. Janaki-Ammal. 1945. Chromosome atlas of cultivated plants. George Allen & Unwin Ltd., London.

de Azkue, D. and A. Martínez. 1990. Chromosome number of *Oxalis tuberosa* alliance (Oxalidaceae). Plant. Syst. Evol. 169:25-29.

Dini, A., G. Migliuolo, L. Rastrelli, P. Saturnino and O. Schettino. 1994. Chemical compositium of *Lepidium meyenii.* Food Chem. 49:347-349.

Espinoza, P., R. Vaca and J. Abad. 1993. Informe sobre la producción de achira en Patate: limitantes y posibilidades. Equipo de Ciencias Sociales:1-22. CIP, Quito.

Ferreyra, R. 1986. Flora del Perú: Dicotiledoneas. Editorial Imprenta Sudamericana S.A., Lima, Perú.

Flores, I. 1991. Estudio del proceso de elaboración de khaya. Pp. 34-53 *in* Generación de tecnología para procesamiento de cultivos andinos. Informe Técnico Final. INIAA-FUNDEAGRO, Huancayo, Perú.

Franco, S., J. Rodríguez and L. Machuca. 1989. Catálogo de colecciones de recursos fitogenéticos de la sierra norte del Perú 1985-1989. INIAA, PRONARGEN, PICA.

Gade, D.W. 1966. Achira, the edible *Canna*, its cultivation and use in the Peruvian Andes. Econ. Bot. 20:407-415.

Gandarillas, H. and J. Luizaga. 1967. Número de cromosomas de la papalisa (*Ullucus tuberosus* Caldas). Sayaña, Revista Boliviana de Agricultura 5(2):8-9.

García-Barriga, H. 1974. Flora medicinal de Colombia. Botánica Médica, Tomo I, Instituto de Ciencias Naturales, Universidad Nacional de Colombia, Bogotá, Colombia. Pp. 212-213.

Giordano, L.B., F.F. Santos and S. Brune. 1994. Breeding Peruvian carrot (*Arracacia xanthorrhiza* Bancroft) by using botanical seed at CNPH- EMBRAPA. P. 29 *in* 10th Symposium of the International Society for Tropical Root Crops (ISTRC), Salvador, Bahia, Brazil.

Gonzales, R. and C. Arbizu. 1995. Niveles de ploidía de las achiras cultivadas en el Perú. *In* Resúmenes del Primer Congreso Peruano de Cultivos Andinos "Oscar Blanco Galdós", Universidad Nacional de San Cristóbal de Huamanga, Facultad de Ciencias Agrarias, Programa de Investigación en Cultivos Andinos, Ayacucho, Perú, 11-16 setiembre, 1995, Cultivos Andinos 5(1):17.

Hata,Y., T. Hara, T. Oikawa, M. Yamamoto, N. Hirose, T. Nagashima, N. Torihama, K. Nakajima, A. Watabe and M. Yamashita. 1983. The effect of oligofructans (neosugar) on hyperpilemia. Geriatr. Med. 21:156-167.

Hawkes, J.G. 1983. The diversity of crop plants. Harvard University Press, Cambridge, Mass.

Hawkes, J.G. 1989. The domestication of roots and tubers in the American tropics. Pp. 481-503 *in* Foraging and Farming: The Evolution of Plant Exploitation (D.R. Harris and B.C. Hillman, eds.). Unwin Hyman, London.

Herrera, F.L. 1934. Botánica etnológica: Filología quechua III (I). Rev. Mus. Nac. 3:37-62.

Herrera, F.L. 1941. Plantas que curan y plantas que matan en la flora del Cusco. Rev. Mus. Nac. (Lima, Perú) 9:102-103.

Herrera, F.L. 1942a. Etnobotánica: Plantas endémicas domesticadas por los antiguos peruanos. Rev. Mus. Nac. (Lima, Perú) 11:25-30.

Herrera, F.L. 1942b. Etnobotánica: Plantas tropicales cultivadas por los antiguos peruanos. Rev. Mus. Nac. (Lima, Perú) 11:179-195.

Hidaka, H., T. Eida, T. Adachi and Y. Saitoh. 1987. Industrial production of fructooligosaccharides, and their applications for humans and animals [in Japanese]. Nippon Nogeikagaku Kaishi 61:915-923.

Hiroe, M. 1979. Umbelliferae of the World. Tokyo, Anake Book Co.

Ho, T.V. and B.T. Hao. 1995. Studies on edible *Canna* in Vietnam. Pp. 47-56 *in* Root Crops Germplasm Research in Vietnam (E. Chujoy, ed). National Institute of Agricultural Sciences (INSA) Hanoi,

Vietnam - International Development Research Centre (IDRC), Singapore - International Potato Center (CIP), Philippines.

Hodge, W.H. 1954. The edible arracacha – a little-known root crop of the Andes. Econ. Bot. 8:195-221.

Icochea, T., H. Torres and W. Pérez. 1995. Etiología del mildew de la maca (*Lepidium meyenii* Walpers). *In* Resúmenes del Primer Congreso Peruano de Cultivos Andinos "Oscar Blanco Galdós", Universidad Nacional de San Cristóbal de Huamanga, Facultad de Ciencias Agrarias, Programa de Investigación en Cultivos Andinos, Ayacucho, Perú, 11- 16 setiembre, 1995, Cultivos Andinos 5(1):7.

Ikeda, T, M. Hirasawa and T. Kurita. 1982. Cariogenesis of nystose as a substrate in vitro. Proceedings of neosugar meeting [in Japanese], I:77-86. Meiji Seika, Ltd., Tokyo.

International Potato Center. 1995. Program Report: 1993-1994. Lima, Peru.

Johns, T. and G.H.N. Towers. 1981. Isothiocyanates and thioureas in enzyme hydrolysates of *Tropaeolum tuberosum*. Phytochemistry 20(12):2687-2689.

Johns, T., W.D. Kitts, F. Newsome and G.H.N. Towers. 1982. Anti-reproductive and other medicinal effects of *Tropaeolum tuberosum*. J. Ethnopharmacol. 5:149-161

Khoshoo, T.N. and I. Mukherjee. 1970. Genetic evolutionary studies on cultivated Cannas. VI. Origin and evolution of ornamental taxa. Theor. Appl. Genet. 40:204-217.

King, S.R. 1988. Economic botany of the Andean tuber crop complex: *Lepidium meyenii*, *Oxalis tuberosa*, *Tropaeolum tuberosum*, and *Ullucus tuberosus*. PhD Thesis, The City University of New York, New York.

Kjaer, A., J. Ogaard and Y. Maeda. 1978. Seed volatiles within the family Tropaeolaceae. Phytochemistry 17:1285-1287.

Knuth, R. 1930. Oxalidaceae. *In* A. Engler, Das Pflanzenreich 4(130):1-41.

Kranzlin, Fr. 1912. Cannaceae. *In* A. Engler, Das Pflanzenreich IV(47):1-77.

Larkka, J. 1991. Citogenética. Pp. 10-13 *en* Investigaciones sobre ulluku (L. Pietilá y M.E. Tapia, eds.). Turku, Finland.

Larkka, J., P. Jokela, L. Pietilá and Y. Viinikka. 1992. Karyotypes and meiosis of cultivated and wild ulluco. Caryologia 45(3-4): 229-235.

León, J. 1964. Plantas alimenticias andinas. Boletín Técnico No. 6, Instituto Interamericano de Ciencias Agrícolas Zona Andina. Lima, Perú.

León, J. 1967. Andean tuber and root crops: origin and variability. Pp. 118-123 *in* Proceedings of the International Symposium on Tropical Root Crops, University of West Indies, St. Augustine, Trinidad, 2-8 April 1967.

Lescano, J.L. 1985. Investigaciones en tubérculos andinos en la Universidad Nacional del Altiplano-Puno, Perú. Pp. 94-114 *in* Avances en las investigaciones sobre tubérculos andinos en los Andes, 2nd. ed. (M.E. Tapia, ed). Prog. Nac. de Sistemas Andinos de Producción Agropecuaria, INIPA, Perú.

Liesner, R.L. 1993. Nictaginaceae. Pp. 750-754 *in* Catalogue of Flowering Plants and Gymnosperms of Peru (L. Brako and J.L. Zarucchi, eds.). Missouri Botanical Garden.

Lipp, F.J. 1971. Ethnobotany of the Chinantec Indians, Oaxaca, México. Econ. Bot. 25(3):234-244.

Maas, P.J.M. and H. Maas. 1988. Cannaceae. *In* Harling and Andersson: Flora of Ecuador 32:1-9.

Macbride, J.F. 1936. Cannaceae. *In* Flora of Perú, Field Mus. Nat. Hist. Bot. 13(3):738-741.

Macbride, J.F. 1937. Mirabilis L. *In* Flora of Peru, Field Mus. Nat. Hist. Bot. 13(2):539-546.

Macbride, J.F. 1949. Oxalidaceae. *In* Flora of Perú, Field Mus. Nat. Hist. Bot. 13(2):544-602.

Martins, R. 1976. New archeological techniques for the study of ancient root crops in Perú. PhD thesis, University of Birmingham, England.

Mathias, M.E. and L. Constance. 1962. *Arracacia* Bancroft. Pp. 13-19 *In* Flora of Peru, XIII (1), part V-A (M.E. Mathias and L. Constance, eds.).

Mathias, M.E. and L. Constance. 1976. *Arracacia* Bancr. Pp. 42-47 *in* Flora of Ecuador, No. 5, 145. Umbelliferae (G. Harling and B. Sparre, eds.). Lund, Denmark.

Méndez, M.L. 1995. Determinación de la ploidía del ulluco (*Ullucus tuberosus* Caldas). Tesis Ing. Agr., Universidad Nacional Agraria La Molina, Lima, Perú.

Méndez, M., C. Arbizu y M. Orrillo. 1994. Niveles de ploidía de los ullucos cultivados y silvestres. *En* Resúmenes de trabajos presentados al VIII Congreso Internacional de Sistemas Agropecuarios Andinos... y su proyección al tercer milenio, Universidad Austral de Chile, Valdivia, Chile, 21-26 marzo 1994, Agro Sur 22:12.

Meza, G. 1995a. Variedades nativas de virraca (*Arracacia xanthorriza* Bancroft) en Cusco. Centro de Investigación en Cultivos Andinos, Facultad de Agronomía y Zootecnia, Universidad Nacional de San Antonio Abad del Cusco, Cusco, Perú.

Meza, G. 1995b. Variedades nativas de achira (*Canna edulis* Ker Gawler) en el Valle del Apurímac. Centro de Investigación en Cultivos Andinos, Facultad de Agronomía y Zootecnia, Universidad Nacional de San Antonio Abad del Cusco, Cusco, Perú.

Morales, R. 1969. Características físicas, químicas y organolépticas del almidón de "achira"(*Canna edulis* Ker var.). Revista de la Academia Colombiana de Ciencias Exactas, Físicas y Naturales XIII(51):357-369.

Motial, V.S. 1982. The *Canna*, p. 1-18. Indian Council of Agricultural Research, New Delhi.

Mukherjee, I. and T.N. Khoshoo. 1970. Genetic evolutionary studies on cultivated Cannas. II. Pollination mechanism and breeding system. Proc. of the Indian National Science Academy Section B, 36(4):271-274.

Nakanishi, T. 1992. Use of composite plants for food and industry, and research of yacón [in Japanese]. Agric. Technol. 47(6):241-246.

NRC (National Research Council). 1989. Lost Crops of the Incas: Little-known Plants of the Andes with Promise for Worldwide Cultivation. National Academic Press, Washington, DC.

O'Neal, L.M. and T.W. Whitaker. 1947. Embroideries of the early Nazca period and the crops depicted on them. Southw. J. Anthropol. 3(4):294-321.

Ørting, B., W.J. Gruneberg and M. Sørensen. 1996. Ahipa (*Pachyrhizus ahipa* (Wedd.) Parodi) in Bolivia. Genet. Res. Crop Evol. (in press).

Palmer, J. 1982. Some lesser known temperate root crops. J. R. New Zealand Inst. Hort. 10:98-101.

Pietilá, L. 1995. Pollination requirements for seed set in ulluco (*Ullucus tuberosus*). Euphytica 84:127-131.

Pietilá, L. and A. Rousi. 1991. Morfología y variabilidad. Pp. 26-34 *in* Investigaciones sobre ulluku (L. Pietilá y M.E. Tapia, eds.). Turku, Finland.

Pool, A. 1993. Oxalidaceae. Pp. 867-875 *in* Catalogue of the Flowering Plants and Gymnosperms of Peru (L. Brako and J.L. Zarucchi, eds.). Missouri Botanical Garden.

Quiros, C., A. Epperson and M. Holle. 1994. Growth and development response to soil pH and photoperiod, chromosome number and polymorphism level in maca *Lepidium meyenii*. Progress report, Department of Vegetable Crops, University of California, Davis, California, USA / International Potato Center, Lima, Perú.

Quiros, C., A. Epperson, J. Hu and M. Holle. 1996. Physiological studies and determination of chromosome number in maca, *Lepidium meyenii* (Brassicaceae). Econ. Bot. 50(2):216-223.

Ramos, V., C. Ruiz and J. Hilfiker. 1976. Ensayo preliminar del estudio de la mashua (*Tropaeolum tuberosum*) como suplemento comparado a un concentrado en la alimentación de becerros. Pp. 121-133 *in* Memorias de trabajos presentados a la V Reunión de Especialistas e Investigadores Forrajeros del Perú, Universidad Nascional de San Cristóbal de Huamanga-Cooperación Técnica Suiza, Programa de Pastos, Ayacucho, Perú.

Rea, J. 1980. Mutaciones somáticas espontaneas en ulluco y oca. Pp. 215-219 *in* Memorias del II Congreso Internacional sobre Cultivos Andinos (L. Corral and J.H. Cáceres, eds.). Riobamba, Ecuador, 4-8 junio 1979.

Rea, J. 1984. *Arracacia xanthorrhiza* en los países andinos de Sud América. Pp. 387-396 *in* Memorias del IV Congreso Internacional de Cultivos Andinos, Pasto, Nariño, Colombia, 22-25 mayo 1984.

Rea, J. 1992. Raíces andinas. *In* Cultivos marginados, otra perspectiva de 1492 (J.E. Hernández Bermejo and J. León, eds.). Colección FAO: Producción y protección vegetal no. 26, Roma.

Rea, J. 1995. Informe técnico sobre conservación in situ de raíces y tubérculos andinos. Programa Colaborativo Biodiversidad de Raices y Tubérculos Andinos, Centro Internacional de la Papa-Cooperación Técnica Suiza, Lima, Perú.

Rea, J. and J. León. 1965. La mauka (*Mirabilis expansa* Ruiz & Pavón), un aporte de la agricultura andina prehispánica de Bolivia. Anales Científicos (Universidad Agraria, La Molina, Perú) 3(1):38-41.

Rea, J. and D. Morales. 1980. Catálogo de tubérculos andinos. Ministerio de Asuntos Campesinos y Agropecuarios, Instituto Boliviano de Tecnología Agropecuaria, Programa de Cultivos Andinos, La Paz, Bolivia.

Robinson, H. 1978. Studies in the Heliantheae (Asteraceae). XII. Re-establishment of the genus *Smallanthus*. Phytologia 39(1):47-53.

Rousi, A., P. Jokela, R. Kalliola, L. Pietilá, J. Salo and M. Yli-Rekola. 1989. Morphological variation among clones of ulluco (*Ullucus tuberosus*, Basellaseae) collected in Southern Peru. Econ. Bot. 43(1):58-72.

Safford, W.E. 1917. Food plants and textiles of ancient America. Pp. 12-30 *in* Proceedings of the 19th International Congress of Americanists, Washington, DC.

Santos, F.F. and M. Hermann. 1994. The processing of arracacha (*Arracacia xanthorrhiza* Bancroft) in Brazil. P. 90 *in* 10th Symposium of the International Society for Tropical Root Crops (ISTRC), Salvador, Bahia, Brazil. 13-19 November 1994.

Santos, F.F., L.B. Giordano and S. Brune. 1994. Avaliaçao de clones de mandiquinha-salsa (*Arracacia xanthorrhiza* Bancroft), visando producao e selecao de genótipos mais precoces. P. 4 *in* 10th Symposium of the International Society for Tropical Root Crops (ISTRC), Salvador, Bahia, Brazil. 13-19 November 1994.

Sauer, J. 1952. Agricultural Origins and Dispersals. Am. Geograph. Soc., New York.

Segeren, W. and P.J.M. Maas. 1971. The genus *Canna* in northern South America. Acta Bot. Neerl. 20(6):663-680.

Seminario, J. 1993. Aspectos etnobotánicos del chago, miso o mauka (*Mirabilis expansa* R. y P.) en el Perú. Boletín de Lima 86:71-79.

Sørensen, M. 1988. A taxonomic revision of the genus *Pachyrhizus* (Fabaceae-Phaseoleae). Nord. J. Bot. 8(2):167-192.

Sørensen, M. 1990. Observations on distribution, ecology and cultivation of the tuber-bearing legume genus *Pachyrhizus* Rich. ex DC. Wageningen Agric. Univ. Papers 90, 3:1-38.

Soto, R.C. 1976. Diccionario quechua Ayacucho-Chanka. Min. de Educación, Instituto de Estudios Peruanos, Lima, Perú.

Soukup, J. 1970. Vocabulario de los nombres vulgares de la flora peruana. Colegio Salesiano, Lima, Perú.

Sparre, B. 1973. Tropaeolaceae. Pp. 4-30 *in* Flora of Ecuador (G. Harling and B. Sparre, eds.). Opera Bot. ser. B. no. 2:4-30.

Sparre, B. and L. Anderson. 1991. A taxonomic revision of the Tropaeolaceae. Opera Bot. 108:1-140.

Sperling, C. 1987. Systematics of the Basellaceae. PhD dissertation. Harvard University, Cambridge, Mass.

Tapia, C., R. Castillo and N. Mazón. 1996. Catálogo de recursos genéticos de raíces y tubérculos andinos en Ecuador. Instituto Nacional Autónomo de Investigaciones Agropecuarias, Departamento Nacional de Recursos Fitogenéticos y Biotecnología, Quito, Ecuador.

Tello, J., M. Hermann y A. Calderón. 1992. La maca (*Lepidium meyenii* Walp): cultivo alimenticio potencial para las zonas altoandinas. Boletín de Lima 81:59-66.

Terrazas, F. 1995. Informe técnico sobre conservación *in situ* de raíces y tubérculos andinos. Programa Colaborativo Biodiversidad de Raíces y Tubérculos Andinos, Centro Internacional de la Papa-Cooperación Técnica Suiza, Lima, Perú.

Towle, M.A. 1961. The Ethnobotany of pre-Columbian Peru. Aldine Publishing Co., Chicago.

Ugent, D., S. Pozorski and T. Pozorski. 1984. New evidence for ancient cultivation of *Canna edulis* in Perú. Econ. Bot. 38(4):417-432.

Ugent, D., S. Pozorski and T. Pozorski. 1986. Archeological manioc (*Manihot*) from Coastal Perú. Econ. Bot. 40(1):78-102.

Uphof, J.C.Th. 1968. Dictionary of Economic Plants. 2nd. ed. Verlag von J. Cramer, New York.

Valdez, L.M. 1994. Investigaciones arqueológicas en Gentilar, Acarí. Boletín de Lima XVI(91-96):351-361.

Valladolid, A. 1996. Niveles de ploidía de la oca (*Oxalis tuberosa* Mol.). MSc thesis, Universidad Nacional Agraria, La Molina, Perú.

Valladolid, A., C. Arbizu and D. Talledo. 1994. Niveles de ploidía de la oca (*Oxalis tuberosa* Mol.) y sus parientes silvestres. *In* Resúmenes de trabajos presentados al VIII Congreso Internacional de Sistemas Agropecuarios Andinos... y su proyección al tercer milenio, Universidad Austral de Chile, Valdivia, Chile, 21-26 marzo, 1994, Agro Sur 22:11.

Vallenas, M. 1995. Vigencia del cultivo de mauka (*Mirabilis expansa* R. & P.) en Puno-Perú. *In* Resúmenes del Primer Congreso Peruano de Cultivos Andinos "Oscar Blanco Galdós", Universidad Nacional de San Cristóbal de Huamanga, Facultad de Ciencias Agrarias, Programa de Investigación en Cultivos Andinos, Ayacucho, Perú, 11-16 setiembre 1995, Cultivos Andinos 5(1):72-73.

Vavilov, N.I. 1951. The origin, variation, immunity and breeding of cultivated plants. Chronica Botanica 13:1-366. [Translated from Russian by K. Starr, Chester].

Vivanco, F. and C. Arbizu. 1995. Variación morfológica del ulluco (*Ullucus tuberosus* Caldas), oca (*Oxalis tuberosa* Mol.) y mashua (*Tropaeolum tuberosum* R. & P.). *In* Resúmenes del Primer Congreso Peruano de Cultivos Andinos "Oscar Blanco Galdós", Universidad Nacional de San Cristóbal de Huamanga, Facultad de Ciencias Agrarias, Programa de Investigación en Cultivos Andinos, Ayacucho, Perú, 11- 16 setiembre 1995, Cultivos Andinos 5(1):18.

Weberbauer, A. 1945. El mundo vegetal de los andes peruanos. Estación Experimental Agrícola La Molina, Ministerio de Agricultura, Lima, Perú.

Wells, J.R. 1965. A taxonomic study of *Polymnia* (Compositae). Brittonia 17:144-159.

Willis, J.C. 1973. A Dictionary of Flowering Plants and Ferns. 8th edition. Cambridge University Press.

Yacovleff, E. 1933. La jiquima, raiz comestible extinguida en el Perú. Rev. Mus. Nac. (Lima, Perú) 2(1):51-66.

Yacovleff, E. and F.L. Herrera. 1934-35. El mundo vegetal de los antiguos peruanos. Rev. Mus. Hist. Nat. (Lima, Perú) 3(3):243-322; 4(1):31-100.

Zardini, E. 1991. Ethnobotanical notes on "yacón," *Polymnia sonchifolia* (Asteraceae). Econ. Bot. 45(1):72-85.

Zúñiga, E. 1995. Insectos dañinos del cultivo de la maca (*Lepidium meyenii* Walp) en el Perú. *In* Resúmenes del Primer Congreso Peruano de Cultivos Andinos "Oscar Blanco Galdós", Universidad Nacional de San Cristóbal de Huamanga, Facultad de Ciencias Agrarias, Programa de Investigación en Cultivos Andinos, Ayacucho, Perú, 11-16 setiembre 1995, Cultivos Andinos 5(1):45.

Chapter 5

Yams

R. Asiedu, N.M. Wanyera, S.Y.C. Ng and N.Q. Ng

BOTANY AND DISTRIBUTION

Several sections have been described under the genus *Dioscorea* of family Dioscoreaceae. The main food yams have been grouped as follows:

Section *Enantiophyllum*

This is the largest section with respect to number of species and food importance (Degras 1993). Members may be further grouped in terms of geography as: Asian - Oceanian species, e.g. *D. alata* L. (water yam, greater yam, white yam), *D. glabra* Roxb., *D. nummularia* Lam., *D. transversa* Br.; Sino-Japanese species (or species complex), e.g. *D. japonica* Thumb. (igname de Chine, Chinese yam), *D. opposita* Thumb., and African species or species complex, e.g. *D. cayenensis* Lam. (yellow yam), *D. rotundata* Poir. (white Guinea yam, white yam).

Section *Lasiophyton*

D. pentaphylla L., *D. hispida* Dennsdest, *D. dumetorum* (Knuth) Pax (bitter yam)

Section *Opsophyton*

D. bulbifera L. (aerial yam)

Section *Combilium*

D. esculenta (Lour.) Burk. (Chinese yam, lesser yam)

Section *Macrogynodium*

D. trifida L. (cush-cush yam)

The many species of yams (*Dioscorea* sp.) have various unique or peculiar characteristics that distinguish them from each other. The principal food species have been described in a series of monographs (Martin 1974a, 1974b, 1976; Martin and Degras 1978a, 1978b; Martin and Sadik 1977). Generally the yam plant comprises a shoot portion made up of a vine with branches, leaves and sometimes bulbils in the axils of the leaves, fibrous roots and an underground storage organ, the tuber. The vine twines clockwise or anticlockwise depending on the species. The growth pattern has sequential phases in which the roots, the vine, the leaves and finally the tuber become the focus of growth and development and therefore the main sink.

For most species the tuber is the organ for propagation and perennation. Some evidence suggests that the tuber originates from the hypocotyl, a region between the stem and the root. In the annual species it remains dormant during the unfavourable agroclimatic period between one harvest and the next planting season. It is the source of food and therefore the economic part of the plant.

Reproductive Biology

The challenges posed to breeding have led to many studies of the reproductive biology of the cultivated species (Akoroda 1983; Bai and Jos 1986; Abraham and Nair 1990; Zoundjihekpon 1993). Many *Dioscorea* species do flower even though some accessions of the cultivated forms do not or only flower irregularly or sparingly. Flowering genotypes of species like *D. alata, D. bulbifera, D. cayenensis* and *D. rotundata* are generally dioecious. Female flowers tend to be more limited than male, especially for *D. alata* (Martin 1976) and sterility is quite common. Akoroda (1985a) reported a range of *in vitro* pollen germination from 0.3 to 85% for *D. rotundata*. Bai and Jos (1986) reported a pollen fertility range of 20-98% for some genotypes of *D. alata*.

At Ibadan, Nigeria, flowering occurs between June and September for *D. rotundata, D. cayenensis, D. praehensilis, D. abyssinica, D. dumetorum, D. burkilliana* and *D. bulbifera*, and from August to November for *D. alata*. Flowering intensity, fruit set and seed set are influenced by shoot vigour which in turn depends on factors like initial tuber sett size, plant health status and leafiness (Akoroda 1983, 1985b) in interaction with edapho-climatic factors.

Origin of the Species, Centres of Diversity and Areas of Domestication

The genus *Dioscorea* is well demarcated as Old and New World species. Southeast Asia (for *D. alata* and *D. esculenta*), West Africa (mainly for *D. rotundata, D. cayenensis* and *D. dumetorum*) and pre-Columbian tropical America (for *D. trifida*) are recorded as centres of origin of the major species. These regions have also been the major areas of domestication and diversity of the important food species. Hanson (1985) described the origin and major areas of cultivation for 13 food and seven medicinal species of *Dioscorea*.

Wild species believed to have produced cultivated forms in West Africa include *D. burkilliana, D. abyssinica* and *D. praehensilis* (Hamon *et al.* 1995). From an investigation of the domestication of *Dioscorea* spp. in Cameroon, Dumont *et al.* (1994) reported 16 wild and 7 cultivated species. They attributed the current diversity in the traditional landraces to the availability of wild yams with cropping potential, different selection pressures, successive domestication, culture-derived modifications and somatic mutations. The origin of *D. cayenensis* or the relationship between it and *D. rotundata* has been the subject of much controversy (Martin and Rhodes 1978; Akoroda and Chheda 1983; Terauchi *et al.* 1992). Some workers classify them as a single species complex. From a survey of ribosomal and chloroplast DNA, Terauchi *et al.* (1992) suggested that *D. rotundata* was domesticated from either *D. abyssinica, D. liebrechtsiana* or *D. praehensilis* or their hybrid but that *D. cayenensis* originates from hybridization between *D. burkilliana, D. minutiflora* or *D. smilacifolia* as male parent and *D. rotundata, D. abyssinica, D. liebrechtsiana* or *D. praehensilis* as female parent.

An east-to-west movement of species is acknowledged to have occurred bringing *D. alata* and *D. esculenta* from Asia to Africa and America, and *D. rotundata* from Africa to America (Hahn *et al.* 1987).

Present geographic distribution of traditional landraces and local cultivars

Approximately 93% of world yam is produced in the 'yam belt' of West and Central Africa which stretches from the Bandama river in central Ivory Coast through Ghana, the Republics of Togo, Benin, Nigeria, down to the Western parts of Cameroon (Coursey 1967; Ayensu and Coursey 1972; Hahn *et al.* 1987). *Dioscorea rotundata* and *D. cayenensis* account for most of this production in Africa. They are the most preferred yams in West Africa owing to the organoleptic properties of the tubers which suit the most prevalent food use for the crop in the region. *Dioscorea alata*, introduced from Asia during the 16th century, is second in terms of volume of production and extent of utilization. However, it has the widest geographic distribution among the food yams (Martin 1976), having been introduced from Asia to many parts of the tropics. This broad distribution has made accurate estimates of its contributions to world yam production rather difficult. Martin (1976) estimated world production of the species at not less than 10 million t per year. Considering its wide distribution, agronomic flexibility, high nutritive value and wide acceptance, he declared it to be the most important cultivated species of *Dioscorea*.

GERMPLASM CONSERVATION AND USE

Genetic Resources
Diversity for, and geographic distribution of, important traits
Dioscorea species show diversity in terms of cycles of aerial and underground parts of the plant, geographical distribution, usage, modes of multiplication and ploidy levels (Hamon *et al.* 1995). Studies in different parts of the world have documented a significant proportion of this diversity (Lawton 1967; Rhodes and Martin 1972; Dumont 1977, 1982; Essad 1984; Hamon *et al.* 1992; Dumont *et al.* 1994; Muzac-Tucker and Ahmad 1995). Data from characterization and evaluation have been reported from the Philippines (Pido 1986, 1987), the South Pacific (Jackson and Firmin 1987), India (Bindroo and Bhat 1988), Jamaica (Muzac-Tucker *et al.* 1995) and Côte d'Ivoire (Hamon *et al.* 1986; Hamon and Ahoussou 1988; Hamon and Toure 1990a, 1990b).

Conservation
The Genetics Resources Unit of IITA conserves over 2600 accessions of the major food yams (Table 5.1) in a field collection with over 1780 of these also conserved *in vitro* (Ng and Ng 1994). These were assembled mainly from countries of the West African yam belt and serve as a valuable resource for genetic improvement of the cultigens.

A field genebank has been the traditional method for preserving yam germplasm at IITA (Ng 1993). They are vegetatively propagated through planting setts from underground tubers or aerial tubers. Plants are grown in the fields for the entire growing season, which lasts 6-9 months, depending on species and genotype. Then the mature underground tubers are dug up, or aerial tubers are plucked, and stored for several months in a traditional yam barn, under shade in open air before the next planting.

To reduce the risk of genetic erosion in the fields and in tuber storage, the Genetic Resources Unit has adopted the yam minisett technique (using 30-45 g setts) coupled with very rigorous control of the prevailing insect pests and diseases by good cultural practices and the application of chemicals (Ng 1993). The establishment of over 2300 accessions in 1992 from minisetts was good. About 39% of the accessions had more than 75% plant establishment; no accessions had less than 25% establishment from minisetts. Tubers produced from the minisett-derived plants are smaller than ware yams produced by the traditional method. These smaller tubers were easier to handle and store than the traditional ware yams (Ng 1993). After harvest, tubers are treated with both fungicide and insecticide before storage in a traditional yam barn or in a controlled store room at 18°C and 50-60% RH. During storage, regular checking of tubers, to remove any rotten ones to prevent them from spoiling other healthy tubers, is essential to ensure that the tubers under storage remain in good condition. Tubers of some accessions have stored well up to a year under the temperature- and humidity-controlled facilities at IITA and all accessions could be safely stored for 6 months. Nonetheless, all accessions still need to be planted out annually.

For *in vitro* conservation, apical and axillary buds obtained from accessions in the field have been used as explants. After surface disinfection, meristems excised from these are cultured on modified Murashige and Skoog medium (Ng and Ng 1994) at 25-30°C, 16-h photoperiod and 4000 lux light intensity. The resulting plantlets are subcultured using nodal cuttings. Plantlets from these are kept under reduced-growth conditions (Ng and Ng 1991) for 1-2 years, depending on the genotype, before subculturing becomes necessary. Malaurie *et al.* (1993) reported the creation of an *in vitro* collection of accessions of 14 *Dioscorea* species from Africa and Asia.

Table 5.1. Germplasm of major yam species maintained at IITA, Ibadan[†].

Country of origin	Number of accessions[‡]						
	D.r.	D.a.	D.b.	D.c.	D.d.	D.e.	Total
Rep. of Benin	31	37	3	5	–	–	76
Côte d'Ivoire	126	23	–	6	–	2	157
Ghana	156	74	2	15	4	2	253
Equatorial Guinea	3	1	2	–	–	–	6
Nigeria	696	124	7	10	2	3	842
Togo	862	392	19	7	13	14	1307
Total	1874	651	33	43	19	21	2641

Source: Ng and Ng 1994.

[†] Also holds 41 accessions of minor yams and wild species.

[‡] D.r. = D. rotundata, D.a. = D. alata, D.b. = bulbifera, D.c. = D. cayenensis, D.d. = D. dumetorum, D.e. = D. esculenta.

In 1985 Hanson described 16 of the major *Dioscorea* collections around the world. However, there are many difficulties in maintaining collections of *Dioscorea* species (Degras 1993). Losses from various national collections over time have been severe (Ng and Ng 1994). *In situ* conservation is fraught with many risks. In Brazil, Pedralli (1991) reported significant reduction in the numbers of accessions found during an expedition in 1987 compared with 1982 when 66 accessions of wild *Dioscorea* and 115 of cultivated species were found. He attributed the losses to habitat destruction and agricultural activities. Hanson (1985) discussed the options for conservation including field, *in vitro* ('active' or medium term and 'base' or long term involving cryopreservation) and seed. Butenko *et al.* (1984) reported successful cryopreservation of *D. deltoidea*. Botanic seeds are of limited usefulness for preservation of species of *Dioscorea* owing to limitations in flowering and problems in maintaining viability.

Use of Germplasm

Wide crosses among *Dioscorea* species have not produced high levels of fruiting or seed set. Akoroda (1985b) reported 46% fruit set from a *D. rotundata* x *D. praehensilis* cross which resulted in more seeds than the *D. rotundata* x *D. cayenensis* crosses. The latter produced 2-8% fruits which were seedless except where bulked pollen was used from a few accessions of *D. cayenensis*. Other crosses reported include *D. japonica* x *D. opposita* (Araki *et al.* 1983), *D. praehensilis* x *D. cayenensis-rotundata* (Hamon *et al.* 1992), *D. floribunda* x *D. composita* and *D. floribunda* x *D. freidrichsthalii*.

Hybridization blocks at IITA have included varieties selected from the IITA base germplasm collection and a working collection comprising a recent collection of the most popular varieties from the major yam-growing areas of Nigeria, Ghana, Togo and the Republic of Benin in addition to the best-performing breeder's lines in IITA clonal trials. The resulting seeds mainly go into the annual seedling nurseries of the Institute but a significant proportion is shared with collaborators (Table 5.2). Parental selection has largely been on the basis of phenotypic attributes of the genotypes. It is expected that as more information is gathered in terms of varietal grouping of the base germplasm collection, the choice of parents will take the groupings into consideration. Furthermore, parental selection needs to be based more on the breeding values of the clones which will be established through studies of combining abilities.

The seeds from controlled hybridization are supplemented with those collected from popular varieties on farmers' fields in Nigeria and other African countries to set up seedling nurseries which form the base of the selection cycle. For instance about 20 000 fruits of *D. rotundata* were collected from 18 farms in seven states of Nigeria in 1974 which contributed to the seedling nursery of 1975 (IITA 1975). Similarly, the 1989 and 1990 nurseries comprised 12 760 seedlings from eight families and 5990 seedlings from 26 families, respectively, from farmers' fields (IITA 1993). Germination was 44% in the 1990 nursery for seeds originating from farmers' fields. Furthermore, analysis of entries in the first clonal evaluations of 1989 and 1990 show 2900 clones tracing back to 33 Nigerian local varieties in 1989 and 1400 clones originating from 12 local varieties in 1990 (IITA 1993). Table 5.3 details the 1990 Clonal Evaluation.

On the basis of good performance in trials over a number of years, some of the local landraces have been selected, subjected to meristem culture for virus elimination and certified for international distribution as plantlets (Table 5.4) since 1988 and also as minitubers since 1991 (Table 5.5).

Properties and Uses

In West Africa the preferred method of preparation of the tuber is boiling and pounding into a thick paste which is then consumed with soup. Tubers may also be consumed directly after boiling or cooked into pottage with added protein sources and oils. Frying in oil or roasting are also important cooking methods. In some parts of the region the peeled tubers are dried and later ground into flour which is reconstituted into paste for consumption. A few commercial products based on dry flakes from the tuber are marketed, especially in Nigeria and Côte d'Ivoire. There is considerable export of yams to Europe and North America, principally targeted at residents originating from Africa and the Caribbean. Table 5.6 summarizes the composition of three major food species. In addition to the food yams, species such as *D. floribunda* are cultivated for their medicinal value.

The impact on regional diversity and distribution of local and international trade in fresh yams, part of which serve as planting material in the destination countries, is yet to be documented. It is pertinent to determine the extent of genetic erosion as the dominant commercial cultivars keep spreading at the expense of minor ones and some traditional yam-producing areas become committed to other crops. On the other hand, it would also be useful to document more fully the positively counteractive effects of domestication and traditional selection practices by farmers.

Breeding

Selection and breeding procedures

The principal objectives for most yam improvement programmes include high and stable yield of marketable tubers, good tuber quality (e.g. in terms of dry matter content, cooking quality/texture, taste, dormancy period, rate of enzymatic browning), resistance to biotic stresses in the field and during post-harvest storage, tolerance to abiotic stress (e.g. drought and low soil fertility) and suitability for prevalent cropping systems (e.g. plant architecture, vigour and maturity period).

Degras (1993) has summarized activities in clonal selection programmes in various institutions around the world. Edem (1975) provided a summary of yield trials of *D. rotundata* in Nigeria for the period 1957-64 which were used as the basis for varietal recommendations. Over the years IITA has identified a number of high-performing landraces from evaluation of its germplasm collection. Some of these materials have been shared with collaborators in various countries.

The genetic improvement scheme at IITA begins with evaluation of germplasm from various sources which leads to the identification of materials with desirable traits relevant to objectives of the programme. Such genotypes go through further evaluation toward varietal development and/or become parental clones for hybridization. Botanic seeds are generated through biparental crosses and open-pollination among selected clones planted in isolation from the main yam fields. Additional seeds are obtained through natural hybridization in clonal trials and on farmers' fields. Seedlings from these seeds are evaluated in nurseries from which selections go through a series of trials towards the selection of superior clones.

These clonal trials start with unreplicated (observational) trials with variable numbers of clones and stands per clone, hill trials (or clonal evaluation) and preliminary yield trials. At the Advanced Yield Trial (AYT) and Uniform Yield Trial (UYT) stages a randomized complete block design has been used with 3 to 6 replications.

Table 5.2. Delivery of yam seeds (*Dioscorea rotundata*) to collaborators in various countries (1992-94).

Year	Country	No. of families	No. of seeds
1992	Malawi	11	14 986
	South Korea	6	6 425
1993	Benin	18	2 127
	Gambia	11	2 104
	Rwanda	12	2 758
	Uganda	9	1 859
1994	Guinea	10	3 199
	Ghana	13	8 641
	Togo	2	1 000
	Uganda	14	9 183
	Gambia	22	10 418
	UK	8	4 972
	Benin	35	20 698
	Total delivery for the period		88 370

Table 5.3. Distribution of entries in 1990 Clonal Evaluation according to maternal parent.

Maternal parent	No. of clones	Maternal parent	No. of clones
Abakaleke	171	Ori	44
Unegbe	201	Erefu	79
Gbakunmo	198	Gbongi	25
Boki ex Iwo	205	Elentu	9
Okunmado	227	Odo	9
Iyawo	230	Kpaki	5

Table 5.4. Distribution of virus-tested white yam genotypes as *in vitro* plantlets.

Year	No. of countries	No. of genotypes	Plantlets
1988	1	5	20
1989	19	5	500
1990	13	5	540
1991	14	5	238
1992	12	9	390
1993	14	16	400
1994	12	16	1772

Table 5.5. Distribution of virus-tested white yam genotypes as minitubers.

Year	No. of countries	No. of genotypes	No. of tubers
1991	8	5	1020
1992	10	5	1520
1993	7	5	1400
1994	3	5	1001

By the fifth year of evaluation the materials can be subjected to a series of tests for their cooking and processing attributes in addition to evaluation for yield and reaction to pests and diseases. Owing to the slow multiplication rate of yams, it is necessary to make a special effort after the first advanced yield trial in the sixth year to multiply planting materials before extensive multilocational testing of selected materials. Simultaneously, the process of virus elimination, micropropagation and certification is carried out by the Tissue Culture Unit in collaboration with the Plant Quarantine Service in Nigeria.

In addition to a broad-based population targeting the Guinea Savanna zone of West Africa, there are specific populations being developed for special traits, e.g. resistance to anthracnose disease in *D. alata*.

The first indication of potential value from hybridization of *D. alata* at IITA was the selection of new hybrid clones with good tuber characteristics in 1985 (Hahn and

Table 5.6. Composition of yam tubers of three *Dioscorea* species from various sources (summarized from Bradbury and Holloway 1988).

	D. alata	*D. rotundata*	*D. esculenta*
Moisture %	65 - 78.6	60 - 71.2	67 - 80
Energy (kJ/100 g)	311 - 452	538 - 550	395 - 482
Protein %	1.1 - 3.05	1.1 - 2.34	1.20 - 2.30
Starch %	15.9 - 28	26.8 - 30.2	17 - 7 - 25
Sugar %	0.5 - 1.39	0.32	0.32 - 0.78
Carbohydrate % (diff.)	21.3 - 29	–	17 - 25
Dietary fibre %	1.19 - 2.36	0.63 - 0.87	0.94 - 1.40
Crude fibre %	0.21 - 1.4	0.34 - 0.8	0.2 - 1.51
Fat %	0.03 - 0.57	0.09 - 0.11	0.04 - 0.11
Ash %	0.7 - 2.1	0.7 - 2.6	0.50 - 1.24
Minerals (mg/100 g)			
Ca	5.7 - 22	4.6 - 6.0	3.2 - 18.9
P	4.8 - 44	6.0 - 31	5.2 - 42
Mg	6.6 - 17.8	11.18	9.9 - 30
Na	2.2 - 108	4.3 - 143	1.16 - 108
K	224 - 451	284 - 361	175 - 393
S	9.8 - 14.4	12	13.9 - 18.4
Fe	0.14 - 1.15	0.60 - 1.8	0.28 - 1.4
Cu	0.055 - 0.21	0.10 - 0.12	0.05 - 0.34
Zn	0.24 - 0.49	0.30 - 0.43	0.32 - 0.63
Mn	0.02 - 0.64	0.03 - 0.77	0.06 - 0.58
Al	0.10 - 1.18	0.63	0.35 - 0.64
B	0.08 - 0.09	0.08	0.07 - 0.08
Vitamins (mg/100 g)			
Vitamin A (ret. + ß - car./6)	0 - 0.018	0.8	0.017
Thiamin	0.031 - 0.09	–	0.044 - 0.084
Riboflavin	0.024 - 0.04	–	0.016 - 0.030
Nicotinic acid	0.07 - 0.47	–	0.41 - 1.07
Pot. Nic. acid = Trp/60	0.28 - 0.60	–	0.66
Ascorbic acid (AA)	10.0	–	13.0
Dehydroascorbic acid (DAA)	17.6	–	7.3
Vitamin C	–	6 - 12	–
Total vitamin C (AA + DAA)	2 - 8.2	–	17 - 20.3

Akoroda 1986). The superior tuber yield of some of the new materials generated in the breeding effort has been maintained over the years.

Significant interactions were found for year by genotype in *D. alata* and *D. rotundata* with respect to yield and anthracnose disease in staked Advanced Yield Trials of 1980-81 at Ibadan, Nigeria. In Uniform Yield Trials involving six clones of *D. alata* from 1980 to 1982 it was established that 30% of the variation in clonal performance was attributable to clone by year interaction and 25% to the clones (IITA 1993).

IITA has placed a lot of emphasis on collaboration with national agricultural research systems (NARS) in Africa and elsewhere in its efforts at genetic improvement of yams. Close collaboration with NARS enables a decentralized approach and end-user participation. This leads to removal of bottlenecks in the delivery system, an increase in the pace of multiplication of propagules and draws on the capacities and opportunities in the NARS. In the end there would be maximization of impact potential, relevance of output to end-users and use of available local experience and knowledge with the potential benefit of building local capacity.

Major breeding activities in progress at IITA and the Central Tuber Crops Research Institute, Trivandrum, India could benefit from exchange of promising genotypes. Many nations could also benefit from introduction of improved germplasm. Lack of appropriate indexing for *D. alata* is a major bottleneck to the process.

Prospects

Yams have tremendous sink capacity. Individual tubers may weigh as much as 20-30 kg. The relatively long tuber dormancy of the dominant cultigens ensures longer shelf-life of the fresh tuber than for other root crops such as cassava.

The initial growth of the planted setts ensures tolerance of drought conditions. The sprout draws on water from the tuber as it emerges, and the vine remains leafless for a while (xerophytic growth) coupled with a rapid establishment of a root system. This growth pattern, combined with dormancy, makes it possible to plant yam setts for the next season's crop many months ahead, which allows some flexibility in labour use.

The value of yams in local trade as well as the current and potential revenue from their export to Europe and the Americas is often underestimated (Asuming-Brempong 1994). Studies in southeastern Nigeria (1984-85), where about 42% of world yams are produced, showed a positive expenditure elasticity of demand for yams at all expenditure levels. It also showed yams to be elastic in price (Nweke *et al.* 1992). It was concluded that yams would continue to have a high market potential and that production research that will increase supply is likely to increase quantities consumed at low-expenditure levels.

It is believed that the wild and cultivated species of *Dioscorea* are in a state of flux with each group benefiting from the other. It is also possible that yam cultivation has survived so many years of neglect with respect to serious research because of this situation coupled with the efforts of various ethnic groups who depend on the species. The full potential contribution of the various species is not yet known and therefore their protection is both urgent and important.

Limitations of the crop

Planting materials for production of ware yams (large tubers for market or home consumption) are derived from the edible portion, the tuber, which is expensive and bulky to transport. The multiplication ratio in the field is very low (1:10) compared, for instance, with some cereals (1:300). These propagules could also serve as sources of virus diseases, nematodes and fungi unless appropriate measures are taken.

In many cultivars, emergence after planting is slow and staggered, especially where a mixture of tuber portions (head, middle and tail) is planted. Ground cover during the first 4 months is slow.

Perishability of the tuber, which is both the edible portion and the organ for field propagation, poses a considerable challenge. Losses during storage have negative impact at several stages in the cycle of yam production and utilization. Even more critical for the producers is the quality of seed yams which influences heavily the performance of the next season's crop.

Many aspects of production – planting, weeding, staking, and harvesting – are very labour-intensive and some are difficult to mechanize. Staking has implications for deforestation.

Dioscorea alata exhibits more agronomic flexibility than *D. rotundata* and *D. cayenensis* in the major production area of West Africa, especially in multiplication ratio and availability of planting material. However, the texture of its cooked tubers renders it not as suitable as the others for preparation of the most preferred food form for yam in the region, which is a paste resulting from pounding of the freshly boiled tuber.

Methods that have been developed, with differing levels of efficiency, for extension of tuber dormancy include application of gibberellic acid, irradiation and refrigeration. Most of these are not available for small-scale operators. It would seem that selection for long dormancy in breeding programmes would be useful. This would, however, exacerbate another problem. The other side of dormancy is with respect to production. Irrespective of when yam is planted, the critical starting point for the growing season is when the dormancy ends and sprouts are initiated. Thus there is a compulsory break from harvesting to the next planting during which period even multiplication of planting materials can not be carried out. This is serious when one considers the painfully slow multiplication of yam. Moreover, losses incurred during storage of seed yams could largely be obviated if there were more flexibility in control of sprouting date.

REFERENCES

Abraham, K. and S.G. Nair. 1990. Floral biology and artificial pollination in *Dioscorea alata*. Euphytica 48:45-51.

Akoroda, M.O. 1983. Floral biology in relation to hand pollination of white yam. Euphytica 32:831-838.

Akoroda, M.O. 1985a. Pollination management for controlled hybridisation of white yam. Scientia Hort. 25:201-209.

Akoroda, M.O. 1985b. Sexual seed production in white yam. Seed Sci. Technol. 13:571-581.

Akoroda, M.O. and H.R. Chheda. 1983. Agro-botanical and species relationships of Guinea yams. Trop. Agric. 60:242-248.

Araki, H., T. Harada and T. Yakwa. 1983. Some characteristics of interspecific hybrids between *D. japonica* and *D. opposita*. J. Jpn. Soc. Hort. Sci. 52:153-158.

Asuming-Brempong, S. 1994. Yams for foreign exchange: potentials and prospects in Ghana. Pp 382-386 *in* Tropical Root Crops in a Developing Economy (F. Ofori and S. K. Hahn, eds.). Proceedings, IXth Symposium of the International Society for Tropical Root Crops, Accra, Ghana, 20-26 October 1991.

Ayensu, E.S. and D.G. Coursey. 1972. Guinea yams: the botany, ethnobotany, use and possible future of yams in West Africa. Econ. Bot. 26:301-318.

Bai, K.V. and J.S. Jos. 1986. Female fertility and seed set in *Dioscorea alata* L. Trop. Agric. (Trin.) 63(1):7-10.

Bindroo, B.B. and B.K. Bhat. 1988. Natural variation, heritability and genotype x year interaction in *Dioscorea deltoidea* Wall. Herba Hungarica 27:11-26.

Bradbury, J.H. and W.D. Holloway. 1988. Chemistry of Tropical Root Crops: significance for nutrition and agriculture in the Pacific. ACIAR Mono. No. 6.

Butenko, R.G., A.S. Popov, L.A. Volkova, N.D. Chernyak and A.M. Nosov. 1984. Recovery of cell cultures and their biosynthetic capacity after storage of *Dioscorea deltoidea* and Panax ginseng cells in liquid nitrogen. Plant Sci. Letters 33:285-292.

Coursey, D.G. 1967. Yams. Longmans, London.

Degras, L. 1993. The Yam: A Tropical Root Crop. The MacMillan Press Ltd., London and Basingstoke, UK.

Dumont, R. 1977. Étude morphobotanique des ignames *Dioscorea rotundata* et *D. cayenensis* cultivées au Nord-Bénin. L'Agronomie Tropicale 32(3):225-241.

Dumont, R. 1982. Ignames spontanées et cultivées au Bénin et en Haute-Volta. Pp. 31-36 *in* Yams - Ignames (J. Miège et S.N. Lyonga, eds.). Clarendon Press, Oxford.

Dumont, R., P. Hamon and C. Seignobos. 1994. Les ignames au Cameroun. Repères, Cultures annuelles. CIRAD-CA, France.

Edem. 1975. White yam (*Dioscorea rotundata*) yield trials: summary of results 1957-1964. Federal Dept. of Agric. Research, Ibadan, Nigeria

Essad, S. 1984. Variation géographique des nombres chromosomiques de base et polyploidie dans le genre *Dioscorea* à propos du dénombrement des espèces *transversa* Brown, *pilosiuscula* Bert. et *trifida* L. Agronomie 4(7):611-617.

Hahn, S.K. and M.O. Akoroda. 1986. First new water yam clones from seeds. Annual Report, Root and Tuber Improvement Programme, IITA, Ibadan, Nigeria.

Hahn, S.K., D.S.O. Osiru, M.O. Akoroda and J.A. Otoo. 1987. Yam production and its future prospects. Outlook on Agric. 16:105-110.

Hamon, P. and N. Ahoussou. 1988. Les ignames *Dioscorea* spp. de Cote d'Ivoire. Plant Genet. Resources Newsl. 72:20-23.

Hamon, P. and B. Toure. 1990a. Characterization of traditional yam varieties belonging to the *Dioscorea cayenensis-rotundata* complex by their isozymic patterns. Euphytica 46:101-107.

Hamon, P. and B. Toure. 1990b. The classification of the cultivated yams (*Dioscorea cayenensis-rotundata* complex) of West Africa. Euphytica 47:179-187.

Hamon, P., J.P. Brizzard, C. Duperey, J. Zoundjihekpon and A. Borgel. 1992. Etude de la teneur en ADN de 8 espèces d'ignames (*Dioscorea* spp.) par cytofluorimétrie en flux. Can. J. Bot. 70:996-1000.

Hamon, P., R. Dumont, J. Zoundjihekpon, B. Tio-Toure and S. Hamon. 1995. Wild Yams in West Africa: Morphological Characteristics. ORSTOM.

Hamon, P., S. Hamon and B. Touré. 1986. Les ignames du complexe *Dioscorea cayenensis-rotundata* de Côte d'Ivoire. Inventaire et descriptions des cultivars traditionnels. Italie, Rome, IBPGR-FAO.

Hanson, J. 1985. Methods for storing tropical root crop germplasm with special reference to yam. Plant Genet. Resources Newsl. 64:24-32.

IITA (International Institute of Tropical Agriculture). 1975. Annual Report for 1975. IITA, Ibadan, Nigeria.

IITA. 1993. Root and Tuber Improvement Program. Archival Report 1989-1992. Part III. Yams (*Dioscorea* spp.). Crop Improvement Division, IITA, Ibadan, Nigeria.

Jackson, G.V.H. and I.D. Firmin. 1987. Characterisation and preliminary evaluation of root crop germplasm in the South Pacific with special reference to pests and diseases. Newsletter, IBPGR Regional Committee for South-east Asia. Special Issue 19-23.

Lawton, J.R.S. 1967. A key to the *Dioscorea* species in Nigeria. J. West African Sci. Assoc., 12(1):3-9.

Malaurie, B., O. Pungu, R. Dumont and M-F. Trouslot. 1993. The creation of an *in vitro* germplasm collection of yam (*Dioscorea* spp.) for genetic resources preservation. Euphytica 65:113-122.

Martin, F.W. 1974a. Tropical Yams and their Potential. Series - Part 1. *Dioscorea esculenta*. USDA Agriculture Handbook No. 457.

Martin, F.W. 1974b. Tropical Yams and their Potential. Series - Part 2. *Dioscorea bulbifera*. USDA Agriculture Handbook No. 466.

Martin, F.W. 1976. Tropical Yams and their Potential. Series - Part 3. *Dioscorea alata*. USDA Agriculture Handbook No. 495.

Martin, F.W. and L. Degras. 1978a. Tropical Yams and their Potential. Series - Part 5. *Dioscorea trifida*. USDA Agriculture Handbook No. 522.

Martin, F.W. and L. Degras. 1978b. Tropical yams and their Potential. Series - Part 6. Minor cultivated *Dioscorea* species. USDA Agriculture Handbook No. 538.

Martin, F.W. and A.M. Rhodes. 1978. The relationship of *Dioscorea cayenensis* and *D. rotundata*. Trop. Agric. 55:195-206.

Martin, F.W. and S. Sadik. 1977. Tropical Yams and their Potential. Series - Part 4. *Dioscorea rotundata* and *Dioscorea cayenensis*. USDA Agriculture Handbook No. 502.

Muzac-Tucker, I. and M.H. Ahmad. 1995. Rapid detection of polymorphism in yams (*Dioscorea* sp.) through amplification by polymerase chain reaction and rDNA variation. J. Sci. Food and Agric. 67:303-307.

Muzac-Tucker, I., H.N. Asemota and M.H. Ahmad. 1995. Diversity in protein, lipid fatty acid and phenolic content of yams (*Dioscorea* sp.) grown in Jamaica. J. Sci. Food and Agric. 62:219-224.

Ng, N.Q. 1993. Annual Report 1992, Genetic Resources Unit, Crop Improvement Division, International Institute of Tropical Agriculture. IITA, Ibadan, Nigeria.

Ng, Q. and S.Y.C. Ng. 1994. Approaches for yam germplasm conservation. Pp. 135-140 *in* Root crops for food security in Africa (M.O. Akoroda, ed.). Proceedings of the 5th Triennial Symposium of the International Society for Tropical Root Crops - Africa Branch, Kampala, Uganda, 22-28 November 1992.

Ng, S.Y.C. and N.Q. Ng. 1991. Reduced growth storage of germplasm. *In* Tissue Culture for Conservation of Plant Genetic Resources (J.H. Dodds, ed.). Croom Helm Ltd., Kent.

Nweke, F.I., E.C. Okoroji, J.E. Njoku and D.J. King. 1992. Elasticities of demand for major food items in a root and tuber-based food system: emphasis on yam and cassava in south eastern Nigeria. RCMP Research Monograph No. 11, IITA, Ibadan, Nigeria.

Pedralli, G. 1991. Collecting yam germplasm in Brazil. Plant Genet. Resources Newsl. 83-84:22.

Pido, N.L. 1986. Yam germplasm collection in Philippines. Newsletter, IBPGR Regional Committee for South-east Asia 10 (2):4-7.

Pido, N.L. 1987. Maintenance, characterization and preliminary evaluation of yam (*Dioscorea* spp.) in the Philippines.

Rhodes, A.M. and F.W. Martin. 1972. Multivariate studies of variations in yams. (*Dioscorea alata* L.). J. Am. Soc. Hort. Sci. 97(5):685-688.

Terauchi, R., V.A. Chikaleke, G. Thottapilly and S.K. Hahn. 1992. Origin and phylogeny of Guinea yams as revealed by RFLP analysis of chloroplast DNA and nuclear ribosomal DNA. Theor. Appl. Genet. 83:743-751.

Zoundjihekpon, J. 1993. Biologie de la reproduction et génétique des ignames de l'Afrique de l'Ouest, *Dioscorea cayenensis-rotundata*. TDM No. 127. ORSTOM, Paris.

Chapter 6

Banana and Plantain

J.-P. Horry, R. Ortiz, E. Arnaud, J.H. Crouch,
R.S.B. Ferris, D.R. Jones, N. Mateo, C. Picq and
D. Vuylsteke

BOTANY AND DISTRIBUTION

The genus *Musa* belongs to the Musaceae family, and 'banana plant' describes all wild species, landraces and cultivars. 'Plantain' describes landraces whose fruit is eaten cooked, mainly in Central and West Africa (Simmonds 1962; Stover and Simmonds 1987). Cultivated banana exhibits parthenocarpic fruit development, a marked degree of sterility, and vegetative propagation. Wild banana (Table 6.1) exhibits sexual reproduction and vegetative propagation by ratooning. Common names are numerous (Jarret 1990), and in Uganda alone Karamura and Karamura (1994) have identified 566 names used just for the subgroup Mutika/Lujugira of the AAA group.

Virtually all the cultivars derive from the Eumusa section, which Simmonds and Weatherup (1990) have divided into two subsections, *Eumusa* (1) and *Eumusa* (2). Edible banana plants derived from the *Eumusa* section are mainly based on the recognition of a single source for most varieties, *Musa acuminata* (A genome) and *M. balbisiana* (B genome) (Cheesman 1947, 1948; Dodds and Simmonds 1948). The *Australimusa* section is important as a genetic source of domestication for textile fibres (Abaca or Manila hemp) and the edible Fe'i or Fehi cultivars, found only in the Pacific Islands. *Callimusa* and *Australimusa* have a basic chromosome number of 10 ($2n=20$) while *Eumusa* and *Rhodochlamys* have 11 ($2n=22$). The Simmonds and Shepherd (1955) classification associates the ploidy level ($2n=2x$, $3x$ or $4x$) with a different contribution of genomes of two species. They use AA, AB, AAA, AAB, ABB, BBB, AAAA, AAAB, etc. to indicate the ploidy level (see Jarret and Gawel 1995, for a review of molecular studies). Landraces that share a common set of morphological features are placed in subgroups, which comprise a set of cultivars with little genetic diversity and derived from each other by somatic mutations (Table 6.2). Natural hybrids of *M. acuminata* and *M. schizocarpa* (*Eumusa* (1) section; S genome) (Argent 1976; Shepherd and Ferreira 1982) also occur in areas where the two species overlap, and artificial hybrids of *M. balbisiana* and *M. textilis* (*Australimusa* section; T genome) have been used to improve fibre (Brewbaker and Umali 1956). Surprisingly, landraces with the three genomes A, B and T have been found in Papua New Guinea (Carreel 1995).

According to Simmonds (1962) the *Rhodochlamys* section is probably a young section originating from *M. acuminata* and *M. flaviflora* (*Eumusa* section). Shepherd (1990) questions placing *Rhodochlamys* and *Eumusa* in separate sections because of the interfertility of certain species. Although their morphological features suggest that they are separate (Simmonds and Weatherup 1990), cytoplasm and nuclear genomes do not confirm it (Gawel *et al.* 1992; Carreel 1995).

Table 6.1. Systematics of the wild bananas.

Family	Genus	Sections	Species	Subspecies
Musaceae	*Musa*	*Callimusa* (2n=20)	*beccarii*	
			borneensis	
			coccinea	
			gracilis	
		Australimusa (2n=20)	*textilis*	
			angustigemma	
			bukensis	
			jackeyi	
			lolodensis	
			maclayi	
			peekelii	
		Eumusa (2n=22)	*acuminata*	a. *burmannica/*
				burmannicoides
				a. *siamea*
				a. *malaccensis*
				a. *truncata*
				a. *microcarpa*
				a. *banksii*
				a. *errans*
				a. *zebrina*
				a. 'Pemba' f.
			balbisiana	
			basjoo	
			cheesmani	
			flaviflora	
			halabanensis	
			itinerans	
			nagensium	
			schizocarpa	
			sikkimensis	
			sumatrana	
		Rhodochlamys (2n=22)	*laterita*	
			ornata	
			sanguinea	
			velutina	
		also *M. ingens* (2n=14), *M. lasiocarpa* (2n=?), *M. boman* (2n=?)		

Table 6.2. Main groups and subgroups among *Eumusa*-derivated edible bananas.

Group	Subgroup	Use[†]	Group	Subgroup	Use[†]
AA		d/c	AAB	Silk	d
AB	Ney Poovan	d		Maia Maoli/Popoulu	c
AAA	Cavendish	d		Laknao	c
	Mutika/Lujugira	c/b		Mysore	d
	Gros Michel	d		Iholena	c
	Ibota	d/c	ABB	Bluggoe	c
	Red/Green Red	d		Pisang Awak	d/c/b
AAB	Plantain	c		Saba	c
	Pome	d	BBB	Lep Chang Kut	c

† Most frequent use: b = beermaking; c = cooking; d = dessert.

Origin, Domestication and Diffusion

Southeast Asia is the cradle of the genus *Musa* (Simmonds 1962). *Rhodochlamys* is restricted to Assam, Thailand, Burma; *Callimusa* to Borneo, Malaysia and Indochina; *Australimusa* to Borneo and Indonesia, The Philippines, Australia and the Pacific; and *Eumusa* to the whole area (Fig. 6.1).

Musa acuminata has by far the widest distribution. Simmonds (1962) placed its centre of diversity in the Malayan area. Shepherd (1990) pointed out that the status of the species in the Indonesian archipelago and in Irian Jaya was unknown and revised the geographic subdivision of the species. Nasution *et al.* (1989) consider the centres of origin and diversification of *M. acuminata* to be in Indonesia. Nasution (1991) later collected and described 15 forms of *M. acuminata* recognized as members of the classical subspecies *malaccensis, truncata, microcarpa, zebrina* and *banksii*. An examination of the polymorphism of the anthocyanins led to a similar conclusion that the Indonesian *zebrina* holotype is the present most primitive form (Horry and Jay 1990).

The use of molecular markers has confirmed the genetic isolation of *M. acuminata* subsp. *truncata* (Carreel *et al.* 1994) which Horry (1989) related to *M. acuminata* subsp. *microcarpa*. However, the nuclear and cytoplasmic profiles of *M. acuminata* subsp. *microcarpa* are very similar to those of *M. acuminata* subsp. *banksii* (Gawel and Jarret 1991a; Carreel 1995). From an analysis of the isozymes, Lebot *et al.* (1993) found that there was great homology between *zebrina* and *sumatrana*. These observations led to the breakdown shown in Figure 6.2.

Musa acuminata is found on a number of islands that are very remote from the distribution area. Most fascinating is the presence of an accession of *M. acuminata* on the island of Pemba, in Tanzania (Simmonds and Shepherd 1952). Its chromosome structure sets it apart from the other *M. acuminata* accessions (Shepherd 1990). *Musa schizocarpa* has only been found in New Guinea and the neighbouring islands (Argent 1976; Sharrock 1990). The distribution of *Australimusa*-derived edible bananas has been discussed by Macdaniels (1947), Stover and Simmonds (1987), Tezenas du Montcel (1990), Jarret *et al.* (1992) and Carreel (1995), and the *Eumusa*-derived edible bananas by Simmonds (1962), Horry and Jay (1988), Tezenas du Montcel (1990) and Lebot *et al.* (1993).

Another species, *M. balbisiana*, is also widespread, distributed throughout the whole of the northern and eastern fringe of the *Eumusa* area. Nevertheless, few different accessions are available in the collections, and weak relative diversity has been observed (Horry 1989; Carreel *et al.* 1994). This species probably was introduced very early in many regions from original populations of southern India, northern Burma and the Philippines (Cheesman 1948). Simmonds (1956) claimed that it was endemic to Papua New Guinea, but Argent (1976) thought that it was introduced. The 'Butuhan' clone, considered to be a typical *M. balbisiana* endemic to the Philippines, was originally an ancient hybrid of *M. balbisiana* and *M. textilis* (Carreel 1995).

The factors that led to the establishment of cultivated varieties from wild species and primitive landraces and their dispersal outside Asia have been amply described by Simmonds (1962).

Access to chloroplast and mitochondrial genomes, and the discovery of a different inheritance for each of these genomes in bananas (maternal transmission of chloroplast DNA and paternal transmission of mitochondrial DNA) make them ideal tools for phylogenetic studies (Gawel and Jarret 1991a, 1991b; Faure *et al.* 1994). The key discoveries made by Carreel (1995) coupled with the joint analysis of nuclear, chloroplast and mitochondrial genomes suggest that the origin of the edible varieties is directly linked to the subspecies *banksii* and *errans*.

In some zones, somatic mutations led to an extraordinary diversity of morphotypes on a very small genetic base, as was the case for the AAB-Plantain. Based on inflorescence morphology, four plantain groups have been defined: French, French Horn, False Horn and True Horn (Tezenas du Montcel *et al.* 1983). Bunch weight increases, while average fruit weight decreases, from the sterile True Horn to the partially fertile French type. The two largest groups are the False Horn and French plantains (Swennen *et al.* 1995). Giant

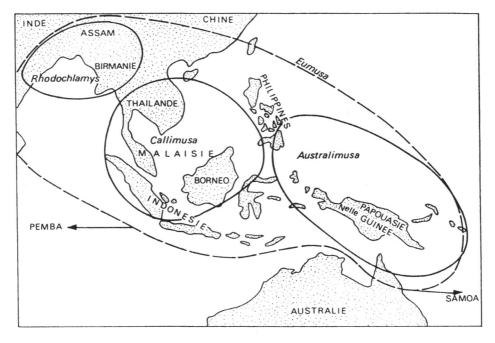

Fig. 6.1. Geographic distribution of the *Musa* sections (source: Horry 1989, adapted from Champion 1967).

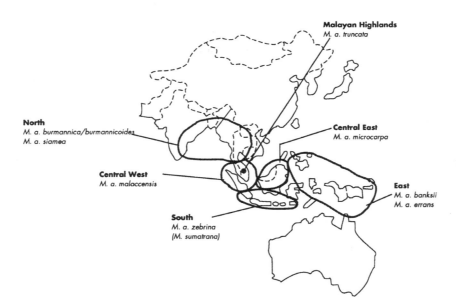

Fig. 6.2. Geographic distribution of the *Musa acuminata* subspecies.

cultivars produce more foliage, resulting in heavier bunches with more hands and fruits (Swennen *et al.* 1995). Groupings that resulted following principal component analysis (PCA) based on time to flowering, pseudostem height and number of fruits agreed with taxonomic groupings based on inflorescence type and plant size (Swennen *et al.* 1995).

Reproductive Biology

The banana plant is a gigantic herb whose development has been well documented. Propagation in landraces and cultivars is vegetative. A review of the variations in the flower features of the genus can be found in Champion (1967). In wild species that have flowers of one sex, the female and male phases are asynchronous and pollination occurs exogenously. A RFLP (restriction fragment length polymorphism) study of 82 wild accessions of *Eumusa* confirms a very low heterozygosity rate (below 10%) in the hermaphrodite types (*M. acuminata* subsp. *banksii*, *M. acuminata* subsp. *errans*), but even among the *acuminata* non-hermaphrodite subspecies the rate only averages 25%. In *M. balbisiana*, a non-hermaphrodite species, the heterozygosity rate is between 0 and 25% (Carreel *et al.* 1994; Carreel 1995). In some species, such as *M. itinerans* or *M. laterita*, the horizontal shoots spread for several metres from the parent before emerging from the ground; root development studies, however, have mainly been conducted on cultivars for cropping on a large scale (Price 1995).

GERMPLASM CONSERVATION AND USE

The bases for coordinating the conservation of banana plants at the international level were laid down at a seminar organized in 1989 in Belgium under the aegis of INIBAP and IBPGR (INIBAP/IBPGR 1990). Panis (1995) summarized the advantages and drawbacks of the four conservation methods recommended in this seminar (Table 6.3). The conservation of banana plant seeds would naturally be limited to the wild species. Work carried out at the Pertanian University of Malaysia (Williams 1987) indicates that this option would be possible with *Musa*. Strotzky *et al.* (1962) emphasized the difficulties of seed germination, but this method should not be discarded (Darjo and Bakry 1990).

Field collections are spread throughout the whole of the intertropical zone (Table 6.4). INIBAP is developing a *Musa* Germplasm Information System (MGIS), sponsored by IDRC. The core of MGIS is an international database on *Musa* germplasm which will record all information generated by research activities on *Musa*. The system will work on a network mode with all relevant data on *Musa* biodiversity being fed directly by the genebank curators into their own files within the database. The database will act as a focal point for information exchange between researchers and curators and will enable the optimization of conservation and use of germplasm. Recognition of Regional Field Collections has made it possible to gather on one site in each geographic region the main diversity known in each continent (INIBAP/IBPGR 1990) (Fig. 6.3).

The INIBAP Transit Centre (INIBAP TC) at the Katholieke Universiteit Leuven (KUL), Belgium, contains 1054 *in vitro* accessions, making it the largest collection of banana plants in the world. Setting up basic *in vitro* collections intended for conservation purposes has been possible through an agreement that TBRI will host all the original accessions from Southeast Asia. In 1994, 414 accessions were therefore duplicated in this centre, protecting 39% of the collection kept at INIBAP TC. IITA in Nigeria and CATIE in Costa Rica have accepted the principle of joining in this effort to conserve the *Musa* genetic heritage (INIBAP 1994).

The growth conditions for rapid propagation at INIBAP TC are 30±2°C for a photosynthetic flow of 63 μmol m^{-2} s^{-1} (5000 lux). This method is not popular for conservation purposes because of its cost, the risks of human error and contamination, and the danger of genetic drift due to somaclonal variations inherent in multiplying the *in vitro* subcultures (Cote *et al.* 1993). To minimize these risks, Banerjee and de Langhe (1985) reduced the speed of growth of the apices being multiplied by reducing the storage temperature and the light intensity to 15±2°C and 2000 lux (25 μmol m^{-2} s^{-1}). This increased the culture period without renewing the media from 4-6 weeks to 13-17 months on average. At the present time these are the best conditions available for banana plants. They are routinely applied for the medium-term conservation of germplasm in the INIBAP collection at KUL and for the 'basic' collections. However, storage times are certainly not uniform: some accessions must be subcultured after only 60 days while others can be kept for over 600 days (Van den Houwe *et al.* 1996). Furthermore, if growth is slowed down, the risks (and the costs) are also reduced, but this does not guarantee

Table 6.3. Necessary conditions for the storage of 1000 banana accessions.

	Field collection	*In vitro* collection		Cryopreservation
		Normal growth	Limited growth	
Amount per accession	5 plants	20 multiple buds	20 multiple buds	millions of cells[†]
Area	30 000 m²	15m²	15m²	2m²
Labour per annum (no. technicians only)	6	9	1.5-2	<0.5
Conditions				
Ambient temperature	+25 to +30°C	+30°C	+16°C	−196°C
Risk of losing germplasm	high	medium	low	nil
Pests and diseases	high risk	no risk	no risk	no risk
Human error	moderate	moderate	low	extremely low
Maintenance	replanting every 3-5 years, application of fertilizers, pesticides and weed control	6 subcultures/year	1 subculture /year	no subculture, refill with liquid nitrogen

Source: Panis 1995.

[†] In the case of meristem cryopreservation the amount per accession will depend on the regeneration frequency after thawing

Table 6.4. Number of accessions in *Musa* field collections.

Country	No. of accessions	Country	No. of accessions
Latin America and Caribbean		**West Africa**	
Brazil	264	Nigeria	200
Colombia	81	Cote d'Ivoire	185
Costa Rica	83	Guinéa	19
Cuba	129	**East Africa**	
Honduras	463	Burundi	225
Jamaica	356	Kenya	22
Guadeloupe	>350	Rwanda	75
Mexico	>40	Tanzania	362
Peru	>20	Uganda	81
Panama	>40	**South East Asia and Pacific**	
Puerto Rico	>50	Australia	?
Venezuela	>50	India	303
Windward Islands	>50	Indonesia	93
Central Africa		Malaysia	73
Cameroon	350	Papua New Guinea	416
Gabon	50	Philippines	245
Congo	24	Thailand	323
Zaire	92		

Source: INIBAP/IBPGR 1990.

long-term conservation. The development of cryopreservation methods at very low temperatures (-196°C) has therefore been recommended. These have been established for isolated cells (zygotic embryos of wild species: Abdelnour-Esquivel *et al.* 1992; somatic embryos and cell suspensions: Abdelnour and Escalant 1994; Panis *et al.* 1994) or meristem cultures (Panis 1995). Only one of the six varieties tested by Panis has been able to regenerate through somatic embryogenesis. The cryopreservation of proliferating meristems after encapsulation in sodium alginate and dehydration seems to be the most simple, the most rapid and the most easily applicable method for the widest range of genotypes (Panis 1995). The development of increasingly more effective molecular marking techniques to make varietal differentiation possible (Jarret *et al.* 1993; Lagoda and Noyer 1995) should lead soon to the availability of early assessment tests for detecting the emergence of somaclonal variants.

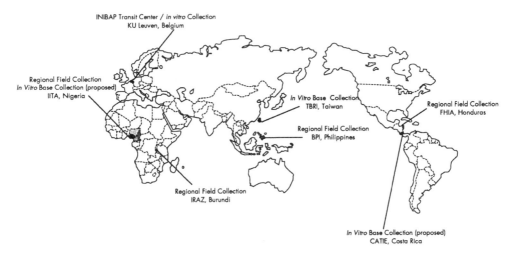

Fig. 6.3. *Musa* germplasm conservation network: regional field collections and *in vitro* collections.

At IITA, shoot-tip culture is routinely used for propagating, exchanging and conserving *Musa* germplasm. More than 300 accessions were introduced as shoot-tip culture from the mid-1980s to the early 1990s with the concurrence of INIBAP and the Nigerian Plant Quarantine Service (Vuylsteke *et al.* 1990). All the plantain field genebank is duplicated *in vitro* using standard shoot-tip culture methodology. Also, plant regeneration by direct somatic embryogenesis in cell suspensions from meristematic tissues (Dhed'a *et al.* 1991) is an efficient means of mass propagation, for long-term cryopreservation of genetic resources and for transformation by biolistic technology (the particle gun) or electroporation. Somaclonal variation in plantains derived from shoot-tip culture has been described in detail (Vuylsteke *et al.* 1991). The narrow spectrum of somaclonal variation and the generally inferior agronomic performance of off-types demonstrated that it is of limited use in plantain improvement (Vuylsteke *et al.* 1996). Therefore, molecular markers for early detection of somaclonal variants arising from *in vitro* propagation are being explored by *Musa* biotechnologists (Howell *et al.* 1994).

Properties and Uses
Banana has a high energy value (from 80 to 135 kcal per 100 g of fresh fruit). The fruit has a water content of between 60 and 76%. It is a major source of carbohydrates (over 90% of the dry matter of the fruit) in the form of starch and sugars, potassium (400 mg per 100 g of pulp), vitamins A (5-60 mg/100 g), B (0.5-0.9 mg/100 g) and C (4 to 16 mg/100 g). It is poor in protein (3-5 g/100 g of dry matter) and fats (about 1 g/100 g) (Stover and Simmonds 1987; Chandler 1995; John and Marchal 1995). The plantain landraces have a dry matter content 2-3% higher than artificially bred plantain hybrids, about 5% higher than cooking bananas and more than 10% higher than dessert export bananas (Ferris 1993).

The main banana-consuming countries are Papua New Guinea, with an estimated annual per capita consumption of 247 kg, followed by Uganda (222.5 kg), Rwanda (180 kg), the Dominican Republic (147 kg), Gabon (143 kg), Ecuador (140 kg), Cameroon (109 kg) and Haiti (102 kg). Bananas may be eaten raw or cooked. Their main uses when processed are powder and flour, purées, meal and chips, juice and jams and alcoholic beverages in East Africa (Chandler 1995; Thompson 1995). Banana beer, obtained by natural fermentation and combined with a cereal (sorghum), is a source of income (152 and 179%) for farmers (Davies 1993). Virtually the whole plant can be used, and some parts are traditionally used in Southeast Asia and in India, where the banana plant is associated with the Tree of Heaven (Stover and Simmonds 1987; Shanmugavelu *et al.* 1992).

Breeding

The genetic improvement of banana aims mainly to develop disease and pest-resistant cultivars, depending on the priorities in different regions (Table 6.5). Detailed information is available (Persley and de Langhe 1987; Ploetz 1990; Valmayor *et al.* 1991, 1994; Ganry 1993; Jones 1994a). Most of the present cultivars are highly sterile triploids that are improved by hybridization or use of somaclonal variation and induced mutation, but biotechnology has received strong research effort and applications may be foreseen within a decade (Jones 1994a; Ortiz *et al.* 1995a).

Genetic improvement using hybridization began at the Imperial College of Tropical Agriculture in Trinidad and Jamaica, by the development of tetraploid varieties through crosses between triploid and diploid varieties. While the triploid parent is supposed to be fixed, it is the diploid parent that produces the improving features, particularly resistance (Fullerton and Stover 1990), and it has then been targeted for improvement. This emphasis has been challenged by the observation at IITA that segregation occurs during the formation of $2n$ $(=3x)$ eggs in plantains (Vuylsteke *et al.* 1993e; Ortiz and Vuylsteke 1994d), a finding that enlarges the prospect of triploid x diploid breeding. Production of tetraploid hybrids is being successfully applied in several breeding programmes (FHIA, Honduras; Banana Board, Jamaica; EMBRAPA/CNPMF, Brazil; IITA, Nigeria and Uganda; CRBP, Cameroon, CIRAD-FLHOR, Guadeloupe FWI; Tamil Nadu Agricultural University, India) and hybrids developed at FHIA, EMBRAPA/CNPMF and IITA are already distributed and adopted by smallholders. A return to the triploid level was envisaged by crossing tetraploid hybrids with diploid individuals. The search for valuable secondary triploids continues at FHIA, EMBRAPA/CNPMF and IITA (Ortiz *et al.* 1995a). To combine favourable characters, FHIA has embarked on the production of secondary tetraploids from tetraploid x tetraploid crosses or secondary triploid x diploid crosses. The most widely distributed improved hybrid at present (FHIA 03) results from this latter scheme.

A different hybridization strategy has been developed by CIRAD-FLHOR attempting to synthesize triploid varieties solely from natural or improved diploid varieties: the first step is the selection and improvement of superior diploid clones. The somatic chromosome stock of one of the selected diploid parents is doubled by treating it with colchicine, and the induced tetraploid is then crossed with another diploid parent. The final product is a triploid hybrid (Horry *et al.* 1993). The feasibility of the scheme was established by the development of improved triploid hybrids (Bakry and Horry 1994).

The development of *in vitro* culture techniques has paved the way to improvement by specifically modifying the genome. Meristem cultures, but moreover embryogenic cell suspensions, somatic embryos and protoplasts are usable material for genetic engineering. *Agrobacterium*-mediated transformation of apical meristems and corm slices has been applied with some success at Texas A&M University, USA (May *et al.* 1995). Biolistics (particle bombardment) is currently applied on somatic embryos (CATIE, Costa Rica) or cell suspensions (KUL, Belgium, CIRAD-FLHOR in collaboration with the Université Paris XI-France, Queensland University of Technology, Australia). Electroporation of protoplasts is investigated at Université Paris XI. The main objective is the incorporation of genes that should bring resistance/tolerance to diseases and pests: coding genes for antifungal proteins against Sigatoka and Panama diseases, virus coat protein genes against virus diseases, and protease inhibitors and toxin genes for insect and nematode control (INIBAP 1993b).

The induction of somaclonal variants through *in vitro* culture as well as induction of heritable variation through irradiation of shoot-tip cultures of banana is currently performed by several research teams (Jones 1994a). However, the spectrum of

Table 6.5. Improvement of *Musa* – INIBAP's perception of needs.

Musa type	Region	Desirable traits needed (small stature is ideal trait for all[1])
Plantain (AAB)	West Africa, Latin America, Southeast Asia, South Asia (India)	BLS/BS resistance[2] at low altitude, S/YS resistance[2] at high altitude, Nematode resistance[3], Weevil borer resistance[4]
Silk (AAB)	Latin America (Brazil), South Asia (India), Southeast Asia	Fusarium wilt resistance[5] BLS/BS resistance[2]
Pome (AAB)	Latin America (Brazil), South Asia (India), Australia	Fusarium wilt resistance[5], BLS/BS resistance[2], S/YS resistance[2]
East African Highland (AAA)	East Africa	Nematode resistance[3], Weevil borer resistance[4], BLS/BS resistance[2], Fusarium wilt resistance[5]
Pisang Awak (ABB)	Asia and East Africa	Fusarium wilt resistance[5]
Bluggoe (ABB)	Worldwide in marginal zones	Fusarium wilt resistance[5], Nematode resistance[3]
Cavendish/Gros Michel (AAA)	Asia, Pacific, Latin America, Africa	BLS/BS resistance[2], S/YS resistance[2], Nematode resistance[3], Weevil borer resistance[4], Virus resistance[6], Delayed ripening[7] (for export)
Miscellaneous dessert types such as Pisang mas (AA), Ney poovan (AB) and Pisang Berangan (AAA)	Southeast Asia, South Asia (India), East Africa and Pacific	BLS/BS resistance[2], S/YS resistance[2], Nematode resistance[3], Weevil borer resistance[4], Fusarium wilt resistance[5]

Source: Jones 1994b.

1   Small stature can be obtained by using a dwarf mother cultivar in conventional breeding or by mutation/somaclonal-induced variants.

2   Mono/oligogenic (vertical) resistance may be acceptable in polyclonal production systems, but polygenic (horizontal) resistance may be more durable in monoclonal production systems. Ideally, both types should be incorporated into the genome. Resistances from different sources should be utilized to increase diversity of resistance mechanisms in breeding programmes. Conventional breeding may be sufficient.

3   Resistance to *Radopholus similis*, which is spread by Cavendish, has been identified in Pisang Jari Buaya. More research needs to be undertaken to identify other sources of resistance to *R. similis* and resistance to *Pratylenchus* subsp. *P. goodeyi* is important over 1000 m asl in tropical Africa. Improvement by transformation is a possibility.

4   Little is known about sources of resistance to weevil borer. Mechanical resistance in corm tissue may be important. All available Bt toxins should be screened to see if any are effective. Gene(s) coding for effective toxin(s) could be inserted into the *Musa* genome.

5   The pathogenic variability of *Fusarium oxysporum* f.sp. *cubense* needs to be considered. Natural sources of resistance/tolerance exist in wild species, cultivars and synthetic diploids. Conventional breeding may be sufficient.

6   Not much work has been undertaken to screen *Musa* for resistance to virus diseases, but evidence to date suggests most germplasm is susceptible to most viruses so conventional breeding may be of little use. Resistance may be possible through the insertion of genes for virus coat proteins or replicase inhibition into the *Musa* genome.

7   Ripening may be delayed by transformation of gene(s) that code for antisense polygalacturanoses and inhibition of ethylene production.

Note: BLS/BS: black leaf streak/black Sigatoka. S/YS: Sigatoka/yellow Sigatoka (cercosporioses)

somaclonal variation seems to be limited. Nevertheless, a *Fusarium* race 4 resistant somaclonal variant of the AAA-Cavendish subgroup was selected in Taiwan (TBRI) while IITA scientists developed a black Sigatoka-resistant Plantain using a somaclonal variant as female parent (Vuylsteke *et al.* 1995).

Mutation breeding applied to banana *in vitro* cultures was first developed at the International Atomic Energy Agency (IAEA), Austria. This technology was transferred later on to other laboratories in the world through interregional, regional and national training courses or individual training organized or supported by IAEA. As a result, an early flowering mutant has been selected in Malaysia (S.H. Jamaluddin, MARDI, Malaysia, pers. comm.) from a mutant Cavendish clone induced at IAEA's Seibersdorf Laboratory (N. Roux, FAO/IAEA Austria, pers. comm.).

Recent advances in breeding and genetics at IITA

Since its creation in 1991, the core activity of the Plantain and Banana Improvement Programme of IITA has been the production of resistant cultivars, which is the most cost-effective and appropriate component intervention. In 1994, IITA won the King Baudouin Award for International Agricultural Research in recognition of breeding plantains for black Sigatoka resistance and advances in *Musa* genetics.

More than 30 sources of black Sigatoka resistance and 37 seed-fertile plantain cultivars (29 French and 8 False Horn plantains) have been identified in IITA *Musa* germplasm collection (Vuylsteke *et al.* 1993c, 1993e). Up to 200 seeds can be produced in one bunch of plantain when pollinated with a wild banana (Swennen and Vuylsteke 1993). Several hundred plantain hybrids have been generated through hand-pollination and embryo rescue (Vuylsteke *et al.* 1993e).

Determination of fruit quality in plantain, banana and their hybrids has been a major activity in support to the breeding programme (Eggleston *et al.* 1992). Fruit size and dry matter content seem to be the key components of consumer acceptability as revealed by taste panels and physicochemical measurements (Vuylsteke *et al.* 1997).

Tetraploids

Fourteen black Sigatoka-resistant tetraploid hybrids of plantain (TMPx) were registered in 1993 in the journal *HortScience* to place these in the public domain (Vuylsteke *et al.* 1993f). One black Sigatoka-resistant tetraploid hybrid, PITA 9 (Vuylsteke *et al.* 1995), was derived from the almost sterile False Horn pool through crosses between the somaclonal French reversion variant of the cultivar Agbagba and the wild banana Calcutta 4.

The breeding of East African banana has just started. So far, 24 cooking/beer bananas have set seed at IITA's breeding station in southeastern Nigeria (4 ABB and 20 AAA) (Ortiz and Vuylsteke 1995). East African highland banana hybrids have been established in the breeding stations for field testing and two promising black Sigatoka-resistant high-yielding cooking banana hybrids (BITA 1 [Fig. 6.4] and BITA 2) have been selected.

Diploids

Plantain-derived diploids (TMP2x) have been obtained from triploid-diploid crosses. These have been released (Vuylsteke and Ortiz 1995) because they are a source of plantain alleles for germplasm enhancement and genetic analysis at the diploid level (Ortiz *et al.* 1995a). Also ten diploid banana hybrids (TMB2x) have been selected at IITA breeding station because they could be potential parents of new East African highland banana hybrids (Vuylsteke *et al.* 1993a). TMB2x hybrids are resistant to black Sigatoka, have adequate agronomic traits and may be resistant to Fusarium wilt. Their breeding value will be assessed in interspecific interploidy crosses with East African cooking/beer bananas.

Gaining insight into the Musa genome

The genetics of resistance were established for black Sigatoka (Ortiz and Vuylsteke 1994d) and banana weevil (Ortiz *et al.* 1995b), susceptibility to virus(es) (Ortiz 1996), dwarfism (Ortiz and Vuylsteke 1995), albinism (Ortiz and Vuylsteke 1994b), suckering behaviour (Ortiz and Vuylsteke 1994c), fruit parthenocarpy (Ortiz and Vuylsteke 1994a), pseudostem waxiness (Ortiz *et al.* 1995c), male/female fertility (Ortiz 1995), bunch

Fig. 6.4. BITA-1, a black Sigatoka resistant cooking banana hybrid, obtained by crossing Bluggoe (left) and Calcutta 4 (right).

orientation (Ortiz 1995), bunch weight and other yield components (Ortiz 1995; Vandenhout *et al.* 1995; Vuylsteke *et al.* 1993d). Other studies (Craenen and Ortiz 1996; Ortiz and Vuylsteke 1995; Vandenhout *et al.* 1995) have established the effects of conventional genetic markers and ploidy on quantitative variation of important economic descriptors in euploid plantain-banana hybrids.

The triploid plantain genome is not fixed (Ortiz and Vuylsteke 1994d), hence much more variability can be recovered from primary 3x-2x crosses than earlier anticipated (Vuylsteke *et al.* 1993e). This finding suggests that the success of tetraploid hybrids does not depend entirely on the qualities transmitted by the pollen as previously believed by banana breeders. Furthermore, Calcutta 4 (the wild non-edible diploid banana) is not the sole source of alleles for black Sigatoka resistance (Ortiz and Vuylsteke 1994d). Other alleles influence resistance in plantain genotypes but they are not expressed phenotypically because of their additive/recessive nature. Similarly, Calcutta 4 donates alleles for non-parthenocarpic fruits (Ortiz and Vuylsteke 1994a); however, its tetraploid hybrids exhibit parthenocarpic fruits and their bunch weight was not inferior to that of the maternal French plantain genotype (Vuylsteke *et al.* 1993e). Moreover, it seems that most morphological traits in *Musa*, including components of yield, are under the control of epistatic genetic systems (Ortiz 1995). This form of inheritance has commonly been reported in vegetatively propagated crops (Peloquin and Ortiz 1992).

Based on this genetic information and materials available, a new breeding strategy was developed (Ortiz *et al.* 1995a; Ortiz and Vuylsteke 1994e; Vuylsteke *et al.* 1993c). This strategy entails the production of improved triploid germplasm, the final target in plantain breeding, by intercrossing tetraploid (TMPx) and diploid (TMP2x) selected hybrids with black Sigatoka resistance.

Prospects

The utilization of interspecific hybridization, ploidy manipulations (2n gametes), embryo culture, rapid *in vitro* multiplication, field testing and selection, has resulted in the identification of improved *Musa* germplasm at IITA in the last 7 years. The breeding goal has been driven by an ideotype which requires the integration of several genes into a selected genotype. The improved plantain or banana cultivar should combine alleles for high and stable yield, disease/pest resistance, wide adaptation and fruit quality. Nevertheless, the ultimate impact of this work will depend on how acceptable the new breeding materials are to farmers. Hence, on-farm trials are currently being carried out in cooperation with extension services of public and private organizations in the region to test not only disease- and pest-resistant *Musa* materials but also potential improved banana and plantain cultivars for the African farming community.

However, sustainable plantain production and utilization systems require a holistic approach because single-component interventions do not provide solutions to the different constraints affecting plantain and banana production in sub-Saharan Africa (Vuyslteke *et al.* 1993b). Therefore, an IITA multidisciplinary team works in collaboration with leading laboratories and NARS on the integration of host plant resistance, enhanced utilization, improved cropping systems and crop husbandry techniques for long-term productivity.

INIBAP, in its network mode of operation, has put a lot of emphasis on collaboration with national and international research agencies. A number of valuable improved cultivars are already being produced under various breeding programmes. These improved cultivars, selected at the national level, have been entrusted to INIBAP to carry out multisite assessment worldwide under its International *Musa* Testing Programme sponsored by the United Nations Development Program (UNDP), which will then disseminate the improved varieties obtained. The first phase was limited to evaluation for agronomic characters and resistance to black Sigatoka disease and resulted in the selection of three tetraploid hybrids from the Hondurean programme now being disseminated worldwide. The second phase widened the evaluation to include Fusarium wilt and Sigatoka diseases and a third phase under development will include an assessment of nematode tolerance.

REFERENCES

Abdelnour, A. and J.V. Escalant. 1994. Cryopreservation of somatic embryos of *Musa* Grande Naine (AAA). *In* Proceedings of the XIth Reunion Meeting of ACORBAT, San José, Costa Rica, 13-18 February 1994. (in press).

Abdelnour-Esquivel, A., A. Mora and V. Villalobos. 1992. Cryopreservation of zygotic embryos of *Musa acuminata (AA)* and *Musa balbisiana* (BB). Cryo-Letters 13:159-164.

Argent, G.C.G. 1976. The wild bananas of Papua New-Guinea. Notes R. Bot. Gard. Edinb. 35:77-114.

Bakry, F. and J.P. Horry. 1994. *Musa* Breeding at CIRAD-FLHOR. Pp. 169-175 *in* The Improvement and Testing of *Musa*: a Global Partnership. Proceedings of the first Global Conference of the International *Musa* Testing Programme held at FHIA, Honduras, 27-30 April 1994. (D.R. Jones ed.) INIBAP, Montpellier, France.

Banerjee, N. and E.A. de Langhe. 1985. A tissue culture technique for rapid clonal propagation and storage under minimal growth conditions of *Musa* (banana and plantain). Plant Cell Reports 4:351-354.

Brewbaker, J.L. and D.L. Umali. 1956. Classification of Philippine *Musae* I. The genera *Musa* L. and *Ensete* Horan. Philipp. Agric. 40:231-241.

Carreel, F. 1995. Etude de la diversité génétique des bananiers (genre *Musa*) à l'aide des marqueurs RFLP. Thèse, INA Paris-Grignon, Paris, France.

Carreel, F., S. Fauré, D. Gonzalez De Leon, P.J.L. Lagoda, X. Perrier, F. Bakry, H. Tenezas Du Montcel, C. Lanaud and J.P. Horry. 1994. Evaluation de la diversité génétique chez les bananiers diploïdes (*Musa* sp.). Genet. Sel. Evol. 26, Suppl 1:125s-136s.

Champion, J. 1967. Les bananiers et leur culture - tome 1 - Botanique et génétique. SETCO, Paris, France.

Chandler, S. 1995. The nutritional value of bananas. Pp. 468-480 *in* Bananas and Plantains (S. Gowen, ed.). Chapman and Hall, London, UK.

Cheesman, E.E. 1947. Classification of Bananas. II: the Genus *Musa* L. Kew Bull. 2:106-117.

Cheesman, E.E. 1948. Classification of Bananas III: critical notes on the species *M. paradisiaca*, *M. sapientum*. Kew Bull. 2:145-153.

Cote, F. X., S. Perrier and C. Teisson. 1993. Somaclonal variation in *Musa* sp.: theoretical risks and risk management. Future research prospects. Pp. 192-199 *in* Biotechnology Applications for Banana and Plantain Improvement. Proceedings of INIBAP workshop, San José, Costa Rica, 27-31 Jan 1992. INIBAP, Montpellier, France.

Craenen, K. and R. Ortiz. 1996. Effect of the black Sigatoka resistance gene *bs*, and ploidy level in fruit and bunch traits of plantain-banana hybrids. Euphytica 87:97-101.

Darjo, P. and F. Bakry. 1990. Conservation et germination des graines de bananiers *(Musa* sp.). Fruits 45:103-113.

Davies, G. 1993. Domestic banana-beer production in Uganda. Infomusa 2(1):12-15.

Dhed'a, D., F. Dumortier, B. Panis, D. Vuylsteke and E. de Langhe. 1991. Plant regeneration in cell suspension cultures of the cooking banana cv. 'Bluggoe' *(Musa* spp. ABB group). Fruits 46:125-135.

Dodds, K.S. and N.W. Simmonds. 1948. Genetical and cytological studies of *Musa* IX: the origin of edible diploid and the significance of interspecific hybridization in the banana complex. J. Genet. 48:285-296.

Eggleston, G., R. Swennen and S. Akoni. 1992. Physicochemical studies on starch isolated from plantain cultivars, plantain hybrids and cooking bananas. Starch/Starke 44:121-128.

Faure, S., J.L. Noyer, F. Carreel, J.P. Horry, F. Bakry and C. Lanaud. 1994. Maternal inheritance of chloroplast genome and paternal inheritance of mitochondrial genome in bananas *(Musa acuminata).* Curr. Genet. 25:265-269.

Ferris, R.S.B. 1993. Dry matter content in plantains and banana and their hybrids. MusAfrica 2:3-4.

Fullerton, R.A. and R.H. Stover (eds). 1990. Sigatoka leaf spot diseases of bananas. Proceedings of an international workshop held at San José, Costa Rica, March 28-April 1, 1989. INIBAP, Montpellier, France.

Ganry, J. (ed.). 1993. Breeding banana and plantain for resistance to diseases and pests: Proceedings of the International Symposium on genetic improvement of bananas for resistance to diseases and pests organized by CIRAD-FLHOR, Montpellier, France, 7-9 September 1992. CIRAD, Montpellier, France.

Gawel, N. and R.L. Jarret. 1991a. Cytoplasmic genetic diversity in bananas and plantains. Euphytica 52:19-23.

Gawel, N. and R.L. Jarret. 1991b. Chloroplast DNA restriction fragment length polymorphisms (RFLPs) in *Musa* species. Theor. Appl. Genet. 81:783-786.

Gawel, N., R. Jarret and A. Whittemore. 1992. Restriction fragment length polymorphism (RI LP)-based phylogenetic analysis of *Musa*. Theor. Appl. Genet. 82:286-290.

Horry, J.P. 1989. Chimiotaxonomie et organisation génétique dans le genre *Musa*. Fruits 44:455-474; 509-520; 573-578.

Horry, J.P. and M. Jay. 1990. An evolutionary background of bananas as deduced from flavonoids diversification. in Identification of genetic diversity in the genus *Musa*. Pp. 41-55 *in* Proceedings of an international workshop held at Los Banos, Philippines, 5-10 September 1988 (R.L. Jarret, ed.). INIBAP, Montpellier, France.

Horry, J.P., F. Bakry and J. Ganry. 1993. Creation of varieties through hybridization of diploids. Pp. 293-300 *in* Breeding Banana and Plantain for Resistance to Diseases and Pests. Proceedings of the International Symposium on Genetic Improvement of Bananas for Resistance to Diseases and Pests, Montpellier, France, 7-9 September 1992. CIRAD/INIBAP, Montpellier, France.

Howell, E.C., H.J. Newbury, R.L. Swennen, L.A. Withers and B.V. Ford-Lloyd. 1994. The use of RAPD for identifying and classifying *Musa* germplasm. Genome 37:328-332.

INIBAP. 1993b. Biotechnology applications for banana and plantain improvement. Proceedings of INIBAP workshop, San José, Costa Rica, 27-31 Jan 1992. INIBAP, Montpellier, France.

INIBAP. 1994. Annual Report. Montpellier, France.

INIBAP/IBPGR. 1990. *Musa* Conservation and documentation: Proceedings of a workshop held in Leuven, Belgium, 11-14 December 1989. INIBAP, Montpellier, France.

Jarret, R.L. (ed.). 1990. Identification of genetic diversity in the genus *Musa*: Proceedings of an international workshop held at Los Banos, Philippines, 5-10 September 1988. INIBAP, Montpellier, France.

Jarret, R.L. and N. Gawel. 1995. Molecular markers, genetic diversity and systematics in *Musa*. Pp. 66-83 *in* Bananas and Plantains (S. Gowen, ed.). Chapman and Hall, London, UK.

Jarret, R.L., N. Gawel, A. Whittemore and S. Sharrock. 1992. RFLP-based phylogeny of *Musa* species in Papua New Guinea. Theor. Appl. Genet. 84:579-584.

Jarret, R.L., D.R. Vuylsteke, N.J. Gawel, R.B. Pimentel and L.J. Dunbar. 1993. Detecting genetic diversity in diploid bananas using PCR and primers from a highly repetitive DNA sequence. Euphytica 68:69-76.

John, P. and J. Marchal. 1995. Pp. 434-467 *in* Bananas and Plantains (S. Gowen, ed.). Chapman and Hall, London, UK.

Jones, D.R. (ed.). 1994a. The improvement and testing of *Musa*: a global partnership: Proceedings of the first global conference of the international *Musa* testing programme held at FHIA, Honduras, 27-30 April 1994. INIBAP, Montpellier, France.

Jones, D.R. 1994b. Report on the meeting. Pp. 7-15 *in* Banana and Plantain Breeding: Priorities and Strategies (INIBAP, ed.). Proceedings of the First Meeting of the *Musa* Breeders Network, Lima, Honduras, 2-3 May 1994. INIBAP, Montpellier, France.

Karamura, D.A. and E.B. Karamura. 1994. A provisional checklist of banana cultivars in Uganda. INIBAP, Montpellier, France.

Lagoda, P. and J.L. Noyer. 1995. Update on *Musa* genome mapping at CIRAD-AGETROP. Infomusa 3(2):4.

Lebot, V., K.M. Aradhya, R. Manshardt and B. Meilleur. 1993. Genetic relationships among cultivated bananas and plantains from Asia and the Pacific. Euphytica 67:163-175.

MacDaniels, L.H. 1947. A study of the Fe'i banana and its distribution with reference to Polynesian migrations. Bernice P Bishop Museum Bull 190, Bishop Museum Press, Honolulu.

May, G.D., R. Afza, H.S. Mason, A. Wiecko, F.J. Novak and C.J. Arntzen. 1995. Generation of transgenic banana (*Musa acuminata*) plants via *Agrobacterium*-mediated transformation. Biotechnology 13:486-492

Nasution, R.E. 1991. A taxonomic study of the species *Musa acuminata* Colla with its intraspecific taxa in Indonesia. Memoire of Tokyo Univ. of Agriculture 32:1-122.

Nasution, R.E., T. Nakamura, K. Izumi, S. Iyama, M. Amano and Y. Hirai. 1989. Taxonomy study of monkey bananas (*Musa acuminata* Colla) of Indonesia. Pp. 167-169 *in* Proc. of 6th Internatl. Congr. of SABRAO.

Ortiz, R. 1995. *Musa* genetics. Pp. 84-109 *in* Bananas and Plantains (S. Gowen, ed.). Chapman and Hall, UK.

Ortiz, R. 1996. The potential of AMMI analysis for field assessment of *Musa* genotypes to virus infection. HortScience 31:829-832.

Ortiz, R. and D. Vuylsteke. 1994a. Trisomic segregation ratios and genome differentiation in AAB plantains. InfoMusa 3(1):21.

Ortiz, R. and D. Vuylsteke. 1994b. Inheritance of albinism in banana and plantain (*Musa* spp.) HortScience 29:903-905

Ortiz, R. and D. Vuylsteke. 1994c. Genetic analysis of apical dominance and improvement of suckering behaviour in plantain. J. Am. Soc. Hort. Sci. 119:1050-1053.

Ortiz, R. and D. Vuylsteke. 1994d. Inheritance of black sigatoka resistance in plantain-banana (*Musa* spp.) hybrids. Theor. Appl. Genet. 89:146-152.

Ortiz, R. and D. Vuylsteke. 1994e. Future strategy for *Musa* improvement at IITA. Pp. 40-42 *in* Banana and Plantain Breeding: Priorities and Strategues, Proceedings of the first meeting of the *Musa* Breeder's Network, La Lima, Honduras, 2-3 May, 1994. INIBAP, Montpellier, France.

Ortiz, R. and D. Vuylsteke. 1995. Factors influencing seed set in triploid *Musa* spp. L. Ann. Botany 75:151-155.

Ortiz, R., R.S.B. Ferris and D. Vuylsteke. 1995a. Banana and plantain breeding. Pp. 110-146 *in* Bananas and Plantains (S. Gowen, ed.). Chapman and Hall, UK.

Ortiz, R., D. Vuylsteke, B. Dumpe and R.S.B. Ferris. 1995b. Banana weevil resistance and corm hardness in *Musa* germplasm. Euphytica 86:95-102.

Ortiz, R., D. Vuylsteke and N.M. Ogburia. 1995c. Inheritance of pseudostem waxiness in banana and plantain (*Musa* spp.). J. Heredity 86:297-299.

Panis, B. 1995. Cryopreservation of banana (*Musa* spp.) germplasm. Doktoraatsproefschrift Nr. 472, Katholieke Universiteit Leuven, Leuven, Belgium.

Panis, B., K. de Smet, I. Van Den Houwe and R. Swennen. 1994. *In vitro* conservation of *Musa* germplasm: Prospects of cryopreservation. in Proceedings of the XIth Reunion Meeting of ACORBAT, San José, Costa Rica, 13-18 February 1994. (in press).

Peloquin, S.J. and R. Ortiz. 1992. Techniques for introgressing unadapted germplasm to breeding populations. Pp. 485-507 *in* Plant Breeding in the 1990s (H.T. Stalker and J.P. Murphy, eds.). CAB International, UK.

Persley, G.J. and E.A. de Langhe (eds.). 1987. Banana and plantain breeding strategies: Proceedings of an international workshop held at Cairns, Australia, 13-17 October, 1986. ACIAR proceedings No.21. Canberra, Australia.

Ploetz, R.E. (ed.). 1990. Fusarium wilt of banana. APS Press, St. Paul, Minnesota, USA.

Price, N.S. 1995. Banana morphology-part I: roots and rhizomes. Pp. 179-189 *in* Bananas and Plantains (S. Gowen, ed.). Chapman and Hall, London, UK.

Shanmugavelu, K.G., K. Aravindakshan and S. Sathiamoorthy. 1992. Banana. Taxonomy, Breeding and Production Technology. Metropolitan Book Co. PVT.Ltd. New Delhi, India.

Sharrock, S. 1990. Collecting *Musa* in Papua New Guinea. Pp. 140-157 *in* Identification of genetic diversity in the genus *Musa* (R.L. Jarrett, ed). Proceedings of an international workshop held at Los Banos, Philippines, 5-10 September 1988 INIBAP, Montpellier, France.

Shepherd, K. 1990. Observations on *Musa* Taxonomy. Pp. 158-165 *in* Identification of genetic diversity in the genus *Musa* (R.L. Jarrett, ed). Proceedings of an international workshop held at Los Banos, Philippines, 5-10 September 1988 INIBAP, Montpellier, France.

Shepherd, K. and F.R. Ferreira. 1982. The PNG Biological Foundation at Laloki, Port Moresby, Papua New Guinea. IBPGR/SEAN Newsl. 8(4):28-34.

Simmonds, N.W. 1956. Botanical results of the banana collecting expedition, 1954-1955. Kew Bull. 11:463-489.

Simmonds, N.W. 1962. Evolution of the Bananas. Longman, London, UK.

Simmonds, N.W. and K. Shepherd. 1952. An Asian banana (*Musa acuminata*) in Pemba, Zanzibar Protectorate. Nature 169:507

Simmonds, N.W. and K. Shepherd. 1955. The taxonomy and origins of the cultivated bananas. J. Linn. Soc. (Bot.) 55:302-312.

Simmonds, N.W. and S.T.C. Weatherup. 1990. Numerical taxonomy of the wild bananas. New Phytol. 115:567-571.

Stover, R.H. and N.W. Simmonds. 1987. Bananas. Longman, London, UK.

Strotzky, G., E.A. Cox and R.D. Goos. 1962. Seed germination studies in *Musa.I.* Scarification and aseptic germination of *Musa balbisiana.* Am. J. Botany 49:515- 520.

Swennen, R. and D. Vuylsteke. 1993. Breeding black sigatoka resistant plantains with a wild banana. Trop. Agric. (Trin.) 70:74-77.

Swennen, R., D. Vuylsteke and R. Ortiz. 1995. Phenotypic diversity and pattern of variation in West and Central African plantains. Econ. Botany 49:320-327.

Tezenas du Montcel, H. 1990. *M. acuminata* subspecies *banksii:* status and diversity. Pp. 211-218 *in* Identification of genetic diversity in the genus *Musa* (R.L. Jarrett, ed). Proceedings of an international workshop held at Los Banos, Philippines, 5-10 September 1988 INIBAP, Montpellier, France.

Tezenas du Montcel, H., E. de Langhe and R. Swennen. 1983. Essai de classification des bananiers plantains (AAB). Fruits 38:461-474.

Thompson, A.K. 1995. Banana processing. Pp. 481-492 *in* Bananas and Plantains (S. Gowen. ed.). Chapman and Hall, London, UK.

Valmayor, R.V., R.G. Davide, J.M. Stanton, N.L. Treverrow and V.N. Roa (eds.). 1994. Banana Nematodes and Weevil Borers in Asia and the Pacific: Proceedings of a conference-workshop on nematodes and weevil borers affecting bananas in Asia and the Pacific, Serdang, Selangor, Malaysia, 18-22 April 1994. INIBAP/ASPNET, Los Banos, Laguna, Philippines.

Valmayor, R.V., B.E. Umali and C.P. Bejosano (eds). 1991. Banana diseases in Asia and the Pacific: Proceedings of a regional technical meeting on diseases affecting banana and plantain in Asia and the Pacific, Brisbane, Australia, 15-18 April 1991. INIBAP, Montpellier, France.

Van den Houwe, I., K. De Smet, H. Tezenas du Montcel and R. Swennen. 1996. Variability in storage potential of banana meristem cultures under medium term storage conditions. Plant Cell Tiss. Org. Cult. (in press).

Vandenhout, H., R. Ortiz, D. Vuylsteke, R. Swennen and K.V. Bai. 1995. Effect of ploidy on stomatal and other quantitative traits in plantain and banana hybrids. Euphytica 83:117-122.

Vuylsteke, D. and R. Ortiz. 1995. Plantain derived diploid hybrids (TMP2x) with black sigatoka resistance. HortScience 30:147-149.

Vuylsteke, D., R. Ortiz and R.S.B. Ferris. 1993a. Genetic and agronomic improvement for sustainable production of plantain and banana in sub-Saharan Africa. African Crop Sci. J. 1:1-8.

Vuylsteke, D., R. Ortiz, R.S.B. Ferris and J.H. Crouch. 1997. Plantain Improvement. Plant Breed. Rev. 14:267-320.

Vuylsteke, D., R. Ortiz, R.S.B. Ferris and R. Swennen. 1995. PITA-9: A black sigatoka resistant hybrid from the 'False Horn' plantain gene pool. HortScience 30:395-397.

Vuylsteke, D., R. Ortiz, C. Pasberg-Gauhl, F. Gauhl, C. Gold, S. Ferris and P. Speijer. 1993b. Plantain and banana research at the International Institute of Tropical Agriculture. HortScience 28:873-874; 970-971.

Vuylsteke, D., R. Ortiz and R. Swennen. 1993c. Genetic improvement of plantains and bananas at IITA. InfoMusa 2(1):10-12.

Vuylsteke, D., R. Ortiz and R. Swennen. 1993d. Genetic improvement of plantains at IITA. Pp. 266-282 *in* Breeding Banana and Plantain for Resistance to Diseases and Pests (J. Ganry, ed.). CIRAD in collaboration with INIBAP, Montpellier, France.

Vuylsteke, D., J. Schoofs, R. Swennen, G. Adejare, M. Ayodele and E. de Langhe. 1990. Shoot-tip culture and third country quarantine to facilitate the introduction of *Musa* germplasm into West Africa. Plant Genet. Resources Newsl. 81/82:5-11.

Vuylsteke, D., R. Swennen and E. de Langhe. 1991. Somaclonal variation in plantains (*Musa* spp. AAB group) derived from shoot-tip culture. Fruits 46:429-439.

Vuylsteke, D., R. Swennen and E. de Langhe. 1996. Field performance of somaclonal variants of plantain (*Musa* spp., AAB group). J. Am. Soc. Hort. Sci. 121:42-46.

Vuylsteke, D., R. Swennen and R. Ortiz. 1993e. Development and performance of black sigatoka-resistant tetraploid hybrids of plantain (*Musa* spp., AAB group). Euphytica 65:33-42.

Vuylsteke, D., R. Swennen and R. Ortiz. 1993f. Registration of 14 improved Tropical *Musa* plantain hybrids with black sigatoka resistance. HortScience 28:957-959.

Williams, R.J. 1987. Banana and plantain germplasm conservation and movement and needs for research. Pp. 177-181 *in* Banana and Plantain Breeding Strategies (G.J. Persley and E.A. de Langhe, eds). Proceedings of an international workshop held at Cairns, Australia, 13-17 October 1986. ACIAR proceedings No. 21. ACIAR Canberra, Australia.

Chapter 7

Cowpea

N.Q. Ng and B.B. Singh

The major cowpea-growing countries are Nigeria, Niger, Mali, Burkina Faso, Senegal, Togo, Benin, Ghana, Chad and Cameroon in West and Central Africa; Tanzania, Somalia, Kenya, Zambia, Zimbabwe, Botswana and Mozambique in East and Southern Africa; India, Pakistan, Sri Lanka, the Philippines, Bangladesh, Indonesia and China in Asia; and Brazil, the West Indies, Cuba and southern USA in the Western hemisphere. The estimated area under cowpea cultivation in the world is over 10 million hectares with about 70% in West and Central Africa and 12% in Brazil (Singh *et al.* 1996).

BOTANY AND DISTRIBUTION

Taxonomy

Cowpea belongs to the family Leguminosae, subfamily Papilionoideae, tribe Phaseoleae and genus *Vigna*.

Vigna is a large and immensely variable genus consisting of more than 85 species, divided into seven subgenera: *Vigna*, *Sigmoidotropis*, *Plectotropis*, *Macrorhyncha*, *Ceratotropis*, *Haydonia* and *Lasiocarpa* (Marechal *et al.* 1978). Seven species are cultivated. Five Asiatic domesticated species – *V. radiata* (L.) R.Wilczek, *V. mungo* (L.) Hepper, *V. umbellata* (Thumb.) Ohwi & Ohashi, *V. angularis* (Wild.) Ohwi & Ohashi and *V. aconitifolias* (L.). (Jacq.) M. M. & S. – falling under the subgenus *Ceratotropis* are genetically highly isolated from cowpea. Bambara groundnut *(V. subterranea* L.), like cowpea, is also an African domesticated species. It is classified under the same subgenus *Vigna* with cowpea, but in a different section *Vigna*. It has very little in common with cowpea.

According to Marechal *et al.* (1978), cultivated cowpea and their closely related wild species are classified under a single botanical species *V. unguiculata* (L.) Walp. This species and *V. nervosa* Mokota are members of the section *Catiang* of the subgenus *Vigna*. Recently, Ng (1995) proposed to reinstate *V. rhomboidea* Burtt Davy as a distinct species which was classified as a variety of *V. unguiculata* by Marechal *et al.* (1978). Thus the section *Catiang* consists of three botanical species. On this classification, all cultivated cowpea are grouped under the subspecies *unguiculata* which is subdivided into four cultigroups, namely Unguiculata, Biflora, Sesquipedalis and Textilis (Westphal 1974; Marechal *et al.* 1978; Ng and Marechal 1985). Verdcourt's earlier work (1970), however, had recognized three cultivated subspecies of *V. unguiculata*, namely subspp. *unguiculata* (L.) Verd., *sesquipedalis* (L.) Verd. and *cylindrica* (L.) Van Eseltine. The synonyms and vernacular names of the cultivar groups are shown in Table 7.1.

The classification and nomenclature of wild *V. unguiculata* is very complex, so it is useful to discuss them here in greater detail to avoid confusion. More than 20 epithet names have been used to designate the wild taxa within *V. unguiculata*. Recent work on wild *V. unguiculata* by several researchers (Pienaar and van Wyk 1992; Padulosi 1993;

Table 7.1. Synonyms and common names of cultivated *Vigna unguiculata*.

Cultigroup	Synonyms	Common or local names
Unguiculata	*V. unguiculata* subsp. *unguiculata* (L.) (Walp.) Verdc.; *V. sinensis* (L.) Hassk. subsp. *sinensis*; *Dolichos unguiculata* L.; *D. biflorus* L.; *D. sinensis* L.; *Phaseolus unguiculata* (L.) Piper; *P. sphaerospermus* L.	Cowpea, Southern pea, Black-eye pea, Crowder pea, China pea, Kaffir pea, Marble pea, Ewa, Nyebee, Wake, Niebe, Aiku vovo, Kondi
Biflora	*V. unguiculata* subsp. *cyllindrica* (L.) Van Eseltine; *V. catjan* (Burm. F.) Walp.; *V. sinensis* (L.) Hassk. var. *catjang* (Burm. f.) Chiov.; *V. cylindrica* (L.) Skeels; *V. unguiculata* (L.) Walp. subsp. *catjang* (Burm. f.) Chiov; *V. unguiculata* var. *cylindrica* (L.) Ohashi, *Dolichos* catjang Burm. F.; *D. tranquebaricus* Jacq; *D. monochalis* Brot.; *D. biflorus* L.; *D. cylindrica* L.	Catjan bean, Hindu sowpea
Sesquipedalis	*V. unguiculata* subsp. *sesquipedalis* (L.) Verdc.; *V. sinensis* (L.) Hassk. var. *sesquipedalis* (IL.) Ascherson and Schweinf.; *V. sinensis* (L.) Hassk. subsp. *sesquipedalis* (L.) Van Eseltine; *V. sesquipedalis* (L.) Fruhw.; *V. unguiculata* var. *sesquipedalis* (L.) Ohashi, *Dolichos sesquipedalis* L.	Yard-long bean, Asparagus bean, Bodi bean, Snake bean

Pasquet 1993; Ng 1995) aims to improve the system of classification earlier proposed by Marechal *et al.* (1978).

This system had recognized three wild subspecies of *V. unguiculata*: subsp. *dekindtiana* (that includes var. *dekindtiana*, var. *mensensis*, var. *pubescens* and var. *protracta*), subsp. *stenophylla* and subsp. *tenuis*. Pienaar and van Wyk (1992) in their study of wild *V. unguiculata* species complex in South Africa, retained the three subspecies described by Marechal *et al.* (1978), but they raised the ranking of the variety *protracta* to a subspecies level and considered the variety *pubescens* to be a conspecies of the subspecies *protracta*. A recent study by Pasquet (1993) has added three new subspecies: *baoulensis*, *letouzeyi* and *burundiensis* and raised the var. *pubescens* to subspecies ranking. Padulosi (1993) also retained the three subspecies recognized by Marechal *et al.* (1978), but he raised the varieties *protracta* and *pubescens* to two distinct subspecies, and added eight new epithet names to describe the variants found within the five subspecies. Table 7.2 shows the synonyms of the various classification systems of wild *V. unguiculata* species complex by the various authors.

Wild taxa within section *Catiang* are closely related to cowpea. Cytogenetic studies show that all taxa within *V. unguiculata* are intercrossable and produce fertile hybrids (Ng and Apeji 1988; Sakupwanya *et al.* 1990; Ng 1990, 1995). Partial incompatibility exists between var. *pubescens* and the cultivated cowpea (Fatokun and Singh 1987). All wild and cultivated taxa within the species *V. unguiculata* belong to the primary genepool of cultivated cowpea. *Vigna rhomboidea* shows marked incompatibility with all other taxa within *V. unguiculata* studied so far (Ng 1995). So far, experimental studies show that crossing between *V. unguiculata* and *V. rhomboidea* is unidirectional and is successful only when *V. unguiculata* is used as the female. Their hybrids are partially sterile. *Vigna rhomboidea* can be considered to be a secondary genepool of cowpea. It could have evolved from a common wild ancestor of *V. unguiculata* through isolation or divergence, in mid-altitude regions in the Transvaal region and Swaziland. Crosses made so far between *V. unguiculata* and *V. nervosa*, another distinct species within section *Catiang*, have not been successful. No species outside the section *Catiang* (over 35 species have been tried so far) has ever been crossed successfully with

Table 7.2. Classification and nomenclature of the wild *Vigna unguiculata* species complex.

Verdcourt (1970)	Marechal *et al.* (1978)	Pasquet (1993)	Padulosi (1993)	Ng (1995)
V. unguiculata	V. unguiculata	V. unguiculata ssp. unguiculata var. spontanea	V. unguiculata	V. unguiculata
ssp. dekindtiana	ssp. dekindtiana var. dekindtiana	ssp. dekindtiana var. dekindtiana	ssp. dekindtiana var. dekindtiana var. huliensis var. congolensis	ssp. dekindtiana var. dekindtiana
ssp. mensensis	var. mensensis	ssp. letouzeyi ssp. burundiensis	var. ciliolata var. grandiflora	var. mensensis
V. unguiculata var. protracta[†]	var. protracta	ssp. stenophylla	ssp. protracta var. protracta var. kgalagadiensis	ssp. stenophylla var. protracta
V. pubescen[†]	var. pubescens	ssp. pubescens	ssp. pubescens	var. pubescens
V. angustifoliolata[†]	ssp. stenophylla	ssp. stenophylla	ssp. stenophylla	var. stenophylla
V. tenuis[†]	ssp. tenuis	ssp. tenuis	ssp. tenuis var. tenuis var. oblonga var. parviflora ssp. protracta var. rhomboidea	var. tenuis V. rhomboidea
		ssp. baoulensis		

[†] It was remarked by Verdcourt as variants of *V. unguiculata.*

members of the section *Catiang*. The study of hybrid embryo development in crosses between cowpea and *V. vexillata* and between cowpea and several other wild species within section *Vigna* of the subgenus *Vigna* has shown that fertilization did occur (Ng 1990; Agwaranze 1992; Barone *et al.* 1992). However, hybrid pods usually aborted within a few days after pollination and attempted embryo rescue has so far not been successful, owing to very early embryo abortion. The chromosome number of all species in the section *Catiang* is diploid $2n=2x=22$. Most *Vigna* species have the same number $2n=22$, but some have $2n=20$, while no natural polyploids in the genus have ever been reported, except for *V. glabrescens* in section *Ceratotropis*.

Origin, Domestication and Diffusion

The centre of diversity of wild *V. unguiculata* and *V. rhomboidea* is in southern and southeastern Africa, whereas the centre of diversity of the cultivated *V. unguiculata* is in West Africa, in an area encompassing the savanna region of Nigeria, Southern Niger, part of Burkina Faso, Northern Benin, Togo and part of Northern Cameroon (Ng 1995).

The wild annual subsp. *dekindtiana* var. *dekindtiana* is the probable progenitor of the cultivated cowpea. It occurs all over Africa south of the Sahara, including Madagascar. It is a creeping or climbing herb with a broad habitat tolerance, most frequently found in sandy grassland and woodland, in open disturbed habitats, particularly near the edges of farmlands, at roadsides, on the banks of small streams or in fallows. It could have been derived from any variety of the subsp. *stenophylla* (Ng 1995). Its growth habit and morphology are very much the same as that of cowpea landraces, except that its pods are usually black, scabrous, dehisive at maturity and much smaller than the cultivated cowpea. The surface of the pods of cultivars is glabrous. Seeds of this wild species are tiny with dark speckles. In most of the African centres of production today, cowpea landraces are cultivated as a component in a mixed cropping system, particularly millet- or sorghum-based farming systems in the semi-arid and subhumid tropics. Cowpea is highly resistant to drought. Its leaves and stems remain green at the end of the cropping season when most standing crops have wilted. The haulms are gathered to feed cattle, particularly in Northern Nigeria, Niger, Mali, Burkina Faso, Northern Cameroon and Senegal. It is equally important as a pulse in these regions.

Wild cowpea (subsp. *dekindtiana*) could have been gathered as fodder to feed cattle and later domesticated as early as 4000 BP in West Africa. During the process of domestication and selection of cowpea from its wild progenitor, characters lost and gained included seed dormancy together with a reduction of pod dehiscence on the one hand, and an increase in pod and seed size on the other.

The selection of the cowpea as a pulse as well as for fodder might have resulted in the establishment of the cultigroup Unguiculata (Ng 1995). Selection for types with long peduncle for fibre as well as for fodder or seed has resulted in the cultigroup Textilis (Ng and Marechal 1985). Once the cultigroup Unguiculata was established in West Africa, diversity developed and accumulated through mutation. Recombination also resulted from occasional hybridization between predominantly inbred cultivars or between cultivars and sympatric populations of the wild progenitor subsp. *dekindtiana*. Some weedy forms are morphologically indistinguishable from truly wild *dekindtiana*, except that their seed is slightly larger. They could be hybrid derivatives resulting from hybridization between cowpea and the ancestor species, or an escape of the most primitive cultivated form in the early stage of domestication, or intermediates between truly wild var. *dekindtiana* and the domesticated species. These weedy forms are commonly found in West Africa (Rawal and Roberts 1975). Taxonomically, they should be classified under var. *dekindtiana*. Pasquet (1993) prefers to call this var. *spontanea* of the subsp. *unguiculata*.

Through centuries of cultivation, short-day cowpea cultivars became adapted to the cereal farming system, while day-neutral cultivars later evolved from these short-day cultivars and became adapted to the yam-based farming system in the humid zone of West Africa (Steele and Mehra 1980). From West Africa the cultigroup Unguiculata was introduced to East Africa. From there it was brought to Europe, where it was known to the Romans about 2300 BP, and to India about 2200 BP (Ng and Marechal 1985). The cowpea underwent further diversification in India and Southeast Asia, producing the cultigroup Biflora for its grain and for use as a cover crop, and the cultigroup Sesquipedalis with its long pods used as a vegetable (Steele and Mehra 1980). Cowpea was probably brought to the Americas during the 17th century by the Spanish and Portuguese traders. The centre of diversity and dispersal route of cowpea are shown in Figure 7.1.

Botany

Seed germination in *V. unguiculata* is epigeal. The cultivated species has deep taproots which are stout with numerous well-nodulated, spreading, lateral roots in the surface soil. The first pair of leaves is simple and opposite, and is succeeded by alternate, pinnately trifolate leaves on stout and grooved petioles 4-15 cm long. The leaflets are ovate-rhomboid to lanceolate, sometimes hastate, 5-18 cm long and 3-16 cm wide, usually entire, but sometimes lobed and subtended by inconspicuous stipels. At the base of each petiole is appended a large stipule which is spurred, the characteristic of the species in the section *Catiang*.

Inflorescence axillary is a compound raceme with several racemes (usually 2-4 flowered) crowded near the tip in alternate pairs on thickened nodes carried on a grooved peduncle 2.5-50 cm long, occasionally up to 1 m long in cultivar group Textilis. Each flower is substended by a deciduous bract and has a campanulate calyx with five acute lobes (triangular teeth) 2-16 mm long. Cowpea flowers are large, with a standard petal 2-3 cm wide, two symmetric wing petals and a boat-shaped keel with beak. The wings are adherent to the keel enclosing the androecium and gynoecium. The petals are either white or with anthocyanin pigmentation in shades of pale mauve or pink to

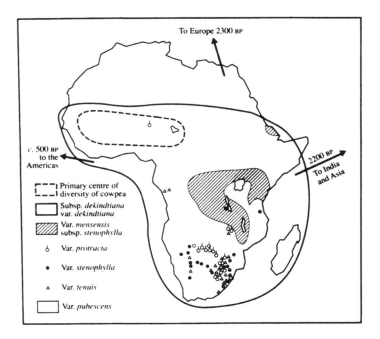

Fig. 7.1. Centre of diversity and distribution of cowpea and its closely related wild species (from Ng 1995).

dark purple. The flowers open in the early morning and then wither, fade and are shed the same day. The extra-floral nectaries at the base of the corolla attract ants, flies and bees. The stamens are diadelphous (9+1) with the vexillary stamen free and 9 fused, forming a tubular sheath around the gynoecium. The ovary is sessile with many ovules and its style is bent, bearded on the inner curve immediately below oblique stigma.

Reproductive Biology

Cowpea is basically an inbreeding species but some outcrossing (about 10%) probably occurs (Duke 1981). Its pollen is sticky and heavy. The anthers of cowpea flowers are usually held closely around the stigma. However, in some cases the position of the stigma is well above the anthers. In this case insects may be needed to help pollinate the flowers. When a heavy insect such as a bee or bumblebee lands on the petal to collect nectar at the base of the corolla, it depresses the keel, thereby releasing the stigma and stamen out of the keel. In this tripping process, the pollen may be rubbed onto the stigma, or may be collected by an insect or may stick in the hair on the bee. In the same way, the stigma may be in contact with the insect and be pollinated by pollens collected from other plants. This process leads to outcrossing. Other insects such as ants and flies may help to pollinate the flowers.

In many wild cowpea species, there is a wide spatial separation between the stigma and anthers, with the former being a few millimeters above the latter. When this spatial separation is pronounced, it may form an effective barrier to self-pollination. Tripping is necessary in this case for pollination. Outcrossing in wild cowpea appears to be very common, as the progenies derived from seeds collected from plants in their natural habitats or multiplication plots frequently segregate.

GERMPLASM CONSERVATION AND USE

Until the establishment of IITA's Cowpea Improvement Programme in 1970, only a few national programmes, notably in the USA, Nigeria, Senegal, Tanzania and Uganda, had initiated breeding activities. By the time IITA's Cowpea Improvement Programme was established, a collection of more than 1000 accessions had been assembled from many parts of the world in the Agricultural Research Service of the Department of Agriculture in the USA. The then Nigerian Grain Legume Genebank at Samaru, Zaria had another colletion of 939 accessions of Nigerian local cowpea cultivars (Ebong 1971; IITA 1972). Nonetheless, the exchange of germplasm between countries was very limited, and most national programmes had limited access to the diverse germplasm resources scattered throughout Africa and other parts of the world. That situation has improved since IITA assumed the global responsibility in cowpea breeding, collecting and germplasm conservation.

From the beginnning of work on cowpea at IITA, germplasm collecting and evaluation have received high priority on its research agenda. Soon after the programme was established, an immediate task was to assemble germplasm from existing collections held by institutions worldwide, such as those in the USA and in other places like Nigeria. Scientists at IITA also initiated plant exploration in West Africa in 1971. Around 4000 accessions of cowpea germplasm had been assembled by 1974.

Systematic plant exploration for the germplasm of cowpea for utilization and conservation expanded greatly after the Genetic Resources Unit (GRU) was established at IITA in 1975. Exploration for wild relatives of cowpea and other wild *Vigna* species was intensified from 1985 (Ng 1990). IITA works in close collaboration with many national programmes and the International Plant Genetic Resources Institute in collecting cowpea and *Vigna* germplasm. To date it has conducted more than 60 exploration missions to 31 African countries. Together with the materials that IITA's scientists collected during exploration and from acquisition or exchange, the GRU now preserves a collection of 14 816 accessions of cultivated cowpea from over 80 countries throughout the world and 1651 accessions of wild *Vigna*. More than 50 wild *Vigna* species from over 50 countries are represented in this collection (Tables 7.3 and 7.4). This collection of cowpea germplasm has been extensively used at IITA for cowpea improvement and related research. Several thousand accessions of germplasm are screened each year by IITA's scientists and their collaborators. Accessions that are preserved at IITA are held in trust under the auspecies of FAO, and so are freely accessible to the public. During the period between 1978 and 1995 GRU/IITA distributed 23 458 samples of cowpea and wild *Vigna* on request, to researchers and workers in over 80 countries (Table 7.5).

Cowpea germplasm is preserved at IITA in both active and base collection conservation seed banks. For the active collection, up to about 500 g of seed of each accession is stored in a plastic screw-top jar in a cold room conditioned at 5°C and 30±5% relative humidity. Requests for germplasm samples from IITA scientists and researchers around the world are filled from this collection. Another lot of seed, usually more than 3000 grains, of the same accession, with seed germination rate known to be greater than 90%, is dried to 5±1% moisture content in a drying cabinet or room conditioned at less than 10% RH and 20°C, and then sealed into aluminium cans or aluminium foil envelopes and stored in a cold room at −20°C for base collection conservation. Under such conditions cowpea seed can remain viable (with 85% germination) for over 100 years (Ng 1991). The regeneration standard for cowpea at IITA is set at 85% germination.

Evaluation

By 1974, about 4000 accessions of cowpea assembled at that time had been evaluated and characterized and a germplasm catalogue was published (Porter *et al.* 1974). By 1996, about 14 000 accessions of the existing collection have been characterized for up to

Table 7.3. Existing germplasm accessions at the Genetic Resources Unit of IITA.

Country or region of origin	Advanced breeding lines	Landraces/ old cultivars	Total
Central and West Africa			
Burkina Faso	37	227	264
Cameroon	–	524	524
Central African Rep.	–	183	183
Chad	–	124	124
Congo	–	44	44
Côte d'Ivoire	2	132	134
Gabon	–	6	6
Gambia	–	4	4
Ghana	1	316	317
Guinea	–	65	65
Liberia	–	9	9
Mali	–	294	294
Mauritania	–	2	2
Niger	8	846	854
Nigeria	635	2908	3543
Republic of Benin	13	354	367
Senegal	280	280	–
Sierra Leone	–	13	13
Togo	1	102	103
Zaire	–	16	16
East and South Africa			
Angola	–	1	1
Botswana	–	507	507
Burundi	–	1	1
Ethiopia	–	7	7
Kenya	11	146	157
Lesotho	–	41	41
Madagascar	–	34	34
Malawi	–	401	401
Rep. of South Africa	–	111	111
Somalia	–	71	71
Sudan	–	52	52
Swaziland	–	18	18
Tanzania	7	409	416
Uganda	1	70	71
Zambia	–	529	829
Zimbabwe	–	118	118
North Africa			
Algeria	–	3	3
Egypt	–	343	343
Asia			
Afghanistan	–	65	65
Bangladesh	–	1	1
China	–	29	29
India	1	2074	2075
Indo China	–	1	1
Indonesia	–	4	4
Japan	–	2	2
Laos	–	2	2
Nepal	–	3	3
Papua New Guinea	–	12	12
Philippines	–	114	114
Republic of Yemen	–	60	60
Sri Lanka	–	2	2
Thailand	–	1	1
USSR	–	42	42

Country or region of origin	Advanced breeding lines	Landraces/ old cultivars	Total
West Pakistan	–	7	7
Middle East			
Iran	–	29	29
Israel	–	8	8
Oman	–	72	72
Syria	–	6	6
Turkey	–	47	47
United Arab Republic	–	4	4
Americas			
Argentina	–	1	1
Brazil	–	171	171
Colombia	–	2	2
Cuba	–	1	1
El Salvador	–	1	1
Guatamala	–	11	11
Honduras	–	1	1
Jamaica	–	1	1
Mexico	–	23	23
Nicaragua	–	1	1
Paraguay	–	12	12
Peru	–	3	3
Surinam	–	17	17
USA	3	825	828
Venezuela	–	3	3
Europe			
France	–	1	1
Greece	–	36	36
Hungary	–	1	1
Italy	–	160	160
Portugal	–	5	5
UK	–	13	13
Australia		24	24
New Caledonia	–	1	1
Unknown	–	886	886
Total	**720**	**14 096**	**14 816**

30 agrobotanical descriptors, following the cowpea decriptors published by IITA (Porter *et al.* 1974) and updated by IPGRI (IBPGR 1982).

Presently the database for cowpea in GRU/IITA contains 11 passport data, 30 agrobotanical characters, and 17 pests and diseases and nutritional values (N and S). The information on passport data is being updated in 1996 to include collaborators. The documentation system will be linked with the Internet through the SINGER project, for public access. Information is also available on request as a computer print-out or a floppy diskette for personal computers.

Growth habits of cowpea vary greatly and they are influenced by both genotype and environment interaction. They range from acute erect to erect (usually photo-insensitive and determinate growth habit), semi-erect, intermediate, semi-prostrate, prostrate or climbing. The length and number of their branches are also quite variable. Erect and determinate plant types have a few short branches measuring less than 1 m. The prostrate and spreading type or sometimes climbing type have five or more orders of branching, and the first and second order branches are up to 5 m long. Testa colour of cowpea is variable, and 42 eye patterns and 62 eye colours have been described (Porter *et al.* 1974), but the common seed colours are cream, buff, brown, red and black. Eye colour or patterns refers to the pigmented area around the hilum, in an otherwise white seed; the colour may spread over the seed from the hilum to varying extents and in various combinations of speckled, mottled or solid distribution. Seed sizes vary from 2 to 12 mm long and individual seed weight can be between 50 and 340 mg. Seed shape may vary from square or globular to kidney-shaped. Seed texture is smooth, rough or wrinkled. Pods vary in length (6-90 cm) and shape (straight, curved or coiled) with between 8 and 20 seeds. Pod pigmentation varies from green or green with a purple tip and/or suture and valves, to purple or brown at the immature stage; and from straw or straw with dull black splashes to deep purple or brown at maturity. Seeds of wild cowpea are tiny, about 2-7 mm long and weigh about 20 mg (3-90 mg). Their coat colour is usually black, speckled or mottled on a grey background, sometimes solid black. Wild cowpea pods are scabrous and sometimes hairy (as in *V. rhomboidea*). They are small, 8 mm (2-18 mm) long and 4 mm (1.7-7.4 mm) wide, usually black and shatter at maturity.

Through intensive screening of germplasm in Ibadan in the early 1970s, superior materials identified were selected and constituted two international disease nursery trials, the Cowpea Disease Nursery and the Cowpea Mosaic Virus Nursery. These were sent to scientists in several countries for evaluation. Some of the germplasm evaluated in the early 1970s was found to have high yield potential as well as resistance to multiple diseases. These are TVu 201(S), TVu 1190, TVu1977 and TVu 4557, which were subsequently described as VITA 1, VITA 3, VITA 4 and VITA 5, respectively. They have been released by several national programmes as direct introductions (IITA 1992, 1994). These elite germplasm accessions formed a major part of the basic source materials for use in the Cowpea Breeding Programme at IITA in the 1970s.

Through intensive screening of the existing collections at IITA, many sources of insect pest resistance to cowpea aphids (*Aphis cracivara*), leafhoppers (*Empoasca signata* and *E. dolichi*), legume bud thrips (*Megalurothrips sjostedti*), pod borers (*Maruca testulalis*), pod-sucking bugs (*Clavigralla tomentosicollis* and *C. shadabi*) and bruchids (*Callosobruchus maculatus*) have been identified (Singh 1977; IITA 1978; Singh and Allen 1979; Singh *et al.* 1983, 1989; Ng 1987; Jackai and Singh 1988; IITA 1989; Ng and Padulosi 1991). Some of the resistant sources are listed in Table 7.6 for those who may be interested in obtaining these germplasm materials for their research.

Sources of resistance against many major diseases, viruses and parasitic weeds have been found, and some of the sources are resistant to multiple diseases (Williams 1975, 1977a, 1977b; Ladipo and Allen 1979; IITA 1978, 1983, 1993; Allen 1980, 1983; Allen *et al.* 1981a, 1981b; Ng and Padulosi 1991; Singh and Ntare 1985; Singh *et al.* 1983, 1989). Germplasm accessions with better drought resistance were also successfully identified and are being used in breeding programmes. Sources of resistance to major viruses, diseases and parasitic weeds and drought are listed in Tables 7.7, 7.8 and 7.9, respectively.

Properties and Uses

Cowpea is important both as a food legume and vegetable for human consumption and as fodder for animal feed. Cowpea seeds can be shelled green or dried, but they are mostly consumed as dried seed. They can be used in plain cooking (soups, stews, boiled or roasted) or in processed dishes (boiled, steamed, dried or baked). Detailed accounts of the use of dried seed in West Africa and the recipes can be found in Dovlo *et al.* (1976). Young pods of the Sesquipedalis cultivars are succulent and are consumed as a vegetable like French beans. Young leaves of all cultivars can be used as leafy vegetables. Long peduncles, particularly of Textilis cultivars, are retted for fibre in the northern part of Nigeria (Duke 1981). The crop can also be used as a green manure. Prostrate varieties with good vegetative growth are suitable for use as a cover crop to reduce soil erosion and suppress weed growth.

Mature cowpea seeds are very nutritious. They contain most of the essential amino acids, although deficient in cystine, methionine and tryptophan. Most traditional cowpea cultivars have an average protein content of 22-24%, ranging from 18 to 29%, with a potential of as high as 34% (Porter *et al.* 1974). The protein content of IITA's improved varieties also falls within this average. The proteins consist of 90% water-

Table 7.4. Existing wild *Vigna* germplasm accessions available at the Genetic Resources Unit of IITA.

Country or region of origin	No. of accessions	Country or region of origin	No. of accessions
Central and West Africa		**East and South Africa (cont.)**	
Cameroon	115	Tanzania	72
Central African Rep.	43	Uganda	1
Chad	28	Zambia	63
Congo	108	Zimbabwe	53
Côte d'ivoire	3	**Asia**	
Equatorial Guinea	4	China	1
Gabon	77	India	11
Ghana	83	Indonesia	1
Kenya	31	Japan	1
Liberia	2	Laos	1
Mali	5	Philippines	2
Niger	75	**North and South America**	
Nigeria	112	Argentina	2
Republic of Benin	26	Brazil	18
Senega	15	Colombia	45
Sierra Leone	1	Costa Rica	8
Zaire	35	Guatemala	2
East and South Africa		Honduras	1
Angola	2	Mexico	6
Botswana	97	Paraguay	1
Burundi	4	Peru	1
Ethiopia	1	Puerto Rico	1
Madagascar	2	Surinam	1
Malawi	86	Uruguay	1
Mozambique	12	USA	1
Namibia	23	Venezuela	1
Rep. of South Africa	117	**Others**	
Rwanda	10	UK	1
Swaziland	45	Australia	3
Uganda	1	Unknown	189
Somalia	7	**Total**	1651

Table 7.5. Cowpea germplasm distributed to researchers and agricultural workers[†] in over 80 countries worldwide by the Genetic Resources Unit, IITA, 1978-95.

Region	No. of samples Cowpea	No. of samples Wild *Vigna*	No. of samples Total
Africa	12 007	737	12 744
Asia	4 741	89	4 830
Europe	1 997	402	2 399
America	3 252	233	3 485
Total	**21 997**	**1461**	**23 458**

[†] These figures exclude numerous advanced breeding lines distributed by IITA cowpea breeders.

soluble globulins and 10% water-soluble albumins. Amino acid content (mg/g N) is: isoleucine 239, leucine 440, lysine 427, methionine 73, cystine 68, phenylalanine 323, tyrosine 163, threonine 225, tryptophan 68, valine 283, arginine 400, histidine 204, alanine 257, aspartic acid 689, glutamic acid 1027, glycine 234, proline 244 and serine 268. Dried mature seeds contain about 12% moisture, 60% carbohydrate, 1.5% fat and 3.7% ash. Each 100 g of seed contains about 338 calories, 5.4 g fibre, 104 mg Ca, 416 mg P, 0.9 mg thiamine, 0.15 mg riboflavin, 2 mg nicotinic acid, 20 IU Vitamin A and a trace of ascorbic acid (0-2 mg) (Duke 1981; Dovlo *et al.* 1976). Total sugars range from 13.7 to 19.7%. Starch composition contains 20-49% amylose and 11-37% amylopectin.

Table 7.6. Sources of cowpea germplasm resistance to some major pests.

Pest	Germplasm accessions[†]	Advanced breeding lines[‡]
Aphid (*Aphis craccivora*)	TVu 18, TVu 36, TVu 42	IT81D-1020, IT82E-1-108
	TVu 301, TVu 408	IT84S-2246, IT90K-59
	TVu 801, TVu 1037	IT86D-534, IT81D-1020
	TVu 2755, TVu 2896	IT90K-76, IT87D-15211
	TVu 3000, TVu 3273	IT82E-1-108, IT88D-807-11
	TVu 3629, TVu 9836	IT89KD-256, IT90K-77
	TVu 9914, TVu 9929	IT90K-101
	TVu 9930, TVu 9944	
Beanfly (*Ophiomyia phaseoli*)	TVu 1433, TVu 3192	IT81D-1205-174, IT82D-644
Cowpea storage weevil	TVu 625, TVu 2027	IT81D-1007, IT81D-1137
(*Callosobruchus maculatus*)	TVu 1192, TVu 11953	IT81D-1157, IT81D-1157
	TVu 4200 (Pod wall)	IT82D-716, IT84S-2246-4
	(all moderate resistance)	IT86D-499, IT86D-534
	TVNu 72, TVNu 73	IT87D-1521, IT87D-1521
	TVNu 70, TVNu 71	IT90K-277-2, IT90K-59
	TVNu 74, TVNu 75	IT90K-76, IT90K-77
	TVNu 121, TVNu 137	IT89KD-457, IT89KD-288
	TVNu 160	IT89KD-256, IT90K-261-3
	(all high resistance)	(all moderate resistance)
Flower bud	TVu 1509, TVu 2870	TVx 3236, IT82D-713
(*Megalurothrips sjostedti*)	(moderate resistance)	IT82D-716, IT84S-2246-4
		IT86D-534, IT90K-76
		IT90K-59, IT90K-77
		IT89KD-457, IT89KD-288
		IT89KD-256 (all moderate)
Legume pod borer	TVu 946, TVu 15556	
(*Maruca testulalis*)	(low resistance)	
	TVNu 64, TVNu 72, TVNu 73,	
	TVNu 77; TVNu 200, TVNu 201;	
	TVNu 97, TVNu 459, TVNu 119,	
	TVNu 84, TVNu 238[‡]	
	TVNu 863[‡]	
	(all high resistance)	
Leafhoppers	TVu 59, TVu 123	
(*Empoasca* sp)	TVu 662, TVu 1190	
	(moderate resistance)	
Pod-sucking bugs	TVu 1, TVu 1890	
(*Clavigralla tomentosicollis*)	TVu 3172, TVu 3354	
	TVu 3199, TVu 3364	
	TVu 3156 (all low level)	
	TVNu 64, TVNu 72	
	TVNu 73, TVNu 85	
	TVNu 167, TVNu 168	
	TVNu 95, TVNu 87	
	TVNu 88, TVNu 168	
	TVNu 95, TVNu 87	
	TVNu 88, TVNu 33	
	TVNu 133 (high resistance)	
C. shadabi	TVu 1977, TVu 7274 (low level)	
Root-knot nematodes	TVu 1283, TVu 1560	
(*Meloidogyne javanica*)		
M. incognita	TVu 401, TVu 1190	IT84S-2246-4, IT89KD-288
	TVu 1962, TVu 1560	IT90K-59, IT90K-76

[‡] *V. unguiculata* subsp. *dekindtiana*.   [†] TVu: Prefix for IITA's cowpea (*V. unguiculata*) accession no.; TVNu: Prefix for IITA's wild *Vigna* accession no.; TVx or IT: Prefix for IITA's cowpea advanced breeding lines. Sources: Singh 1977; IITA 1978, 1983, 1989, 1993; Singh and Allen 1979; Karel and Malinga 1980; Zannou 1981; Jackai and Singh 1988; Ng and Padulosi 1991; Singh *et al*. 1996.

Table 7.7. Sources of cowpea germplasm resistance to some major viruses

Viruses[†]	Germplasm accessions[‡]	Advanced breeding lines[†]
CYMV	TVu 113, TVu 274, TVu 310, TVu 410, TVu 433, TVu 470, TVu 866, TVu 746, TVu 1190, TVu 2769, TVu 2563, TVu 3650, TVu 7941	
SBMV	TVu 79, TVu 113, TVu 238, TVu 310, TVu 347, TVu 393, TVu 445, TVu 470, TVu 470, TVu 486, TVu 493, TVu 697, TVu 746, TVu 1185, TVu 1851, TVu 1851, TVu 1878, TVu 1888, TVu 1948, TVu 1985, TVu 1986, TVu 2672, TVu 2755, TVu 2896, TVu 6365	
CMeV	TVu 36, TVu 42, TVu 205, TVu 471, TVu 488, TVu 566, TVu 947, TVu 1171, TVu 1261, TVu 1477, TVu 2269, TVu 2885, TVu 2886, TVu 2887, TVu 2933, TVu 2939, TVu 2962, TVu 2964, TVu 2968, TVu 2971, TVu 3043, TVu 6433, TVNu 64, TVNu 65, TVNu 66, TVNu 72, TVNu 73, TVNu 84, TVNu 180	
CMeV + CYMV	TVu 274, TVu 346, TVu 393, TVu 493, TVu 1888	
CMeV + CYMV + CAbMV	TVu 266-1, TVu 433, TVu 393, TVu 493, TVu 1888, TVu 1947, TVu 2755	
CMeV + CYMV + SBMV	TVu 238, TVu 393, TVu 470, TVu 493, TVu 2755, TVu 1185, TVu 1849, TVu 2896	
CMeV + CYMV + SBMV + CAbMV	TVu 493, TVu 697, TVu 746, TVu 1185, TVu 1888, TVu 1948, TVu 2755, TVu 393	IT85F-867, IT85F-2687, IT84D-449, IT82D-442
CMeV + CYMV + SBMV + CAbMV + CGMV	TVu 393, TVu 493, TVu 2755, TVu 1185	
CYMV + CAbMV + CGMV		IT82D-716, IT84S-2246, IT90K-59, IT90K-76
CAbMV + CYMV + CUMV + CMMV		IT86D-880, IT86F-2089-5, IT86D-1010
CYMV + CAbMV + CGMV + CUMV + SBMV		IT82D-889, IT83S-818, IT83D-442, IT85F-867-5
CYMV + CAbMV + CMeV + CUMV + SBMV		IT83D-442, IT85F-867-5, IT85F-2687
CYMV + CAbMV + CUMV + CMeV + SBMV + CGMV		IT82E-16, IT82D-889, IT82D-950

[†] CYMV: cowpea yellow mosaic virus; SBMV: southern bean mosaic virus; CMeV: cowpea mottle virus; CAbMV: cowpea aphid-borne mosaic virus; CGMV: cowpea golden mosaic virus; CUMV: cucumber mosaic virus; CMMV: cowpea mild mottle virus.

[‡] TVu: Prefix for IITA's cowpea (*Vigna unguiculata*) accession no.; TVNu: Prefix for IITA's wild *Vigna* accession no.; IT: Prefix for IITA's cowpea advanced breeding lines.

Sources: Williams 1975, 1977b; Ladipo and Allen 1979; Allen 1980; IITA 1978, 1983; Singh and Ntare 1985; Singh *et al.* 1989; Ng and Padulosi 1991; Thottappilly *et al.* 1994.

Table 7.8. Sources of cowpea germplasm resistance to multiple diseases.

Disease	Germplasm accessions[†]	Advanced breeding lines[†]
Ascochyta	TVu 4536, TVu 4569, TVu 4557 (moderate level of resistance)	
Brown blotch	TVu 201, TVu 1977	IT82E-16, IT86D-719, IT86D-715, IT84S-2246-4
Scab	TVu 843, TVu 1404, TVu 1433, TVu 1977	IT86D-1056, IT88S-501-8, IT90K-81-4, TVx 3236, IT84S-2246-4, IT90K-59, IT90K-76
Bacterial blight	TVu 347, TVu 410, TVu 456, TVu 483-2, TVu 726, TVu 745, TVu 1190, TVu 1977	
Bacterial blight + Cercospora		IT90K-284-2, IT90K-277-2, IT86D-719IT85D-3517-2, IT89KD-374-57, IT89KD-391, IT89KD-109, IT89KD-109, IT88D-867-11
Septoria	TVu 456, TVu 483, TVu 486, TVu 726, TVu 853, TVu 1433, TVu 1583, TVu 2455, TVu 1977	IT86D-1056, IT88S-501-8, IT81D-994, IT90K-82-2, IT90K-81-4, IT85D-3577, IT86D-885, IT90K-284-2
Fusarium wilt	TVu 109, TVu 347, TVu 984, TVu 1000, TVu 1016	
Fusarium root rot	TVu 202, TVu 231, TVu 243, TVu 266, TVu 274, TVu 316, TVu 320, TVu 408, TVu 1563	
Phakospora rust	TVu 612, TVu 1258, TVu 1962, TVu 2455, TVu 4540	
Phytophthora stem rot	Ku 235	
Web blight	TVu 317, TVu 2483, TVu 4539 (all are moderately resistant)	
Synchytrium fast rust	TVu 43, TVu 222, TVu 612, TVu 4535, TVu 4537, TVu 4569, TVu 6666	
Black spot		IT87D-509-5, IAR 48, IT88S-524-7, IT84D-666
Target spot	TVu 1190	
Bacterial blight+ scab + septoria+ brown blotch	TVu 1977	
Scab + septoria	TVu 853, TVu 1433	IT86D-1056, IT88S-501-8, IT90K-81-4
Bacterial blight+ septoria	TVu 456, TVu 726, TVu 4832	IT90K-284-2
Combined resistance to Anthracnose, cercospora leaf spot, rust, bacterial pustule, target spot	TVu 8, TVu 62, TVu 64, TVu 176, TVu 201, TVu 317, TVu 374, TVu 1190, TVu 2657, TVu 2757, TVu 2785, TVu 2846, TVu 2847, TVu 2939, TVu 3273, TVu 3408, TVu 3415, TVu 3430, TVu 3511, TVu 3521, TVu 3552, TVu 3847, TVu 4536, TVu 4546, TVu 4549, TVu 4557, TVu 4558, TVu 4562	
Anthracnose + bacterial pustule + Cercospora leaf spot + rust + cowpea yellow mosaic virus	TVu 201, TVu 310, TVu 345, TVu 347, TVu 393, TVu 408, TVu 410, TVu 537, TVu 645, TVu 697, TVu 746, TVu 990, TVu 1190, TVu 1283, TVu 1452, TVu 1452, TVu 1980, TVu 2430, TVu 2755, TVu 3415	

[†] TVu: Prefix for IITA's cowpea (*V. unguiculata*) accession no.; IT: Prefix for IITA's cowpea advanced breeding lines.

Source: Williams 1975, 1977a, 1977b; Allen *et al.* 1981a, 1981b; Patel 1981; Oyekan *et al.* 1976; IITA, 1978, 1989, 1993; Singh and Allen 1979, Singh *et al.* 1989.

Table 7.9. Sources of cowpea germplasm resistance to *Striga*, *Alectra* and drought.

Stress	Germplasm accessions[†]	Advanced breeding lines[†]
Striga gesnerioides	TVu 1271, TVu 1272	IT88D-867-11
	TVu 1330, TVu 1331	IT90K-59
	TVu 1332, TVu 4642	IT90K-76
	TVu 8337, TVu 12415	IT90K-77
	TVu 12430, TVu 12431	IT90K-82-2
	TVu 12432, TVu 12449	IT81D-994
	TVu 12470, TVu 11788	IT82D-849
	TVu 9238, TVu 13035	
	TVu 13035, TVu 8453	
	TVu 13950, TVu 15553	
Alectra vogelii	TVu 12470, TVu 9238	IT90K-59
	TVu 12432, TVu 12415	IT90K-76
	TVu 13950, TVu 11788	
Drought	TVu 11979, TVu 14914	IT90K-59-2
	TVu 15553, TVu 8713	
	TVu 6914, TVu 7841	

[†] TVu: Prefix for IITA's cowpea (*V. unguiculata*) accession no.; IT: Prefix for IITA's cowpea advanced breeding lines.
Sources: IITA 1989, 1993.

The crop (almost entirely of the cultigroup Unguiculata) is grown extensively in the lowlands and midaltitude regions of Africa, particularly in the dry savannas, sometimes as a sole crop but more often in mixtures with cereals, typically sorghum and millet. In these associations, two distinct types of cowpea are often found. One is an early, photo-insensitive type that matures long before the cereals and provides an important source of protein at a time when food supplies have dwindled. The other is a late photosensitive cowpea that is planted in alternate rows with the early more erect type, and produces abundant vegetation as it spreads across the ground. This latter type, which matures after the cereals, is an important source of livestock feed during the long dry season, mostly in West and Central Africa. In the eastern and southern parts of the continent, farmers value the tender, green leaves of cowpea, which are harvested when succulent, then boiled or blanched, sun-dried and compressed into balls or ground into powder for use in stews and soups. In this region, the traditional varieties are grown as intercrops or sole crops for leaves as well as dry grain. They belong to medium (85 days) maturity and late photosensitive groups.

In more humid areas of Southeast Asia and Southern China, the young succulent pods of the yard-long type (cultigroup Sesquipedalis) are used as a green vegetable and are more important than the pulse type. This cultigroup is also grown in parts of Japan, South Korea, the Indian subcontinent, southern Italy, Africa, Central and South America. Cultivars of this group are mostly climbing, and the pods became deflated and flabby when mature. They are grown in small patches in vegetable gardens around cities and villages or in backyard gardens. Biflora is like cultigroup Unguiculata, which is used as both dry seed and fodder. It is also used as a cover crop in plantations. It is cultivated mainly in India and Sri Lanka for its seed and used as fodder. Cultigroup Unguiculata is grown more in the drier regions in India than in other Asian countries. It is also used as seed and fodder.

In many places in the tropical lowland ecology in Latin America it is grown as a substitute for lima bean (*Phaseolus vulgaris*), especially in the drier regions and on less fertile soil where it performs better than lima bean. In the drought-prone region in Northeastern Brazil, cowpea is frequently sown by farmers who practise shifting cultivation on marginal lands. In South America cowpea is grown as a pure crop (Ferreira and Silva 1987) but is more frequently intercropped with maize and perennial cotton (Beltrao *et al.* 1986). The preferred plant type for this ecology with erratic rainfall is one with medium duration (80-100 days) and prostrate growth habit; it covers the ground, reducing evaporation and controlling weed growth (Watt *et al.* 1985). In the savanna ecologies with intermediate rainfall (600-1500 mm) in Brazil, Venezuela, Colombia, Honduras, Nicaragua and Cuba, cowpea is frequently grown as a monocrop but occasionally with maize. In areas of high rainfall, climbing Sesquipedalis cultivars, whose young pods are used as a green vegetable, are frequently grown but on a limited scale in the Amazon River basin, Peru, the coastal regions of Ecuador, Surinam and Guyana and also in Trinidad.

Cowpea has been cultivated in the southern parts of the USA since the early 18th century. Grain-type cowpea cultivars are mainly grown. Although it was once an important agronomic crop, it is grown and consumed on a limited basis at present but is still relatively important horticulturally. Dry seed is produced for the canning industry as well as fresh seed for market. An extensive canning industry exists to supply the processed and dry seeds that are marketed nationwide in the USA (Fery 1985).

Breeding

Cowpea varietal requirements vary greatly in terms of plant type, seed type, maturity and use pattern in the diverse cowpea-growing ecologies throughout the tropics. The seed colour preference and use patterns differ from region to region and the maturity, growth habit and photosensitivity requirements depend upon the cropping systems (IITA 1992; Singh *et al.* 1996). No single cultivar is suitable for all the needs of the cowpea-growing regions. The general objectives and strategies to meet these requirements were described by Singh and Ntare (1985) but these have now been enlarged. The breeding objectives of IITA thus evolve over the years (Ng 1995). Prior to 1987, IITA directed almost 100% of its efforts toward developing cowpea varieties for pure cropping which are early to medium maturity and resistant to pest and diseases. Since then, IITA has diversified its objectives to include breeding for intercropping as an important component of its overall cowpea improvement programme since the bulk of cowpea in West and Central Africa is still grown as an intercrop (IITA 1992, 1993). The current focus of IITA's cowpea breeding programme is to develop the following types of varieties (Singh *et al.* 1996):

1. Extra-early maturity (60-70 days), photo-insensitive grain type for use as sole crop in multiple cropping systems and short rainy seasons.
2. Medium maturity (75-90 days) photo-insensitive grain type for use as sole crop and intercrop.
3. Late-maturing (85-120 days) photo-insensitive dual-purpose (grain + leaf) types for use as sole crop and intercrop.
4. Photosensitive early maturing (70-80 days) grain types for intercropping.
5. Photosensitive medium-maturing (75-90 days) dual-purpose (grain + fodder) types for intercropping.
6. Photosensitive late-maturing (85-120 days) fodder type for intercropping.
7. High-yielding non-staking Sesquipedalis varieties for use as a vegetable.
8. Desirable seed types and seed colours with high protein content and short cooking time.
9. Varieties resistant to major diseases, insect pests and parasitic weeds.
10. Varieties tolerant to drought, low pH and adapted to sandy soil and low fertility.

In 1975, IITA's breeding programme initiated the Cowpea International Testing (CIT) programme, in which trials were sent to many countries in the tropics (Goldsworthy and Redden 1982). This testing programme has continued until today, as a regular feature of IITA's Cowpea Improvement Network and has served as IITA's main vehicle for distributing improvement germplasm to national programmes.

Limitations and Prospects

Cowpea is very susceptible to numerous insect pests and diseases, which attack the plants at all stages of their growth, beginning from seedling stages to post-flowering and at storage. Remarkable progress in cowpea improvement has been made during the last two decades in Africa, the USA and Asia (Singh and Rachie 1985; IITA 1992; Singh *et al.* 1996). The most spectacular achievements are the multiple disease- and pest-resistant (aphid and bruchid), high-yielding, short- and medium-duration strains suitable to a wide range of ecologies and cropping seasons made by IITA. However, the crop is still facing serious problems, especially for its susceptibility to *Maruca* pod borers, coreid buds and flowering thrips in West Africa. No sources of germplasm of the cultivated cowpea have been found to possess high levels of resistance to those three major pests. Under high population pressure, the crop may be stripped bare of pod and grain. Current improved cowpea varieties could not achieve their yield potential of over 2.5 t/ha at farm level without the use of insecticides for protection against these pests.

Cowpea is very nutritious, about 72% of its total seed protein being digestible. However, like most other food legumes, cowpea is deficient in sulphur-containing amino acids. This fact is very important for the region where the main diet is based on root and tuber crops, and starchy food (Bressani 1985). It also contains high oligosaccharides which have flatulent effects that may make cowpea food less acceptable.

Cowpea plants are sensitive to waterlogging and cold temperature, which limits the extension of the crop to these ecologies.

There are ample opportunities for further improvement of cowpea through the exploitation in breeding of currently available germplasm. The existing collection of about 1600 accessions of wild *Vigna* including about 500 accessions from the section *Catiang* should be exploited. Sources of resistance to *Maruca* pod borers and pod sucking bugs have been identified from wild *Vigna* (Table 7.6). Currently, efforts are being made by IITA scientists and their collaborators in advanced laboratories to try to find ways of exploiting this genepool. These efforts include the identification of sources of resistance, isolation of genes, interspecific hybridization and genetic transformation (Ng 1995). At the same time, efforts are continuing to identify and use the best available sources of resistance to the various abiotic and biotic stresses from the diversity found in the cultivated cowpea for further improvement of existing cultivars to suit different cropping systems, needs and ecologies. Efforts are also being made by IITA's scientists and their collaborators in advanced laboratories to utilize foreign genes for cowpea improvement through genetic transformation.

Research on the integrated pest management of cowpea that includes cultural practices and biocontrol agents may help reduce the pest problems of cowpea in West Africa and in other parts of the world.

REFERENCES

Agwaranze, N.F. 1992. Morphological variability, inheritance of pubescence in *Vigna vexillata* and the histology of hybrids between wild *Vigna* and cultivated cowpea (*Vigna unguiculata* (L) Walp). PhD Thesis submitted to the University of Ibadan, Ibadan, Nigeria.

Allen, D.J. 1980. Identification of resistance to cowpea mottle virus. Trop. Agric. (Trin.) 57:325-332.

Allen, D.J. 1983. The Pathology of Tropical Food Legumes. Disease Resistance in Crop Improvement. John Wiley and Sons, New York.

Allen, D.J., A.A. Emechebe and B. Ndimande. 1981a. Identification of resistance in cowpea to diseases of the African savannas. Trop. Agric. (Trin.) 58:267-274.

Allen, D.J., C.L.N. Nebane and J.A. Raji. 1981b. Screening for resistance to bacterial blight in cowpea. Trop. Pest Manage. 27:218-224.

Barone, A., A. Del Giudice and N.Q. Ng. 1992. Barriers to interspecific hybridization between *Vigna unguiculata* and *Vigna vexillata*. Sex Plant Reprod. 5:195-200.

Beltrao, N.E. de M., J.C.F. de Santana, J.R. Crisos-Tomo, J.P.P. de Araujo and R.P. de Sousa. 1986. Evaluation of cowpea cultivars under intercropping system with annual cotton. Pesquisa Agropecuaria Brasileira 21 (11) 1147-1153.

Bressani, R. 1985. Nutritive value of cowpea. *In* Cowpea Research, Production and Utilization (S.R. Singh and K.O. Rachie, eds.). John Wiley and Sons, Chichester.

Dovlo, F.E., C.E. Williams and L. Zoaka. 1976. Cowpeas Home Preparation and Use in West Africa. International Development Research Centre, Ottawa, Canada.

Duke, J.A. 1981. Handbook of Legumes of World Economic Importance. Plenum Press, New York.

Ebong, U.U. 1971. The Nigerian Grain Legume Gene Bank. (Paper presented at the Ford Foundation/IITA/IRAT Grain Legume Seminar, Ibadan, Nigeria, 1970.) Samaru Agric. Newsl. 13:21-24.

Fatokun, C.A. and B.B. Singh. 1987. Interspecific hybridization between *Vigna pubescens* Wilcz. and *Vigna unguiculata* (L.) Walp. through embryo culture. Plant Cell Tissue Organ Cult. 9:229-233.

Ferreira, J.M. and P.S.L.E. Silva. 1987. 'Green bean' yield and other characteristics of cowpea cultivars. Pesquisa Agropecuaria Brasileira:55-59.

Fery, R.L. 1985. Improved cowpea cultivars for the horticultural industry in the USA. Pp. 129-136 *in* Cowpea Research, Production and Utilization (S.R. Singh and K.O. Rachie, eds.). John Wiley and Sons, Chichester.

Goldsworthy, P.R. and R. Redden. 1982. Cowpea improvement at IITA: 1970-1980. Trop. Grain Legume Bull. 24:9-15.

IBPGR. 1982. Descriptors for cowpea. International Board of Plant Genetic Resources, Rome.

IITA (International Institute of Tropical Agriculture). 1972. IITA Annual Report for 1971. IITA, Ibadan, Nigeria.

IITA. 1978. IITA Annual Report for 1977. IITA, Ibadan, Nigeria.

IITA. 1983. IITA Annual Report for 1982. IITA, Ibadan, Nigeria.

IITA. 1989. Grain Legume Improvement Program Annual Report 1987. IITA, Ibadan, Nigeria.

IITA. 1992. Sustainable food production in sub-saharan Africa. 1. IITA's contributions, pp. 113-123. IITA, Ibadan, Nigeria.

IITA. 1993. Cowpea breeding Archival Report (1988-1992). Grain Legume Improvement Program. IITA, Ibadan, Nigeria.

IITA. 1994. IITA Annual Report for 1993. IITA, Ibadan, Nigeria.

Jackai, L.E.N. and S.R. Singh. 1988. Screening techniques for host plant resistance to cowpea insect pest. Trop. Grain Legume Bull. 35:2-18.

Karel, A.K. and Y. Malinga. 1980. Leaf hopper and aphid resistance in cowpea. Trop. Grain Legume Bull. 20:10-11.

Ladipo, J.L. and D.J. Allen. 1979. Identification of resistance to southern bean mosaic virus in cowpea. Trop. Agric. (Trin.) 56:33-40.

Marechal, R., J.M. Mascherpa and F. Stainer. 1978. Etude taxonomique d'un groupe complexe d'espèces des genres *Phaseolus* et *Vigna* (Papilionaceae) sur la base de données morphologiques et polliniques, traitées par l'analyse informatique. Boissiera 28:1-273.

Ng, N.Q. 1987. The genebank at IITA and its significance for crop improvement in Africa, with special reference to cowpea germplasm. *In* Proceedings of International Symposium on Conservation and Utilization of Ethiopian Germplasm (J.M.M. Engels, ed.). Addis Ababa, October 1986. PGRC/E, Addis Ababa, Ethiopia.

Ng, N.Q. 1990. Recent developments in cowpea germplasm collection, conservation, evaluation and research at the genetic resources unit, IITA. Pp. 13-28 *in* Cowpea Genetic Resources (N.Q. Ng and L.M. Monti, eds.). IITA, Ibadan.

Ng, N.Q. 1991. Long-term seed conservation. Pp. 135-148 *in* Crop Genetic Resources of Africa. Vol. II. (N.Q. Ng, P. Perrino and F. Attere, eds.). IITA, Ibadan, Nigeria.

Ng, N.Q. 1995. Cowpea *Vigna unguiculata* (Leguminosae - Papilionoideae). Pp. 326-332 *in* Evolution of Crop Plants, Second edition (J. Smartt and N.W. Simmonds, eds.). Longman Scientific & Technical, Essex, England.

Ng, N.Q. and J. Apeji. 1988. Interspecific crosses between cowpea and wild Vigna. Genetic Resources Annual Report for 1988. IITA, Ibadan, Nigeria.

Ng, N.Q. and R. Marechal. 1985. Cowpea taxonomy, origin and germplasm. Pp. 11-12 *in* Cowpea Research, Production and Utilization (S.R. Singh and K.O. Rachie, eds.). John Wiley and Sons, Chichester.

Ng, N.Q. and S. Padulosi. 1991. Cowpea gene pool distribution and crop improvement. Pp. 161-174 *in* Crop Genetic Resources of Africa. Vol. II. (N.Q. Ng, P. Perrino and F. Attere, eds.). IITA, Ibadan, Nigeria.

Oyekan, P.O., P.T. Onesirosan and R.J. Williams. 1976. Screening for resistance in cowpea to web blight. Trop. Grain Legume Bull. 3:6-8.

Padulosi, S. 1993. Genetic diversity, taxonomy and ecogeographic survey of the wild relatives of cowpea (*Vigna unguiculata* (L.) Walpers). Thesis, Universite Catholique de Louvain-La-Neuve. Louvain, Belgique.

Pasquet, R.S. 1993. Classification infraspecifique des formes spontanées de *Vigna unguiculata* (L.) Walp. (Fabaceae) a partir de données morphologiques. Bull. Jard. Bot. Nat. Bel. 62:127-173.

Patel, P.N. 1981. Pathogen variability and host resistance in bacterial pustule disease of cowpea in Africa. Trop. Agric. (Trin.) 58:275-280.

Pienaar, B.J. and A.E. van Wyk. 1992. The *Vigna unguiculata* complex (Fabaceae) in southern Africa. S. Afr. J. Bot. 58 (6):414-429.

Porter, W.M., K.O. Rachie, K.M. Rawal, H.C. Wien, R.J. Williams and R.A. Luse. 1974. Cowpea germplasm catalogue no. 1. IITA, Ibadan.

Rawal, K.M. and L.M. Roberts. 1975. Natural hybridization among weedy and cultivated *Vigna unguiculata* (L.) Walp. Euphytica 24:69-707.

Sakupwanya S., R. Mithen and T. Mutangandura-Mhlanga. 1990. Studies on the African *Vigna* genepool. II. Hybridization studies with *Vigna unguiculata* var. *tenuis* and var. *stenophylla*. Plant Genet. Resour. Newslett. 78/79:5-9.

Singh, B.B. and B.R. Ntare. 1985. Development of improved cowpea varieties in Africa. Pp. 105-115 *in* Cowpea Research, Production and Utilization (S.R. Singh and K.O. Rachie, eds.). John Wiley and Sons, Chichester.

Singh, B.B., O.L. Chambliss and B. Sharma. 1996. Recent advances in cowpea breeding. *In* Proceedings of the Second World Cowpea Conference (in press).

Singh, S.R. 1977. Cowpea cultivars resistant to insect pests in world germplasm collection. Trop. Grain Legume Bull. 9:3-7.

Singh, S.R. and D.J. Allen. 1979. Cowpea pests and diseases. Manual Series No. 2. IITA, Ibadan, Nigeria.

Singh, S.R. and K.O. Rachie (eds.). 1985. Cowpea Research, Production and Utilization. John Wiley and Sons, Chichester.

Singh, S.R., B.B. Singh, L.E.N. Jackai and B.R. Ntare. 1983. Cowpea Research at IITA. Information Series No. 14. IITA, Ibadan, Nigeria.

Singh, S.R., L.E.N. Jackai, B.B. Singh, B.R. Ntare, H.W. Rossel, G. Thottappilly, N.Q. Ng, M.A. Hossain, K. Cardwell, S. Padulosi and G. Myers. 1989. Cowpea Research at IITA. GLIP Research Monograph No. 1, IITA, Ibadan, Nigeria.

Steele, W.M. and K.I. Mehra. 1980. Structure, evolution and adaptation to farming system and environment in *Vigna*. *In* Advances in Legume Science (R.J. Summerfield and A.H. Bunting, eds.). Her Majesty's Stationery Office, London, UK.

Thottappilly, G., N.Q. Ng and H.W. Rossel. 1994. Screening germplasm of *Vigna vexillata* for resistance to cowpea mottle virus. Int. J. Trop. Plant Diseases 12:75-80.

Verdcourt, B. 1970. Studies in the Leguminosae-Papilionoideae for the "Flora of Tropical East Africa". IV. Kew Bull. 24:507-569.

Watt, E.E., E.A. Kueneman and J.P.P. de Araujo. 1985. Achievements in breeding cowpea in Latin America. Pp. 125-135 *in* Cowpea Research, Production and Utilization (S.R. Singh and K.O. Rachie, eds.). John Wiley and Sons, Chichester.

Westphal, E. 1974. Pulses in Ethiopia: their taxonomy and agriculture significance. Agric. Res. Rep. 815:1-261. Wageningen, Centre Agric. Publ. Document.

Willliams, R.J. 1975. International testing programme: aims, progress and problems seen from IITA. Pp. 66-72 *in* Proceedings of IITA collaborators meeting on grain legume improvement, 9-13 June, 1975 (R.A. Luse and K.O. Rachie, eds.). IITA, Ibadan, Nigeria.

Williams, R.J. 1977a. Identification of multiple disease resistance in cowpea. Trop. Agric. (Trin.) 54:53-60.

Williams, R.J. 1977b. Identification of resistance to cowpea yellow mosaic virus. Trop. Agric. (Trin.) 56:61-67.

Zannou, T.A. 1981. Resistance to root-knot nematodes *Meloidogyne javanica* (Teub, 1885) Chitwood, 1949, and *M. incognita* (Kofoid and White, 1919) Chitwood, 1949, in cowpea, *Vigna unguiculata* (L.) Walp. Thesis, University of Ibadan, Nigeria.

Chapter 8

Chickpea

K.B. Singh, R.P.S. Pundir, L.D. Robertson,
H.A. van Rheenen, U. Singh, T.J. Kelley,
P. Parthasarathy Rao, C. Johansen and N.P. Saxena

Cultivated chickpea (*Cicer arietinum* L.) is a diploid, self-pollinated, leguminous crop that ranks second in area and third in production among the pulses. It is cultivated primarily for its protein-rich seed, and the plant is an efficient symbiotic nitrogen-fixer, playing an important role in farming systems. Two types of chickpea are grown: desi, with angular and coloured seeds, primarily grown in South Asia; and kabuli, with large, owl-head shape and beige-coloured seeds, grown in the Mediterranean region. Germplasm is maintained at ICRISAT (Patancheru: 18°N, 78°E) and ICARDA (Tel Hadya: 35°5'N, 36°55'E). Before 1970, wild *Cicer* species were scarce, but a number of annual species accessions are now available.

BOTANY AND DISTRIBUTION

The name *Cicer* is of Latin origin and probably derives from the pre-Indo German *kichere* in the Pelagian language of the tribes populating north Greece before Greek-speaking tribes took over. Chickpea belongs to subfamily Papilionoideae, tribe Viceae Alef, but its position is sufficiently distinct to consider the *Cicer* genus a tribe of its own, the Cicereae Alef. (Kupicha 1977). Van der Maesen (1987) dealt with this genus in detail and listed 43 species, including 34 wild perennial species, 8 wild annual and the cultivated annual, *C. arietinum*. The study on chromosome count in *Cicer* species has been limited because of rare availability of living materials. The chromosome number 2*n*=16 can be generalized for *Cicer* species because some taxa which were reported to have other numbers (Iyenger 1939; Contandriopoulos *et al.* 1972; van der Maesen 1972; Santos Guerra and Lewis 1986) were confirmed to have only 2*n*=16 (Ladizinsky and Adler 1976; Ocampo *et al.* 1992; Pundir *et al.* 1993).

Ladizinsky (1975) has described *C. reticulatum*, which has been suggested as the wild progenitor (Ladizinsky and Adler 1976). Since this species crossed normally with chickpea and produced fertile hybrids, the category 'variety' seemed appropriate (van der Maesen 1987). Four sections are recognized in the intrageneric classification of the *Cicer* genus.

First, section *Cicer* (=*Monocicer*) M.G. Popov, the most relevant section for applied research, encompasses eight species with annual, erect or prostrate growth, imparipinnate leaves or the rachis ending in a tendril. Second, section *Chamaecicer* M.G. Popov encompasses one each of annual and perennial species, shrubs bearing thin creeping branching, and leaves with 3-7 leaflets. Third, section *Polycicer* M.G. Popov contains 23 perennial species with large flowers, imparipinnate leaves, or the rachis ending in a tendril. And fourth, *Acanthocicer* M.G. Popov accommodates nine perennial species with large flowers, paripinnate leaves, rachis ending in a spine and spiny calyx teeth.

Don (1882) suggested that *Cicer* was derived from the Greek *kikus*, meaning strength. The Latin epithet *arietinum* probably derived as a translation of the Greek *krios*, another name for both ram and chickpea, alluding to the seed shape, similar to a ram's (Aries) head. In

Turkey, Romania, Bulgaria, Iran, Afghanistan, Russia and CIS, chickpea is called *nakhut* or its derivation such as *nut, naut* or *nohot*. *Hommos* is the name of chickpea in Arabic. In English, chickpea is derived from Cicer pea (=chich-pea). In most Sanskrit-derived languages in the Indian subcontinent, chickpea is called *chana*. The Anglo-Indian name *Bengal gram* was derived from the Portuguese *grao*, meaning grain (=*gram*), and in Spanish *garbanzo*; in Mexico and the USA it is the garbanzo bean.

Cicer is hypogeal. The first two nodes on the main stem have simple, scale-like leaves, fused with two lateral scale-like stipules (Westphal 1974). Leaves are borne single at each node arranged in alternate phyllotaxy. A unipinnate compound leaf is the rule in the *Cicer* genus, although simple and multipinnate leaves are also found (Pundir *et al.* 1990). Chickpea possesses a deep taproot system with three to four defined rows of lateral roots. In the early stages of growth, most of the roots are confined to the surface layer of soil from 0 to 30 cm depth. As surface soil dries out, development continues to deeper layers. In deep vertisols, the roots have penetrated below 120 cm by the time fruiting begins (Sheldrake and Saxena 1979). The primary and secondary roots may develop large, lobed nodules containing rhizobia that fix atmospheric nitrogen symbiotically. Practically all nodules are confined to the top 30 cm and 90% are within 15 cm of the surface (Sheldrake and Saxena 1979).

Five growth habit classes are recognized (Pundir *et al.* 1985). The angle of branches from the vertical is taken as the basis of this grouping (Fig. 8.1). Although named primary, secondary and tertiary branches, these terms lack a botanical and logical basis (L.J.G. van der Maesen and R.P.S. Pundir, unpublished). A sounder pattern is illustrated in Figure 8.2. Flowers are typically papillionaceous. The twin flower characteristic is a useful trait in cultivated chickpea and is monogenic recessive. This trait can offer an advantage of about 6-11% in yield under conditions in which the trait is well expressed (Sheldrake *et al.* 1978). One carpel a flower is the rule in the *Cicer* genus, but a natural mutant has 1-3 carpels a flower, producing 1-3 pods and well-formed seeds (Pundir *et al.* 1988). Pod size is an important yield trait, but difficult to measure and categorize. It has been estimated by equivalent volume of water replacement (Pundir *et al.* 1992). In a set of 83 diverse chickpea lines, the pod volume varied from 0.30 to 2.77 ml per pod. Pod-filling varies from 8.97 to 56.53%. Twenty-one different seed colours and shades have been recognized (Pundir *et al.* 1988).

Origin, Domestication and Diffusion

Chickpea most probably originated in the area of present-day southeast Turkey, where three wild species of *Cicer* (*C. bijugum* K.H. Rech., *C. echinospermum* P.H.Davis and *C. reticulatum* Ladiz) are found. Seed remnants from Hacillar near Burder dated to 5450 BC (Helbaek 1970) and the most closely related wild species occur there (Ladizinsky and Adler 1976). De Candolle (1883) traced the origin of chickpea to an area south of the Caucasus and northern Persia (now Iran). Vavilov (1926) designated two primary centres of origin (now centres of diversity) – southwest Asia and the Mediterranean – and a secondary one, Ethiopia. In 1951 he recognized four centres of diversity for cultivated chickpea: (i) the Mediterranean, (ii) Central Asia, (iii) the Near East, and (iv) India, and a secondary centre in Ethiopia (Vavilov 1951).

Hopf (1969) found proof of chickpea cultivation at Jericho, where large quantities were found in layers dating back to the early Bronze Age (3200 BC). The earliest record of chickpea in India (Uttar Pradesh) dates from 2000 BC (Chowdury *et al.* 1971; Vishnu-Mittre 1974) and seed remnants dating from 300 to 100 BC were found at Nevasa (near Aurangabad), India. Chickpea spread with human movement toward the west to the Mediterranean basin and south toward the Indian subcontinent via the Silk Route (Afghanistan). Records in Ethiopia date to 1520 BC (van der Maesen 1972). The Spanish and Portuguese introduced chickpea to the New World around 1500 AD, and kabuli is of recent introduction to India (1700 AD). Domestication of chickpea began in the USA and Australia only in the last three decades.

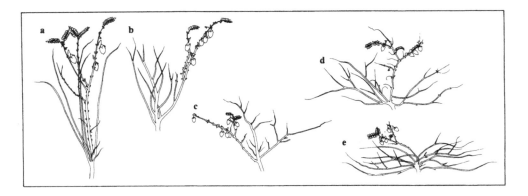

Fig. 8.1. Growth patterns in chickpea plants: (a) erect (ICC 8923); (b) semi-erect (ICC 5003); (c) semi-spreading (ICC 552); (d) spreading (ICC 11888); (e) prostrate (ICC 5434). (Source: ICRISAT Chickpea Germplasm Catalog, Pundir *et al.* 1988).

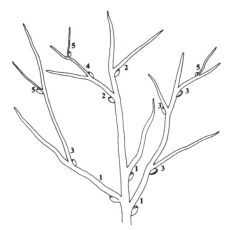

Fig. 8.2. Branch groups in a chickpea plant.
1. Basal primary branches – emerge from the leaf axils on the lower half of the main stem.
2. Apical primary branches – emerge from the leaf axils on the upper half of the main stem.
3. Basal secondary branches – emerge from the leaf axils of basal primary branches.
4. Apical secondary branches – emerge from the leaf axils of apical primary branches.
5. Tertiary branches – emerge from the leaf axils of basal and apical secondary branches. (Source: ICRISAT Chickpea Germplasm Catalog, Pundir *et al.* 1988).

Reproductive Biology

Chickpea is a quantitative long-day plant, but in short days vernalization may replace this requirement in some cultivars (Angus and Moncur 1980). The response to cold also appears to be quantitative (Saxena and Siddique 1980). Flowering begins sooner in warm weather and at long photoperiods (Roberts *et al.* 1980). At ICRISAT, flowering continues for 20-30 days, depending on soil moisture. Only 20-50% of the flowers set pods (Eshel 1968; Pundir *et al.* 1983) and pod set is inhibited at temperatures below 5°C (Saxena and Sheldrake 1980). High mean temperature of 25°C or cloudy weather also prevent production of flowers and pods. It takes about 5-6 days from appearance of the bud and flower to bloom. Temperature is more important than photoperiod in determining the length of the reproductive period.

GERMPLASM CONSERVATION AND USE

Moreno and Cubero (1978) revealed the existence of two complexes in chickpea: *macrosperma* and *microsperma*. These complexes differ in a cluster of traits associated with seed, pod and leaf morphology, but no taxonomic basis exists for treating them as subspecies.   The researchers postulated that *macrosperma* derived from *microsperma* through mutation and was selected during relatively recent times.

The frequency by country of origin of the cultivated chickpea collections of ICARDA and ICRISAT is given in Table 8.1.   To facilitate information exchange, chickpea descriptors were developed and jointly published by IBPGR, ICRISAT and ICARDA (1993).   At ICRISAT, a total of 12 018 accessions representing 41 countries were evaluated for 23 botanical and agronomic traits, and those with resistance to diseases or insects listed. ICARDA evaluated 3400 germplasm accessions of the kabuli type for 29 descriptors during the spring season of 1980 and published their first chickpea catalogue (Singh *et al.* 1983).   Similarly, 6467 accessions of kabuli type were evaluated for 24 characters at ICARDA during winter, and a catalogue lists desirable traits (Singh *et al.* 1991a).   The range of variability recorded is given in Tables 8.2 and 8.3.

Geographic Distribution of the Major Traits

The biomass and grain yield of chickpea are the ultimate results but not the major traits.   Furthermore, some traits, such as large-seeded chickpea (ca. 30 g per 100 seeds), may not be major traits universally.   The large seed of this crop has high consumer preference and adaptation in the Mediterranean countries, whereas small seed (ca. 15 g per 100 seeds) is a well-accepted and adapted trait in northern India and Pakistan.   A list of traits that are optimum in different environments would be too extensive, but Table 8.4 summarizes the geographic distribution of major traits.

Wild Genetic Resources

Wild *Cicer* species were scarce in collections before 1970, but now a reasonable number of accessions of annual species are available at ICRISAT and ICARDA (Table 8.5) and at the Western Plant Introduction Division/USDA at Pullman, Washington, USA. Availability of perennial species is still rare, and limited information on their potential use is available.

Ladizinsky and Adler (1976) proposed grouping *Cicer* species into three genepools, namely GP 1, GP 2 and GP 3, following the scheme suggested by Harlan and De Wet (1971).   As a result of their own research work, Ladizinsky and Adler included *C. reticulatum* in GP 1 of chickpea because gene exchange between these species was regular.   The species *C. echinospermum* was placed in GP 2 because gene exchange of this species with chickpea was impaired owing to high sterility of their F_1 hybrids.   All remaining species were placed in GP 3. Subsequent results reported by Pundir *et al.* (1992) and Singh and Ocampo (1993) showed that *C. echinospermum* crosses produced F_1 hybrids with 50% fertility.   The remaining six annual and 34 perennial *Cicer* species show no possibility of gene exchange with cultivated chickpea; hence they should be placed in GP 3.   Wild species have been extensively evaluated for resistance to biotic and abiotic stresses (Table 8.6).   Efforts are underway to transfer genes for resistance to cyst nematode and cold (ICARDA 1994).   Also the possibility of transfer of yield genes from wild to cultivated species is being explored (Singh and Ocampo 1993).   More accessions of wild annual *Cicer* species need to be collected, particularly the progenitor species, *C. reticulatum*, because they possess genes for resistance to almost all stresses for which evaluations have been made.   Introgression of genes from wild species can upgrade the genetic potential for seed yield.

Gaps in germplasm collections

Gaps in the genetic resources of crops include geographic, taxonomic, administrative and special-purpose traits (van der Maesen *et al.* 1988).   For cultivated chickpea, geographic gap areas include Russia and CIS, Colombia, Peru, Western Nepal, Eritrea and Myanmar.   The gaps for wild species are Russia and CIS, Turkey, Afghanistan, Iraq and Ethiopia. Known entity gaps include the types of chickpea that have been documented or described as herbarium material but are no longer available as live seed.   A limited number of accessions are assembled for wild *Cicer* species.   For example, only one

Table 8.1. Frequency by country of origin of the ICARDA and ICRISAT cultivated chickpea germplasm collections.

Country	ICARDA	ICRISAT	Country	ICARDA	ICRISAT
Afghanistan	889	686	Malawi	3	81
Algeria	50	16	Mexico	117	396
Armenia	3	–	Moldova	6	–
Australia	4	3	Morocco	225	249
Azerbaijan	13	–	Myanmar	–	129
Bangladesh	1	170	Nepal	4	80
Bulgaria	191	9	Nigeria	1	3
Chile	346	139	Pakistan	254	445
China	25	24	Palestine	36	–
Colombia	1	1	Peru	4	3
Cyprus	46	44	Portugal	121	84
Czechoslovakia	9	8	Romania	2	–
Ecuador	1	–	Russia	22	–
Egypt	55	53	Spain	283	121
Ethiopia	58	932	Srilanka	–	3
France	18	2	Sudan	11	12
Georgia	1	–	Soviet Union	106	133
Germany	1	11	Syria	2097	203
Greece	18	25	Tanzania	–	97
Hungary	2	4	Tajikistan	8	–
India	395	7180	Tunisia	263	33
Iran	1737	4856	Turkey	804	449
Iraq	30	18	Uganda	–	1
Israel	–	48	Ukrania	9	–
Italy	66	45	UK	8	–
Jordan	143	25	USA	120	82
Kazakhstan	1	–	Uzbekistan	11	–
Kenya	–	1	Yugoslavia	6	6
Kyrgyzstan	1	–	Unknown	240	179
Lebanon	28	19	Breeding lines	972	–
Libya	2	–	**Total**	**9628**	**17108**

Table 8.2. Range of variation observed in chickpea germplasm at ICRISAT.

Traits	Mean	Minimum	Maximum
Time to 50% flowering (days)	64	28	96
Time to maturity (days)	117	84	169
Plant canopy height (cm)	38	14	96
Canopy width (cm)	40	13	124
Pods per plant (number)	39	3	238
Seeds per pod (number)	1.2	1.0	3.2
Grain yield (kg/ha)	1286	70	5130
100-seed weight (g)	16.1	3.8	59.1
Seed protein content (%)	19.8	12.1	29.6

accession of *C. cuneatum* is known to be available anywhere, and of the 34 perennial *Cicer* species known, live seed is available only for nine species.

Seed conservation

Chickpea has orthodox seed that can be dried and stored for long periods with minimum loss of viability. In the genebanks at ICRISAT and ICARDA, chickpea germplasm has been stored in medium-term (4°C, 20% RH) and long-term (–20°C) facilities for about 15 years with full seed viability. Ellis (1988) has given practical advice on seed viability in storage. For example, chickpea seed having 99% initial

Table 8.3. Ranges among kabuli chickpea accessions in the ICARDA collection grown at Tel Hadya, Syria during spring 1980 and winter of 1987/88.

Character	Spring			Winter		
	Mean	Min.	Max.	Mean	Min.	Max.
Time to flower (days)	81	70	94	137	115	156
Flower duration (days)	23	11	36	29	12	83
Time to maturity (days)	–	114	124	182	174	206
Plant height (cm)	30	15	50	54	25	85
Canopy width (cm)	40	15	60	57	20	96
Pods per plant (number)	25	4	100	–	–	–
Seeds per pod (number)	1.1	0.1	3.1	–	–	–
Biological yield (g/m^2)	643	35	533	574	28	1200
Grain yield (g/m^2)	311	23	921	272	1	567
Harvest index (%)	49	7	76	48	1	78
100-seed weight (g)	25.1	8.7	59.1	30.0	8.4	70.1
Protein content (%)	20.1	16.0	24.8	23.0	13.5	28.2

Table 8.4. Geographical distribution of major chickpea traits.

Trait		Countries/regions[†]
Crop duration	Short	EGY, ETH, IND, MEX, SDN
	Medium	BGD, IND, MEX, MYN
	Long	ESP, IND, IRN, MAR, NPL, PAK, RUS, SYR, TUR
Branches	High number	AFG, IND, IRQ, ITA
	Low number	CHL, IND, TUN, TUR
Growth habit	Erect	GRC, IND, ITA, RUS
	Erect (+tall)	GRC, ITA, RUS
Seed number	High	AFG, EGY, IND, MEX, NPL, PAK
	Low	Mediterranean countries
Seed mass	High	Mediterranean countries, IND, MEX
	Low	Indian subcontinent, ETH, MYN, TZA
Seed type	Desi	Indian subcontinent, Eastern Africa, MYN
	Kabuli	Mediterranean countries, Western Asia, CHL
	Intermediate	ETH, IRN
Resistance to fusarium wilt		BGD, ETH, IND, IRN, PAK
Resistance to dry root rot		IND, IRN
Resistance to ascochyta blight		IND, IRN, MEX, TUR
Resistance to gray mould		IRN
Tolerance to *Helicoverpa* pod borer		IND
High seed protein content		PAK, SDN

† AFG=Afghanistan; IND=India; NPL=Nepal; BGD=Bangladesh; IRN=Iran; PAK=Pakistan; CHL=Chile; IRQ=Iraq; RUS=Russia; EGY=Egypt; ITA=Italy; SDN=Sudan; ESP=Spain; MAR=Morocco; SYR=Syria; ETH=Ethiopia; MEX=Mexico; TZA=Tanzania; GRC=Greece; MYN=Myanmar; TUN=Tunisia; TUR=Turkey.

Table 8.5. Wild annual *Cicer* species collections of ICARDA and ICRISAT.

Taxa	ICARDA	ICRISAT
bijugum	43	9
chorassanicum	8	3
cuneatum	1	1
echinospermum	15	4
judaicum	76	21
pinnatifidum	54	11
reticulatum	61	6
yamashitae	5	3
Total	**263**	**58**

Table 8.6. Economic traits present in the wild *Cicer* species.

Trait	*Cicer* species	Reference[†]
Resistance to fusarium wilt	*judaicum*	2,4
	bijugum	2,3,4
	echinospermum	3,4
	canariense, chorassanicum,]	2
	cuneatum, pinnatifidum]	
	reticulatum	4
Resistance to combined soilborne	*bijugum, cuneatum,*]	5
diseases	*judaicum, pinnatifidum*]	
Resistance to gray mould disease	*bijugum*	3
Resistance to ascochyta blight	*bijugum*	3,7
	judaicum	1,3,7
	pinnatifidum	1,3,7
	montbretii	1
Resistance to cyst nematode	*bijugum, pinnatifidum, reticulatum*	8
Tolerance to cold	*bijugum, echinospermum*	10
	judaicum, pinnatifidum, reticulatum	
Higher seed protein	*bijugum, reticulatum*	6
Higher biomass	*cuneatum,* most perennial species	–
Resistance to leaf miner	All wild annual species	9
Twin pods	*anatolicum, bijugum,*	12,11
	chorassanicum, cuneatum, microphyllum,	
	pinnatifidum, oxyden, soongaricum	
Multiple seeds	*cuneatum, montbretii*	12,11

[†] 1. Singh *et al.* 1981; 2. Kaiser *et al.* 1994; 3. Haware *et al.* 1992; 4. Infantino *et al.* 1996; 5. Reddy *et al.* 1991; 6. Singh and Pundir 1991; 7. Singh and Reddy 1993b; 8. Di Vito *et al.* 1996; 9. Singh and Weigand 1994; 10. Singh *et al.* 1990; 11. Robertson *et al.* 1995; 12. van der Maesen 1987.

viability, 10% moisture content and stored under 4°C temperature, after a storage span of 20 years has an accepted viability of 80%.

Use and Distribution

Chickpea improvement research gained momentum with the establishment of ICRISAT and ICARDA in 1972 and 1977, respectively. These centres have distributed a large number of germplasm samples (Table 8.7). Besides their use in research, many lines have been found promising and worth cultivating in specific areas of adaptation. For example, 10 accessions supplied from the germplasm collection at ICRISAT and ICARDA were found superior to local cultivars and released for cultivation in Algeria, China, Cyprus, Egypt, India, Iran, Iraq, Italy, Jordan, Lebanon, Morocco, Myanmar, Nepal, Oman, Sudan, Syria, Turkey, Tunisia and the USA.

Properties and Uses

Crude protein content in chickpea seed ranges from 12.6 to 30.5% (Singh 1985) and contains non-protein nitrogen. Total seed carbohydrates vary from 52.4 to 70.9% (Williams and Singh 1987). Starch, the principal carbohydrate, is 20-30% amylose and the remainder amylopectin. Chickpea seeds contain 4.0-10.0% fat and 4.8-8.5% soluble sugars (Singh 1985). The seed coat contributes about 70% of the total seed calcium.

Chickpea proteins are deficient in the sulphur-containing amino acids, methionine and cystine. Next to sulphur, the amino acids tryptophan, threonine and valine appear important in chickpea because in many cases the chemical scores for these amino acids are below satisfactory levels (Williams and Singh 1987). Supplementation with 10, 20, 30, 40 and 50% chickpea flour significantly enhances the nutritive value of Arabic bread, but 20% supplementation is most acceptable organoleptically (Hallab *et al.* 1974). The biological value was considerably higher for the kabuli than for the desi type because the kabuli type contains more utilizable protein and may be nutritionally better than desi (Singh *et al.* 1991b). Although it depends on the methods of cooking, protein quality of chickpea is

Table 8.7. Distribution of seed samples from the chickpea germplasm collection at ICRISAT and ICARDA, from 1974 to 1994.

Year	ICRISAT[†]	ICARDA[‡]	Year	ICRISAT[†]	ICARDA[‡]
1974[2]	3070	–	1985	4808	–
1975	7020	–	1986	3104	–
1976	2687	–	1987	6268	–
1977	800	–	1988	5095	–
1978	2318	–	1989	8825	–
1979	1454	–	1990	2860	462
1980	8336	–	1991	4745	83
1981	10202	–	1992	1945	714
1982	5861	–	1993	2624	165
1983	10548	–	1994	1166	3344
1984	6596	–	Total	100332	4768

[†] Germplasm samples from ICRISAT were supplied to various institutes in 80 countries and from ICARDA to 24 countries.
[‡] Germplasm samples from ICARDA were supplied from 1977 to 1989, but proper records were not maintained.

improved more by moist heat than by dry heat treatment, as available lysine was less in roasted chickpea than in boiled or pressure-cooked chickpea (Geervani and Theophilus 1980).

In comparison with soybean, pea and common bean, chickpea offers less problem with protease (trypsin and chymotrypsin) inhibitors so far as a group of antinutritional factors are concerned (Singh 1988). Although not at a toxic level, chickpea produces a certain amount of agglutinating activity in cow erythrocytes, which indicates the presence of phytolectins as toxic factors in it (Contreras and Tagle 1974). The polyphenolic compounds that interfere with protein digestibility are higher in chickpea cultivars with dark seed-coat colour (Singh 1988). The cooking time of whole seed of chickpea cultivars generally ranges between 70 and 90 minutes. However, some genotypes may require more than 2 hours to cook. Cooking time of chickpea genotypes can be significantly reduced by soaking the seeds in sodium bicarbonate solution before cooking (Singh *et al.* 1988). Cooking time is a heritable character (Williams and Singh 1987).

Breeding
Although chickpea breeding has a relatively short history (Singh 1987), mean world chickpea yields in 1961-63, 1976-78 and 1991-93 have registered a steady increase from 613 to 682 to 704 kg/ha. Conventional techniques of crop improvement that are widely used for this self-pollinator are: mass selection, pure line selection and the pedigree, bulk and backcross methods. The latter three methods are followed singly or in combination. Mutation breeding has been mainly followed for creating new variability, e.g. to generate ascochyta blight resistance (Haq *et al.* 1988). Some annual wild *Cicer* species have been used in hybridization programmes to broaden the crossing spectrum (Singh 1993). To create variability for wide adaptation, chickpea breeding often adopts a 'polygon' programme (van Rheenen 1991; van Rheenen and Haware 1994). Some breeding programmes have used multilocational, early generation, bulk-yield testing (Dahiya *et al.* 1984). However, Geletu (1987) concluded that correlations of yield between generations were low and did not justify continuation of the elaborate method of multilocational F_2/F_3 yield testing. Attempts to develop and make use of molecular markers in selection of desirable traits, such as resistance to fusarium wilt and ascochyta blight, are being pursued strongly, although their practical application has not yet been achieved (Muehlbauer *et al.* 1990; ICARDA 1994; Kahl *et al.* 1994).

Smithson *et al.* (1985) suggested that stagnation in chickpea productivity can be attributed to yield instability from abiotic and biotic constraints. Breeding efforts have been directed toward alleviating these constraints at ICRISAT and ICARDA by identifying resistance sources and using them in breeding programmes. The results of 17 years of evaluation of germplasm accessions for resistance to various stresses at ICARDA are summarized in Table 8.8. Similar efforts have been made at ICRISAT to identify sources of resistance to biotic and abiotic stresses. Out of more than 19 000 germplasm accessions screened at ICARDA, 14 genotypes were identified as resistant to ascochyta blight (*Ascochyta rabiei* (Pass.) Lab.) (Singh and Reddy 1993a). Use of these genotypes in breeding programmes resulted in yield increases from 60 to 75% (Nene and Reddy 1987; ICARDA 1994). Freezing cold also endangers the winter-sown crop. Again the germplasm collection was screened under low winter temperatures, and 16 out of 5781 accessions were rated as resistant (Singh *et al.* 1989; Singh 1993). These and the wild species *C. reticulatum* have been used successfully in breeding programmes.

The earliest-flowering chickpea cultivar in ICRISAT's genebank of about 17 000 accessions is ICCV 2, a released kabuli cultivar. It effectively escapes drought stress, and yield increases of 20% over control have been reported (van Rheenen 1991). Saxena (1987) identified two drought-resistant lines, ICC 4958 and ICC 10448. The former showed a particularly large root volume and is being used in breeding programmes to enhance drought resistance. ICCV 10 has also shown drought resistance, and reportedly exceeded controls by 16-20% in seed yield (van Rheenen 1991). Singh *et al.* (1996) have identified FLIP 87-59C as a resistant line from an evaluation of more than 4000 accessions.

In the West Asia and North Africa (WANA) region, temperatures lower than −10°C may occur. Biodiversity, both intra- and interspecific, has enabled successful enhancement of cold tolerance, which showed high heritability estimates in genetic studies (Singh *et al.* 1989, 1994). At latitudes from 25 to 30°, temperatures close to 0°C can be harmful in causing flower drop. A continued search for tolerance to chilling finally resulted in identification of a few lines and an active breeding programme is underway to capitalize on chilling tolerance (Saxena *et al.* 1988; ICRISAT 1990). Chickpea often has to cope with above-optimal temperature conditions. By sowing chickpea in January at Patancheru and exposing flowering plants to maximum temperature ranges of 33-39°C, heat-tolerant genotypes were identified.

In chickpea, 47 diseases and 54 insect pests have been reported, but two diseases and two insect pests are of major importance. Fusarium wilt and ascochyta blight are the major disease threats. Fusarium wilt, caused by *F. oxysporum* f.sp. *ciceri*, is probably the most widespread disease of chickpea. Inheritance studies have contributed significantly to the current success in breeding resistant varieties (Singh *et al.* 1987). Several sources of resistance were identified at both ICRISAT and ICARDA (Table 8.8) and are being utilized in the breeding programme. Many sources of resistance to ascochyta blight are being used successfully in breeding programmes, although the inheritance of the resistance has been a controversial issue (Malik *et al.* 1988; Dey and Singh 1993; van Rheenen and Haware 1994). Many researchers find this trait to be simply inherited (Singh and Reddy 1991). Good progress has been made in breeding ascochyta blight resistant cultivars worldwide.

About 15 000 germplasm accessions have been screened to find resistant genotypes to pod borer (*Helicoverpa* and *Heliothis* species), the best among them being ICC 506 (Reed *et al.* 1987; Lateef and Pimbert 1990). More lines were recommended to be used as resistant parents in crossing programmes (Sachan 1990). Progress has been made in development of lines with pod borer resistance and high yield at ICRISAT. About 7000 germplasm accessions have been evaluated at ICARDA to find resistant lines to leaf miner (*Liriomyza cicerina* Rond.), and three lines (ILC 3800, ILC 5900 and ILC 7738) have been identified (Singh and Weigand 1996). Little progress has been made in incorporating genes for resistance to leaf miner in high-yielding lines.

Although breeding has been largely defensive and aimed to incorporate factors that alleviate stresses, proactive research has been conducted on quality factors such as seed appearance (ICRISAT 1991).

Table 8.8. Reaction of chickpea germplasm accessions to some biotic and abiotic stresses at Tel Hadya between 1978 and 1992.

Scale	Ascochyta blight[t]	Fusarium wilt[‡]	Leaf miner[§]	Seed beetle[§]	Cyst nematode[¶]	Cold[tt]	Drought[‡‡]
1	0	7	0	0	0	0	0
2	0	3	0	0	0	0	0
3	10	6	3	0	0	3	0
4	22	31	5	0	0	10	3
5	9	68	485	0	20	1191	202
6	1444	174	710	164	0	1023	2007
7	1833	277	1269	185	494	1014	1172
8	1185	636	13	1551	1104	2284	607
9	14867	1817	3540	3253	7639	3570	77
Total	19370	3019	6025	5153	9257	9095	4168

[t] Singh and Reddy 1993a; [‡] Jimenez-Diaz *et al.* 1991; [§] Weigand and Tahhan 1990; [¶] Di Vito *et al.* 1996; [tt] Singh *et al.* 1989, [‡‡] Singh *et al.* (unpublished).

Table 8.9. Chickpea area, production, yield and compound growth rates, by region.

Region	1989-91 average			Compound growth rates (1971-91)		
	Area	Production	Yield	Area	Production	Yield
South Asia	8180	5562	670	−0.67**	−0.19	0.48
WANA	1346	1115	828	4.76**	5.54**	0.75**
NCS America	161	194	1205	−2.92**	−1.26	1.70**
Oceania	160	173	1081	64.09**	61.98**	−1.28
Europe	94	68	723	−4.53**	−3.49**	1.08**
World	10078	7116	706	0.19	0.42	0.62*

* Significant at 0.10 *P* level, ** Significant at 0.05 *P* level.
Sources: FAO, unpublished data, and Government of India, DES, various years.

Table 8.10. Chickpea imports and exports ('000 t), by region.

Region	1975-77 average		1989-91 average	
	Imports	Exports	Imports	Exports
South Asia	1	2	168	5
WANA	28	56	102	290
NCS America	12	36	27	59
Europe	32	6	106	9
Oceania	0	0	0	110
World	78	103	411	474

Sources: FAO 1988, 1992a, 1992b, 1993.

Production and Trade

Chickpea accounts for 12% (or 7.1 million t) of world pulse production. This figure, however, understates the importance of this commodity in specific geographical regions of the world. The area of cultivation is often under-reported because of its secondary status as a companion crop in intercropped systems (Jodha and Subba Rao 1987).

Compared with significant gains in world pulse production during the last two decades, chickpea share has declined (Table 8.9), and the overall declining trend in chickpea area in India and South Asia will likely continue, barring any major breakthrough in chickpea productivity to enhance its competitiveness (Kelley and Parthasarathy Rao 1994). Production increases in Turkey have come about mainly through area expansion spurred on by strong demand from importers and an attractive export incentive policy of the government. Oceania (mainly Australia) shows a positive trend in area and production (Table 8.9); this growth was in response to the relaxation of import restrictions on pulses in India during recent times (Rees *et al.* 1992).

The WANA region ranks first in exports of chickpea with 290 000 t exported annually (Table 8.10). The world market for chickpea is thin: about 6.5% of the total chickpea produced. This is low compared with all other pulses, where exports represent about 11% of world production (Oram and Agcaoili 1994). Should a free exchange of commodities come to pass, world exports of chickpea are expected to increase as production shifts to areas of greater comparative advantage. Domestic production declines in India are already resulting in increased imports from Australia.

REFERENCES

Angus, J.F. and M.W. Moncur. 1980. Photoperiodic and vernalisation effects on phasic development in chickpea. Int. Chickpea Newsl. 2:8-9.

Chowdury, K.A, K.S. Saraswat, S.N. Hasan and R.G. Gaur. 1971. 4000-3500 year old barley, rice and pulses from Atranjikhera. Science and Culture 37:531-533.

Contandriopoulos, J., A. Pamukuoglu and P. Quezel. 1972. A propos des *Cicer* vivaces du pourtour mediterranee oriental. Biologia Gallo-Hellenica 4(1):1-18.

Contreras, S. and M.A. Tagle. 1974. Toxic factors in Chilean legumes. Arch Latinoa mer Nutr. 24:191-199.

Dahiya, B.S., R.S. Waldia, L.S. Kaushik and I.S. Solanki. 1984. Early generation yield testing versus visual selection in chickpea (*Cicer arietinum* L.). Theor. Appl. Genet. 68:525-529.

De Candolle, A. 1883. Origine des Plantes Cultivées. Paris:208-260.

Dey, S.K. and G. Singh. 1993. Resistance to ascochyta blight in chickpea-Genetic basis. Euphytica 68:147-153.

Di Vito, M., K.B. Singh, N. Greco and M.C. Saxena. 1996. Sources of resistance to cyst nematode in cultivated and wild *Cicer* species. Genetic Resources and Crop Evolution 43(2):103-107.

Don, G. 1882. General history of dichlamydeous plants. 2:311.

Ellis, R.H. 1988. The viability equation, seed viability nomographs, and practical advice on seed storage. Seed Sci. and Technol. 16:29-50.

Eshel, Y. 1968. Flower development and pollen viability of chickpea (*Cicer arietinum* L.). Israel J. Agric. Res. 18:31-33.

FAO. 1988. Food Balance Sheets 1983-1985 average. FAO, Rome.

FAO. 1992a. FAO production tapes 1992. FAO, Rome.

FAO. 1992b. FAO trade tapes 1992. FAO, Rome.

FAO. 1993. Production yearbook. Vol. 47. FAO, Rome, Italy.

Geervani, P. and F. Theophilus. 1980. Effect of home processing on nutrient composition of certain high yielding legume varieties. Ind. J. Nutr. Diet. 17:443-446.

Geletu, B. 1987. Relationships among the F_2 to F_6 generations, and effect of spacing and selection in F_4 on performance in F_5 generation in chickpea. PhD Thesis, Andhra Pradesh Agricultural University, Hyderabad, India.

Hallab, H., H.A. Khatchadourian and I. Jabr. 1974. The nutritive value and organoleptic properties of white arabic bread supplemented with soybean and chickpea. Cereal Chem. 51:106-112.

Haq, M.A., M. Sadiq and M. Hasan. 1988. Improvement of chickpea through induced mutations. Pp. 75-78 *in* Improvement of Grain Legume Production using Induced Mutations. Proceedings of a workshop, Pullman, Washington, USA, 1-5 July 1986. International Atomic Agency, Vienna, Austria.

Harlan, J.R. and J.M.J. de Wet. 1971. Toward a rational classification of cultivated plats. Taxon 20(4):509-517.

Haware, M.P., J. Narayan Rao and R.P.S. Pundir. 1992. Evaluation of wild *Cicer* species for resistance to four chickpea diseases. Int. Chickpea Newsl. 27:16-18.

Helbaek, H. 1970. The plant husbandry at Hacillar. Pp. 189-244 *in* Excavation at Hacillar (J. Mellaart, ed.). Edinburg University Press. Gerald Duckworth and Co, London.

Hopf, M. 1969. Plant remains and early farming in Jericho. Pp. 355-358 *in* the Domestication of Plants and Animals (P.J. Ucko and W.G. Dimbleby, eds.). Gerald Duckworth and Co., London.

IBPGR/ICRISAT/ICARDA. 1993. Descriptors for chickpea (*Cicer arietinum* L.). ICRISAT, Patancheru, India.

ICARDA. 1994. Legume Program Annual Report for 1993. ICARDA, Aleppo, Syria.

ICRISAT. 1990. Annual Report 1989. Pp. 86-87. ICRISAT, Patancheru, India.

ICRISAT. 1991. Proceedings of the Consultancy meeting on uses of grain legumes, 27-28 March 1989, ICRISAT Centre, India (R. Jambunathan, ed.). ICRISAT, Patancheru, India.

Infantino, A., A. Porta-Puglia and K.B. Singh. 1996. Screening wild *Cicer* species for resistance to Fusarium wilt. Plant Dis. 80:42-44.

Iyenger, N.K. 1939. Cytological investigations on the genus *Cicer*. Ann. Botany, N. S. 3:271-305.

Jimenez-Diaz, R.M., K.B. Singh, A. Trapero-Casas and J.L. Trapero-Casas. 1991. Resistance in kabuli chickpea to Fusarium wilt. Plant Dis. 75:914-918.

Jodha, N.S. and K.V. Subba Rao. 1987. Chickpea: World importance and distribution Pp. 1-10 *in* The Chickpea (M.C. Saxena and K.B. Singh, eds.). CAB International, Wallingford, UK.

Kahl, G., D. Kaemmer, K. Weising, S. Kost, F. Weigand and M.C. Saxena. 1994. The potential of gene technology and genome analysis for cool season food legume crops: theory and practice. Pp. 705-725 *in* Expanding the Production and Use of Cool Season Food Legumes (F.J. Muehlbauer and W.J. Kaiser, eds.). Kluwer Academic Publishers, Dordrecht, The Netherlands.

Kaiser, W.J., A.R. Alcala-Jimenez, A. Hervas-Vargas, J.L. Trapero-Casas and R.M. Jimenez-Diaz. 1994. Screening of wild *Cicer* species for resistance to races 0 and 5 of *Fusarium oxysporum* f.sp. *ciceris*. Plant Dis. 78:962-967.

Kelley, T.G and P. Parthasarathy Rao. 1994. Chickpea competitiveness in India. Economic and Political Weekly, Vol 29, No 26, pp. 89-100, June 25.

Kupicha, F.K. .1977. The delimitation of the tribe *Vicieae* (Leguminosae) and the relationship of *Cicer* L. Botanical J. Linnean Soc. 74:131-162.

Ladizinsky, G. 1975. A new *Cicer* from Turkey. Notes Roy. Bot. Gard. Edin. 34:201-202.

Ladizinsky, G. and A. Adler. 1976. Genetic relationships among the annual species of *Cicer* L. Theor. Appl. Genet. 48:197-204.

Lateef, S.S. and M.P. Pimbert. 1990. The search for host-plant resistance to *Helicoverpa armigera* in chickpea and pigeonpea at ICRISAT. Pp. 14-18 *in* Summary proceedings of the First Consultative Group Meeting on the Host Selection Behaviour of *Helicoverpa armigera*, 5-7 Mar 1990, ICRISAT Centre, India. ICRISAT, Patancheru, India.

Malik, B.A., M.M. Verma, M.M. Rahman and A.N. Bhattarai. 1988. Production of chickpea, lentil, pea and faba bean in South-East Asia. Pp. 1095-1111 *in* World Crops: Cool Season Food Legumes (R.J. Summerfield, ed.). Proceedings of the International Food Legume Research Conference on Pea, Lentil, Faba bean and Chickpea, 6-11 Jul 1986, Spokane, Washington, USA. Kluwer Academic Publishers, Dordrecht, The Netherlands.

Moreno, M.T. and J.I. Cubero. 1978. Variation in *Cicer arietinum* L. Euphytica 27:465-485.

Muehlbauer, F.J., C.J. Simon, S.C. Spaeth and N.I. Haddad. 1990. Genetic improvement of chickpea: key factors to be considered for a breakthrough in productivity. Pp 209-216 *in* Chickpea in the Nineties (H.A. van Rheenen and M.C. Saxena, eds.). Proceedings of the Second International Workshop on Chickpea Improvement, 4-8 Dec 1989, ICRISAT Centre, India. ICRISAT, Patancheru, India.

Nene, Y.L. and M.V. Reddy. 1987. Chickpea diseases and their control. Pp. 233-270 *in* The Chickpea (M.C. Saxena and K.B. Singh, eds.). CAB International, Wallingford, UK.

Ocampo, B; G. Venora, A.Errico, K.B. Singh and F. Saccardo. 1992. Karyotype analysis in the genus *Cicer*. J. Genet & Breed. 46:229-240.

Oram, P.A. and M. Agcaoili. 1994. Current status and future trends in supply and demand of cool-season food legumes. Pages 3-49 *In* Expanding the Production and Use of Cool Season Food Legumes (F.J. Muehlbauer and W.J. Kaiser, eds.). Kluwer Academic Publishers, Dordrecht.

Pundir, R.P.S., M.H. Mengesha and K.N. Reddy. 1990. Leaf type and their genetics in chickpea (*Cicer arietinum* L.). Euphytica 45:197-200.

Pundir, R.P.S., M.H. Mengesha and G.V. Reddy. 1993. Cytomorphology of *Cicer canariense*, a wild relative of chickpea. Euphytica 69:73-75.

Pundir, R.P.S., N.K. Rao and L.J.G. van der Maesen. 1983. Induced tetraploidy in chickpea (*Cicer arietinum* L.). Theor. Appl. Genet. 65:119-122.

Pundir, R.P.S., N.K. Rao and L.J.G. van der Maesen. 1985. Distribution of qualitative traits in the world germplasm of chickpea (*Cicer arietinum* L.). Euphytica 34:697-703.

Pundir, R.P.S., K.N. Reddy and M.H. Mengesha. 1988. ICRISAT Chickpea Germplasm Catalog: evaluation and analysis. ICRISAT, Patancheru, India.

Pundir, R.P.S., K.N. Reddy and M.H. Mengesha. 1992. Pod volume and pod filling as useful traits of chickpea. Int. Chickpea Newsl. 17:18-20.

Reddy, M.V., T.N. Raju and R.P.S. Pundir. 1991. Evaluations of wild *Cicer* accessions for resistance to wilt and root rots. Indian Phytopathol. 44:389-391.

Reed, W., C. Cardona, S. Sithanantham and S.S. Lateef. 1987. Chickpea insect pests and their control. Pp. 283-318 *in* The Chickpea (M.C. Saxena and K.B. Singh, eds.). CAB International, Wallingford, UK.

Rees, R.O., J.B. Brower, J.E. Mahoney, G.H. Walton, R.B. Brinsmead, E.J. Knights and D.F. Beech. 1992. Pea and chickpea production in Australia. Pp. 412-425.*in* Expanding the Production and Use of Cool Season Food Legumes (F.J. Muehlbauer and W.J. Kaiser, eds.). Kluwer Academic Publishers, Dordrecht.

Roberts, E.H., J.R. Summerfield, F.R. Minchin and P. Hadley. 1980. Phenology of chickpeas (*Cicer arietinum*) in contrasting aerial environments. Exp. Agric. 16:343-360.

Robertson, L.D., K.B. Singh and B. Ocampo. 1995. A catalog of annual wild *Cicer* species. 1995. ICARDA, Aleppo, Syria.

Sachan, J.N. 1990. Progress in host-plant resistance work in chickpea and pigeonpea against *Helicoverpa armigera* (Hubner) in India. Pp. 14-18 *in* Summary Proceedings of the First Consultative Group Meeting on the Host Selection Behaviour of *Helicoverpa armigera*, 5-7 Mar 1990, ICRISAT Centre, India. ICRISAT, Patancheru, India.

Santos Guerra, A. and G.P. Lewis. 1986. A new species of *Cicer* (Leguminosae - Papilionoideae) from the Canary Islands. Kew Bulletin 41(2):459-462.

Saxena, M.C. and M.H. Siddique. 1980. Response of some diverse kabuli chickpea genotypes to vernalisation. Int. Chickpea Newsl. 2:7-8.

Saxena, M.C., N.P. Saxena and A.K. Mohamed. 1988. High temperature stress. Pp. 845-856 *in* World Crops: Cool Season Food Legumes (R.J. Summerfield, ed.). Kluwer Academic Publishers, Dordrecht, The Netherlands.

Saxena, N.P. 1987. Screening for adaptation to drought: case studies with chickpea and pigeonpea. Pp. 63-76 *in* Adaptation of Chickpea and Pigeonpea to Abiotic Stresses (N.P. Saxena and C. Johansen, eds.). Proceedings of the Consultants' Workshop, 19-21 Dec 1984, ICRISAT Centre, India. ICRISAT, Patancheru, India.

Saxena, N.P. and A.R. Sheldrake. 1980. Physiology of growth, development and yield of chickpea in India. *in* Proceedings of the International Workshop on Chickpea Improvement, 28 Feb-2 Mar 1979, Hyderabad, A.P., India. ICRISAT, Patancheru, India.

Sheldrake, A.R. and N.P. Saxena. 1979. The growth and development of chickpea under progressive moisture stress. Pp. 465-485 *in* Stress Physiology in Crop Plants (H. Mussel and R.C. Staples, eds.). John Wiley and Sons, Chichester.

Sheldrake, A.R., N.P. Saxena and L. Krishnamurthy. 1978. The expression and influence on yield of the double-podded character in chickpea (*Cicer arietinum* L.). Field Crops Res. 1:243-253.

Singh, H., J. Kumar, J.B. Smithson and M.P. Haware. 1987. Complementation between genes for resistance to race 1 of *Fusarium oxysporum* f.sp. *ciceri* in chickpea. Plant Pathol. 36:539-543.

Singh, K.B. 1987. Chickpea breeding. Pp. 127-162 *in* The Chickpea (M.C. Saxena and K.B. Singh, eds.). CAB International, Wallingford, UK.

Singh, K.B. 1993. Problems and prospects of stress resistance breeding in chickpea. Pp. 17-35 *in* Breeding for Stress Tolerance in Cool Season Food Legumes (K.B. Singh and M.C. Saxena, eds.). John Wiley and Sons, Chichester, UK.

Singh, K.B. and B. Ocampo. 1993. Interspecific hybridization in annual *Cicer* species. J. Genet. and Breed. 47:199-204.

Singh, K.B. and M.V. Reddy. 1991. Advances in Disease Resistance Breeding in Chickpea. Adv. Agron. 45:191-222.

Singh, K.B. and M.V. Reddy. 1993a. Resistance to six races of *Ascochyta rabiei* in the world germplasm collection of chickpea. Crop Sci. 33:186-189.

Singh, K.B. and M.V. Reddy. 1993b. Sources of resistance to ascochyta blight in wild *Cicer* species. Neth. J. Plant Pathol. 99:163-167.

Singh, K.B. and S. Weigand. 1994. Identification of resistant sources in *Cicer* species to *Liriomyza cicerina*. Genet. Resour. and Crop Evolution 41:75-79.

Singh, K.B. and S. Weigand. 1996. Registration of three leafminer-resistant chickpea germplasm lines: ILC 3800, ILC 5901, and ILC 7738. Crop Sci. 36:472.

Singh, K.B., W. Erskine, L.D. Robertson, H. Nakkoul and P.C. Williams. 1988. Influence of pretreatment on cooking quality parameters of dry legumes. J. Sci. Food Agric. 44:135-142.

Singh, K.B., G.C. Hawtin, Y.L. Nene and M.V. Reddy. 1981. Resistance in chickpea to *Ascochyta rabiei*. Plant Dis. 65:586-587.

Singh K.B., L. Holly and G. Bejiga. 1991a. A catalog of kabuli chickpea germplasm: An Evaluation Report of Winter Sown Kabuli Chickpea, Land Races, Breeding Lines and Wild *Cicer* Species. ICARDA, Aleppo, Syria.

Singh, K.B., R.S. Malhotra and J.E. Witcombe. 1983. Kabuli chickpea germplasm catalog. ICARDA, Aleppo, Syria.

Singh, K.B., R.S. Malhotra and M.C. Saxena. 1989. Chickpea evaluation for cold tolerance under field conditions. Crop Sci. 29:282-285.

Singh, K.B., R.S. Malhotra, M.H. Halila, E.J. Knights and M.M. Verma. 1994. Current status and future strategy in breeding chickpea for resistance to biotic and abiotic stresses. Pp. 572-591 *in* Expanding the Production and Use of Cool Season Food Legumes (F.J. Muehlbauer and W.J. Kaiser, eds.). Kluwer Academic Publishers, Dordrecht, The Netherlands.

Singh, K.B., R.S. Malhotra and M.C. Saxena. 1990. Sources of tolerance to cold in *Cicer* species. Crop Sci. 30:1136-1138.

Singh, K.B, M. Omar, M.C. Saxena and C. Johansen. 1996. Registration of FLIP 87-59C, a drought-tolerant chickpea germplasm line. Crop Sci. 36:472.

Singh, U. 1985. Nutritional quality of chickpea (*Cicer arietinum* L.): current status and future research needs. J. Plant Food Hum. Nutr. 35:339-357.

Singh, U. 1988. Antinutritional factors of chickpea and pigeonpea and their removal by processing. Plant Foods Hum. Nutr. 38:251-261.

Singh, U. and R.P.S. Pundir. 1991. Amino acid composition and protein content of chickpea and its wild relatives. Int. Chickpea Newsl. 25:19-20.

Singh, U., N. Subrahmanyam and J. Kumar. 1991b. Cooking quality and nutritional attributes of some newly developed cultivars of chickpea. J. Sci. Food Agric. 55:37-46.

Smithson, J.B., J.A. Thompson and R.J. Summerfield. 1985. Chickpea (*Cicer arietinum* L.). Pp. 312-390 *in* Grain Legume Crops. Collins, London, UK.

van der Maesen, L.J.G. 1972. *Cicer* L., A monograph of the genus with special reference to chickpea (*Cicer arietinum* L.), its ecology and cultivation. Maded. Landbou. Wageningen, 72-10.

van der Maesen, L.J.G. 1987. Origin, history and taxonomy of chickpea. Pp. 11-34. *in* The Chickpea (M.C. Saxena and K.B. Singh, eds.). CAB International, UK.

van der Maesen, L.J.G., W.J. Kaiser, G.A. Marx and M. Worede. 1988. Genetic basis of pulse crop improvement: Collection, preservation and genetic variation in relation to needed traits. Pp. 55-66 *in* World Crops: Cool Season Food Legumes (R.J. Summerfield, ed.). Kluwer Academic Publishers.

van Rheenen, H.A. 1991. Production aspects and prospects of chickpea. Pp. 31-35 *in* Proceedings of the consultancy meeting on uses of grain legumes (R. Jambunathan, ed.)., 27-28 Mar 1989, ICRISAT Centre, India. ICRISAT, Patancheru, India.

van Rheenen, H.A. and M.P. Haware. 1994. Mode of inheritance of resistance to ascochyta blight (*Ascochyta rabiei* Pass. Labr.) in chickpea (*Cicer arietinum* L.) and its consequences for resistance breeding. Int. J. Pest Manage. 40:166-169.

Vavilov, N.I. 1926. [Studies on the origin of cultivated plants.] Leningrad 129-238.

Vavilov, N.I. 1951. The origin, variation, immunity and breeding of cultivated plants. Chronica Botanica, New York 13-1/6:26-38, 75-78, 151 (1949-50).

Vishnu-Mittre, A. 1974. The beginnings of agriculture: Paleobotanical evidence in India Pp. 23-24 *in* Evolutionary Studies on World Crops: Diversity and Change in the Indian Sub-continent (J.B. Hutchinson, ed.). Cambridge University Press.

Weigand, S. and O. Tahhan. 1990. Chickpea insect pests in the Mediterranean zones and new approaches to their management. Pp. 169-175 *in* Chickpea in the Ninties (H.A. van Rheenen and M.C. Saxena, eds.). ICRISAT, Patancheru, India.

Westphal, E. 1974. Pulses in Ethiopia, their taxonomy and agricultural significance. Agric. Res. Rep. 815. Wageningen, The Netherlands.

Williams, P.C. and U. Singh. 1987. The chickpea - Nutritional quality and the evaluation of quality in breeding programs. *In* The Chickpea (M.C. Saxena and K.B. Singh, eds.). CAB International, Wallingford, UK.

Chapter 9

Groundnut[1]

A.K. Singh and S.N. Nigam

Groundnut, *Arachis hypogaea* L. (also called peanut in English, *mani* in Spanish, *amondoim* in Portuguese, *pistache* in French, *mungphali* in Hindi and *ying zui dou* in Chinese), ranks 13th among food crops and annual oilseed crops (FAO 1995). Its high oil and protein contents serve important needs for food, energy and industrial uses. Although a native of South America, the crop is now cultivated in tropical, subtropical and warm temperate regions of the world extending from 40°N to 40°S.

BOTANY AND DISTRIBUTION
Arachis hypogaea L. is a member of family Leguminoseae-Papilionoideae, tribe Aeschynomeneae and subtribe Stylosanthinae. It is a tetraploid with $2n=40$. Krapovickas and Gregory (1994) divided the genus *Arachis* into nine sections. Section *Arachis* contains cultivated groundnut, *A. hypogaea*, another tetraploid species *A. monticola* Krapov. & Rigoni and a number of wild diploid species. Gregory *et al.* (1973) earlier divided *A. hypogaea* into two subspecies, *fastigiata* Waldron and *hypogaea* Krap. et Rig., and each subspecies into two botanical varieties. According to the new classification, subsp. *fastigiata* is subdivided into four botanical varieties, *fastigiata, peruviana* Krapov. & W.C. Gregory, *aequatoriana* Krapov. & W.C. Gregory and *vulgaris* C. Harz. The two botanical varieties in subsp. *hypogaea* are *hypogaea* and *hirsuta* Kohler. The key for identification of different botanical varieties is given in Box 9.1.

Origin, Domestication and Diffusion
The genus *Arachis* is naturally restricted to Argentina, Bolivia, Brazil, Paraguay and Uruguay in South America. Both Krapovickas (1969, 1973) and Gregory *et al.* (1980) postulated a planalto profile from Corumba to Joazeiro, Brazil as the centre from which distribution of *Arachis* occurred. Cultivated groundnut most probably originated in the region of southern Bolivia and northwestern Argentina (Krapovickas 1969), which is an important centre of diversity of subsp. *hypogaea*. A few forms of subsp. *fastigiata*, certain wild diploid annuals such as *A. duranensis* Krapov. & W.C. Gregory and *A. batizocoi* Krapov. & W.C. Gregory and *A. monticola*, considered to be the probable ancestors of *A. hypogaea* (Singh 1988), also occur naturally in this area. It has been suggested that *A. duranensis* (with A genome) and *A. batizocoi* (with B genome) initially evolved into the wild tetraploid *A. monticola* through amphidiplodization, which on domestication gave rise to the cultivated *A. hypogaea* (Smartt *et al.* 1978; Singh 1986, 1988), although RFLP results do not show *A. batizocoi* to be close to *A. hypogaea*. Subsequent spread of the crop to different agroclimatic zones brought further diversification and variability in growth habit and seed and pod characteristics (Singh 1995).

---

[1] Submitted as Journal Article No. 1851 by ICRISAT.

Box 9.1. Key to distinguish the taxa of *Arachis hypogaea*

A.   Central axis without flowers and lateral branches, the   **subsp. *hypogaea***
     vegetative and reproductive branches alternate regularly
     (alternate ramification).

     B. Leaflets with a glabrous dorsal surface or with some   **var. *hypogaea***
        hair along the midrib.

     B'. Leaflets with hairy (1-2 mm) dorsal surface, entire   **var. *hirsuta***
        surface is hairy.

A'. Central axis with flowers and lateral branches, the   **subsp. *fastigiata***
     reproductive and vegetative branches show no order
     (sequential ramification).

C.   Fruits with more than two seeds. Open/widespread
     fruiting.

D.   Leaflets with a glabrous dorsal surface and hair only on
     the midrib.
     E. Fruits with smooth or lightly marked reticulation,   **var. *fastigiata***
        without highlighting of the longitudinal ribs.
        Reproductive branches mostly short and thin.

     E'. Fruits with very marked reticulation, and with   **var. *peruviana***
        prominent longitudinal ribs.
        Long, strong, reproductive branches (5-10 cm), with
        strong central axis and lateral branches.

     D'. Leaflets with a hairy (1-2 mm) dorsal surface, entire   **var. *aequatoriana***
        surface hairy.
        Long reproductive branches, mainly the lateral
        branches.
        Central axis mostly with short inflorescence and
        reproductive branches.

     C'. Fruits mostly 2-seeded. Bunched fruits, pointing to   **var. *vulgaris***
        the base of the plant. Frequently with compact ears.

Source: Krapovickas and Gregory 1994 (translated from Spanish).

Domestication probably first took place in the valleys of the Parana and Paraguay river systems in the Gran Chaco area of South America. Early European explorers found native Indians cultivating this crop in many islands in the Antilles, on the northeastern and eastern coasts of Brazil in all warm regions of the Rio de la Plata basin; extensively in Peru and sparsely in Mexico (Hammons 1994).

In South America, where the greatest diversity is found, Krapovickas (1969) and Gregory and Gregory (1976) recognized the Chaco region between southern Bolivia and northwestern Argentina as the primary centre of diversity and another six regions as secondary centres of diversity for cultivated groundnut (Fig. 9.1).

On the basis of presence of distinct landraces found during further exploration in Ecuador, Singh and Simpson (1994) recently have added Ecuador as another secondary centre of diversity. Most authorities believe that in the late 15th century the Portuguese carried two-seeded groundnut varieties from the east coast of South America (Brazil) to Africa, to the Malabar coast of southeastern India and possibly to the far east. The Spaniards in the early 16th century took three-seeded Peruvian types (including

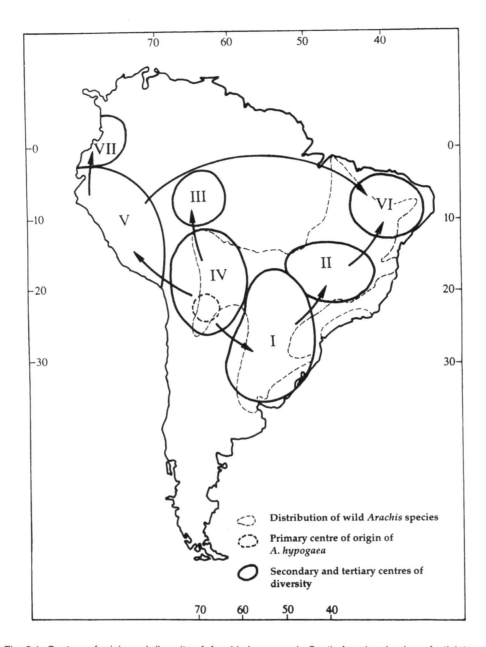

Fig. 9.1. Centres of origin and diversity of *Arachis hypogaea* in South America. I subsp. *fastigiata* var. *fastigiata* and var. *vulgaris*; II subsp. *fastigiata* var. *fastigiata*; III subsp. *hypogaea* var. *hypogaea*; IV subsp. *hypogaea* var. *hypogaea*, subsp. *fastigiata* var. *fastigiata*; V subsp. *hypogaea* var. *hypogaea* and var. *hirsuta*, subsp. *fastigiata* var. *fastigiata* and var. *peruviana*; VI subsp. *fastigiata* var. *fastigiata* and var. *vulgaris*; VII subsp. *fastigiata* var. *aequatoriana*.

hirsuta) to Indonesia and China up to Madagascar from the west coast of South America via the western Pacific. By the middle of the 16th century, groundnut made its way to North America from Africa as well as from the Caribbean islands, Central America and Mexico and was distributed worldwide. By the 19th century, groundnut became an important crop in West Africa, India, China and the USA. Among these new areas of introduction, Africa is considered a tertiary centre of diversity. Although various types of groundnut introduced into Africa came from a single centre in South America, near Bolivia, there exists a significant variability in the continent. Similarly, India and China, with long histories of groundnut cultivation and landraces, are considered other important centres of diversity.

In addition to *A. hypogaea*, the wild *Arachis* species, some of which are used for edible seeds (e.g. *A. villosulicarpa* Hoehne) or forage (e.g. *A. glabrata* Benth., *A. pintoi* Krapov. & W.C. Gregory and *A. repens* Handro) constitute another genetic reservoir of useful characteristics for the improvement of cultivated groundnut. They are notable as sources of host-plant resistance to diseases and insect pests and perhaps also of agronomic traits (Gouk *et al.* 1986).

Genetic diversity in genus *Arachis* has been classified into four genepools by Singh and Simpson (1994).

1. Primary genepool consisting of landraces of *A. hypogaea* and its wild form *A. monticola* (although some consider *A. monticola* as a separate genepool).
2. Secondary genepool consisting of diploid species from section *Arachis* that are cross-compatible with *A. hypogaea*.
3. Tertiary genepool consisting of species of section *Procumbentes* that are weakly cross-compatible with *A. hypogaea*.
4. The fourth genepool consisting of the remaining wild *Arachis* species classified into seven other sections.

Reproductive Biology

Groundnut is an annual or weakly perennial herb that may flower as early as 17-18 days from the date of emergence. Most flowers are self-pollinated before or as they open (cleistogamy) and cross-pollination is rare, but some wild species, such as *A. lignosa*, may also require insects for pollination (Banks 1990). Sporogenesis or gametogenesis occurs 2 days prior to anthesis, when bud length is around 5 mm. Unlike other legumes, groundnut antipodals degenerate several hours before fertilization. The pollen tube takes around 10-18 hours after pollination to reach the ovary and effect fertilization. After fertilization the flowers wither rapidly and the intercalary meristematic cells that comprise the basal tissue of the ovary produce a geotropic stalk-like structure called a peg (carpophore). Initially associated with embryo development, the peg grows at first slowly and then rapidly. The tip of the peg usually contains two (sometimes 3-5, depending on variety) fertilized ovules. At the time of peg growth, the embryo is at the 8-12 cell stage and becomes quiescent. Peg growth continues until penetration into the soil (after 8-14 days of fertilization), and when it receives mechanical stimulus the peg transforms into a pod. Ovules and embryos then start growing, mature to form seeds within the pod, which later becomes dry and brittle to form the shell.

GERMPLASM CONSERVATION AND USE

We have made significant progress in the collection of groundnut germplasm from various centres of diversity in the last two decades (Table 9.1). Gaps in genetic diversity and geographical representation still exist. *Arachis hypogaea* subsp. *hypogaea* var. *hirsuta*, one among the six botanical varieties of *A. hypogaea*, remains unrepresented in the ICRISAT collection of 14 000 accessions of groundnut. Similarly, traditional groundnut areas in subsistence agriculture, areas of early introduction in countries like Laos and China in Asia, Angola, Malagasay Republic, Namibia and South Africa in Africa, and the areas of secondary centres of diversity in South America, Peru, Ecuador, Uruguay and Paraguay have not been fully explored. Information on each accession is available for most of the important features indicated in Groundnut Descriptors (IBPGR and ICRISAT 1992), but data on several descriptors in passport data, and for some characteristics of regional or local importance in evaluation data, are still far from complete.

Most of the traditional landraces, which constitute around 33% of the total world collection conserved at ICRISAT, originate from different countries of South America, Africa and Asia (Tables 9.1 and 9.2). Variability analysis of the world collection has shown comparatively greater amounts of variation in landraces than other germplasm (Singh *et al.* 1992). Most variability, particularly for resistance to diseases and insect pests and also for some agronomic characters like seed mass, is mainly available in the landraces originating from the primary and secondary centres of diversity in South America (Table 9.3).

Genus *Arachis* presents a considerable amount of botanical diversity. The basic plant structure in wild *Arachis* species and *A. hypogaea* is similar. In growth habit, genotypes can be procumbent runner type with short or long main axis and laterals growing horizontally to various lengths. Other genotypes may be decumbent, where laterals have an ascending tendency or are erect with shortened internodes. The angle between the main axis and secondary branches may vary and consequently the growth habit, classified into decumbent types (IBPGR and ICRISAT 1992). Table 9.3 summarizes the range of variability recorded at ICRISAT for various plant, pod and seed characters.

Three foliar diseases – late leaf spot [*Phaeosariopsis personata* (Berk. & Curt.) V. Arx.], early leaf spot (*Cercospora arachidicola* Hori) and rust (*Puccinia arachidis* Speg.) – are the most widely distributed and economically important diseases of groundnut. At ICRISAT 143 rust-resistant lines have been identified (Mehan *et al.* 1994b; Subrahmanyam *et al.* 1995). Extensive screening for leaf spot resistance has resulted in identification of several resistance sources (Foster *et al.* 1980, 1981; Melouk *et al.* 1984; Subrahmanyam *et al.* 1982). Fifty-four lines resistant to late leaf spot have been identified, 29 of which are also resistant to rust (Subrahmanyam *et al.* 1995). Screening of more than 2000 accessions in Malawi for early leaf spot over seasons has resulted in identification of five promising lines (Subrahmanyam, pers. comm.). Many wild *Arachis* species have been evaluated against these three foliar diseases and high levels of resistance have been identified in a large number of species/accessions for early leaf spot (Gibbons and Bailey 1967; Abdou *et al.* 1974; Foster *et al.* 1981), late leaf spot and rust (Abdou *et al.* 1974; Subrahmanyam *et al.* 1985).

Six important groundnut virus diseases are groundnut rosette (GRV) in Africa, peanut bud necrosis virus (PBNV) in India, tomato spotted wilt virus (TSWV) in the USA, peanut mottle (PMV) worldwide, peanut stripe (PStV) in East and Southeast Asia and peanut clump (PCV) in West Africa and India. Resistance to GRV disease was found in landraces from Burkina Faso (de Berchoux 1960) and also in wild *Arachis* species, *A. glabrata* and *A. repens* (Gibbons 1969). Recently, wild *Arachis* species, *A. appressibila* (30003), *A. chacoensis* (now *A. diagoi* Hoehne) (K.R. Bock, pers. comm.) and an interspecific derivative involving *A. diagoi* have shown high levels of resistance to GRV disease (Moss *et al.* 1993). Numerous lines with consistently less than 20% PBNV disease incidence in the field – such as ICGs 848, 851, 852, 862, 869, 885, 2271, 2306, 3806, 5030, 6135, 7676, 7892 – have been identified at ICRISAT (Dwivedi *et al.* 1995). Among wild *Arachis* species, *A. diagoi* showed no infection after mechanical or vector-effected inoculation (Subrahmanyam *et al.* 1985).

Screening of around 9000 accessions for PStV in Indonesia did not result in identification of any resistant line in *A. hypogaea*. However, screening of wild *Arachis* species has resulted in identification of several accessions with negative reaction to PStV (Culver *et al.* 1987; Prasada Rao *et al.* 1991). For PMV some germplasm lines, such as NC Ac 2240 and NC Ac 2243, have shown consistently low yield losses due to this disease (ICRISAT 1983). For PCV, screening of 7000 accessions did not result in identification of any resistant line. For both PMV and PCV, a number of wild *Arachis* species have shown promise (ICRISAT 1985; Subrahmanyam *et al.* 1985). Considerable variability for apparent resistance to TSWV has been reported in breeding lines in the USA (Culbreath *et al.* 1994).

Variation in reaction to several soilborne diseases has been reported in the germplasm. Resistance to bacterial wilt caused by *Pseudomonas solanacearum* (Smith)

Table 9.1. Number of accessions of cultivated groundnut and wild *Arachis* species from different centres of diversity available at ICRISAT (December 1994).

Collections	Centres of diversity					
	Primary	Secondary	Tertiary	India/China	Others	Total
Accessions	601	945	1194	88	1974	4802
Landraces	408	786	1943	1422	587	5146
Breeding lines	135	81	987	1558	1801	4562
Named cultivars	10	1	28	135	146	320
Interspecific derivatives	–	–	–	167	8	175
Wild *Arachis* spp.	132	146	–	–	31	309

Table 9.2. Status of groundnut germplasm accessions by botanical variety (December 1994).

Botanical group	Number of accessions				
	Landraces	Breeding lines	Released cultivars	Others[†]	Total
vulgaris	1738	1494	158	1384	4774
fastigiata	978	529	24	575	2106
peruviana	325	5	0	10	390
acquatoriana	3	4	0	7	14
hypogaea (bunch)	1128	1574	75	856	3633
hypogaea (runner)	1064	956	63	656	2739
hirsuta	0	0	0	0	0

[†] Others includes 165 interspecific derivatives and other accessions with status not known, doubtful, or not clear.

Smith was identified as early as 1920. Around 5000 germplasm accessions and breeding lines have been screened in wilt-sick plots in China and Indonesia, resulting in identification of about 54 lines (Mehan *et al.* 1994a). Most of these belong to the Chinese dragon type (subsp. *hypogaea* var. *hirsuta* ?). Resistance to black rot disease caused by *Cylindrocladium crotalariae* (C.A. Loos) D.K. Bell & Sobers has been identified in several Virginia and Spanish genotypes (Green *et al.* 1983) and *A. monticola* (Fitzner *et al.* 1985). NC 3033, a line resistant to black rot, was found resistant to *Sclerotium rolfsii* Sacc. Genotypes resistant to *Pythium* pod rot, *Sclerotinia minor*, have been identified (Smith *et al.* 1989).

Aflatoxin contamination in groundnut is a serious concern. Many sources have been found with resistance to pre-harvest seed infection, *in vitro* seed colonization and aflatoxin production. These include PI 337409, PI 337394F, UF 71513 (resistant to seed invasion and colonization), Doran and Shulamit (resistant to pod infection), U-4-477, 55-437, 73-30 and J 11 (resistant to seed infection in the field), and U 4-7-5 and VRR 245 (low production of aflatoxin B_1) (Mehan 1989).

Sources of resistance to most insect pests have been identified in both *A. hypogaea* and wild *Arachis* species (Lynch *et al.* 1981; Stalker and Campbell 1983; Stalker *et al.* 1984; Wightman *et al.* 1989; Lynch 1990; Wightman and Ranga Rao 1994). Some wild *Arachis* are cross-compatible with *A. hypogaea*. Resistance in wild *Arachis* spp. has been identified for the plant-parasitic nematodes *Meliodogyne arenaria* and *M. hapla* (Baltensperger *et al.* 1986; Nelson *et al.* 1988; Holbrook and Noe 1990). Eleven *A. hypogaea* have been reported resistant to these two nematode species (Anonymous 1985). A number of genotypes – such as ICGs 1697, 4110, 6322, 7889, 7897 – have been identified as resistant to a severe nematode disease popularly called Kalahasti Malady caused by *Tylenchorhynchus brevilineatus* in Andhra Pradesh, India (Mehan *et al.* 1993). Recently groundnut germplasm has been evaluated for crop growth rate, water use efficiency and partitioning (Nageswara Rao *et al.* 1994). The number of accessions identified with resistance to various biotic and abiotic stresses, variation in reaction, and the total number of accessions screened are summarized in Tables 9.4 and 9.5.

Table 9.3. Range of variation in cultivated groundnut observed at ICRISAT Asia Center.

Character	Minimum	Maximum	Intermediate(s)
Life form	Annual	–	–
Growth habit	Erect	Procumbent	Decumbent
Branching pattern	Sequential	Alternate	Irregular
Stem pigmentation	Absent	Present	–
Stem hairiness	Glabrous	Woolly	Hairy, very hairy
Reproductive branch length	> 1 cm	10 cm	Continuous
No. of flowers/ inflorescence	1	5	2,3,4
Peg colour	Absent	Present	–
Standard petal colour	Yellow	Garnet	Lemon yellow, light orange, orange, dark orange
Standard petal markings	Yellow	Garnet	Lemon yellow, light orange, orange, dark orange
Leaf colour	Yellowish green	Dark green	Light green, green, bottle green
Leaflet length (L)	17 mm	94 mm	Continuous
Leaflet width (W)	7 mm	52 mm	Continuous
Leaflet L/W ratio	1	6	Continuous
Leaflet shape	Suborbicular	Linear lanceolate	Elliptic, ovate, obovate, oblong
Hairiness of leaflet	Subglabrous	Profuse and long	Scarce and short, scarce and long, profuse and short
No. of seeds/pod	1	5	2,3,4
Pod beak	Absent	V. prominent	Slight, moderate, prominent
Pod constriction	Absent	Very deep	Slight, moderate, deep
Pod reticulation	Smooth	Prominent	Slight, moderate
Pod length	14 mm	65 mm	Continuous
Pod width	7 mm	20 mm	Continuous
Seed colour pattern	One	Variegated	–
Seed colour	White	Dark purple	Yellow, shades of tan, rose shades of red, grey-orange, shades of purple
Seed length	4 mm	23 mm	Continuous
Seed width	5 mm	13 mm	Continuous
100-seed weight	14 g	140 g	Continuous
Days to emergence	4	18	Continuous
Days to 50% flowering	15	54	Continuous
Days to maturity	75	> 155	Continuous
Fresh seed dormancy	0 days	> 66 days	Continuous
Oil content	31.8%	55.0%	Continuous
Protein content	15.8	34.2	Continuous

Groundut germplasm is conserved as pods or seeds, except for some wild *Arachis* species, mostly in section *Rhizomatosae*, which are conserved as live plants in concrete rings under contained conditions. The following facilities are used for processing and *ex situ* conservation of seed.

1. **Short-term storage.** This facility at ICRISAT is maintained at 18°C and 30% RH. Pods/seeds in these chambers remain viable for a few years without much loss in viability.

Table 9.4. Number of accessions identified with resistance to different biotic and abiotic stresses and high biological nitrogen fixation capacity (BNF)† at ICRISAT.

Stress/Factor	Status‡					Botanical type§					
	LR	BL	RC	Wild	Others	Vul	Fst	Hyb	Hyr	Pru	Aeq
Biotic stresses											
Late leaf spot	49	6	–	27	4	2	17	2	–	37	1
Rust	135	15	–	57	4	3	19	19	6	105	2
Seed invasion and colonization by A. flavus in the laboratory	18	11	1	–	9	35	2	–	–	2	–
Seed infection by A. flavus in the field	2	2	1	–	2	6	1	–	–	–	–
Peanut bud necrosis	1	19	–	–	3	–	–	13	10	–	–
Aphids	–	2	–	–	2	–	–	2	2	–	–
Leaf miner	6	7	–	–	1	–	3	9	2	–	–
Jassids	70	48	7	–	11	13	14	38	42	29	–
Thrips	–	14	1	3	–	–	–	9	6	–	–
Abiotic stresses											
Drought	20	12	7	–	7	28	11	7	–	–	–
High BNF	3	4	1	–	–	3	3	1	1	–	–

† Source: ICRISAT published and unpublished data.

‡ LR=Landrace, BL=Breeding line, RC=Released cultivar, Wild=Arachis spp.

§ Vul=vulgaris, Fst=fastigiata, Hyb= hypogaea bunch, Hyr=hypogaea runner, Pru=peruviana, Aeq=aequatoriana.

Table 9.5.   Variation in reaction of groundnut accessions to various stresses and for nutritional factors at ICRISAT.

Stress/Factor	Level of reaction[1]			Susceptible/ Average	Access. screened
	HR/H	R/M	MR/L		
Fungal disease					
Early leaf spot	–	–	5	2084	2089
Late leaf spot	–	59	39	10103	10201
Rust	79	75	23	10024	10201
Aflatoxin production	–	4	–	578	582
Aspergillis flavus seed invasion	21	8	10	539	580
Pod rot	–	6	–	3216	3222
Viral disease					
PBNV	–	–	23	7377	7400
PMV	–	–	2	6942	6944
Pest					
Thrips	–	15	–	5330	5345
Jassids	105	28	3	6709	6845
Termites	–	9	–	511	520
Aphids	2	2	–	596	600
Leaf miner	14	–	–	10187	10201
Abiotic stress					
Drought	–	38	8	774	820
Nutritional quality					
High oil	20	5247	632	8849	8868
High protein	117	3119	97	8751	8868

1. HR=Highly Resistant, R=Resistant, MR=Moderately Resistant; H=50-58%, M=40-50%, L=31.8-40% for oil and H=>31%, M=21-30%, L=< 20% for protein and Average=average oil and protein contents.

2. **Medium-term chambers.** These modules are maintained at a temperature around 4°C and 20% RH.   The pods can remain for 25-35 years without much loss in seed viability.

3. **Long-term chambers.**   These modules are maintained at –18°C without any control over RH and host 1000-1500 seeds of base or duplicate collections.   The seeds are dried to a moisture level of 4-5% in the dryers maintained at a temperature of 15°C and 15% RH and are hermetically sealed in aluminium pouches before being transferred to long-term chambers. Long-term chambers can hold the seeds for periods in excess of 35-50 years without much loss in viability.

Seed/pod samples have been supplied worldwide for research and use in breeding programmes to improve the genetic potential of existing groundnut cultigens (Table 9.6). Several wild *Arachis* species have been used for transfer of foliar disease resistance into cultivated groundnut.

Properties and Uses

Groundnut is rich in oil and protein, most of which is found in the cotyledons. Chemically a groundnut seed contains around 30% protein, 48% fat, 15.0% carbohydrate, 3.0% crude fibre, 5.0% moisture and 2.0% ash (Natrajan 1980). Groundnut protein is deficient in lysine, methionine and threonine (Pancholy *et al.* 1978). Non-protein or free amino acids are thought to react with glucose and fructose, produced by hydrolysis of sucrose during the browning process, to produce the typical roasted groundnut flavour, colour and aroma (Young *et al.* 1974; Woodroof 1983). The ratio of these amino acids varies with seed size (Young *et al.* 1974).   A methionine-rich protein (MRP) also has been identified in groundnut seed.   Studies have shown considerable variation in MRP composition, thereby suggesting the possibility of improving the nutritional value of groundnut (Basha 1991).

Table 9.6. Number of groundnut germplasm accessions distributed to different regions of the world from ICRISAT Center (1976 to December 1995).

Region	Individual	University	Internat. Programme	National Programme	Others[†]	Total
Asia	59	492	727	14201	10	15489
Europe	8	98	133	771	6	1016
India	11	15732	62311	21956	53	100063
Oceania	0	32	0	598	0	630
N & E Africa	0	25	12	1207	0	1244
S Africa	5	61	11028	983	8	12085
W & C Africa	20	65	5434	924	0	6443
C America	30	57	80	67	0	234
N America	14	207	68	291	12	592
S America	0	43	25	312	98	478

[†] Others includes supplies to commercial companies, non-governmental organizations and regional institutes.

Groundnut shells are used in many ways: as fuel, conditioner for heavy soil, filler in cattle feed, a raw source of activated carbon, combustible gases, organic chemicals, reducing sugars, alcohol and extender resins, a cork substitute and a component of building block and hardboard. The use of shells as mulch or manure is beneficial in areas of scarce rainfall. Residue left after furfurol extraction makes good compost after treatment with H_2SO_4 and neutralization with tricalcium phosphate. Groundnut hay (haulm) is used for livestock feed. Nutritionally it is not superior to alfalfa but is comparable to or better than grasses. Recently there has been some interest in exploitation of wild *Arachis* species for forage. *Arachis glabrata* and *A. pintoi* have been released as forage species in Australia, Brazil and the USA. These species are good sources of protein for livestock in grazing lands.

Breeding

Breeding procedures in use for cultivar development in groundnut are those generally used for self-pollinated crops. A real boost to groundnut breeding came with the perfection of a field hybridization procedure (Nigam *et al.* 1990). A modified pedigree (single-seed descent) procedure has been followed in some countries with good success (Hildebrand 1985). Only limited use has been made of the recurrent selection procedures (Wynne and Gregory 1981), owing to space and hybridization requirements. However, intercrossing of derived breeding lines is commonly resorted to, and it represents a delayed recurrent selection programme. Backcross breeding has been used to a limited extent to transfer simply inherited traits such as resistance to groundnut rosette virus disease into adapted cultivars (Gibbons 1969). Resistance to most of the diseases and insect pests is not simply inherited. Bulk and bulk pedigree methods are extensively used in regional and international programmes to retain variability in breeding populations for exploitation by the breeders in collaborating countries.

Prospects

Most of the cultivars of groundnut stand on a very narrow genetic base, either because of non-availability of varied sources from different centres of diversity or lack of proper characterization of available groundnut genetic resources. Utilization of available genetic resources can be improved with the use of some advanced molecular marker techniques in genetic characterization and identification of uniqueness of genotypes.

Poor partitioning, particularly in runner-type landraces of subsp. *hypogaea*, has resulted in the erosion of these types from traditional production systems all over the world. Besides causing ecological imbalance, this loss also has resulted in soil degradation because of lack of soil-binding capacity in the introduced erect types.

Correction is required, either through simple *in situ* conservation of runner types or an on-farm *in situ* conservation with a mixture of local and introduced high-yielding runner types in these areas.

With the development of new mixed production systems and agroforestry, the lack of tolerance to shade, heat and cold could become a limitation in the further spread of groundnut. South America has extensive areas where groundnut can be cultivated but is not, because of acid soils and deficiency of micronutrients. Evaluation of genetic diversity for resistance to these constraints could help overcome them and spread groundnut to new production systems and areas.

Biotic and abiotic stresses reduce yield, and some of these are amenable to chemical and cultural controls but such approaches are not always possible in low-input rain-fed agriculture. Abiotic stresses such as drought, nutrient toxicity, nutrient deficiency and low pH have received very little attention in breeding. Genetic amelioration to stresses, through exploitation of host-plant resistance by conventional methods, has been successful in many cases, such as rust, late leaf spot and groundnut rosette virus diseases, but much more remains to be done. The wild *Arachis* species are resistant to many stresses such as early leaf spot, peanut stripe virus, peanut bud necrosis virus, *Spodoptera* and leafminer where the genetic variation in *A. hypogaea* is limited. Advancement in molecular techniques related to the groundnut crop is expected to overcome some of the barriers to gene transfer between wild and domesticated species.

Lack of adapted cultivars is often cited as one of the major constraints in increasing groundnut production and further spread of groundnut cultivation to new areas. Development of location-specific, improved germplasm with stable performance will require more attention by breeders. In sub-Saharan Africa, which is characterized by increased frequency of a shorter and less reliable wet season, extra-early genotypes are required if groundnut is to maintain its position there.

Limitations

Compared with the leading agricultural crops of the world the groundnut crop remains poorly researched and as such suffers from many limitations. In spite of significant developments in crop improvement research, benefits have been very slow to reach small farmers because of low seed multiplication rates and the bulky nature of the seed of this crop. These limitations have made the crop unattractive to the commercial seed sector. Public sector seed-producing agencies have not been able to meet the demand for improved seed.

Since most of the Ca requirement of the developing pod and seed is met by direct absorption, the availability of moisture in the first 8-10 cm of topsoil at the pod-developing stage is crucial for high yield. The crop is not able to make use of soil moisture in deeper layers in a productive manner in spite of its availability at the pod-developing stage. In certain soil types and under conditions of end-of-season drought, the harvesting of pods becomes very difficult. Because the crop is indeterminate the crop lacks uniformity in maturity of pods at the time of harvest. Consequently, the economic yield is reduced owing to discarding of immature pods.

The current world trade of groundnut is mainly in edible types. With increasing health consciousness, the high oil content of groundnut becomes a limitation in food trade. Susceptibility of groundnut to aflatoxin contamination is a serious health hazard. With better crop husbandry, tolerant genotypes and appropriate post-harvest technology, this problem can be overcome to a large extent. However, groundnut remains predominantly a rain-fed crop making it vulnerable to aflatoxin contamination in the field due to drought stress at the pod-developing stage. Groundnut as an oilseed crop is losing its competitiveness with other oilseed crops in many countries. If groundnut is to maintain its position as a leading oilseed crop, its productivity under low-input rain-fed agriculture will have to increase to meet the requirements and expectations of small-scale farmers in developing countries, where it continues to be a labour-intensive crop.

To exploit its full potential on the world food market, the crop will have to move away from the subsistence level. As a food crop, complete freedom from aflatoxin and chemical residues in the produce will be an essential requirement. To meet this requirement, crop husbandry including curing and drying will require more attention. Crop husbandry will have to be more environmentally friendly with less dependence on agrochemicals. Quality considerations also will become more important: seed shape, size and colour, low oil content, better taste and flavour and increased shelf-life.

Diversification in crop uses and development of new groundnut products will help increase groundnut demand in the world market. Despite relatively low nutritional value, groundnut protein has unique functional properties, such as low solution viscosity and relatively high concentration (5-10%), good compatibility with bread dough systems, white colour and bland flavour. In view of this, opportunities exist for the food industry to manufacture defatted groundnut flours, groundnut protein isolates and concentrates as well as a wide range of food products, which might include vitamin-fortified infant food, precooked dehydrated foods, groundnut bread, groundnut cheese and groundnut milk. Texturized groundnut protein can provide an excellent substitute for expensive animal protein to meet the food requirements of developing countries in Africa and Asia. Many less-industrialized countries do not produce enough vegetable oil to meet domestic demand. It is usually in rural areas where deficit occurs because of the cost of transportation and distribution. The development and introduction of technologies for processing of groundnut oil on a small scale for use in rural areas will help alleviate the short supply of edible oil and generate employment, adding value to agricultural production and developing local engineering skills for rural agro-industrial development.

Groundnut shells can be processed for economically useful purposes such as in the manufacture of activated charcoal, biogas, alcohol, extender resins, cork substitute and hardboards. The manufacture of adhesive glues, fire-extinguishing liquid and water-resistant powder from groundnut press cake has not yet been exploited commercially. Groundnut has good potential to move into new and non-traditional areas of cultivation. However, suitable cultivars will have to be tailored for such areas. In the rice- and wheat-based cropping systems, groundnut can play a significant role together with other legumes in restoring the balance in soil fertility and arresting the decline in productivity of the system. The use of wild *Arachis* species, which have great diversity for growth forms and adaptation, in forage production is another potential area for future research and conservation in forage germplasm banks. To date only three species have been recognized and commercialized for forage production. These species have demonstrated high yields and high quality of forage, high palatability, excellent haymaking quality, persistence under intensive grazing, tolerance to low fertility and high aluminium and manganese, good drought tolerance and minimal loss due to pest and diseases, which are essential for good forage and successful animal production.

REFERENCES

Abdou, Y.A.M., W.C. Gregory and W.E. Cooper. 1974. Sources and nature of resistance to *Cercospora arachidicola* and *Cercosporidium personatum* in *Arachis* spp. Peanut Sci. 1: 6-11.

Anonymous. 1985. Report of co-operative research between the International *Meloidgyne* project (IMP) and the International Crops Research Institute for the Semi-Arid Tropics (ICRISAT), North Carolina State University, Raleigh.

Baltensperger, D.D., G.M. Prine and R.A. Dunn. 1986. Root-knot nematode resistance in *Arachis glabrata*. Peanut Sci. 13:78-80.

Banks, D.J. 1990. Hand-tripped flowers promoted seed production in *Arachis lignosa*, a wild peanut. Peanut Sci. 17:22-24.

Basha, S.M. 1991. Deposition pattern of methionine-rich protein in peanuts. J. Agric. Food Chem. 39:88-91.

Culbreath, A.K., J.W. Todd, W.D. Branch, D.W. Gorbet, C.C. Holbrook, W.F. Anderson and J.W. Demski. 1994. Variation in susceptibility to Tomato Spotted Wilt Virus among Peanut genotypes. *In* Proceedings of American Peanut Research and Education Society Inc. 26:49.

Culver, J.N., J.L. Sherwood and H.A. Melouk. 1987. Resistance to Peanut Stripe Virus in *Arachis* germplasm. Plant Dis. 71:1080-1082.

de Berchoux, C. 1960. La rosette de l'arachide en Haute Volta. Comportement des lignesresistantes. Oleagineux 15:237-239.

Dwevedi, S.L., S.N. Nigam, D.V.R. Reddy, A.S. Reddy and G.V. Ranga Rao. 1995. Progress in breeding groundnut varieties resistance to peanut bud necrosis virus and vector. Pp. 35-40 *in* Recent Studies on Peanut Bud Necrosis Disease (A.M.M. Buiel, J.E. Parlevliet and J.M. Lenne, eds.). Dept. of Plant Breeding, University of Wageningen / ICRISAT Center, Patancheru, India.

FAO. 1995. Year Book: Production 1994. 48:108-109.

Fitzner, M.S., S.C. Alderman and H.T. Stalker. 1985. Greenhouse evaluation of cultivated and wild peanut species for resistance to Cylindrocladium black rot. *In* Proceedings of American Peanut Research and Education Society Inc. 17:28.

Foster, D.J., H.T. Stalker, J.C. Wynne and M.K. Beute. 1981. Resistance of *Arachis hypogaea* L. and wild relatives to *Cercospora arachidicola* Hori. Oleagineux 36:139-143.

Foster, D.J., J.C. Wynne and M.K. Beute. 1980. Evaluation of detached leaf culture for screening peanuts for leaf spot resistance. Peanut Sci. 7:98-100.

Gibbons, R.W. 1969. Groundnut rosette research in Malawi. Pp. 1-8 *in* Third African Cereals Conference Zambia and Malawi. Mimeo Report.

Gibbons, R.W. and B.E. Bailey. 1967. Resistance to *Cercospora arachdicola* in some species of *Arachis*. Rhodesia, Zambia and Malawi. J Agric. Res. 5:57-59.

Gouk, H.P., J.C. Wynne and H.T. Stalker. 1986. Recurrent selection within a population from an interspecific peanut cross. Crop Sci. 26:249-253.

Green, C.C., M.K. Beute and J.C. Wynne. 1983. A comparison of methods of evaluating resistance to *Cylindrocladium crotalariae* in peanut field tests. Peanut Sci. 10:66-69.

Gregory, W.C. and M.P. Gregory. 1976. Groundnut. Pp. 151-154 in Evolution of Crop Plants (N.W.Simmonds ed.). Longman Group Ltd., London.

Gregory, W.C., M.P. Gregory, A. Krapovickas, B.W. Smith and J.A. Yarbrough. 1973. Structures and genetic resources of peanuts. Pp. 47-133 *in* Peanuts - Culture and Uses (C.T. Wilson, ed.). American Peanut Research and Education Association Inc., Stillwater, Oklahoma.

Gregory, W.C., A. Krapovickas and M.P. Gregory. 1980. Structures, Variation, Evolution and Classification in *Arachis*. Pp. 469-481 *in* Advances in Legume Science (R.J. Summerfield and A.H. Bunting, eds.). Royal Botanic Gardens, Kew.

Hammons, R.O. 1994. The origin and history of the groundnut. Pp. 24-42 *in* The Groundnut Crop: A Scientific Basis for Improvement (J. Smartt, ed.). Chapman & Hall, London.

Hildebrand G. 1985. Use of single-seed descent method of selection in groundnut breeding in Zimbabwe. Pp. 137-140 *in* Proceedings of the Regional Groundnut Workshop for Southern Africa, 26-29 March, 1984 Lilongwe Malawi. ICRISAT, Patancheru, India.

Holbrook, C.C. and J.P. Noe. 1990. Resistance to *Meloidogyne arenaria* in *Arachis* spp. and the implications on development of resistant peanut cultivars. Peanut Sci. 17:35-38.

IBPGR and ICRISAT. 1992. Descriptors for groundnut. International Board for Plant Genetic Resources, Rome, Italy; International Crops Research Institute for the Semi-Arid Tropics, Patancheru, India.

ICRISAT. 1983. Annual Report 1982. ICRISAT, Patancheru, India.

ICRISAT. 1985. Annual Report 1984. ICRISAT, Patancheru, India.

Krapovickas, A. 1969. The origin, variability, and spread of the groundnut (*Arachis hypogaea*). Pp. 427-440 *in* The Domestication and Exploitation of Plants and Animals (R.J. Ucko and C.W. Dimbleby, eds.). Duckworth, London.

Krapovickas, A. 1973. Evolution of the genus *Arachis*. Pp. 135-157 *in* Agricultural Genetics: Selected Topics (R. Moav, ed.). National Council for Research and Development, Jerusalem.

Krapovickas, A. and W.C. Gregory. 1994. Taxonomia del genero *Arachis (Leguminosae)*. Bonplandia VIII:1-187.

Lynch, R.E. 1990. Resistance in peanut to major arthropod pests. Florida Entomol. 73:422-445.

Lynch, R.E., W.D. Branch and J.W. Garner. 1981. Resistance of *Arachis* species to the Fall Armyworm, *Spodeptera frugiperd* A. Peanut Sci. 8:106-109.

Mehan, V.K. 1989. Screening of groundnuts for resistance to seed invasion by *Aspergillus flavus* and to aflatoxin production. Pp. 323-334 in Aflatoxin Contamination of Groundnut: Proceedings of the International Workshop, 6-9 October 1987, ICRISAT Center, India. ICRISAT, Patancheru, India.

Mehan, V.K., B.S. Lio, Y.J. Tan, A. Robinsmith, D. McDonald and A.C. Hayward. 1994a. Bacterial wilt of groundnut [In English, Summaries in English, French]. Information Bulletin no. 35. ICRISAT, Patancheru, India.

Mehan, V.K., D.D.R. Reddy and D. McDonald. 1993. Resistance in groundnut genotypes to kalahasti malady caused by the stunt nematode, *Tylenchorhynchus brevilineatus*. Int. J. Pest Manage. 39:201203.

Mehan, V.K., P.M. Reddy, K. Vidyasagar Rao and D. McDonald. 1994b. Components of rust resistance in peanut genotypes. Phytopathology 84:1421-1426.

Melouk, H.A., D.J. Banks and M.A. Fanous. 1984. Assessment of resistance to *Cercospora arachidicola* in peanut genotypes in field plots. Plant Dis. 68:395-397.

Moss, J.P, A.K. Singh, P. Subrahmanyam, G.L. Hildebrand and A.F. Murant. 1993. Transfer of resistance to groundnut rosette disease from wild *Arachis* species into cultivated groundnut. Int. *Arachis* Newsl. 13:22-23.

Nageswara Rao, R.C., A.K. Singh, L.J. Reddy and S.N. Nigam. 1994. Prospects for utilization of genotypic variability for yield improvement in groundnut. J. Oilseeds Res. 11:259-268.

Natrajan, K.R. 1980. Peanut protein ingredient, preparation, properties and food uses. Adv. in Food Res. 26:215-273.

Nelson, S.C., J.L. Starr and C.E. Simpson. 1988. Resistance to *Meloidogyne arenaria* in exotic germplasm of the genus *Arachis*. J. Nematology 20:651

Nigam, S.N., M.J. Vasudeva Rao and R.W. Gibbons. 1990. Artificial hybridization in Groundnut. Information Bulletin no. 90. ICRISAT, Patancheru, India.

Pancholy, S.K., A.S. Deshpande and S. Krall. 1978. Amino acid, oil and protein content of some selected peanut cultivars. *In* Proceedings of American Peanut Research and Education Association Inc. 10:3-37.

Prasada Rao, R.D.V.J., A.S. Reddy, S.K. Chakrabarty, D.V.R. Reddy, V.R. Rao and J.P. Moss. 1991. Identification of Peanut Stripe Virus resistance in wild *Arachis* germplasm. Peanut Sci. 18:1-2.

Singh, A.K. 1986. Utilization of wild relatives in genetic improvement of *Arachis hypogaea* L. 8. Synthetic amphidiploids and their importance in interspecific breeding. Theor. Appl. Genet. 72:433-439.

Singh, A.K. 1988. Putative genome donors of *Arachis hypogaea (Fabaceae)*, evidence from crosses with synthetic amphidiploids. Plant Syst. Evol. 160: 143-151.

Singh, A.K. 1995. Groundnut, *Arachis hypogaea* Lin. (Leguminosae-Papilionoideae). Pp. 246-250 *in* Evolution of Crop Plants (N.W. Simmonds and J. Smartt, eds.). 2nd ed. Longman Group Ltd., London.

Singh, A.K. and C.E. Simpson. 1994. Biosystematics and genetic resources. Pp. 96-137 *in* The Groundnut Crop: A scientific basis for improvement (J. Smartt ed.). Chapman & Hall London.

Singh, A.K., V.R. Rao and M.H. Mengesha. 1992. Groundnut genetic resources: progress and prospects. Pp. 297-309 *in* Groundnut - A Global Perspective (S.N. Nigam ed.). Proceedings of an International Workshop 25-30 Nov. 1991, ICRISAT Center, India. ICRISAT, Patancheru, India.

Smartt, J., W.C. Gregory and M.P. Gregory. 1978. The genome of *Arachis hypogaea* L. Cytogenetic studies of putative genome donors. Euphytica 27:665-675.

Smith, O.D., T.E. Boswell, W.J. Gricher and C.E. Simpson. 1989. Reaction of selected peanut (*Arachis hypogaea* L.) lines to southern stem rot and *Pythium* pod rot under varied disease pressures. Peanut Sci. 16:9-13.

Stalker, H.T. and W.V. Campbell. 1983. Resistance of wild species of peanut to an insect complex. Peanut Sci. 10:30-33.

Stalker, H.T., W.V. Campbell and J.C. Wynne. 1984. Evaluation of cultivated and wild peanut species for resistance to the lesser cornstalk borer (*Lepidoptera: pyralidae*). J. Econ. Entomol. 77:53-57.

Subrahmanyam, P., A.M. Ghanekar, B.L. Nolt, D.V.R Reddy and D. McDonald. 1985. Resistance to groundnut diseases in wild *Arachis* species. Pp. 49-55 *in* Proceedings of the International Workshop on Cytogenetics of *Arachis*, 31 October-2 November 1983, ICRISAT Center, India. ICRISAT, Patancheru, India.

Subrahmanyam, P., D. McDonald, F. Waliyar, L.J. Reddy, S.N. Nigam, R.W. Gibbons, V. Ramnath Rao, A.K. Singh, S. Pande, P.M. Reddy and P.V. Subba Rao. 1995. Screening methods and sources of resistance to rust and late leaf spot of groundnut. ICRISAT Information Bulletin No. 47. ICRISAT, Patancheru India.

Subrahmanyam, P., R.W. Gibbons, S.N. Nigam and V.R. Rao. 1982. Resistance to rust and late leaf spot diseases in some genotypes of *Arachis hypogaea*. Peanut Sci. 9:6-10.

Wightman, J.A. and G.V. Ranga Rao. 1994. Groundnut pests. Pp. 395-469 *in* The Groundnut Crop: A Scientific Basis for Improvement (J. Smartt, ed.). Chapman & Hall, London.

Wightman, J.A., K.M. Dick and G.V. Ranga Rao. 1989. Pests of groundnut in the semi-arid tropics. Pp. 243-322 *in* Insect Pests of Food Legumes (S.R. Singh, ed.). John Wiley and Sons, New York.

Woodroof, J.G. 1983. Peanut Production, Processing, Products, 3rd edn. Avi Publishing Company Inc. Westport, Connecticut.

Wynne, J.C. and W.C. Gregory. 1981. Peanut Breeding. Adv. Agron. 34:39-72

Young, C.T., R.S. Matlock, M.E. Mason and G.R. Waller. 1974. Effect of harvest date and maturity upon free amino acid levels in three varieties of peanuts. J. Am. Oil Chemist's Soc. 51:269-273.

Chapter 10

Lentil

L.D. Robertson and W. Erskine

Lentil (*Lens culinaris* Medikus) is a dietary mainstay and one of the principal pulse crops in the drier regions of the Middle East, North Africa and the Indian subcontinent. The seed provides an important source of protein to people of these regions, where lentil straw is valued for animal production. Lentil is grown to a lesser extent in southern Europe and the Americas, and as a field legume it is usually grown in rotation with cereals. The major factor in the domestication of lentil has been selection pressure for an appropriate phenology (Erskine *et al.* 1989). This force still drives the ICARDA breeding strategy. Most accessions of lentil in the ICARDA collection came from the West Asia and North Africe (WANA) region, which is the centre of origin and primary diversity (Zohary and Hopf 1988). The strategy has led to the successful use of landraces from the collection for direct release as cultivars for the WANA region and beyond. Separate programmes target improvements for the diverse environments in which lentil is grown in the developing world.

BOTANY AND DISTRIBUTION
Lentil is derived from the genus *Lens*, which describes the shape of the cultivated lentil seed. The genus *Lens* Miller belongs to the order Rosales, suborder Rosinae, family Leguminosae and subfamily Papilionaceae, in the tribe Vicieae (Kupicha 1981). *Lens* is characterized by small-flowered, low annual herbs. Cultivated lentil is a slender, pilose annual, 20-40 cm tall, long-day plant. All species in the genus are diploid with $2n=14$ and have similar karyotypes. Wild *Lens* has four taxa (Ladizinsky 1993):

  Lens culinaris subsp. *orientalis* (Boiss.) Ponert
  Lens odemensis Ladiz.
  Lens ervoides (Brign.) Grande
  Lens nigricans (M.Bieb) Godr.

Lens culinaris subsp. *orientalis* is fully crossable with the cultigen and is considered to be the lentil progenitor (Zohary 1972; Ladizinsky 1979; Ladizinsky *et al.* 1984). Common names of lentil are *phakos* in Greek; *ads, adas* and *ades* in Arabic; *adis* and *bersim* in Berber; *merjimek* and *mecumeck* in Turkish; *lenteje* in Spanish; *masur dhal* in Hindi; *lentille* in French; *chechevitza* in Russian and *showpindu* in Chinese. Many researchers still use *macrosperma* and *microsperma* as subspecies (Baurlina 1930) to distinguish seed sizes of 6-9 mm and 2-6 mm (Muehlbauer *et al.* 1995).

Origin, Domestication and Diffusion
Baurlina (1930) suggested the Hindu Kush-Himalaya border region as the centre of origin. Because wild species evidence does not support this hypothesis, the border may be a secondary centre of diversity. This mountainous area has many isolated valleys in which

the effects of genetic drift and gene fixation could have led to the high diversity of endemic varieties Baurlina found.

Wild progenitors and relatives are primarily found in the West Asian region (Cubero 1981). *Lens culinaris* subsp. *orientalis* (Boiss.) Ponert is endemic to this region, although it has probably been distributed to some extent as a weed with the cultigen in the Asian Republics of the former Soviet Union.

Archaeological records of carbonized lentil seed reveal the pattern of dissemination from its centre of origin. The crop was cultivated as early as 8000 BC in the Middle East (Zohary and Hopf 1988) and domesticated in the Lidia-Kurdistania region (Cubero 1981; Ladizinsky 1979) along the Near East Arc with einkorn, emmer and barley (Ladizinsky 1979; Vavilov 1992). It spread to Cyprus and southeastern Europe in 6000 BC, to Central Europe via the Danube, with wheat and barley to Egypt and with Hamitic invaders to Ethiopia. To the east, lentil reached Georgia in 5000 BC and early 4000 BC, then India and Pakistan around 2000 BC. It reached the Americas after their discovery.

Reproductive Biology

Lentil is a self-pollinated species with little outcrossing (Wilson and Law 1972), although up to 6.6% outcrossing has been reported (Erskine and Muehlbauer 1991). Flowers are complete and borne singly or in multiples on peduncles from the upper part of the plant.

GERMPLASM CONSERVATION AND USE

The primary lentil genepool consists of the cultigen and its progenitor, *L. culinaris* subsp. *orientalis*. Germplasm available for the cultigen at ICARDA is listed in Table 10.1. The early ICARDA years focused on collecting and assembling a large germplasm base. Today, 21% of the ICARDA lentil accessions are landraces collected by ICARDA and ALAD (Arid Lands Agricultural Development Program, the precursor organization of ICARDA). Breeding programmes have built upon the foundation of the germplasm collections.

Seed Conservation

Because lentil has less than 1% outcrossing (Wilson and Law 1972), it is easily grown for multiplication and rejuvenation in adjacent plots without taking measures to control pollination. The seeds of wild *Lens* species shatter, so these plants are grown in plastic or glasshouses and bagged at flowering. Lentil and its wild relatives have orthodox seeds that are easily maintained in the ICARDA genebank in active and base collections. Seeds are dried to 6% moisture content before storage at -20°C in the base collection and 0°C/16% RH in the active collection. Cultivated lentil germplasm is tested on a schedule to ensure that seed is pathogen- and virus-free (Khare 1981; Bos *et al.* 1988).

ICARDA has signed an agreement for the safety-duplication of its cultivated lentil collection with the National Bureau of Plant Genetic Resources (NBPGR) in New Delhi, India, prepared and stored 91% of its collection there, and will store the remainder after regeneration for seed supply and to obtain proper viability for storage.

Germplasm Characterization and Evaluation

Abiotic and biotic stresses affect lentil, and sources of resistance are being identified (Table 10.2). In addition, resistance to more than one stress is often needed for the farming systems, and several accessions have resistance to two or more diseases (Table 10.3).

Rust (induced by *Uromyces fabae* (Pers.) de Bary) is the most important foliar disease. Epiphytotics of rust are common in Chile, Ecuador, Ethiopia, India, Morocco and Pakistan (Erskine and Saxena 1993). ICARDA screens for rust resistance through joint research with the national programmes of Ethiopia, Morocco and Pakistan. As a result of this effort, rust-resistant cultivars have been released in Chile, Ecuador, Ethiopia, Morocco and Pakistan. An international nursery for rust resistance was initiated with national programmes in 1990 to clarify the host-pathogen relationships in different regions and to assist in

Table 10.1. Current status of the cultivated lentil germplasm collection at ICARDA by country of origin[†].

Country	No. access.	Country	No. access.	Country	No. access.
AFG	142	ETH	380	PAL	2
ALB	2	FRA	9	PER	2
ARG	11	GBR	1	POL	18
ARM	9	GRC	104	PRT	14
AUS	2	GTM	1	ROM	3
AZE	4	HUN	28	RUS	31
BEL	2	IND	1854	SAU	3
BGD	36	IRN	912	SDN	2
BGR	41	IRQ	24	SOM	1
BLR	1	ITA	12	SUN	82
BRA	4	JOR	392	SYR	1206
CAN	4	KAZ	1	TJK	6
CHL	350	LBN	76	TUN	21
CHN	3	LBY	1	TUR	412
COL	8	MAR	89	UKR	14
CRI	2	MEX	45	URY	1
CSK	21	MUS	1	USA	34
CYP	28	NLD	1	UZB	1
DEU	32	NOR	1	YEM	60
DZA	35	NPL	263	YUG	25
EGY	98	NZL	1	Not known	106
ESP	180	PAK	222	**Total**	**7477**

[†] Origin country codes as per ISO (International Standardization Organization) three-letter country codes.

Table 10.2. Resistant sources to biotic and abiotic stresses in lentil.

Stress	Resistant sources
Fusarium wilt	ILL 241, 632, 813, 1712, 4403, 5714, 5871, 5883, 6024, 6025, 6410, 6427, 6458, 6461, 6797, 6976, 6991, 7005, 7012, 7180, 7192, 7193, 7199, 7204
Rust	ILL 358, 4605, 5604, 6002, 6209
Ascochyta blight	ILL 358, 2439, 5244, 5480, 5588, 5597, 5684, 5714, 5725, 5755, 6258
Winter hardiness	ILL 52, 465, 468, 590, 662, 780, 857, 975, 1878, 1918

Table 10.3. Some examples of multiple disease resistance in lentil with combinations of resistance to Ascochyta blight, rust and vascular wilt.

Accession no. (ILL)	Resistance reaction[†]		
	Ascochyta blight	Rust	Vascular wilt
2439	R	S	R
4605	MR	R	S
5588	R	S	R
5714	R	S	R
5871	R	na	R
5883	T	R	R
6024	MR	R	R
6025	R	S	R
6212	MR	R	S
6258	R	MR	R
6264	na	R	R
6458	R	S	R

[†] R=Resistant, MR=Moderately resistant, T=Tolerant, S=Susceptible, na=Data not available.

identifying variation in the fungus. Vascular wilt (induced by *Fusarium oxysporum* f.sp. *lentis* Vasd. and Srin.) is the most important soilborne disease of lentil in the Mediterranean region and also causes major yield losses in the Indian subcontinent. Bayaa and Erskine (1990) have developed an efficient screening method for vascular wilt in lentil and identified several useful sources of resistance. Ascochyta blight (induced by *Ascochyta lentis* Bond. and Vassil.) causes losses in productivity in WANA, Ethiopia, parts of the

Indian subcontinent and Canada. Losses affect not only the standing crop but also seed quality from infection in the swathe. Good sources of resistance to ascochyta blight have been identified in cooperation with the National Agricultural Research Centre, Islamabad, Pakistan and are now being used in the breeding programme. Although more than 1774 accessions of lentil have been screened for *Orobanche crenata* resistance, no useful source of resistance has been found.

At higher altitudes, experiments have shown that changing from spring-sown to winter-sown lentil can increase yields by 50-100% (Sakar *et al.* 1988). Screening for cold tolerance started in 1980 near Ankara, Turkey, in cooperation with the national breeding programme (Erskine *et al.* 1981). Winter survival of lentil often requires tolerance to factors other than cold, such as frost-heaving, waterlogging and diseases caused by root pathogens and ascochyta blight. This screening has resulted in identification of sources of winterhardiness. With the onset of spring, rising temperatures lead to higher evapo-transpiration and drought stress that coincide with the crop's reproductive growth stage, and thus yields are often low. Drought avoidance by earliness is the only effective trait for resistance to drought in the cultigen (Hamdi *et al.* 1992; Silim *et al.* 1993). Erskine *et al.* (1993) have found differences in country of origin for iron deficiency when screening the ICARDA germplasm collection. Lentil accessions from the Fertile Crescent expressed iron deficiency symptoms at much lower frequencies than those from areas with warmer climates.

The cultivated lentil germplasm collection also has been evaluated for agronomic and stress descriptors over the past 15 years. A catalogue describes the evaluation of the first 4550 accessions in the collection (Erskine and Witcombe 1984). Geographic patterns of diversity emerge when the data are analyzed by country of origin. Most significant of these patterns was time to flowering. Germplasm from the Indian subcontinent flowered and matured early. This material was also characterized by small-seeded types, an expected result because *microsperma* types are preferred in this region. Data collected for an additional 1859 accessions in which 20 descriptors were evaluated are being prepared for publication in catalogue form. Summary statistics (Table 10.4) reveal that the largest variation for quantitatively scored descriptors was for the yield descriptors seed yield (SYLD), biomass (BYLD), straw yield (STYLD) and 100-seed weight (HSW). Days to flowering, plant height, seed yield and straw yield showed normal distributions.

Detailed analyses also have been made for quantitative traits in germplasm evaluation trials, with significant differences found for all characters, including straw and seed yield, between countries of origin (Erskine 1983; Erskine *et al.* 1989). This information was used to separate the lentil genepool into three major regional groups:

1. a Levantine group (Egypt, Jordan, Lebanon and Syria)
2. a northern group (Greece, Iran, Turkey and the former USSR)
3. accessions from India and Ethiopia.

Trials have been conducted to characterize the responses to temperature and photoperiod for time to flowering of lentil (Erskine *et al.* 1990a, 1994). Most of the variation among accessions in time to first flowering was attributed to country of origin and was dependent upon latitude of origin.

Seed protein content varied from 18.6 to 30.2% (Erskine and Witcombe 1984). Cooking time has been determined in the world collection and showed a large variation (Singh *et al.* 1988). However, this variation was lost when pre-soaking was done before cooking, which is a common practice. Again, cooking time is not an objective in breeding programmes, although it is periodically determined in elite lines.

Lentil originated in the Fertile Crescent and its wild relatives are most common in this region (Cubero 1981; Ladizinsky 1979, 1993). The progenitor species of lentil, *Lens culinaris* subsp. *orientalis*, is endemic to this region. Collections of wild *Lens* are much smaller than those of the cultigens, but the one at ICARDA is the largest in the world (Table 10.5) and has been evaluated for morphological and agronomic descriptors (Table 10.6). Mean values for the quantitatively scored descriptors for each species are given in Table 10.7. High levels of winterhardiness have been found within *L. culinaris* subsp. *orientalis* (ICARDA 1994). The wild subspecies of *Lens* were found to be more drought resistant

Table 10.4. Summary statistics for 1859 lentil germplasm accessions evaluated at Tel Hadya, Syria during 1991/92.

Descriptor[†]	Check mean	Mean	Minimum	Maximum	CV (%)
DFLR[a] (days)	122.4	122.8	106	147	4.8
DMAT (days)	162.2	162.5	142	184	4.4
PTHT (days)	23.8	22.8	6	41	19.6
HTFP (cm)	12.4	11.1	1	26	33.7
HI (%)	44.3	42.0	2.0	75.8	18.8
HSW (g)	5.5	4.3	1.6	10.1	40.9
SYLD (kg/ha)	1302	976	23	3453	43.8
BYLD kg/ha)	2924	2325	360	7293	40.3
STYLD (kg/ha)	1623	1349	207	4667	43.2

[†] DFLR=days to flowering; DMAT=days to maturity; PTHT=plant height; HTFP=height to 1st pod; HI=harvest index; HSW=100-seed weight; SYLD=seed yield; BYLD=biomass yield; STYLD=straw yield.

Table 10.5. Frequency by country of origin[†] of the ICARDA wild *Lens* germplasm collection.

Taxa	Country of origin	No. of accessions	Taxa	Country of origin	No. of accessions
orientalis	CYP	3	ervoides	ITA	3
	IRN	7		JOR	6
	JOR	4		LBN	15
	LBN	4		PAL	3
	PAL	4		SYR	47
	SUN	4		TUR	36
	SYR	101		UKR	2
	TJK	5		YUG	14
	TKM	2			
	TUR	65	nigricans	ESP	11
	UZB	12		FRA	5
				ITA	3
odemensis	LBY	1		JOR	2
	PAL	2		TUR	13
	SYR	34		UKR	2
	TUR	11		YUG	7
			Total		429

[†] Origin country codes as per ISO (International Standarization Organization) three-letter country codes.

than the cultigens with their low relative reduction in yield with drought stress (ICARDA 1992, 1994). Two accessions of *L. culinaris* subsp. *orientalis* have high levels of resistance to vascular wilt (Bayaa *et al.* 1995). Twenty-four accessions of *L. culinaris* subsp. *orientalis* have high levels of resistance to ascochyta blight (Table 10.8) (Bayaa *et al.* 1994).

Gaps in Germplasm Collections

Gaps remain in the collection, especially for wild accessions of *L. culinaris* subsp. *orientalis*. The cultivated lentil collection of ICARDA has incomplete passport data for some regions, especially Iran and India, and the true origin of many accessions from these countries is questionable. Extensive effort is being made to document the ICARDA genetic resources to clarify the status of the passport data of the collections to make their use more efficient. Within the WANA region, data from Algeria and to a lesser extent Tunisia are incomplete. Outside this region, the major gap is for Bangladesh; however, a collecting mission just completed in April 1995 will probably complete this information. Specific regions within a country may not be adequately covered, and when updating the passport data these regions need to be addressed. A major gap in the germplasm collections of lentil (both cultivated and wild) is the lack of accessions with a high biomass. Low biomass restricts the yield ability of the crop and reduces its competitiveness against weeds. There also is a lack of resistance for pea leaf weevil (*Sitona* sp.) and the parasitic weed broomrape (*Orobanche* sp.).

Table 10.6. Frequency distributions for some discretely scored descriptors for 316 accessions of wild *Lens* evaluated at Tel Hadya, Syria in 1991/92.

Descriptor/	*Lens* species (%)			
score	*orientalis*	*odemensis*	*nigricans*	*ervoides*
Flower colour				
White	0.6	0.0	0.0	0.0
White, blue veins	2.6	0.0	0.0	0.0
Blue	1.3	0.0	0.0	0.0
Violet	91.0	100.0	60.0	93.5
Pink	0.6	0.0	40.0	0.0
Other	3.9	0.0	0.0	6.5
Pod pigmentation				
Absent	64.5	26.5	88.6	93.2
Slight, some pods	27.1	14.7	11.4	5.7
Slight, most pods	7.7	14.7	0.0	1.1
Deep purple	0.6	44.1	0.0	0.0
Growth habit				
Erect	3.4	5.3	3.0	0.0
Semi-erect	45.6	47.3	30.3	72.0
Semi-spreading	38.8	36.8	48.5	26.3
Spreading	12.3	10.5	18.2	1.8
Cotyledon colour				
Yellow	29.7	0.0	14.3	46.7
Orange/red	55.5	8.8	65.7	47.8
Olive-green	14.2	20.6	17.1	4.3
Mixed	0.6	70.6	2.9	1.1
Tendril				
Rudimentary	55.5	55.9	77.1	92.4
Prominent	44.5	44.1	22.9	7.6
Arista				
Present	5.8	11.8	8.6	80.2
Absent	94.2	88.2	91.4	19.8
Number of accessions	**155**	**34**	**35**	**92**

Table 10.7. Means for quantitative descriptors for 316 accessions of wild *Lens* species evaluated at Tel Hadya, Syria in 1991/92.

Descriptor/	*Lens* species[‡]				
score[†]	check	*orien.*[b]	*odem.*	*nigr.*	*ervo.*
DFLR (days)	103.1	106.3	106.2	120.0	112.3
DAP (days)	110.7	120.8	120.8	131.3	121.5
DDFLR (days)	43.2	47.2	47.1	43.1	41.1
DDAP days)	2.9	5.8	7.1	6.5	5.9
PTHT (mm)	241.7	136.8	102.7	99.3	86.3
PWX (mm)	224.4	215.3	143.6	178.4	108.6
PWY (mm)	187.9	209.3	132.2	165.1	101.5
FPP	17.1	11.8	12.1	12.0	10.2
INL (mm)	21.9	15.5	10.9	12.6	13.1
LPL	9.4	7.7	7.0	7.0	4.9
LFTL (mm)	11.6	8.0	6.0	7.5	8.8
LFTW (mm)	3.5	2.7	2.0	2.6	1.8
LFL (mm)	25.0	14.0	10.9	11.7	4.9
LPP	188.3	99.9	46.5	73.4	41.3
LAP (cm²)	338.0	104.7	23.7	58.1	18.4

[†] DFLR=days to flowering; DAP=days to podding; DDFLR=duration of flowering; DDAP=duration of podding; PTHT=plant height; PWX=plant width between rows; PWY=plant width within rows; FPP=flowers per plant; INL=internode length; LPL=leaflets per leaf; LFTL=leaflet length; LFTW=leaflet width; LFL=leaf length; LPP=leaves per plant; LAP=leaf area per plant.

[‡] *Orien.=orientalis; odem.=odemensis; nigr.=nigricans; ervo.=ervoides.*

Use and Distribution

Since 1989, ICARDA has distributed 6230 accessions of lentil and 638 accessions of the wild *Lens* species (Table 10.9), mostly to countries in the developing world. Selection was first made among, then within, locally adapted landraces, and they were distributed through the International Testing Network (ITN). To date, 18 cultivars of lentil have been released either from landraces or by direct selection from the germplasm collections (Table 10.10). Many of these selections also have been purified for such traits as seed size, disease resistance, frost tolerance and non-shattering. Major use of this material has been for screening and sources of resistance to rust, ascochyta blight and fusarium wilt. Other requests have been fulfilled to broaden the genetic base of the crop in particular countries or to supply wild species of *Lens* for research. Several genebanks in the ICARDA region, e.g. Iran, have been supplied with large numbers of wild *Lens* accessions to start their own collections.

Bhatty (1988) and Savage (1988) have reviewed composition and nutritive quality, and Table 10.11 lists dry matter composition. Lentil contains a number of antinutritional factors that are unimportant in human diets because they are heat-labile and deactivated by cooking. The chief food uses of lentil are summarized in Table 10.12. Processing includes cleaning, sizing, dehulling, splitting and polishing (Williams *et al.* 1993). The most common lentil-based foods have been summarized by Hawtin and Sears (1993). In the Middle East, lentil is cultivated for its straw in addition to grain. In Syria, a farmer's revenue from straw is sometimes greater than that from the grain. Straw enters both national and international trade as a livestock feed, particularly for sheep, contributing up to 20% of the diet of ewes from November to February. The straw comes from the traditional threshing process and includes broken branches, pod walls and leaflets. The protein content of lentil straw, whose genetic variation of quality is limited (Erskine *et al.* 1990b), varies from 5 to 7% and its digestible dry matter from 43 to 46%.

Breeding

The geographic distribution of landrace variation in the world lentil collection for morphological characters (Erskine and Witcombe 1984; Erskine *et al.* 1989), responses in flowering to temperature and photoperiod (Erskine *et al.* 1990a, 1994), winterhardiness (Erskine *et al.* 1981) and iron deficiency chlorosis (Erskine *et al.* 1993) collectively illustrate the specificity of adaptation in lentil. The ICARDA breeding programme is linked closely to national efforts aimed toward producing genetic material across regions (Table 10.13). Many cultivar releases by national programmes are selections from landraces in the ICARDA germplasm collection. For example, within the lowland, medium-rainfall Mediterranean region, selections from Jordanian landraces are registered in Iraq, Jordan, Lebanon, Libya and Syria; Lebanese germplasm has been released in southeastern Turkey, and a selection from a Syrian landrace has been released in Algeria and Tunisia (ICARDA 1995). In the highland region, releases in Turkey emanate from germplasm selections from highland Algeria and Iran. These registrations demonstrate the value of direct exploitation of landraces (Stage 1) as well as the underexploitation of the lentil.

A particular combination of characters required for specific regions is often not found 'on the shelf' in the germplasm collection. In such cases, hybridization and the bulk/pedigree breeding method with off-season generation advancement (at Terbol, Lebanon, 950 m asl), single-plant selection in the F_4 generation and selection in Syria and Lebanon have been used to produce Stage 2 material. Today, however, selection at ICARDA in West Asia is limited to adaptation to the home region – the Mediterranean, low-to-medium elevation – and for traits where ICARDA has a comparative advantage. This selection includes breeding for such traints as vascular wilt resistance, because selection locally for adaptation to a specific target area is the most efficient selection method.

For the other regions, Stage 3 crosses are agreed with cooperators and made at ICARDA, Tel Hadya. After off-season generation advance, segregating populations are shipped to national cooperators for local selection. Since 1985, specific crosses have been produced for the national programmes of Algeria, Bangladesh, India, Jordan, Morocco,

Table 10.8. Sources of resistance in *Lens orientalis.*

Disease	Source of resistance
Fusarium wilt	ILWL 79, 113
Ascochyta blight	ILWL 4, 7, 69, 77, 80, 84, 86, 88, 93, 94, 117, 121, 146, 180, 181, 248, 257, 277, 302, 304, 315, 330, 331
Cold	ILWL 89, 91, 97

Table 10.9. Distribution of cultivated lentil and wild *Lens* species from the ICARDA collections, 1990-92.

Country code[†]	Cultivated	Wild	Total	Country code[†]	Cultivated	Wild	Total
1990				**1993**			
ARG	18	–	18	AUT	4	–	4
GBR	33	39	72	BGR	52	–	52
IND	10	12	22	CSK	–	3	3
POL	23	–	23	ETH	66	–	66
RUS	27	4	31	IND	301	–	301
Total	**111**	**55**	**166**	IRQ	20	–	20
1991				ITA	19	–	19
AUS	495	44	539	JPN	–	8	8
PAK	365	–	365	NZL	–	16	16
POL	–	36	36	PAK	1000	–	1000
SYR	1	–	1	SYR	25	6	31
Total	**861**	**80**	**941**	TUR	123	–	123
1992				**Total**	**1610**	**33**	**1643**
AUS	101	–	101	**1994**			
BEL	–	4	4	CSK	11	–	11
BGD	36	–	36	EGY	–	20	20
CAN	96	–	96	ETH	102	–	102
CSK	1	–	1	GBR	1	13	14
EGY	–	20	20	IND	303	–	303
IND	79	30	109	JOR	1549	–	1549
IRN	152	322	474	LBN	–	20	20
ITA	46	–	46	NPL	500	–	500
JOR	78	–	78	NZL	4	–	4
MDA	145	30	175	PAK	300	–	30
MEX	19	–	19	SYR	25	11	36
SYR	50	–	50	**Total**	**2795**	**64**	**2859**
TUR	1	–	1				
ZAF	49	–	49	**Grand**			
Total	**853**	**406**	**1259**	**Total**	**6230**	**638**	**6868**

[†] Origin country codes as per ISO (International Standardization Organization) three-letter country codes.

Table 10.10. Lentil cultivars selected from ICARDA's germplasm collections.

Country/ crop	Cultivar released	Year of release	Country/ Crop	Cultivar released	Year of release
Algeria	ILL 4400	1988	Morocco	ILL 4605 (Precoz)	1990
Canada	ILL 481 (Indian head)	1989	Nepal	ILL 4402	1989
Chile	ILL 5523 (Centinela)	1989	Pakistan	ILL 4605 (Manserha 89)	1990
Egypt	ILL 4605 (Precoz)	1990	Sudan	ILL 813 (Rubatab1)	1993
Ethiopia	ILL 358	1984	Tunisia	ILL 4400 (Neir)	1986
	NEL 2704	1993	Turkey	ILL 942 (Erzurum '89)	1990
Iraq	ILL 5582	1992		ILL 1384 (Malazgirt '89)	1990
Jordan	ILL 5582	1990		ILL 854 (Sazak'91)	1991
Libya	ILL 5582	1993	USA	ILL 784 (Crimson)	1991

Table 10.11. Ranges[†] for proximate composition (% of dry matter) of lentil (Savage 1988).

	Crude protein	Ether extract	Crude fibre	Nitrogen-free extract	Ash
Raw					
Whole seed	19.5-35.5	0.6-3.9	1.4-5.9	52.5-69.7	1.9-5.7
Testa	19.5-32.5	0.9-2.8	0.2-3.0	35.3-68.3	2.2-3.0
Cooked					
Whole seed	20.5-31.4	0.7-1.6	2.7-4.9	61.4-69.4	1.4-3.5
Testa	22.5-29.0	0.6-2.2	0.5-3.9	60.4-66.4	1.0-2.8

[†] All values converted to an oven-dried basis.

Table 10.12. Common lentil-based foods.

Type	Hulled	Dehulled
Macrosperma		
Green seeds	Green vegetable	
Mature seeds	Soup (whole seeds or lentil flour), fried lentil, *mujeddharah* (whole seeds with rice or cracked wheat)	
Microsperma		
Mature seeds	Soup (whole seeds)	Soup (splits), *masur dhal*, *khichri*

Nepal, Syria and Turkey. Random Amplified Polymorphic DNA markers linked to the genes for resistance to these diseases are also being explored to determine the potential of marker-assisted selection (Eujayl *et al.* 1995).

For autogamous small grains, pure lines are the usual product of breeding in the developed world, whereas in the developing world subsistence cropping is characterized by varietal mixtures. From a review of research in temperate regions under modern agriculture, it is clear that improved stability and lower disease severity are common features of mixtures compared with their components in monoculture (Lenné and Smithson 1995). Landraces possess considerable heterogeneity (Erskine and Choudhary 1986). To provide some heterogeneity within individual lines, the breeding programme at ICARDA for the West Asian lowlands is producing advanced-generation bulk lines, formed by bulking the progeny of early generation (F_3 or F_4) single-plant selections. Continued segregation in later generations at the remaining heterozygous alleles produces a low level of heterogeneity within bulked lines parallel to that within landraces.

The wild genepool also has been used to improve the cultivated lentil. After crossing lentil with its wild progenitor *L. culinaris* subsp. *orientalis* at ICARDA, 10 lines were selected from bulk segregating populations for distribution worldwide in the Lentil International Nursery Program in 1984. Among these selections, the small-seeded ILL *5700 ranked 3rd, 1st, 7th and 2nd for average yield among 24 entries of the Lentil International Yield Trial tested in 13-15 countries from 1985 to 1988. This selection with wild parentage has since been widely used in crossing to introgress wild genes into the cultivated plant. The other major current use of wild germplasm in the breeding programme is to introgress winterhardiness genes from *L. culinaris* subsp. *orientalis*.

Prospects

Exciting gains in sustainable production arise from the integration of a change in agronomic practice with using a new cultivar. In highland West Asia, lentil is sown in spring. Sowing in late autumn/early winter, however, coupled with the use of a winter-hardy cultivar gives yield advantages of approximately 50% (Sakar *et al.* 1988). Although agronomic constraints to winter-sown lentil in the Highlands require further research, this substantial yield increase can be expected in the future to augment the area and production in this agro-ecological zone.

In some areas around the Mediterranean, such as south Syria, lentil production has ceased primarily because of vascular wilt disease. The registration of wilt-resistant cultivars ILL 5883 in Syria and Talya 2 in Lebanon offers the prospect of a return of lentil

Table 10.13. Target agro-ecological regions of production of lentil and key breeding aims.

Region	Key traits for recombination
Mediterranean low to medium elevation	
1. 300-400 mm annual rainfall	Biomass (seed + straw), attributes for mechanical harvest and wilt resistance
2. <300 mm annual rainfall	Biomass, drought escape through earliness
3. Morocco	Biomass, attributes for mechanical harvest and rust resistance
4. Egypt	Seed yield, response to irrigation, earliness and wilt resistance
High elevation	
1. Anatolian highlands	Biomass and winter hardiness
2. N. African highlands	Seed yield and low level of winter hardiness
South Asia and E Africa	
1. India, Pakistan, Nepal and Ethiopia	Seed yield, early maturity, resistance to rust, ascochyta and wilt
2. Bangladesh	Seed yield, extra earliness and rust resistance

cultivation to such areas. The further spread of lentil harvest mechanization technology will probably also contribute to increased lentil production in West Asia in the future. In south Asia, because wilt, rust and ascochyta blight are key disease problems, large areas are left fallow over winter following the harvest of late paddy rice. Farmers need early maturing, disease-resistant varieties of lentil with late-sowing potential for such situations or when land becomes available for lentil planting after the monsoon floodwaters have subsided. As extra-early maturing cultivars with combined resistance to diseases become available, the prospect will open up a major expansion of lentil production in the rice-based cultivation system in the Indian subcontinent.

REFERENCES

Baurlina, H. 1930. Lentils of the U.S.S.R. and other countries. Bull. Appl. Genet. Plant Breed. (Leningrad) Suppl. 40:1-319.

Bayaa, B. and W. Erskine. 1990. A screening technique for resistance to vascular wilt in lentil. Arab J. Plant Pathol. 8:30-33.

Bayaa, B., W. Erskine and A. Hamdi. 1994. Geographic distribution of resistance to Ascochyta blight in wild lentil. Genetic Resour. and Crop Evolution 41:61-65.

Bayaa, B., W. Erskine and A. Hamdi. 1995. Evaluation of a wild lentil collection for resistance to vascular wilt. Genet. Resour. Crop Evolution (in press).

Bhatty, R.S. 1988. Composition and quality of lentil (*Lens culinaris* Medik.): A review. Can. Inst. Food Sci. Technol. 21(2):144-160.

Bos, L., R.O. Hampton and K.H. Makkouk. 1988. Viruses and virus diseases of pea, lentil, faba bean and chickpea. Pp. 591-615 *in* World Crops: Cool Season Food Legumes (R.J. Summerfield, ed.). Kluwer, Dordrecht, The Netherlands.

Cubero, J.I. 1981. Origin, taxonomy and domestication. Pp. 15-38 *in* Lentils (C. Webb and G. Hawtin, eds.). CAB, Farnham, UK.

Erskine W. 1983. The relationship between the yield of seed and straw in lentils. Field Crops Res. 7:115-121.

Erskine, W. and M.A. Choudhary. 1986. Variation between and within lentil landraces from Yemen Arab Republic. Euphytica 35:695-700.

Erskine, W. and F. Muehlbauer. 1991. Allozyme and morphological variability, outcrossing rate and core collection formation in lentil germplasm. Theor. Appl. Genet. 83:119-125.

Erskine, W. and M.C. Saxena. 1993. Breeding lentil at ICARDA for southern latitudes. Pp. 207-219 *in* Lentil in S. Asia (W. Erskine and M.C. Saxena, eds.). ICARDA, Syria.

Erskine, W. and J.R. Witcombe. 1984. Lentil Germplasm Catalog. ICARDA, Aleppo, Syria.

Erskine, W., Y. Adham and L. Holly. 1989. Geographic distribution of variation in quantitative characters in a world lentil collection. Euphytica 43:97-103.

Erskine, W., R.H. Ellis, R.J. Summerfield, E.H. Roberts and A. Hussain. 1990a. Characterization of responses to temperature and photoperiod for time to flowering in a world lentil collection. Theor. Appl. Genet. 80:193-199.

Erskine, W., A. Hussain, M. Tahir, A. Baksh, R.H. Ellis, R.J. Summerfield and E.H. Roberts. 1994. Field evaluation of a model of photothermal flowering responses in a world lentil collection. Theor. Appl. Genet. 88:423-428.

Erskine, W., K. Myveci and N. Izgin. 1981. Screening a world lentil collection for cold tolerance. LENS Newsl. 8:5-8.

Erskine, W., S. Rihawe and B.S. Capper. 1990b. Variation in lentil straw quality. Ann. Feed Sci. Technol. 28:61-69.

Erskine, W., N.P. Saxena and M.C. Saxena. 1993. Iron deficiency in lentil: Yield loss and geographic distribution in a germplasm collection. Plant and Soil 151:249-254.

Eujayl, I., M. Baum, W. Erskine, E. Pehu and F.J. Muehlbauer. 1995. Development of a genetic linkage map for lentil based on RAPD markers. *In* Proceedings of the European Association for Grain Legume Research Conference, Copenhagen, Denmark July, 1995. (in press).

Hamdi, A., W. Erskine and P. Gates. 1992. Adaptation of lentil seed yield to varying moisture supply. Crop Sci. 34:987-991.

Hawtin, L. and L. Sears. 1993. Legume Cookbook: Cooking with Chickpeas, Faba beans and Lentils. ICARDA, Aleppo, Syria.

ICARDA. 1992. Legume Program Annual Report for 1991, pp. 105-107. ICARDA, Aleppo, Syria.

ICARDA. 1994. Legume Program Annual Report for 1993, pp. 105-108. ICARDA, Aleppo, Syria.

ICARDA. 1995. Annual report for 1994 of the germplasm program: Legumes. ICARDA, Aleppo, Syria.

Khare, M.N. 1981. Diseases of lentils. Pp. 163-172 *in* Lentils (C. Webb and G. Hawtin, eds.). CAB/ICARDA, London, UK.

Kupicha, F.K. 1981. Pp. 377-381 *in* Advances in Legume Systematics. Part 1 (R.M. Polhiss and P.H. Raven, eds.). Royal Botanic Gardens, Kew, UK.

Ladizinsky, G. 1979. The origin of lentil and its wild genepool. Euphytica 28:179-187.

Ladizinsky, G. 1993. Wild lentils. Critical Rev. in Plant Sci. 12:169-184.

Ladizinsky, G., D. Braun, D. Goshen and F.J. Muehlbauer. 1984. The biological species of the genus *Lens* L. Bot. Gaz. 145:253-261.

Lenné, J.M. and J.B. Smithson. 1995. Varietal mixtures: a viable strategy for sustainable productivity in subsistence agriculture? Ann. Appl. Biol. 39 (in press).

Muehlbauer, F.J., W.J. Kaiser, S.L. Clement and R.J. Summerfield. 1995. Production and breeding of lentil. Adv. Agron. 54:283-332.

Sakar, D., N. Durutan and K. Meyveci. 1988. Factors which limit the productivity of cool season food legumes in Turkey. Pp. 137-146 *in* World Crops: Cool Season Food Legumes (R.J. Summerfield, ed.). Kluwer, the Netherlands.

Savage, G.P. 1988. The composition and nutritive value of lentils (*Lens culinaris*). Nutrition Abstr. and Rev. (Series A). 58(5):320-343.

Silim, S.N., M.C. Saxena and W. Erskine. 1993. Adaptation of lentil to the Mediterranean environment. I. factors affecting yield under drought conditions. Exp. Agric. 29:9-19.

Singh, K.B., W. Erskine, L.D. Robertson, H. Nakkoul and P.C. Williams. 1988. Influence of pretreatment on cooking quality parameters. J. Sci. Food Agric. 44:135-142.

Vavilov, N.I. 1992. Origin and geography of cultivated plants. [English translation from Russian]. Cambridge University Press, Cambridge.

Williams, P.C., W. Erskine and U. Singh. 1993. Lentil processing. LENS Newsl. 20(1):3-13.

Wilson, V.E. and A.G. Law. 1972. Natural crossing in *Lens esculenta* Moench. J. Am. Soc. Hort. Sci. 97:142-143.

Zohary, D. 1972. The wild progenitor and the place of origin of the cultivated lentil: *Lens culinaris*. Econ. Bot. 26:326-332.

Zohary, D. and M. Hopf. 1988. Domestication of Plants in the Old World. Clarendon Press, Oxford, UK.

Chapter 11

Phaseolus Beans

R. Hidalgo and S. Beebe

Phaseolus beans are cultivated in almost all continents. The common bean is produced in a wide range of cropping systems and environments, as diverse as Latin America, Africa, the Middle East, China, Europe and North America. *Phaseolus* beans are grown in more than 90 countries. Total production surpasses 9.6 million t, on approximately 13 million ha; the estimated value of production is more than US$4380 million. The other species of the genus have a more localized regional production. Cultivation ranges from subsistence intercropping or secondary cropping to monoculture on smallholdings in the tropics and subtropics to commercial operations.

BOTANY AND DISTRIBUTION
The genus *Phaseolus* includes about 55 species of annuals and semiperennials throughout the warm regions of both hemispheres. It fits botanically within the order Rosales, family Leguminosae (Fabaceae), subfamily Papilionaideae, tribe Phaseoleae, subtribe Phaseolinae. Only in the past 25 years has Phaseolinae taxonomy been agreed (Maréchal *et al.* 1978; Delgado Salinas 1985; Debouck 1991) and three main sections have been proposed: *Phaseolus, Alepidocalyx* and *Minkelersia* (Appendix 1). Scientific and common names for the cultivated species are given in Box 11.1.

Origin, Distribution and Diffusion
Archaeological evidence for *Phaseolus* origin comes from Mesoamerica and the Andes (Kaplan 1965). The oldest remains of *P. vulgaris*, estimated at 7000 years BP (Tehuacán, Mexico) and 8000 years BP (Ancash, Peruvian, Andes), have been revised downwards (Kaplan 1994). For *P. lunatus*, they are dated from 1400 years BP (Tehuacán, Mexico) and 5300 years BP (Chilca, Peruvian Andes). For the complex *P. coccineus-P. polyanthus*, remains dated from about 7500 years BP (Ocampo, Mexico) and for *P. acutifolius*, from 5000 years (Tehuacán, Mexico). Because all the specimens were of cultivated forms, the sites and dates of domestication have been open to question (Kaplan 1981). Gepts and Debouck (1991), using data from Vanderborght (1982, 1983), compared wild forms of *P. vulgaris* from Mesoamerica with those of Andean origin. They explained that wild common beans are morphologically differentiated according to their geographic origin and, apparently, the derived cultivated forms maintain these differences.

Biochemical and molecular markers are helping us to understand the evolutionary role played by a seed storage protein, phaseolin. Gepts and Bliss (1986, 1988) showed how this single trait can be used as an evolutionary marker in *P. vulgaris*. Wide polymorphism was observed in this protein and certain patterns were exclusive to beans domesticated in Mesoamerica and others to those domesticated in the Andes. A type S (for cultivar Sanilac) phaseolin was found predominantly in wild and cultivated

---

Box 11.1. Scientific and common names for cultivated species of *Phaseolus*.

***Phaseolus vulgaris* L. (1753)**
Synonyms:	*Phaseolus esculentus* Salisb. (1798). *Phaseolus communis* L. ex Pritzel (1855)
Common names:	English: common bean, dry bean, navy bean, snap bean. Spanish: *frijol común, alubia, habichuela, caraota.* French: *haricot.* Portuguese: *feijao*

***Phaseolus lunatus* L. (1753)**
Synonyms:	*Phaseolus inamoemus* L. (1753). *Phaseolus bipunctatus* Jacq. (1770). *Phaseolus puberulus* (1823). *Phaseolus xuaresii* Zuce (1825). *Phaseolus limensis* Macfadyen (1837). *Phaseolus saccharatus* Macfadyen (1837)
Common names:	English: lima bean. Spanish: *pallar, frijol lima, frijol haba.* French: *haricot de lima*

***Phaseolus coccineus* L. (1753)**
Synonyms:	*Phaseolus multiflorus* Sam. (1789)
Common names:	English: scarlet runner bean. Spanish: *ayocote.* French: *haricot d'Espagne*

***Phaseolus polyanthus* Greenman (1907)**
Synonyms:	*Phaseolus coccineus* subsp. *polyanthus* (Greenman) Maréchal, Mascherpa & Stainier (1948). *Phaseolus coccineus* subsp. *darwiniwanus* Hernandez and Miranda Colin (1959)
Common names:	English: runner bean, year bean. Spanish: *piloy, frijol de vida, frijol cacha*

***Phaseolus acutifolius* Asa Gray (1852)**
Synonyms:	None
Common names:	English: tepary. Spanish: *tepari*

---

forms from Mesoamerica, while a type T (for cultivar Tendergreen) was predominant in materials from the Andes. This finding, plus later developments, led to the hypothesis that, most probably, several primary centres of domestication and diversification exist: one in middle America, another in the southern Andes and a possible third in Colombia. Similar findings were obtained for both the wild and cultivated forms of *P. lunatus* (Gutiérrez *et al.* 1995).

Several characteristics of wild species have changed drastically as a consequence of domestication (Smartt 1990; Gepts and Debouck 1991), including a tendency toward giant seeds, suppression of seed dispersal mechanism, modified growth form (Voysest 1983), modified life forms, loss of seed dormancy, changes in physiology (Smartt 1990; O. Voysest, pers. comm.), biochemistry (Vanderborght 1979) and seed colour. *Phaseolus vulgaris* has undergone more apparent evolutionary advance under domestication than the other cultivated species, and *P. acutifolius* the least (Smartt 1990), perhaps because the ecological preferences of ancient civilizations favoured growing this species.

The major bean-growing regions coincide with the distribution of the six proposed races in the primary centres of diversity plus some areas of sub-Saharan Africa, of which about 6.0 million ha are planted with small-seeded bean cultivars of the Mesoamerica race; another 3.5 million ha are planted with large-seed cultivars of the Nueva Granada race, and 1.5 million ha are planted with medium-seeded cultivars of the Durango race (Singh 1992).

Reproductive Biology

Despite the extremely small size of chromosomes in *Phaseolus*, Maréchal (1969) carried out perhaps the most comprehensive karyotype of the genus and found the chromosome number ($2n=2x=22$). The occurrence of polyploids has not yet been reported. Three of the five cultivated species are considered as self-pollinating, with variable amounts of outcrossing; two species, *Phaseolus coccineus* and *Phaseolus polyanthus,* which are usually cross-pollinated by bees, appear to have genetic barriers to self-pollination (Hawkins and Evans 1973).

GERMPLASM CONSERVATION AND USE

The genepool concept (Harlan and De Wet 1971) helps us to understand the genetic diversity potential of a genus and to lay a sound foundation for conserving genetic resources.

A primary genepool (GP1) is equivalent to the biological species. Related species form the secondary pool (GP2), and the least related is the tertiary pool (GP3). Smartt (1990) and Debouck (1991) defined the extent of genepools for *Phaseolus* cultivated species. The former suggested two main subdivisions, GP1A (cultivated) and GP1B (wild), a classification questioned by Debouck (1991) because cultivated and wild forms should not be given specific and separate taxonomical treatment. Smartt (1990) subdivided the GP3 into GP3A for viable but sterile hybrids, and GP3B for inviable hybrids. Table 11.1 summarizes the status of genepools for the cultivated species of *Phaseolus.*

Of the cultivated species, *P. vulgaris* is the most widely distributed, has the broadest range of genetic resources, and is most used as a food crop throughout the world, especially in Latin America and Africa (Pachico 1989). Three hierarchical levels of genepools for common bean developed through the evolution and domestication of this species (Singh *et al.* 1991a). The first level appeared before domestication, and can be identified by molecular markers, morphological traits, reproductive isolation and ecogeographic adaptation. At this level, two major and apparently independent genepools related to distant geographic regions exist for the wild species: Middle American and Andean South American. Smaller groups may exist in the Northern Andes (Debouck *et al.* 1993; S. Beebe, unpublished). The second level of genepool development appeared more as a consequence of domestication with the development of cultivated forms that, for several reasons, can be grouped into races. A third level of genepool development has been suggested by Singh *et al.* (1991b), in which the races can be subdivided into smaller groups according to their growth habit, crop duration and yield potential; this level has an orientation toward breeding objectives. Nine subdivisions of the genepools were proposed: five for Middle America and four for South America. Cultivars have the same morphological and biochemical features as their respective wild ancestrals, suggesting that they, too, derive from the same two genepools (Gepts and Debouck 1991).

Crosses of a certain percentage of cultigroups from different centres are genetically incompatible and expressed in hybrid dwarfism (Singh and Gutiérrez 1984; Gepts and Bliss 1985). Multivariate statistical analysis of several allozymes also supports evidence of the existence of Middle American and Andean cultigroups of common beans (Singh *et al.* 1991b). Singh *et al.* (1991a) also developed the concept of races by using the idea of independent domestication centres; data from multivariate statistical analyses of morphological, agronomic, molecular characteristics and ecogeographic information from the presumed centres of domestication were used. Landraces from Middle America and Andean South American can be distinguished by molecular markers, such as phaseolin and allozymes, and vegetative and reproductive traits. The races also differ for their responses to biotic and abiotic factors such as photoperiod, diseases, pests, temperatures, water stress and soil fertility (Table 11.2). Relatively higher levels of polymorphism for phaseolin seed protein patterns are found in the wild forms of *P. vulgaris* (Gepts and Bliss 1986), however, and the discovery of immunity for bean weevil in wild forms from Mexico (Schoonhoven *et al.* 1983) suggests that useful genes not found in the cultivated species could be explored in wild bean populations (Singh 1992).

Although the neotropical origin and dispersion of the lima bean was referred to briefly by Piper (1926) and Bukasov (1931), it was Mackie (1943) who proposed for the species a specific geographic origin and routes of dispersion in pre-Columbian times. Geographical and botanical evidence, however, do not fully support Mackie's proposals

(Baudoin 1988).   Recent information on the geographic distribution of wild and cultivated forms and their electrophoretic protein patterns shows that wild forms of lima bean can be divided into two groups whose main difference is seed size (Gutiérrez *et al.* 1995). Some beans were apparently selected first for their aesthetic value (e.g. toys for children) and second as an emergency food plant (Debouck 1989). Introgression may have occurred between the two genepools (Maquet *et al.* 1990).

Until recently, the runner bean species, *P. coccineus-P. polyanthus*, had been considered as closely related at the conspecific level, with similar ecological adaptation and geographic distribution. Recent work indicates that the two species should be analyzed separately. The traditional classification system separates the cultivars of the complex into two main groups: twining climbers with short and long vines used in traditional farming, and usually associated with maize; and bushy, non-climbing types that are erect and self-standing (Delgado Salinas 1988). Another system separates the cultivars of this complex into two races: Ayocote and Piloy. Race Ayocote has two forms: the Mexicano or Ayocote, with a shrubby tendency and adapted to drier and hot climates on sandy soil; and the Guatemalteco or Botil, a climber adapted to humid and cooler climates (Freytag 1965). Schmit and Debouck (1991) used protein electrophoretic patterns to explore *P. polyanthus* genetic relationships, geographic distribution, folk descriptions, botanical descriptors, and wild, weedy and feral forms. They found strong similarities in the morphology, ecology and biochemical characterization between a wild form discovered in Guatemala and the cultivated forms of Mesoamerica and northern Andes, and the origin of the species may involve wild forms of *P. vulgaris* and *P. coccineus*. They also concluded that the protein electrophoretic patterns showed much more variability among wild forms than among cultivated forms. Germplasm from Mesoamerica presented more morphological variants than did northern Andean germplasm, suggesting Guatemala as the centre of origin and domestication with two dispersion routes, and *P. polyanthus* appears to be a single species with a single genepool that shows a clinal genetic drift from Mesoamerican materials to northern Andes materials. A final conclusion was that the ready adaptability of *P. polyanthus* to tropical montane rain forest (mainly because of resistance to fungal diseases) and its attractive seeds were the main factors in its domestication and association with maize. The use of chloroplast DNA validated the above conclusions (Schmit *et al.* 1993).

Despite its restricted geographic area of adaptation, doubts still exist about the exact centre of diversification and origin for *Phaseolus acutifolius*. Some authors (Carter 1945; Kaplan 1965 cited in Pratt and Nabhan 1988; Freeman 1992) suggested the Sonora Desert (Mexico) and southwestern United States as two possible centres of origin and/or diversification based on the presence of cultivated and wild forms, and archeological findings. Others (Pratt and Nabhan 1988) proposed the entire arid region, covering northwestern Mexico and southwestern United States (binational arido-American region). Factors leading to their suggestion are the diversity of both tepary domesticates and wild populations found in the area, the strong evidence for the origins of maize and common bean close to the southern extreme of this region, and archeological evidence of tepary domesticates.   Tepary beans might be useful in common bean improvement for fragile moisture-deficient and low soil fertility environments, for their higher levels of tolerance to common bacterial blight and leafhoppers (Singh 1992). The primary centres of diversity of the tepary bean can be classified according to the distribution of wild and cultivated forms (Table 11.3), as ascertained by their morphology, especially leaf shape.   Table 11.4 shows the geographic distribution of several truly wild species. A summary of the geographical distribution of the *Phaseolus* species for both the Mesoamerica and the Andean South America diversity centres is shown in Figure 11.1.

In situ conservation of landraces takes place on small fields that are in a continual adaptation, with or without farmer selection. For *ex situ* conservation, a sample of the variability existing on-farm is stored; for *Phaseolus*, this is readily done by placing seed in cold stores, thanks to their 'orthodox' classification, from the standpoint of seed conservation (Cromarty *et al.* 1982). Diverse germplasm is vulnerable for conservation

Table 11.1. Known genepools of cultivated *Phaseolus* bean species (adapted from Smartt 1990).

| Cultigen | GP1 | | GP2 | GP3 | |
	GP1A	GP1B	GP2	GP3A	PG3B
P. vulgaris (P.v)	+ (P.v. cultivated)	+ (P.v. wild)	+ (Complex P.c. and P.p)	+ (P. acutifolius)	+ (P. lunatus and other species)
P. lunatus (P.l)	+ (P.l. cultivated)	+ (P.l. wild)	−	+ (True wild species: P. maculatus, P. polystachyus)	+
P. coccineus (P.c)	+ (P.c. cultivated)	+ (P.c. wild)	+ (P. vulgaris, P. polyanthus)	+ (True wild species)	+
P. polyanthus (P.p)	+ (P.p. cultivated)	+ (P.p. wild)	+ (P. vulgaris, P. coccineus)	+ (True wild species)	+
P. acutifolius (P.a)	+ (P.a. cultivated)	+ (P.a. wild)	−	+	+ (P. vulgaris)

Table 11.2. Some discriminant characters among the races of cultivated common bean, *Phaseolus vulgaris* L. (adapted from Singh *et al.* 1991b).

Race	Predom. habit	Seed size	Bracteole	Phaseolin	Geographic distribution	Response to abiotic-biotic factors
Middle America						
Mesoamerica	2,3	Small	Large cordate	S,Sb,B	Lowlands of Latin America	Photoperiod insensitivity, resistance to BCMV, tolerance to angular leaf spot, BGMV, high temp., moisture stress, low soil fertility
Durango	3	Medium	Small ovate	S,Sd	Semi-arid highlands of Mexico	Early maturity, drought tolerance, high harvest index, positive general combining ability, tolerance to some virus and to anthracnose
Jalisco	4	Medium	Oval, round, cylindrical	S	Humid highlands of Mexico	High seed yield, high levels of resistance to Apion spp. and anthracnose, tolerance to angular leaf spot, and low soil fertility
Andean South America						
Nueva Granada	1,3	Medium and large	Small, lanceolate or triangular	T	Intermediate altitudes of Andes	Insensitivity to photoperiod, early maturity, resistance to BCMV, halo blight, anthracnose and angular leaf spot
Chile	3	Medium	Small, triangular	C,H	Southern Andes	Adaptation to relatively drier areas
Peru	4	Medium and large	Large lanceolate or cordate	T,C,H	Andean highlands (>2000 m)	Highly photoperiod-sensitive, adapted to moderately wet and cool temperatures, long growing cycles

Table 11.3.   Centres of diversity of the different wild and cultivated forms of *Phaseolus acutifolius*.

Biological form	Region
P. acutifolius var. *tenuifolius* (wild)	Mexico, southwestern USA
P. acutifolius var. *acutifolius* (wild)	Mexico, southwestern USA, Guatemala
P. acutifolius, domesticated landraces of Mesoamerica	Northern Mexico (Sonora, Chihuahua, Durango, Sinaloa, Nayarit, Jalisco)

Table 11.4. Geographic centres of diversity for some truly wild *Phaseolus* species.[†]

Species	Geographic distribution
P. glabellus	Cool highlands of Mexico
P. glaucocarpus	Volcanic axis of Mexico
P. marechalii	Volcanic axis of Mexico
P. xolocotzii	Sierra Madre del Sur, Mexico
P. jaliscanus	Western Mexico
P. salicifolius	Western Mexico
P. sempervirens	Western Mexico
P. leptostachyus	Nuevo León, Mexico, to San José, Costa Rica
P. macrolepis	Mountainous ranges in Central-Western Guatemala
P. minimiflorus	Baja, California
P. ovatifolius	Northern Mexico
P. venosus	Northern Mexico
P. maculatus	Northern Mexico
P. microcarpus	Mountains of Mexico
P. polymorphus	San Luis Potosi, Mexico
P. sinuatus	Southeast USA (Florida)
P. polystachyus	Southeast USA (Florida, Arkansas, Missouri, Kentucky)
P. sonorensis	Sonora, Mexico
P. grayanus	North of Mexico, Arizona, New Mexico
P. angustissimus	Texas, New Mexico, Arizona
P. filiformis	California, Gulf of California
P. pedicellatus	Highlands of Mexico
P. oaxacanus	Oaxaca, Mexico

[†] Adapted from Maréchal *et al.* 1978.

in situ (Debouck 1979, 1986b).   In addition, the bean crop is being displaced and traditional landraces are being replaced by more profitable varieties (Kaplan 1956; Esquinas A. 1983).   Chang (1985) stressed that the primary goal in *ex situ* conservation should be to conserve as many representative samples of existing germplasm as human resources permit.   Bettencourt *et al.* (IBPGR 1989) found more than 106 000 *Phaseolus* bean accessions stored as *ex situ* collections.   Most of these collections are under short- or medium-term storage and are working collections of the National Programmes (Table 11.5).

The *Phaseolus* beans collection at CIAT comprises a total of 27 813 accessions (Debouck 1979, 1986b, 1988) (Table 11.5).   The seed storage facility includes a long-term storage (–20°C), short-medium term storage (5-8°C, 35% RH) and a seed-drying room (20°C, 20% RH).   It can hold 100 000 accessions, half of which have been assigned to *Phaseolus*.   Characterization has been carried out for the most important morphological traits for the cultivated beans: seed type, growth habit and flowering characteristics. The five species present high polymorphism mainly for seed colour and size as well as for growth habit.   Nine groups of seed colours and four prototypes of growth habits have been defined.   More than 270 000 bean samples have been requested by the resident Bean Program, about 80 000 samples have been distributed internationally and many new cultivars have been produced (Tables 11.6 and 11.7).

Properties and Uses

The wider distribution of the common bean, *P. vulgaris*, has made it by far the most important in terms of consumption (Pachico 1989). The estimated per capita consumption in Latin America is 13 kg/year, while in Central and East Africa it is about 31 kg/year (Shellie-Dessert and Bliss 1991). Both dry seed and immature seed are utilized for human consumption, usually by boiling. Because of their protein content and through combination with other foods, especially cereals, *Phaseolus* beans constitute a cheap source of high-value dietary protein in developing countries (Pinchinat 1977). Protein content of dry beans ranges from 18 to 29%, but beans also have a substantial proportion of carbohydrates (24-68%), as well as minerals and vitamins (3.5-4.7%) (Uebersax and Occeña 1991). In developed countries, beans are generally prepared by commercial food-processing operations and consumed as canned beans in brine or sauce. Seeds of *Phaseolus* beans offer numerous opportunities for processing into a variety of food products: sprouts, curds, ferments, bean meals, flours, powders, quick-cooking beans, frozen/dehydrated beans, extruded products, microwaveable products and weaning products (Uebersax and Occeña 1991)

Popping beans called *ñuñas* have been selected by the Incas in the highlands of Peru and Bolivia in the pre-Ceramic era (Debouck 1989) because of the difficulty in cooking beans by boiling at high altitudes. In Peru, beans are toasted 5-10 minutes in a hot frying pan and these bean types may become a widely available, nutritious, tasty and fuel-conserving food (National Research Council 1989). They also possess wide

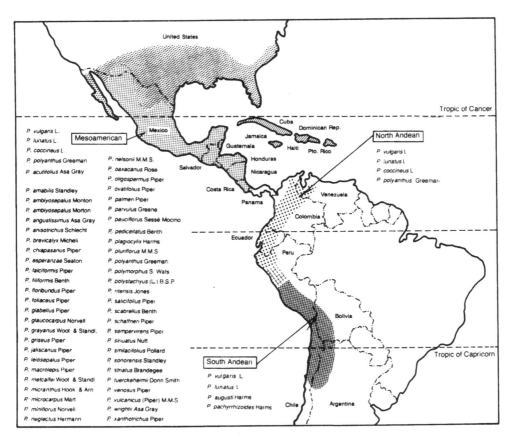

Fig. 11.1. Geographic distribution of the *Phaseolus* species in the three primary centres of diversity, as proposed by Debouck (1986a).

Table 11.5. *Ex situ Phaseolus* beans germplasm collection[†]

Region/ Country	No. of inst.	P. vul	P. lun	P. coc, P. pol	P. acu	P. spp.	Phas. ?	Total access.
Americas								
Argentina	1	561	–	–	–	–	–	561
Brazil	2	1800	373	–	–	–	–	2173
		4202	–	–	–	–	–	4202
Colombia	2	939	–	–	–	–	–	939
CIAT		24563	1548	889	271	164	–	27435
Costa Rica	1	870	51	91	9	–	511	1532
Cuba	1	–	834	–	–	–	–	834
Ecuador	2	860	–	–	–	–	–	860
		1336	–	–	–	–	–	1336
Guatemala	1	600	–	–	–	–	–	600
Mexico	1	8315	580	1335	289	21	–	10540
Peru	1	1500	50	18	–	–	–	1568
USA	3	7965	910	342	61	26	–	9304
		4000	300	–	–	–	–	4300
		4465	567	116	24	5	–	5177
Venezuela	1	827	–	–	–	–	–	827
Subtotal		62803	5213	2791	654	216	511	72188
Europe								
Austria	1	281	–	–	–	–	–	281
Belgium	3	122	94	94	57	99	–	466
		150	–	–	–	–	–	150
		42	–	–	–	–	130	172
Bulgaria	1	2420	–	–	–	–	–	2420
Czechoslov.	2	545	–	–	–	–	–	545
		173	–	–	–	–	–	173
Germany	2	511	–	2	2	–	13	528
		3696	22	148	5	–	–	3871
Greece	2	410	–	–	–	–	–	410
		256	–	–	–	–	–	256
Hungary	3	57	–	–	–	–	–	57
		65	–	–	–	–	–	65
		3331	–	–	–	–	–	3331
Italy	2	39	–	–	–	–	–	39
		31	2	23	–	–	–	56
Netherlands	1	1000	–	20	10	–	–	1030
Poland	1	26	–	–	–	–	–	26
Portugal	2	354	–	14	–	–	–	368
		559	–	28	–	–	190	777
Rumania	1	159	–	–	–	–	52	211
Spain	1	1007	–	6	–	–	–	1013
Russia	1	41	4	–	4	–	–	49
UK	2	4950	–	–	–	10	–	4960
		700	–	100	–	–	–	800
Subtotal		20925	122	435	78	109	385	22054
Africa								
Ethiopia	1	263	–	–	–	–	–	263
Ghana	2	–	8	–	–	–	–	8
		–	193	–	–	–	–	193
Kenya	1	1601	–	–	–	–	–	1601
Morocco	1	315	–	–	–	–	–	315
Tanzania	1	1348	–	–	–	–	–	1348
Subtotal		3527	201	–	–	–	–	3728
Asia								
Afghanistan	1	94	–	–	–	–	–	94
India	1	1700	–	–	–	–	–	1700
Indonesia	1	–	3846?	–	–	–	–	3846

Region/ Country	No. of inst.	P. vul	P. lun	P. coc, P. pol	P. acu	P. spp.	Phas. ?	Total access.
Iran	1	274	–	–	–	–	–	274
Israel	2	–	–	–	10	–	–	10
		350	–	–	–	–	–	350
Philippines	1	495	515	–	–	–	–	1010
Turkey	1	883	–	12	–	–	–	895
Subtotal		3796	4361	12	10	–	–	8179
Total	**55**	**91051**	**9897**	**3238**	**742**	**325**	**896**	**106149**

[†] Adapted from IBPGR 1989.

Table 11.6. Traits evaluated in CIAT's *Phaseolus* collection[†] and number of promising accessions.

Traits evaluated	Number of accessions	
	Evaluated	Promising[‡]
Resistance to disease/pest		
BCMV	21686	3079
BGMV	1660	12
Angular leaf spot	23848	221
Anthracnose	23924	1941
Common bacterial blight	23060	109
Zabrotes subfasciatus	10973	12
Acanthoscelides obtectus	6550	8
Leafhopper	17706	487
Apion spp.	1600	100
Other traits		
Low P tolerance	2178	143
Drought tolerance	5156	118
D1 genes	114	–
Phaseolin types	658	–
Photoperiod response	1655	–

[†] Includes all *Phaseolus* germplasm conserved at CIAT.

[‡] Includes approximate numbers, and some accessions should be rechecked for confirmation.

Table 11.7. Materials released as cultivars through international bean nurseries of the Bean Program, 1979-94[†].

Region	No. of countries	Sources and number of cultivars		
		Genebank (GRU)	Breeding programme	Total
North America	2	2	–	2
Central America	7	7	42	49
Caribbean	2	3	9	12
South America	8	21	70	91
Europe	–	–	–	–
Africa	13	20	23	43
Asia-Oceania	4	2	4	6
Total	37	55	148	203

[†] Source: CIAT Bean Program.

genetic variability (Tohme *et al.* 1995b). In some areas, green pods are consumed, and flat, oval or round-podded cultivars are preferred (Purseglove 1968). A rough estimate of world production is close to 5 million t (Silbernagel *et al.* 1991).

In Africa, young leaves are used as a vegetable and source of vitamin A (Westphal 1974; Silbernagel *et al.* 1991). Young shoots and leaves of *P. coccineus* are reported to be boiled and seasoned with chile and salt (Laughlin 1975; Arias 1980, in Delgado Salinas

Table 11.8. Comparison of nutrient composition (100 g, fresh weight) and/or physical properties[1] of different parts of the common bean plant.

Nutrient and/or physical properties	Leaves (raw)	Snap beans (raw)	Inmature seeds (cooked)	Dry beans (raw)
Energy (calories)	36	36	50	336
Moisture (%)	87	89	88	12
Protein (g)	3.6	2.5	1.7	21.7
Lipid (g)	0.4	0.2	3.2	1.5
Carbohydrate (g)	6.6	7.9	5.5	60.9
Ca (mg)	274	43	42	120
Fe (mg)	9.2	1.4	0.8	8.2
Thiamine (mg)	0.18	0.08	-	0.37
Niacin (mg)	1.3	0.5	-	2.4
Cooking time	10	25	40	120
Post-harvest storability	short	short	short	medium-long

[1] Adapted from Shellie-Dessert and Bliss 1991; Silbernagel *et al.* 1991.

1988). Likewise, *P. coccineus* produces a considerable amount of large tuberous roots, which are dug up by natives, then boiled and used as starchy food (Brücher 1989). Raw roots are used for medicinal prevention of vomiting, malaria and other ailments. Flowers also can be boiled and then fried (Delgado Salinas 1988).

Some physical and biological drawbacks need to be addressed in order to enhance consumption and utilization, such as cooking time, the post-harvest seed storage changes (hardening and discolouration), antinutrients (protease inhibitors and lectins, amylase inhibitors and phytic acid), digestibility and flatulence (Uebersax and Occeña 1991). Table 11.8 is a nutritional comparison of the different parts of the plant.

Breeding

The genetics of common bean has been studied since the beginning of this century (Singh 1992), and research areas reviewed: Yarnell (1965) on bean genetics, Leakey (1988) and Prakken (1970, 1972) on seed, pods and other marker traits, Zaumeyer and Meiners (1975) on bean diseases, Roberts (1982) on major genes, and Basset (1988) on linkage mapping of marker genes. Table 11.9 summarizes the reports. Using molecular techniques of DNA analysis, several genetic maps have been created with Restriction fragment length polymorphisms (RFLP) (Vallejos *et al.* 1991; Nodari *et al.* 1993) and Randomly amplified polymorphic DNA (RAPD) (Jung *et al.* 1994; McClean *et al.* 1994). Also, recent efforts are directed to coordinate several linkage maps of *P. vulgaris*, in order to produce one high-density linkage map in this species (Freyre *et al.* 1995); this map is urgently needed to have a more comprehensive correlation and interpretation of those results.

Plant breeding efforts lead to increased yield potential as well as attempting to stabilize yield potential by overcoming production constraints. Although both objectives have received attention in bean improvement, the alleviation of production constraints often has received more emphasis. Recent reviews have been published on breeding common bean for adaptation to drought (White and Singh 1991), for disease resistance (Beebe and Pastor C. 1991) and for insect resistance (Kornegay and Cardona 1991). Additionally, Singh (1992) has reviewed bean breeding in tropical environments, as well as breeding for seed yield (Singh 1991), and has proposed that breeding programmes be organized around the principle of races and genepools of common bean.   The strategy of utilizing distinct races to improve yield potential of Mesoamerican beans has already proven successful (Kelly and Adams 1987; Singh *et al.* 1992, 1993).   It is likely that this strategy will also be useful for overcoming yield constraints as well. The genepool structure of bean germplasm also has shed light on co-evolution with certain bean pathogens, especially *Colletotrichum lindemutheanum* and *Phaeoisariopsis griseola* (Beebe and Pastor C. 1991).

Table 11.9. Summary of reports on the genetics of the common bean *Phaseolus vulgaris* (adapted from Singh 1991).

Trait	Genes Number	Genes Action
Incompatibility		
Seedling wilt, leaf rolling	2	Recessive
Semilethal chlorosis	1	Recessive
Virescent foliage	3	Recessive
Cripple morphology	2	Recessive
Variegated foliage	2;3	Duplicated, recessive
Leaf distortion	1	Dominant
Dwarfism	2	Complem-dominant
Male sterility (MS)		
Indehiscent anthers	1	Recessive
Gamma induced MS	2	Recessive
Restorer cytoplasmic	3;1	Complem.-dominant
Seed characters		
Size	1;2	Additive
Colour	9	Complementary(?)
Brilliance	1	Dominant
Coat rupture	1	Incomplete dominant
Semihard seed	1	Recessive
Immature white coat	1	Recessive
Coat whiteness	2	Additive-dominant
Total proteins	–	Quantitatively, GCA
Phaseolin types	1	Codominant alleles
Lectins absence	1	Recessive
Arcelin	1	Dominant
Tannins	1	Quantitative
Cooking quality	–	Additive, GCA
Flowering–maturity		
Delayed flowering	1;2;3	Domin., complem., recessive
Early flower maturity	1	Dominant
Flowering behaviour	4	
Early daylength-neutral	2	Recessive
Photoperiod sensitivity	2	Recessive, epistatic
Early flowering	–	Heterosis, GCA, SCA
Time to flowering	–	Addit.-domin.-overdominant
Late maturity	1	Dominant
Flower colour	1	Pleitropic, linked
Purple colour	1;2	Dominant, complem.
White colour	–	Pleitropic
Blotch on standard	1	Dominant
Pod characters		
Weight of dry pods	–	Heterosis, overdominant
Number of pods	–	Domin., addit., epistatic
Length	1;4	Domin., addit., epistatic
Diameter	–	Heterosis, GCA, SCA
Flat shape	1	Incomplete dominance
Round shape	1	Dominant
Curved tip shape	1	Dominant
Interlocular cavitation	Poly	Dominance, additive
Stringy rogue	1	Recessive
Stringless	1	Domin., duplic., recessive
Detachment force	2	Dominant
Yellow wax colour	–	Quantitative
Dry pod colour	1	Dominant
Colour pattern	3	Dominant, complem.
Broken end discolouration	1	Dominant, epistatic
Flavour (alfa-terpinol)	1	Dominant

Trait	Genes Number	Genes Action
Leaf characters		
Colour-shape mutants	9	Independent, recessive
Abaxial pubescence	2	Dominant, complem.
Reclining foliage	1	Recessive
Photonastic movement	2	Dominant, epistatic
Leaf area	–	Additive
Leaf number	–	Dominance
Photosynthethic efficiency	Few	Quantitative
Root characters		
Dry weight	–	Transgressive, additive
Nitrogen fixation	–	GCA
Low nitrogenase activity	1	Dominant
Isozyme systems	6	Independent, codominant
Stem characters		
Indeterminate	1	Dominant
Climbing ability	1; poly	Dominant
Sprawling vs. upright	1	Dominant
Plant height habit	1; poly	Dominant-epist.-modif.-heter.-addit.-GCA
Internode length	6	Dominant-heter.-GCA
Number of branches	–	Heterosis-GCA-SCA
Number of nodes	–	Heterosis-GCA-SCA
Dry weight	–	Heterosis
Inflorescence	–	Heterosis
Harvest index	–	Heter.-addit.-dominant
Abiotic stress		
Phosphorus absorption	–	Addit.-domin.-mix
Iron defficiency	2	Dominant
Low zinc tolerance	2	Dominant, complem.
Ozone tolerance	2	Dominant, interact.
Heat tolerance	–	Addit.-domin.-epistatic
Temp.-drought resistance	1;2	Dominant, complem.
Biotic stress		
Leafhopper resistance		Dominant, GCA, SCA
Nematode resistance	2;3	Recessive
BCMV virus resistance	1	Dominant
BYMV virus resistance	1	Dominant
Rugose virus resistance	1	Allelic dominance
Tobacco mosaic resistance	1	Dominant
Soybean mosaic resistance	1	Incomplete dominance
Subclover stunt resistance	2	Partial dominance
Watermelon mosaic	1	Dominant
BICMV, CAbMV resistance	2	Independent-dominant
Peanut mottle resistance	1	Incomplete-dominant
Common bacterial resistance	Few	Domin.-quant.-additive
Halo blight resistance	–	Domin. race-specific
Bacterial brown spot	1	
Bacterial wilt resistance	–	Duplic-recessive, quantit.
Rust resistance	–	Race specific
Anthracnose	–	Race specific
Angular leaf spot resistance	1	Dominant
White mould	–	Quantitative, additive
Thielaviopsis	3	Quantitative, additive
Fusarium	2	Partly recessive, additive

The major world's germplasm collection of *Phaseolus* beans has been assembled at the Genetic Resources Unit at CIAT. This bank holds more than 27 000 accessions, which includes a wide genetic variability of the five cultivated species, their respective wild ancestral forms plus 22 wild non-cultivated species; the largest proportion of this collection (90%) is filled by *P. vulgaris*. In attempting to understand the genetic value of the sizeable collection, a core collection subset from CIAT's collection has been assembled. About 1500 accessions of this species were selected based on agro-ecological origin, plant and seed morphology, and history of specific regions as areas of traditional bean production (Tohme *et al.* 1995a). Besides detailed morphological characterization and analysis of seed for protein characteristics, the core collection is being evaluated for traits such as yield potential, tolerance to low phosphorus and drought tolerance. It is hoped that these evaluations will serve to identify pockets of useful variability which can be further explored in the reserve collection. The core is also being subjected to molecular analysis to elucidate more fully the genetic structure of genepools and races of bean. Interspecific crosses have been utilized on a very limited scale and in most cases to obtain resistances to specific diseases: with *P. acutifolius* for common blight resistance (Honma 1956; McElroy 1985) and *P. coccineus* for resistance to root rots (Wallace and Wilkinson 1965). The use of molecular markers, especially DNA markers for Marker Assisted Selection (MAS), has created a revolution in plant breeding in recent years. Molecular markers for several economic traits in beans have been reported: resistance to common bacterial blight (Nodari *et al.* 1993; Jung *et al.* 1994), bean rust (Miklas *et al.* 1993), bean golden mosaic virus (Miklas *et al.* 1995), bean common mosaic virus (Johnson and Gepts 1994) and anthracnose (Young and Kelly 1994).

Prospects

Despite the outstanding advances in knowledge of the evolution of the genetic diversity of *P. vulgaris* and, on a minor scale, of the other *Phaseolus* cultivated species, a wide range of research still claims urgent attention. Thus, key genetic variability of cultivated and wild relatives from Southern (Peru) and Northern Andes (Colombia, Venezuela) needs to be collected; there is a high risk of genetic erosion in these areas. Also, more systematic collecting of variability of the non-cultivated wild species is required, because it is lacking in all genebanks. Research on molecular markers as a tool for breeding and for the completion of genetic mapping offers tremendous potential for improvement of the crop. The setting-up of core collections will improve utilization of the germplasm. Interspecific crossing, using wild non-relative species, will be the challenge of the future. Genetic resistance and/or tolerance to several biotic and/or abiotic stresses has been poor or absent in germplasm of the cultivated species, and hence the wild non-cultivated species are a world to explore.

Limitations

The most widespread and endemic production problems of the common bean are diseases, insects, drought and low soil fertility; in fact, many species of pests and diseases attack *Phaseolus* (Schoonhoven and Voysest 1989; Beebe and Pastor C. 1991; Kornegay and Cardona 1991; Singh 1992). Bean yields are also reduced by edaphic and climatic stresses. The most important diseases are bean common mosaic virus (BCMV), bean golden mosaic virus (BGMV), common bacterial blight (*Xanthomonas campestris* pv. *phaseoli* (Smith) Dye), anthracnose (*Colletotrichum*), angular leaf spot (*Phaeoisariopsis griseola* (Sacc) Ferraris), rust (*Uromyces appendiculatus* (Pers) Unger var. *appendiculatus*) and root rots caused by several fungi (Beebe and Pastor C. 1991). Among the insects, the most important are leafhoppers (*Empoasca kraemeri* Ross and Moore), bean pod weevil (*Apion godmani* Wagner), Mexican bean beetle (*Epilachma varivestis* Mulsant), bean fly (*Ophiomyia phaseoli* Triyon), Mexican bean weevil (*Zabrotes subfaciatus* Boheman) and bean weevil (*Acanthoscelides obtectus*). Deficiencies of phosphorus and nitrogen are common in bean production. Agronomical and breeding efforts in most regions must be directed toward recovering yield potential of existing commercial cultivars, stabilizing production and reducing crop losses (Singh 1992).

REFERENCES

Basset, M.J. 1988. Linkage mapping of marker genes in common bean. Pp. 329-353 *in* Genetic Resources of *Phaseolus* Beans: their Maintenance, Domestication, Evolution, and Utilization (P. Gepts, ed.). Kluwer, Dordrecht, Netherlands.

Baudoin, J.P. 1988. Genetic resources, domestication and evolution of lima bean, *Phaseolus lunatus*. Pp. 393-407 *in* Genetic Resources of *Phaseolus* Beans: their Maintenance, Domestication, Evolution, and Utilization (P. Gepts, ed.). Kluwer, Dordrecht, Netherlands.

Beebe, S.E. and M. Pastor C. 1991. Breeding for disease resistance. Pp. 561-617 *in* Common Beans: Research for Crop Improvement (A. van Schoonhoven and O. Voysest, eds.). CAB International, Wallingford, UK/CIAT, Cali, Colombia.

Brücher, Heins. 1989. Useful Plants of Neotropical Origin: and their Wild Relatives, pp. 87-103. Springer-Verlag, Berlin, New York.

Bukasov, S.M. 1931. Las plantas cultivadas de Mexico, Guatemala y Colombia, Pp. 54-66. [Trad. de la versión en Inglés de M.H. Byleveld por Jorge León. 1981.] CATIE.

Carter, G.F. 1945. Plant geography and culture history in the American Southwest. Viking Fund Publications in Anthropology 5:1-140.

Chang, T. 1985. Principles of genetic conservation. Plant genetic resources: key to future plant production. Iowa State J. Res 59(4):325-348.

Cromarty, A.S., R.H. Ellis and E.H. Roberts. 1982. The Design of Seed Storage Facilities for Genetic Conservation, pp. 21-30. IBPGR, Rome, Italy.

Debouck, D.G. 1979. Proyecto de recolección de germoplasma de *Phaseolus* en México, CIAT-INIA, 1978-1979. Centro Internacional de Agricultura Tropical, CIAT (internal report), Cali, Colombia.

Debouck, D.G. 1986a. Primary diversification of Phaseolus in the Americas: three centres? Plant Genet. Resour. Newsl. 67:2-8.

Debouck, D.G. 1986b. *Phaseolus* germplasm collection in Cajamarca and Amazonas, Peru. Trip report. International Board for Plant Genetic Resources, Rome, Italy, AGP/IBPGR 86/161.

Debouck, D.G. 1988. Recoleccion de de *Phaseolus* en Bolivia. Internal Report. CIAT, Cali, Colombia.

Debouck, D.G. 1989. Early beans (*Phaseolus vulgaris* L. and *P. lunatus* L.). Domesticated for their aesthetic value?. Ann. Rept. Bean Improve. Coop. 32:62-63.

Debouck, D. 1991. Systematics and morphology. Pp. 55-118 *in* Common Beans: Research for Crop Improvement (A. van Schoonhoven and O. Voysest, eds.). CAB International, Wallingford, UK/ CIAT, Cali, Colombia.

Debouck, D.G., O. Toro, O. Paredes, W. Johnson and P. Gepts. 1993. Genetic diversity and ecological distribution of *Phaseolus vulgaris* (Fabaceae) in Northwestern South America. Econ. Bot. 47:408-423.

Delgado Salinas, A.O. 1985. Systematics of the genus *Phaseolus* (Leguminosae) in North and Central America. Thesis (PhD). The University of Texas at Austin.

Delgado Salinas, A. 1988. Variation, taxonomy, domestication, and germplasm potentialities. in *Phaseolus coccineus*. Pp. 441-463 *in* Genetic Resources of *Phaseolus* Beans: their Maintenance, Domestication, Evolution, and Utilization (P. Gepts, ed.). Kluwer, Dordrecht, Netherlands.

Esquinas A., J.T. 1983. Los recursos fitogenétiocs: una inversión para el futuro. Instituto Nacional de Investigaciones Agrarias, Madrid, España.

Freeman, G.F. 1992. Southwestern beans and teparies. University of Arizona, Agric. Exp. Stat. Bull. 68:57-619

Freyre, R., P.W. Skroch, A. Adam-Blondon, A. Shirmohamadali, W.C. Johnson, R.O. Nodari, E.M. Koinange, M. Sevignac, H. Bannerot, J. Nienhuis and P. Gepts. 1995. Towards and integrated linkage map of common bean (*Phaseolus vulgaris* L): Coordination of the Davis, Paris and Wisconsin maps. Ann. Rept. Bean Improve. Coop. 38:115-116.

Freytag, G.F. 1965. Clasificación del frijol común (*Phaseolus vulgaris* L) y especies afines. Ceiba (Honduras) 11(1):51-64.

Gepts, P. and F.A. Bliss. 1985. F₁ hybrid weakness in the common bean: differential geographic origin suggests two gene pools in cultivated bean germplasm. J. Hered. 76:447-450.

Gepts, P. and F.A. Bliss. 1986. Phaseolin variability among wild and cultivated common beans (*Phaseolus vulgaris*) from Colombia. Econ. Bot. 40(4):469-478.

Gepts, P. and F.A. Bliss. 1988. Dissemination pathways of common bean (*Phaseolus vulgaris*, Fabaceae) deduced from phaseolin electrophoretic variability, II: Europe and Africa. Econ. Bot. 42(1):86-104.

Gepts, P. and D.G. Debouck. 1991. Origin, domestication , and evolution of the common bean (*Phaseolus vulgaris* L.). Pp. 7-53 *in* Common Beans: Research for Crop Improvement (A. van Schoonhoven and O. Voysest, eds.). CAB International, Wallingford, UK/ CIAT, Cali, Colombia.

Gutiérrez Salgado, A., P. Gepts and D.G. Debouck. 1995. Evidence for two gene pools of the Lima Bean, *Phaseolus lunatus* L., in the Americas. Genet. Resour. and Crop Evolution 42:15-28.

Harlan, J.R. and J.M.J. de Wet. 1971. Toward a rational classification of cultivated plants. Taxon 20:509-517.

Hawkins, C.F. and A.M. Evans. 1973. Elucidating the behaviour of pollen tubes in intra- and interspecific pollinations of *Phaseolus vulgaris* and *P. coccineus*. Euphytica 22:378-85.

Honma, S. 1956. A bean interspecific hybrid. J. Hered. 47:217-220.

IBPGR. 1989. Directory of Crop Germplasm Collections. 1. Food Legumes (E. Bettencourt, J. Konopka and A.B. Damania, eds.). IBPGR, Rome.

Johnson, W.C. and P. Gepts. 1994. Two new molecular markers linked to *bc-3*. Ann. Rept. Bean Improve. Coop. 37:206-207.

Jung, G., D.P. Coyne, P.W. Skroch, J. Nienhuis, E. Arnaud-Santana, J. Bokosi, S.M. Kaeppler and J.R. Steadman. 1994. Construction of a genetic linkage map and location of common blight, rust resistance and pubescence loci in *Phaseolus vulgaris* L. using Random Amplified Polymorphic DNA (RAPD) markers. Ann. Rept. Bean Improve. Coop. 37:37-38.

Kaplan, L. 1956. The cultivated beans of the prehistoric Southwest. Annal. Miss. Bot. Gard. 43:189-251.

Kaplan, L. 1965. Archeology and domestication in American *Phaseolus*. Econ. Bot. 19:358-368.

Kaplan, L. 1981. What is the origin of the common bean?. Econ. Bot. 35(2):240-254.

Kaplan, L. 1994. Accelerator mass spectrometry dates and the antiquity of *Phaseolus* cultivation. Ann. Rept. Bean Improve. Coop. 37:131-132.

Kelly, J.D. and M.W. Adams. 1987. Phenotypic recurrent selection in ideotype breeding of pinto beans. Euphytica 36:69-80.

Kornegay, J. and C. Cardona. 1991. Breeding for insect resistance in beans. Pp. 619-648 *in* Common Beans: Research for Crop Improvement (A. van Schoonhoven and O. Voysest, eds.). CAB International, Wallingford, UK/ CIAT, Cali, Colombia.

Laughlin, R.M. 1975. The great tzotzil Dictionary of San Lorenzo Zinancatan. Smithsonian Contr. to Anthropology 19. Smithsonian Institute Press, Washington, DC.

Leakey, C. 1988. Genotypic and phenotypic markers in common bean. Pp. 245-327 *in* Genetic Resources of *Phaseolus* Beans: their Maintenance, Domestication, Evolution, and Utilization (P. Gepts, ed.). Kluwer, Dordrecht, Netherlands.

Mackie, W.W. 1943. Origin, dispersal and variability of the lima bean, *Phaseolus lunatus*. Hilgardia 15:1-29.

Maquet, A., A. Gutiérrez and D.G. Debouck. 1990. Further biochemical evidence for the existance of two gene pools in lima beans. Ann. Rept. Bean Improve. Coop. 33:128-129.

Maréchal, R. 1969. Données cytologiques sur les espéces de la sous tribu des Papilionaceae - Phaseoleae - Phaseolinae. Premiére serie. Bull. de Jardin BotaniqueNational de Belgique 39:125-165.

Maréchal, R., J.M. Mascherpa and F. Stainier. 1978. Etude taxonomique d'un groupe complexe d'espèces des genres *Phaseolus* et *Vigna* (Papilionaceae) sur la base de données morphologiques et polliniques, traitées par l'analyses informatique. Mémoires des Conservatoire et Jardin Botaniques de la Ville de Genève. Boissiera (Geneva) 28:1-273.

McClean, P., J. Ewing, M, Lince and K. Grafton. 1994. Development of RAPD map of *Phaseolus vulgaris* L. Ann. Rept. Bean Improve. Coop. 37:79-80.

McElroy, J.B. 1985. Breeding dry beans, *Phaseolus vulgaris* L., for common bacterial blight resistance derived from *Phaseolus acutifolius* A. Gray. PhD dissertation, Cornell University, Ithaca, NY, USA.

Miklas, P., E. Johnson and J. Beaver. 1995. RAPD markers for QTLs expressing BGMV resistance in dry bean. Ann. Rept. Bean Improve. Coop. 38:111-112.

Miklas, P.N., J.R. Stavely and J.D. Kelly. 1993. Identification and potential use of a molecular marker for rust resistance in common bean. Theor. Appl. Genet. 85:745-749.

National Research Council. 1989. Lost Crops of the Incas: Little Known Plants of the Andes with Promise for Worldwide Cultivation, pp. 113-179. National Academy Press. Washington, DC, USA.

Nodari, R.O., S.M. Tsai, R.L. Gilbertson and P. Gepts. 1993. Towards an integrated linkage map of common bean. 2. Development of an RFLP-based linkage map. Theor. Appl. Genet. 85 (5):513-520.

Pachico, D. 1989. Trends in world common bean production. Pp. 1-8 *in* Bean Production Problems in the Tropics (H.F. Schwartz and M.A. Pastor-Corrales, eds.). CIAT, Cali, Colombia.

Pinchinat, A.M. 1977. The role of legumes in tropical America. *In* Proceedings of Workshop on Exploring the Legume-*Rhizobium* Symbiosis in Tropical Agriculture, Kahului, Maui, Hawaii, 1976. Department of Agronomy and Soil Science, College of Tropical Agriculture, Honolulu, University of Hawaii. Miscellaneus Publications No. 145:171-182.

Piper, C.V. 1926. Studies in American *Phaseolinae*. Contrib. U.S. Natl. Herb. 22:603-701.

Prakken, R. 1970. Inheritance of colour in *Phaseolus vulgaris* L., II: a critical review. Meded. Landbouwhogesch. Wageningen 70(23):1-40.

Prakken, R. 1972. Inheritance of colour in *Phaseolus vulgaris* L., III: on genes for red seedcoat colour and a general synthesis. Meded. Landbouwhogesch. Wageningen 72(29):1-82.

Pratt, R. C., and G.P. Nabhan. 1988. Evolution and diversity of *Phaseolus acutifolius* genetic resources. Pp. 409-440 *in* Genetic Resources of *Phaseolus* Beans: their Maintenance, Domestication, Evolution, and Utilization (P. Gepts, ed.). Kluwer Academic Publishers, Dordrecht.

Purseglove, J.W. 1968. Tropical Crops. Dicotyledons 1, Pp. 284-285. John Wiley and Sons. Inc., NY.

Roberts, M.H.E. 1982. List of genes-*Phaseolus vulgaris* L. Ann. Rept. Bean Improve. Coop. 25:109-127.

Schmit, V. and D.G. Debouck. 1991. Observations on the origin of *Phaseolus polyanthus* Greenman. Econ. Bot. 45(3):345-364.

Schmit, V., P. Du-Jardin, J.P. Baudoin and D.G. Debouck. 1993. Use of chloroplast DNA polymorphisms for the phylogenetic study of seven *Phaseolus* taxa including *P. vulgaris* and *P. coccineus*. Theor. Appl. Genet. 87 (4):506-516.

Schoonhoven, A. van, C. Cardona and J. Valor. 1983. Resistance to the bean weevil and the Mexican bean weevil (Coleoptera: Bruchidae) in non-cultivated common bean accessions. J. Econ. Entomol. 76:1255-1259.

Schoonhoven, A. van and O. Voysest. 1989. Common beans in Latin America and their constraints. Pp. 33-57 *in* Bean Production Problems in the Tropics. 2nd ed. (H.F. Schwartz and M.A. Pastor-Corrales, eds.). CIAT. Cali, Colombia.

Shellie-Dessert, K.C. and F.A. Bliss. 1991. Genetic improvement of food quality factors. Pp. 649-677 *in* Common Beans: Research for Crop Improvement (A. van Schoonhoven and O. Voysest, eds.). CAB International, Wallingford, UK/ CIAT, Cali, Colombia.

Silbernagel, M.J., W. Janssen, J.H.C, Davis and G. Montes de Oca. 1991. Snap bean production in the tropics: implications for genetic improvement. Pp. 835-862 *in* Common Beans: Research for Crop Improvement (A. van Schoonhoven and O. Voysest, eds.). CAB International, Wallingford, UK/ CIAT, Cali, Colombia.

Singh, S.P. 1991. Bean genetics. Pp. 199-286 *in* Common Beans: Research for Crop Improvement (A. van Schoonhoven and O. Voysest, eds.). CAB International, Wallingford, UK/ CIAT, Cali, Colombia.

Singh, S.P. 1992. Common bean improvement in the tropics. Plant Breeding Reviews 10:199-269.

Singh, S.P. and J.A. Gutierrez. 1984. Geographical distribution of the Dl_1 and Dl_2 genes causing hybrid dwarfism in *Phaseolus vulgaris* L., their association with seed size, and their significance to breeding. Euphytica 33:337-345.

Singh, S.P., A. Molina, C.A. Molina, and J.A. Gutierrez. 1993. Use of interracial hybridization in breeding the race Durango common bean. Can. J. Plant Sci. 73:785-793.

Singh, S.P., P. Gepts and D.G. Debouck. 1991a. Races of common bean (*Phaseolus vulgaris*, Fabaceae). Econ. Bot. 45(3):379-396.

Singh, S.P., R. Nodari and P. Gepts. 1991b. Genetic diversity in cultivated common bean: 1. Allozymes. Crop Sci. 31(1):19-23.

Singh, S.P., C.A. Urrea, A. Molina and J.A. Gutierrez. 1992. Performance of small-seeded common bean from the second selection cycle and multiple cross intra-and interrracial populations. Can. J. Plant Sci. 72:735-741.

Smartt, J. 1990. Grain Legumes: Evolution and Genetic Resources. Cambridge University Press, Cambridge, New York.

Tohme, J., P. Jones, S. Beebe and M. Iwanaga. 1995a. The combined use of agroecological and characterization data to establish the CIAT *Phaseolus vulgaris* core collection. Pp. 95-107 *in* Core Collections of Plant Genetic Resources (T. Hodgkin, A.H.D. Brown, Th.J.L. van Hintum and E.A.V. Morales, eds.). J. Wiley and Sons, Chichester, UK.

Tohme, J., O. Toro, J. Vargas and D. Debouck. 1995b. Variability in Andean Nuña Common Beans (*Phaseolus vulgaris* Fabaceae). Econ. Bot. 49(1):78-95.

Uebersax, M.A. and L.G. Occeña. 1991. Composition and nutritive value of dry edible beans: commercial and world food relief applications. Michigan Dry Bean Digest 15(5):3-12,28.

Vallejos, C.E. and C.D. Chase. 1991. Extended map for the phaseolin linkage group of *Phaseolus vulgaris* L. Theor. Appl. Genet. 82 (3):353-357.

Vanderborght, T. 1979. Le dosage de l'acide cyanohydrique chez *Phaseolus lunatus* L. Annales de Gembloux 85:29-41.

Vanderborght, T. 1982. Seed increase and evaluation of the wild *Phaseolus vulgaris* germplasm. Internal Report. CIAT, Cali, Colombia.

Vanderborght, T. 1983. Evaluation of *Phaseolus vulgaris* wild types and weedy forms. Plant Genet. Resour. Newsl. 54:18-24.

Voysest, O. 1983. Variedades de frijol en América Latina y su origen. CIAT, Cali, Colombia.

Wallace, D.H. and R.E. Wilkinson. 1965. Breeding for *Fusarium* root rot resistance in beans. Phytopathology 55:1227-1231.

Westphal, E. 1974. Pulses in Ethiopia, their taxonomy and agricultural significance, pp. 129-176. Agric. Res. Rep. 815. Wageningen, Netherlands.

White, J.W. and S.P. Singh. 1991. Breeding for adaptation to drought. Pp. 501-560 *in* Common Beans: Research for Crop Improvement (A. van Schoonhoven and O. Voysest, eds.). CAB International, Wallingford, UK/CIAT, Cali, Colombia.

Yarnell, S.H. 1965. Cytogenetics of the vegetable crops, IV: Legumes. Bot. Rev. 31(3):247-330.

Young, R.A. and J. Kelly. 1994. A RAPD marker for the ARE anthracnose gene in beans. Ann. Rept. Bean Improve. Coop. 37:77-78.

Zaumeyer, W.J. and J.P. Meiners. 1975. Disease resistance in beans. Annu. Rev. Phytopathol. 13:313-334.

Appendix 1. Tentative list of *Phaseolus sensu stricto* species (including the three sections *Phaseolus, Alepidocalyx* and *Minkelersia*) (as proposed by Debouck 1991).

Species	Section[†]	Species	Section[†]
P. acutifolius A. Gray	Ph	*P. nelsonii* M.M.S	Mi
P. amabilis Standley	Mi	*P. oaxacanus* Rose	Ph
P. amblyosepalus (Piper) Morton	Al	*P. oligospermus* Piper	Ph
P. angustissimus Asa Gray	Ph	*P. ovatifolius* Piper	Ph
P. augustii Harms	Ph	*P. pachyrrhizoides* Harms	Ph
P. brevicalyx Micheli	Ph	*P. palmeri* Piper	Ph
P. chiapasanus Piper	Ph	*P. parvulus* Greene	Mi
P. coccineus L.	Ph	*P. pauciflorus* Sessé Mociño	Mi
P. costaricensis Freytag & Debouck	Ph	*P. pedicellatus* Bentham	Ph
P. esperanzae Seaton	Ph	*P. pluriflorus* M.M.S.	Mi
P. filiformis Bentham	Ph	*P. polyanthus* Greenman	Ph
P. floribundus Piper	Ph	*P. polymorphus* S. Wats.	Ph
P. foliaceus Piper	Ph	*P. polystachyus* (L.) B.S.P.	Ph
P. glabellus Piper	Ph	*P. ritensis* Jones	Ph
P. glaucocarpus Norvell	Ph	*P. salicifolius* Piper	Ph
P. grayanus Woot. & Standl.	Ph	*P. scabrellus* Bentham	Ph
P. griseus Piper	Ph	*P. schaffneri* Piper	Ph
P. hintonii Delgado	Ph	*P. sempervirens* Piper	Ph
P. jaliscanus Piper	Ph	*P. sinuatus* Nutt	Ph
P. leptostachyus Bentham	Ph	*P. smilacifolius* Pollard	Ph
P. lunatus L.	Ph	*P. sonorensis* Standley	Ph
P. macrolepis Piper	Ph	*P. striatus* Brandegee	Ph
P. maculatus Scheele	Ph	*P. tuerckheimii* Donn. Smitt	Ph
P. micranthus Hook. & Arn.	Ph	*P. venosus* Piper	Ph
P. microcarpus Mart.	Ph	*P. vulgaris* L.	Ph
P. minimiflorus Norvell	Ph	*P. xanthotrichus* Piper	Ph[‡]
P. mollis Hook	Ph	*P. xolocotzii* Delgado	Ph
P. neglectus Hermann	Ph		

[†] Ph = *Phaseolus*; Mi = *Minkolorsia*; Al = *Alepidocalyx*.

[‡] May form another section.

Chapter 12

Pigeonpea[1]

P. Remanandan and L. Singh

BOTANY AND DISTRIBUTION

Pigeonpea, *Cajanus cajan* (L.) Millsp., belongs to the subtribe Cajaninae, tribe Phaseoloideae, subfamily Papilionoideae and family Leguminosae. It is an ancient cultivated crop with many vernacular and trade names in various languages and dialects. In India, it is called *arhar* and *tur*, in Portuguese *guand* and in Spanish *guandu*. The name pigeonpea was first reported from Barbados where the seeds were used to feed pigeons (Plukenet 1692). Van der Maesen (1986) has recorded over 300 names of pigeonpea. The important names include red gram, *tur, arhar, guandul* and *pois d'Angole*.

The plant is a perennial shrub, though in India it is usually cultivated as an annual crop. Pigeonpea has a deep root system which helps to withstand drought. The plant has a C_3 pathway for carbon fixation, and interacts with cowpea strains of *Rhizobium* to fix atmospheric nitrogen. Most traditionally grown pigeonpeas have an indeterminate flowering habit. The inflorescences develop as axillary racemes from the branches and flowering proceeds acropetally. In the determinate types, the apical buds of the main shoots develop into inflorescences. Flowers are yellow with orange or red streaks, and the back of the corolla is fully red or orange. The flower has a typical papilionaceous structure with diadelphous (9+1) stamens, superior, subsessile or short, stalked ovary, long, filiform style with a terminal stigma. Pods are compressed, pubescent with a diagonal depression, 5-8 cm long and 1-5 cm wide. Seed number per pod varies from 2 to 9, but is usually 3 to 5. Seed shape is usually oval; other shapes are square, pea (globular) and elongate. Seed colour varies from white to almost black, and cultivars show remarkable diversity in seed mass (weight), with 100-seed mass range of 3-25 g. Seeds are non-endospermic and contain two large cotyledons. Germination is hypogeal.

Origin, Diffusion and Distribution

Pigeonpea originated in India, and spread to Africa quite early, where a secondary centre of diversity developed (van der Maesen 1986). Seventeen species of *Cajanus*, including the closest species, *Cajanus cajanifolius* (Haines) van der Maesen *comb. nov.*, occur in India. *Cajanus cajanifolius*, the most probable progenitor of pigeonpea, is found only in India. The only distinguishable morphological trait by which it differs from cultivated pigeonpea is the presence of seed strophiole in *C. cajanifolius*. The species readily crosses with pigeonpea. It is extremely rare and is probably on the verge of extinction since its natural habitat consists of open areas which are vulnerable to grazing. The species still occurs in rare patches in the southern parts of Orissa and Madhya Pradesh states, and the genetic diversity found in the pigeonpea landraces

---

[1] Submitted as Journal Article no. 1895 by ICRISAT.

presently grown in this area is remarkable. This leads us to believe that pigeonpea originated in the central and eastern parts of India. Reddy (1973) and De (1974) postulated that the genus *Cajanus* probably originated from an advanced *Atylosia* species through single-gene mutation. It is now apparent that this advanced species is *C. cajanifolius*. Pigeonpea is cultivated in the tropical and subtropical areas between 30°N and 30°S latitudes. It has been introduced into most of the tropical and subtropical countries to which people of Indian origin have migrated. With the slave trade, pigeonpea moved from Africa to the Americas, and is now widely grown in the Caribbean.

Reproductive Biology

Pigeonpea is a quantitative, short-day plant with a critical daylength of 13 hours (Sharma *et al.* 1981). However, there are genotypic variations in sensitivity to photoperiod. No pigeonpea cultivar is truly photoperiod-insensitive and the degree of sensitivity varies quantitatively, the shortest-duration types being the least sensitive. The floral structure of pigeonpea favours self-pollination, which occurs in the flower before the petals open. When the petals are open, insect pollination may take place (van der Maesen 1986). Outcrossing of 0-70% has been reported, depending on the genotype, location, season and the pollinators.

Taxonomy

Following the revision of the taxonomy of *Cajanus* (van der Maesen 1986), its nearest relatives, earlier classified in *Atylosia* W. and A., have been merged with *Cajanus* DC. These include all the species of *Atylosia*, three newly described species from Australia, two species of *Endomallus* Gagnep. and the cultivated *C. cajan*. Primarily on the basis of the success in hybridization between pigeonpea and its wild relatives, van der Maesen (1990) placed all the species interfertile with pigeonpea in the secondary genepool. These are *C. acutifolius* (F. von Muell.) van der Maesen *comb. nov.*, *C. albicans* (W. and A.) van der Maesen *comb. nov.*, *C. cajanifolius* (Haines) van der Maesen *comb. nov.*, *C. lanceolatus* (W. V. Fitzg.) van der Maesen *comb. nov.*, *C. latisepalus* (Reynolds and Pedley) van der Maesen *comb. nov.*, *C. lineatus* (W. and A.) van der Maesen *comb. nov.*, *C. reticulatus* (Dryander) F. von Muell., *C. scarabaeoides* (L.) Thou. van der Maesen *comb. nov.*, *C. sericeus* (Benth. ex Bak.) van der Maesen *comb. nov.* and *C. trinervius* (DC.) van der Maesen *comb. nov.* All these were earlier included in *Atylosia* and most of the crossing efforts were directed toward these species. Those species which do not readily cross with pigeonpea have been placed in the tertiary genepool. These are *C. goensis* Dalz., *C. heynei* (W. and A.) van der Maesen *comb. nov.*, *C. mollis* (Benth.) van der Maesen *comb. nov.*, *C. platycarpus* (Benth.) van der Maesen *comb. nov.*, *C. rugosus* (W. and A.) van der Maesen *comb. nov.* and such other Cajaninae as *Rhynchosia* Lour., *Dunbaria* W. and A., *Eriosema* (DC.) Reichenb.

GERMPLASM CONSERVATION AND USE

Genetic Resources

Some 12 885 accessions of pigeonpea from 72 countries are conserved at the ICRISAT Asia Center, Patancheru, India. This diverse assemblage of germplasm has been systematically characterized and subjected to preliminary evaluation (Remanandan *et al.* 1988a, 1988b) and detailed information on 11 034 accessions originating from 52 countries is now available.

Table 12.1. Range of variability in pigeonpea germplasm held in the ICRISAT genebank.

Character	Minimum	Maximum	No. observations[†]
Days to 50% flowering	52	237	11 968
Days to 75% maturity	97	299	11 956
Plant height (cm)	39	385	11 957
Primary branches (number)	2	66	11 900
Secondary branches (number)	0.3	145.3	11 865
Racemes (number)	6.0	915.0	11 052
Seeds/pod (number)	1.6	7.6	11 888
100-seed mass (g)	2.8	25.8	11 887
Harvest index (%)	0.6	62.7	10 975
Shelling ratio (%)	5.3	87.5	11 006
Seed protein percentage (%)	12.4	29.5	11 601

[†] This indicates the number of accessions in which a specific trait has been measured.

Range of diversity for major characteristics

Pigeonpea germplasm offers a wide range of variation for almost all yield components, quality traits and adaptation (Table 12.1). Genetic male sterility is available and can be transferred easily to any cultivar. This is currently used in hybrid pigeonpea breeding programmes (Lal *et al.* 1989). Use of multiple disease resistant lines and insect-tolerant lines can substantially contribute to yield stability. Information on the origin of accessions from diverse ecological regions can aid in selecting cultivars suitable for introduction to new areas. The ICRISAT Asia Center collection has a fair representation of germplasm from arid areas, acidic soils, high altitudes and other harsh environments. Cultivars have various combinations of desirable traits, such as disease-resistant, large-seeded lines adapted to high altitude, or short-duration, determinate lines.

Some genetic stocks have such unique traits as dwarfism, genetic markers, modified flowers (cleistogamous and/or wrapped flowers) and mutants. The potential for improvement within the germplasm is evident from the remarkable diversity available and the yields attained in experimental fields. Many traditional landraces have been released directly as cultivars without any change in genetic constitution and several have been used as sources for specific characters such as yield components, resistance to biotic and abiotic stresses, quality traits and adaptation (Table 12.2).

Notwithstanding the impressive variation available in the germplasm, limits to genetic enhancement of the species are frequently encountered. The perennial nature of the species, its slow initial growth rate and poor partitioning of biomass resulting in low harvest index all constrain genetic enhancement for yield. Although short-duration pigeonpeas are less perennial, true annuality is not known in pigeonpea.

The reproductive biology of the species favours self-pollination. However, pigeonpea is often cross-pollinated. Therefore, traditional breeding methods of self-pollinated species, particularly pedigree breeding, have not been very effective for traits with low heritability.

The large genotype by environment interaction, in particular, photoperiod by temperature interaction, further reduces the efficiency of selection. There is genotypic variation for sensitivity to photoperiod in the germplasm and there are genotypes which are comparatively less sensitive to photoperiod.

The exploitation of hybrid vigour has recently become possible with the discovery of genetic male sterility, but the cost of hybrid seed production remains high. The search for cytoplasmic male sterility (CMS) was initiated in the 1990s; a stable CMS system has not yet been found in the species.

Insect pests are the major yield constraints in all pigeonpea-growing areas. Tolerant genotypes have been reported, but a stable source of host-plant resistance, an important component of integrated pest management, remains a weak link in the overall effort to improve productivity of pigeonpea. Biotechnological techniques for genetic enhancement of this critical trait have not yet been developed.

Geographic distribution of important traits

The Indian subcontinent is the major area of genetic diversity. Nearly all the important traits are found in the subcontinent, particularly in north and central India. Pigeonpea varies widely in duration, from 3 months to >9 months; this trait is very important in the adaptation of cultivars to various agroclimatic areas and cropping systems. In recent years there has been increased interest in short-duration pigeonpeas. These are relatively less sensitive to photoperiod by temperature interactions and have a short growing season in which to interact with environmental factors. One major limitation in achieving significant impressive genetic enhancement in short-duration pigeonpeas is its narrow genetic base. Out of the 943 short-duration lines presently conserved in the ICRISAT genebank, over 50% are breeding products, sharing many common parents. Nearly all short-duration accessions originate from northern and central India. Most of the medium-duration accessions also originate from India, and, as we have noted above, it is interesting that nearly all the pigeonpeas found outside the Indian subcontinent are long-duration, indeterminate types, many of which are vegetable types with large pods and large, light-coloured seeds.

Pigeonpeas of all durations and plant types are found in various parts of the Indian subcontinent. However, vegetable types are grown mainly in the state of Gujarat. The vegetable types are preferred by tribal people who grow pigeonpea around their houses in the hilly areas of the states of Madhya Pradesh, Orissa and in the Western Ghats of southern India.

Determinate genotypes are generally short in stature and have clusters of pods at the top of the plant canopy; they are predominantly of short or medium duration. Most of the determinate genotypes are also products of breeding programmes. The ICRISAT genebank presently holds 463 determinate accessions, of which 309 accessions are breeding products, 258 from ICRISAT and 51 from the University of Queensland, Australia. A further 34 lines are purified germplasm for photoperiodic insensitivity. The remaining 120 accessions originate from India, Sri Lanka and the Caribbean. The genetic base in this group is not very large.

Host-plant resistance to disease is another important trait in sustainable crop improvement programmes. Systematic screening at ICRISAT of germplasm for host-plant resistance to diseases resulted in the identification and development of resistant genotypes to three diseases of major economic importance: sterility mosaic (SM), fusarium wilt and phytophthora blight.

The collection at ICRISAT Asia Center has 321 accessions resistant to SM, of which 59 are landraces directly identified as resistant and 267 are lines purified through single-plant selection. Most of the accessions resistant to these three diseases are from northern and central India (Table 12.3). *Cajanus platycarpus*, a wild relative, is the only source which is immune to phytophthora blight and this species, which is rather rare now, is distributed in the Western Ghats of India. The species belongs to the tertiary genepool and hence does not cross readily with *Cajanus*. Recently, a successful hybrid was produced at ICRISAT through the embryo-rescue method (Nalini and Moss 1995). Thirteen accessions of *C. platycarpus*, which is on the verge of extinction, have been salvaged from the Western Ghats in the state of Maharashtra, and northern parts of Uttar Pradesh and Himachal Pradesh. It is essential to secure more accessions of this species because it has other desirable traits such as earliness of flowering and high seed protein (Remanandan 1990).

Central and northern India are rich in other important yield components such as number of primary and secondary branches and number of racemes. This area is the home of pigeonpea's progenitor, *C. cajanifolius*, which has vanished from most of its natural habitats but still occurs, albeit rarely, in the Bailadela area in Madhya Pradesh.

Table 12.2. Origin of pigeonpea cultivars released since 1979 by ICRISAT and its collaborators.

ICP no.	Cultivar	Origin	Country, year of release	DF[t]	Remarks
Short duration					
15598	MN 5	Sel.Extra SD Composite	USA, 1995	44	Selection made at Minnesota from Extra SD composites developed at ICRISAT
15599	MN 8	Sel.Extra SD Composite	USA, 1995	49	Selection made at Minnesota from Extra SD composites developed at ICRISAT
15597	MN 1	Sel.Extra SD Composite	USA, 1995	55	Selection made at Minnesota from Extra SD composites developed at ICRISAT
14440	Sarita (ICPL 85010)	ICPL 87 x DL 78-1	India, 1996	63	Short-duration DT cultivar released in H.P. India
14057	Hunt	Prabhat x ICP 7018	Australia, 1983	76	ICP 7018 originates from Madhya Pradesh
14057	Megha	Prabhat x ICP 7018	Indonesia, 1987	76	ICP 7018 originates from Madhya Pradesh
11543	Pragati (ICPL 87)	T21 x ICP 6393	India, 1986; Myanmar, 1992	77	ICP 6393 originates from Madhya Pradesh
13194	Quantum	T21 x ICP 6393	Australia, 1985	78	
11605	Jagriti (ICPL 151)	ICP 6997 x Prabhat	India, 1990	78	ICP 6997 originates from Madhya Pradesh
14421	Durga (ICPL 84031)	ICPL 87 x (Prabhat x UPAS 120)	India, 1995	78	Released in Andhra Pradesh, India
–	ICPH 8	ms Prabhat x ICPL 161	India, 1991	80	Short-duration indeterminate hybrid released in the central zone of India. Parent of ICPL 161 (ICP 6) originates from Andhra Pradesh
–	PPH 4	ms Prabhat x AL 688	India, 1994	–	Short-duration indeterminate hybrid released in Punjab, India
–	CoH 1	ms T 21 x ICPL 87109	India, 1994	–	Short-duration indeterminate hybrid released in Tamil Nadu, India
–	Quest	(Prabhat x HY 3C) x (ICP 7018 x ICP 7035)	Australia, 1988	–	ICP 7018 and ICP 7035 originate from Madhya Pradesh
Medium duration					
15600	Birsa Arhar	Wilt-resistant bulk from BDN 1 x ICP 7620	India, 1992	106	Wilt-resistant population, released in Bihar, India
8863	Maruti	Sel. ICP 7626	India, 1985	111	Wilt-resistant selection from ICP 7626, which originates from Uttar Pradesh, India
14770	Abhaya (ICPL 332)	Sel. ICP 1903	India, 1989	122	Pod borer tolerant selection from ICP 1903, which originates from Andhra Pradesh

ICP no.	Cultivar	Origin	Country, year of release	DF[†]	Remarks
14056	Royes (UQ 50)	Q 8189 West Indies	Australia, 1979; Fiji	129	Medium-duration DT cultivar, released for mechanized production
14722	Asha (ICPL 87119)	C 11 x ICP 1-6	India, 1993	135	Wilt- and SM-resistant; C 11 is a released cultivar; ICP 1-6 is a wilt-resistant selection from another released cultivar 'Sharda'
Long duration					
7035	Kamica	Landrace from Madhya Pradesh	Fiji, 1985	138	Landrace, resistant to wilt and SM, released directly as a variety
13829	Cerro Pelon	Landrace from Grenada	Venezuela, 1991	139	Semi-determinate vegetable type landrace selected from farmers' fields in Grenada (PR 6534), released directly as a variety
6997	Rampur Rhar	Landrace from Madhya Pradesh	Nepal, 1992	140	SM-resistant landrace, released directly as a variety
9905	La Cerrera	Sel. ICP 7217	Venezuela, 1991	144	Selection from a landrace ICP 7217 (Turk-Thogari) which originates from Karnataka, India, released directly as a variety
9145	Nandolo wa nswana	Landrace from Kenya	Malawi, 1988	163	Wilt-resistant landrace, released directly as a variety
11916	Aroa	Landrace from W.Ghats, India	Venezuela, 1991	170	A high-performing ratoonable landrace selected from the fields of ethnic communities of Attapadi, Western Ghats, India (PR 5193) released directly as a variety for crop-livestock system in Venezuela
11384	Bageswari	Landrace from Nepal	Nepal, 1992	176	SM-resistant landrace, selected from farmers' fields in Nepal (PR 5147), released directly as a variety.

† Days to 50% flowering at ICRISAT Asia Center, Patancheru, India.

Table 12.3. Number of accessions resistant to three major diseases of pigeonpea, and their geographical origin.

Area of origin	Sterility mosaic	Fusarium wilt	Phytophthora blight
Uttar Pradesh	59	6	44
Madhya Pradesh	35	7	12
Bihar	80	2	22
Maharashtra	4	3	14
Orissa	21	–	–
Andhra Pradesh	36	3	10
Karnataka	15	2	3
Assam	3	–	2
Tamil Nadu	5	–	2
Rajasthan	–	–	1
Haryana	–	–	1
Gujarat	–	–	2
Meghalaya	–	–	1
West Bengal	2	–	–
Punjab	1	–	–
Others[†]	60[‡]	6	26
Total	321	29	140

[†] These include breeding lines, where the precise origin of the parents which contributed disease resistance could not be traced to their original habitats.
[‡] Out of these, 11 accessions are from Kenya.

Wild relatives as a source of diversity

Information on the desirable traits available in closely related wild species can be found in Remanandan (1981). *Cajanus albicans, C. lineatus, C. sericeus* and *C. crassus* var. *crassus* are resistant to SM. *Cajanus platycarpus* is resistant to phytophthora blight, although such high levels of resistance are not known in cultivated pigeonpeas. This species flowers in 48 days. *Cajanus sericeus* is resistant to both blight and SM. Insect damage is the major constraint in pigeonpea production. *Cajanus scarabaoides* has a higher level of tolerance to insect pests than any other pigeonpea line. Despite this comparatively higher level of tolerance, even this species is not totally resistant to insect pests such as pod borers and pod flies; a truly resistant source is not yet known.

Evaluation

Passport information and characterization data of the world collection of pigeonpea have been documented and computerized using descriptors and descriptor states jointly developed by the International Board for Plant Genetic Resources (IBPGR) and ICRISAT (IBPGR/ICRISAT 1981). Passport information consists of accession identifiers, information on origin and related data. Characterization data include 40 descriptors on morpho-agronomic traits, of which 22 have been entered into a computer-based catalogue. The pigeonpea germplasm passport and characterization information is stored in a Relational Database Management Software known as System 1032 (Compuserve Techologies, USA) on VAX (VMS) environment. The application software developed for this crop allows authorized users to add/update the information. Other users on the system can retrieve the information on all or a desired set of accessions (either with all the data on these accessions, or with information on only a few descriptors). They can also retrieve information on accessions belonging to a particular class. Further, the system facilitates manipulation of the stored data for statistical analysis to examine patterns of variation. It hence serves as a live catalogue, which is frequently revised and updated as new information becomes available.

The utility of the stored information depends largely upon its accessibility to germplasm users the world over. To achieve this, ICRISAT published a catalogue of pigeonpea germplasm (Remanandan *et al.* 1988a, 1988b). In this catalogue, the accessions have been classified into a number of natural and artificial groups, together with several short lists of accessions that have frequently required combinations of morpho-agronomic traits.

Conservation techniques

Pigeonpea has orthodox seeds and can be safely stored under low temperature and low relative humidity after the moisture content of the seed is brought to the desired level. The seeds produced during the post-rainy season are generally of high quality and are used for conservation.

Since pigeonpea is a partially cross-pollinated crop, seed multiplication is carried out under controlled pollination. The branches or the whole plant are covered with muslin bags. Selfed seeds from about 30 plants per accession are bulked to constitute the next generation and to reconstitute the original population as closely as possible.

Sowing pigeonpea close to the shortest day of the year results in reduced plant height, and thus allows whole plants to be conveniently covered with muslin bags. It is also possible to control pollination by using cages made of frames covered with nets. When small numbers of cultivars are to be multiplied for large-scale seed production, geographic isolation of about 100 m is desirable. Accessions are rejuvenated by resowing selfed seed. It is desirable, however, to restrict the number of rejuvenations to the bare minimum to minimize the risk of genetic drift.

Harvested pods are sun-dried and machine-threshed. Since the ambient relative humidity at the time of harvest (Feb-Apr) is generally low at ICRISAT Asia Center (in India), the moisture content of the harvested seed will also be low (10-12%), which is sufficient for medium-term conservation. However, for long-term conservation, the moisture content should be 5±1%, and the conventional technique of sun-drying does not always achieve this level without affecting viability. Therefore, a satisfactory drying system at 15°C and 15% RH is employed to dry seeds to the desired moisture level for long-term conservation. The active germplasm is stored under medium-term conditions, at 4°C and 20% RH. To minimize the frequency of rejuvenation, a relatively large quantity of seed (300-400 g) is stored. Seeds are stored in aluminium cans with screw caps that have rubber gaskets inside to keep them moisture proof.

The base collection is stored at –20°C for long-term conservation using modules procured from Watford Refrigeration (UK). About 5000 seeds per accession are stored in the base collection. We are in the process of transferring a duplicate set of the world collection to the National Bureau of Plant Genetic Resources (NBPGR), New Delhi, India, as an extra measure of safety. The viability of the seed is determined by conducting germination tests at regular intervals (3-5 years) during storage to predict the storage life and time of rejuvenation.

Properties and Uses

Pigeonpea is a unique crop with multiple uses as a source of food, feed, fuel and fertilizer. It is used in more diverse ways than any other legume.

In the Indian subcontinent, pigeonpea is used mainly as human food. Starch and protein are the principal constituents of the seed. Starch content of the cotyledons in diverse cultivars varies from 51.4 to 58.8% (mean 54.7%); protein content ranges between 18.5 and 26.3% (mean 21.5%). Closely related wild species have a seed protein content of up to 30%. Faris and Singh (1990) have listed the distribution of dietary nutrients in different parts of mature pigeonpea seed. In India, pigeonpea is used mostly in the form of dehulled split cotyledons known as *dhal*. In Africa, the dry seed is cooked together with vegetables or meat. The use of pigeonpea to partially substitute for soyabean is being explored in Southeast Asia. Pigeonpea sauce, made by fermenting pigeonpea with *Aspergillus oryzae, A. niger* and *Rhizopus* sp., is an acceptable substitute for soyasauce in Indonesia. Green seeds are consumed as a vegetable in eastern Africa, the Caribbean islands, South/Central America, Indonesia the Philippines and parts of India (Fig. 12.1).

Pigeonpea plants and grain have been used as animal feed for centuries by Indian farmers. After the harvest, plants are often left in the field to be browsed by animals. The crop debris left behind after threshing is widely used as cattle feed. Its value as animal feed will depend on the proportions of stem, leaf and pods in the residue. Pigeonpea is considered an acceptable protein source for all classes of poultry feed (Wallis *et al.* 1986).

Dry pigeonpea stems (Fig. 12.2) are widely used as cooking fuel in the Indian subcontinent and the drier parts of eastern and southern Africa. They are also used to make field fences, huts and baskets. Farmers often grow pigeonpea in poor soils on which it is difficult to grow other crops. The canopy provides extensive ground-cover,

Fig. 12.1.   Pods and seeds of purified germplasm with excellent vegetable-type characteristics; large pods and large, globular seeds are preferred in vegetable pigeonpea.

helping to control soil erosion by wind and water. Being a legume, it fixes nitrogen; leaf fall further improves soil fertility. Dry roots, leaves, flowers and seeds are used to treat a wide range of skin, liver, lung and kidney ailments.

Breeding

Genetic variation

The landraces that have been grown for centuries are the principal primary genepool of pigeonpea. The landraces grown in Asia, Africa and the Americas are largely of medium to long duration (180-250 days), photoperiod-sensitive, tall (2-3 m), with slow early growth and indeterminate flowering habit. They are adapted to inter/mixed cropping with tall cereals; after the cereal is harvested, pigeonpea grows, produces vegetative biomass, flowers and fruits on residual soil moisture. These populations, however, are heterogeneous. Since the early 1900s, breeders have used these landraces for selection, hybridization and improvement for such traits as yield, plant height (genetic dwarfs), growth habit, duration, resistance to fusarium wilt, seed size and colour. More recently, landraces have provided sources for resistance to sterility mosaic disease, shorter duration (100-150 days), relative insensitivity to photoperiod and temperature, and genetic male sterility.

When this primary genepool is grown outside its area of adaptation, e.g. from coastal regions to high altitudes or from 20°N latitude to over 40°N latitude, hidden variability for sensitivity to photoperiod by temperature interactions is expressed, providing a source for selection to increase or reduce this sensitivity.

This primary genepool, a reservoir of apparent and potential variability, has evolved in low-input rain-fed agriculture, retaining such traits as perenniality, low harvest index and slow early growth. These genotypes are adapted to intercropping systems and appear to be in a state of equilibrium with locally occurring diseases and pests.

Fig. 12.2. Harvested long-duration pigeonpea, showing the huge biomass that can be used for fuel and fodder.

Table 12.4. Breeding procedures for different traits in pigeonpea genetic enhancement.

Trait	Breeding procedures
Dry grain yield in all maturity groups	Selection among landraces; hybridization-pedigree or bulk pedigree methods; multilocational testing. Hybrids using genetic male sterility.
Resistance to SM, fusarium wilt, phytophthora blight (single or multiple resistance)	Screening in disease nurseries and other procedures as above. Introgression from wild species and cultivated landraces.
Seed characteristics	Especially for eastern and southern Africa, selections for large (100-grain mass >12 g) white seeds in all maturity groups.
Improved resistance to *Helicoverpa* pod borer and podfly (not much work is being done on tolerance to pod-sucking bugs, *Maruca*, etc.)	Screening and selection under pesticide-free conditions. Strategic studies on traits relating to pest incidence for future application as integrated pest management components. Introgression from wild species and transformation (limited).
Tolerance to drought and waterlogging, insensitivity to photoperiod x temperature interactions	Presently looking for genetic variants and mechanism of tolerance; no specific breeding programme.
Plant type (stature, growth habit, harvest index, biomass)	Characterization of genetic variability and introgression.

With changing patterns in agriculture, e.g. monocropping and mechanization, breeders now look for such traits as short plant stature, short duration, more product-ivity per unit area/time, stability and sustainability of the cropping system or the crop/livestock system. To achieve some of these goals, variability in the primary, secondary and tertiary genepools will be more intensively studied and utilized (Table 12.4).

Prospects

Interest in pigeonpea is growing in many countries because it has multiple uses and is a hardy plant that, when intercropped with a cereal, ensures a measure of income stability in the rain-fed semi-arid tropics. Known for its efficient root system, the pigeonpea plant has been described as a biological plough, because of the improvement its cultivation brings about in soil structure. Substantial organic deposition takes place in the soil through leaf fall and mobilization of unavailable soil phosphorus; this contributes further to the improvement of soil fertility. It is a potential crop for marginal lands both in the plains and on hilly slopes, not only to improve soils, but also to prevent soil erosion. Since the crop does not require high inputs, its importance will increase as a component of sustainable production systems, particularly in the rain-fed areas of tropical Asia, South America and Australia. The potential of pigeonpea as a cover crop in the new rubber plantations of Thailand and Indonesia has been demonstrated (Wallis *et al.* 1986)

The agro-industry for dehulling pigeonpea dry grain to produce *dhal* is dependent on the raw material, which is produced largely in the Indian subcontinent and eastern Africa. The demand for *dhal* currently exceeds production in the Indian subcontinent and in many parts of the world with large Asian populations. The *dhal*-milling industry will, therefore, grow further and generate employment in both the rural and urban sectors. The by-products of dehulling such as seed husk (10%) and broken cotyledons and powder (15-25%) are used mostly as cattle feed. Their nutritive value needs to be investigated (Whiteman and Norton 1981). Because of the high cost of such sources of animal protein as fishmeal, high-quality plant protein is being used increasingly to feed animals, particularly pigs and poultry. Recent ICRISAT studies show that pigeonpea can be successfully grown in semi-arid Venezuela, thus reducing expensive imports of soyabean for poultry and animal feed. Pigeonpea also has high potential for cultivation in many other parts of semi-arid South America.

In recent years, short-duration cultivars developed by the Indian national programme and ICRISAT have begun to change the existing cropping system in some parts of India from intercropping to sole cropping with high planting density. There is also considerable potential to use short-duration genotypes in double-cropping and multiple-harvest systems. Further, the short-duration cultivars are relatively less sensitive to photoperiod; there is therefore the potential to extend the adaptation of pigeonpea from tropical and subtropical areas (up to 30° latitude) to temperate regions (up to 45° latitude). The potential for using pigeonpea in agroforestry/alley cropping systems for fodder, firewood and grain, exploiting its perenniality and ratoonability, has not been exploited fully.

REFERENCES

De, D.N. 1974. Pigeonpea. Pp. 79-87 *in* Evolutionary Studies in World Crops: Diversity and Change in the Indian Subcontinent (J. Hutchinson, ed.). Cambridge University Press, London, UK.

Faris, D.G. and U. Singh. 1990. Pigeonpea: nutrition and products. Pp. 401-433 *in* The Pigeonpea (Y.L. Nene, S.D. Hall and V.K. Sheila, eds.). CAB International, Wallingford, UK.

IBPGR/ICRISAT. 1981. Descriptors for Pigeonpea. IBPGR, Rome, Italy.

Lal, S., A.N. Asthana and C.L.L. Gowda. 1989. Use of chickpea and pigeonpea germplasm and their impact on crop improvement in India. Pp. 101-103 *in* Collaboration on Genetic Resources: Summary Proceedings of a Joint ICRISAT/NBPGR (ICAR) Workshop on Germplasm Exploration and Evaluation in India, 14-15 Nov 1988, ICRISAT Center, India. ICRISAT, Patancheru, India.

Nalini, M. and J.P. Moss. 1995. Production of hybrids between *Cajanus platycarpus* and *Cajanus cajan*. Euphytica 83:43-46.

Plukenet, L. 1692. Phytographia 3.

Reddy, L.J. 1973. Interrelationship of *Cajanus* and *Atylosia* Species as Revealed by Hybridization and Pachytene Analysis. PhD Thesis, Indian Institute of Technology, Kharagpur, India.

Remanandan, P. 1981. The wild gene pool of *Cajanus* at ICRISAT, present and future. Pp. 29-38 *in* Proceedings of the International Workshop on Pigeonpeas, 15-19 Dec 1980, ICRISAT Center, India, Vol. II. ICRISAT, Patancheru, India.

Remanandan, P. 1990. Pigeonpea: genetic resources. Pp. 89-115 *in* The Pigeonpea (Y.L. Nene, S.D. Hall and V.K. Sheila, eds.). CAB International, Wallingford, UK.

Remanandan, P., D.V.S.S.R. Sastry and M.H. Mengesha. 1988a. ICRISAT Pigeonpea Germplasm Catalog: Evaluation and Analysis. ICRISAT, Patancheru, India.

Remanandan, P., D.V.S.S.R. Sastry and M.H. Mengesha. 1988b. ICRISAT Pigeonpea Germplasm Catalog: Passport Information. ICRISAT, Patancheru, India.

Sharma, D., L.J. Reddy and K.C. Jain. 1981. International adaptation of pigeonpea. Pp. 71-81 *in* Proceedings of the International Workshop on Pigeonpeas, 15-19 Dec 1980, ICRISAT Center, India, Vol. I. ICRISAT, Patancheru, India.

van der Maesen, L.J.G. 1986. *Cajanus* D.C. and *Atylosia* W. & A. (Leguminosae). Wageningen Papers, 85-4 (1985). Agricultural University, Wageningen, Netherlands.

van der Maesen, L.J.G. 1990. Pigeonpea: origin, history, evolution, and taxonomy. Pp. 15-46 *in* The Pigeonpea (Y.L. Nene, S.D. Hall, and V.K. Sheila, eds.). CAB International, Wallingford, UK.

Wallis, E.S., D.G. Faris, R. Elliott and D.E. Byth. 1986. Varietal improvement of pigeonpea for smallholder livestock production systems. Pp. 536-553 *in* Proceedings of the Crop Livestock System Research Workshop, 7-11 July 1986, Khon Kaen, Thailand. Farming Systems Research Institute, Dept. of Agriculture, Thailand, and Asian Rice Farming Systems Network, IRRI, Philippines.

Whiteman, P.C. and B.W. Norton. 1981. Alternative uses of pigeonpea. Pp. 365-377 *in* Proceedings of the International Workshop on Pigeonpeas, 15-19 Dec 1980, ICRISAT Center, India, Vol. I. ICRISAT, Patancheru, India.

Chapter 13

Faba Bean

L.D. Robertson

Faba bean (*Vicia faba* L.) is an Old World legume that has been called 'poor man's meat'. China and Ethiopia are the major producers, but it also is cultivated in the Mediterranean region, northern Europe, North Africa and West Asia. Faba beans introduced to South America are mostly grown at higher elevations instead of *Phaseolus* beans. Australia has dramatically increased the area planted to faba bean in the past decade.

BOTANY AND DISTRIBUTION

The genus *Vicia* belongs to the order Rosales, suborder Rosinae, family Leguminosae and subfamily Papilionaceae and is in the tribe Vicieae (Kupicha 1981). *Vicia faba* is in the subgenus *Vicia* and section *Faba*.

Muratova (1931) subdivided *V. faba* into two subspecies: *paucijuga* and *eu-fabae*. Maxted (1993) placed *V. faba* in the monospecific section *Faba* (Miller) Ledeb. and contributed the following infraspecific classification:

V. faba subsp. *faba* L.

    V. faba subsp. *faba* var. *minor* Beck

    V. faba subsp. *faba* var. *equina* Pers.

    V. faba subsp. *faba* var. *faba* L.

    V. faba subsp. *paucijuga* Murat.

The crop is known by many names in the English language: faba bean, broadbean, field bean, tick bean, horse bean. The common names of faba bean in different countries are: *foul*, Arabic; *haba*, Spanish; *feve, feverole*, French; *chechevitza*, Russian; *zeindo*, Chinese.

The varieties *major* (large-seeded), *minor* (small-seeded) and *equina* (intermediate seed size) are also used. Field beans are usually of the *minor* (tick beans) and *equina* (horse bean) types.

Cubero and Suso (1981) used the concept of fertility barriers for defining biological species and concluded that *V. faba* could only be grouped into one subspecies, *Vicia faba* subsp. *faba*. They felt that the *paucijuga* forms were probably the most similar to the wild types and more closely related to the supposed extinct progenitor. Faba bean is diploid species with $2n=12$. *Vicia faba* does not produce fertile hybrids with any other species and to date no gene exchange has been achieved between this and any other species; thus it is the only species in its primary and secondary genepool. Faba bean is a partially outcrossing species, with cross-pollination the result of insect pollination (Bond and Poulsen 1983). Still retaining vestiges of its past, faba bean can loosely be regarded as an incompletely domesticated species, as indicated by the breeding system of the species (which stands between full autogamy and full allogamy), the indeterminate growth habit and dehiscent pods.

Origin, Domestication and Diffusion

Schafer (1973) hypothesized that *V. faba* originated from an extinct ancestor and *Vicia narbonensis* L. (2*n*=14) has received the greatest attention as the putative ancestor of faba bean. *Vicia faba* was most likely domesticated between Afghanistan and the eastern Mediterranean in 7000-4000 BC (Hanelt 1972). Cubero (1972) concluded that culture spread in four directions: north to central Europe, northwest to western Europe, west to the Mediterranean and east to the Far East. The *minor* type faba bean was introduced to China in 100 BC (Tao 1981) and the *major* type in 1200 AD (Hanelt 1972).

Reproductive Biology

Bond and Poulsen (1983) summarized the data on outcrossing from many studies and reported an average of 35% with a range of 4-84%. The reproductive system of faba bean influences improvement for yield and yield stability through both autosterility problems of cultivars and its effects on efficiency of various breeding strategies. Pollination control is a major problem for faba bean breeding programmes, whether they follow pedigree selection, mass selection within crosses or develop synthetics. The major constraint to most breeding programmes is in overcoming problems of pollination control.

GERMPLASM CONSERVATION AND USE

ICARDA has maintained the faba bean resources in two types of germplasm collections (Robertson 1985). The original germplasm accessions were originally maintained as populations (Burton 1970) and are now maintained as composite bulks in the landrace or International Legume Faba Bean (ILB) collection. A Faba Bean Pure Line collection (BPL) has been derived from the ILB collection by a 'pure-breeding' process of single-plant progeny rows to obtain uniformity using insect-proofed screenhouses to ensure selfing (Fig. 13.1). The status of the collections is presented in Table 13.1. Fifteen percent of faba bean accessions held at ICARDA are landraces collected by ICARDA and ALAD (Arid Lands Agricultural Development Program, precursor organization of ICARDA) and most accessions are from the West Asia and North Africa (WANA) region.

Germplasm Characterization and Evaluation

This material was first screened in trials and nurseries in Syria and Lebanon, covering a wide range in rainfall and thermal regimes to include elite lines in the International Testing Network. To date, four cultivars of faba bean have been released from landraces or landrace selections from the germplasm collections (Table 13.2). Many are selections that have been purified for such traits as seed size, disease resistance, frost tolerance, non-shattering, etc.

Selections for disease resistance sources in faba bean for chocolate spot (induced by *Botrytis fabae* Sard.), ascochyta blight (induced by *Ascochyta fabae* Speg.), rust (induced by *Uromyces fabae* (Pers.) de Bary) and stem nematodes (*Ditylenchus dipsaci* (Kühn) Filipjev) have followed a parallel course to that described for the development of homogeneous BPL accessions in Fig. 13.1. The same cyclic single-plant progeny row system was used, with the additional factor of selecting under artificial inoculation for disease resistance, instead of taking random plants, as done for the normal BPL accessions. In this way, many useful sources of resistance were developed for these pathogens (Table 13.3). This material was tested internationally and several sources were found to have durable resistances for chocolate spot and ascochyta blight (Hanounik and Robertson 1987, 1988). Research under drought conditions at ICARDA identified lines ILB 1814, 80S43856 and 80L90121 as more efficient in water use over three seasons (Robertson and Saxena 1993). The BPL germplasm collection has been partially evaluated for descriptors listed by IBPGR/ICARDA. Results for 840 BPL lines are summarized in a published catalogue (Robertson and El-Sherbeeny 1988) which provides listings of useful accessions for various

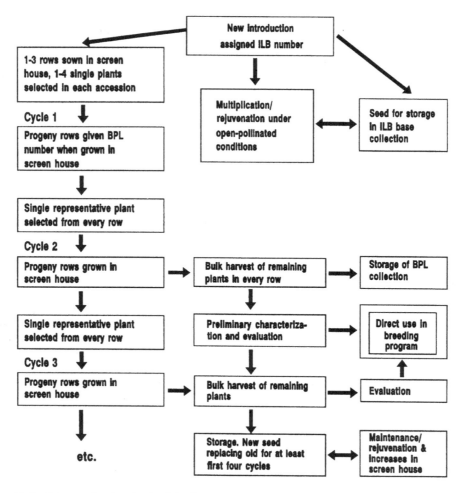

Fig. 13.1. Scheme for developing faba bean pure line (BPL) collection at ICARDA.

descriptors (Table 13.4) along with their summary statistics. Analysis of discretely scored descriptors (Robertson and El-Sherbeeny 1991) for frequency distributions showed eight (stipule spot pigmentation, leaflet size, lodging, intensity of streaks, pod angle, pod shape, pod colour and seed shape) with significant variation (Table 13.5). Summary statistics (Table 13.6) for quantitative descriptors show large variation, except for maturity date. Inbred lines self-pollinated for at least five generations were found to show a normal distribution for seed yield, with yield varying between 0 and 7.5 t/ha, mean 3.6 t/ha (Fig. 13.2a).

The faba bean pure line collection has been screened for autofertility (Robertson and El-Sherbeeny 1995), the ability to self-pollinate in the absence of pollinating insects (Bond and Pope 1987). The mean seed index (SI, measurement for autofertility) among the 840 BPL accessions was 0.54 with a standard error of 0.01 (Table 13.7). The distribution for autofertility clearly demonstrates marked differences among accessions (Fig. 13.2b). A high number of BPL accessions failed to set seeds without tripping and are described as autosterile. A large number of accessions possessed high autofertility (Table 13.4) and marked differences were found among countries of origin, with Egyptian faba beans having the highest autofertility and the former USSR the lowest (Table 13.8).

Table 13.1. Frequency by country of origin[†] of the ICARDA faba bean germplasm collections.

Country	ILB	BPL	Country	ILB	BPL
AFG	94	108	LBY	11	–
ARG	1	4	LKA	2	4
AUS	9	30	LVA	1	1
AUT	1	2	MAR	330	181
BGR	14	2	MEX	7	14
BOL	–	2	NLD	10	10
CAN	172	9	NPL	5	1
CHE	2	–	OMN	4	–
CHN	159	342	PAK	35	50
COL	43	95	PAL	4	10
CSK	12	–	PER	35	47
CYP	104	251	POL	40	20
DEU	68	64	PRT	109	15
DZA	41	41	ROM	33	14
ECU	97	235	RUS	2	2
EGY	74	127	SDN	115	44
ESP	339	779	SUN	52	34
ETH	384	682	SWE	6	11
FRA	15	15	SYR	330	176
GBR	85	98	TUN	31	65
GRC	29	54	TUR	147	212
HUN	15	12	UKR	4	7
IDN	1	1	URY	1	2
IND	11	10	USA	7	11
IRN	15	17	YEM	13	27
IRQ	64	114	YUG	14	30
ITA	57	73	ZAF	–	3
JOR	23	33	Not known	1134	951
JPN	6	12			
LBN	36	88	**Total**	**4453**	**5248**

[†] Origin country codes as per ISO (International Standardization Organization) three-letter country codes.

Table 13.2. Release of faba bean varieties based on ICARDA-supplied genetic resources.

Country	Variety	Year of release
Australia	ICARUS (BPL 710)	1992
Egypt	Giza Blanka (ILB 1270)	1991
	Giza 461 (ILB 938)	1993
Iran	Barkat (ILB 1269)	1986

Table 13.3. The most important faba bean inbred resistant sources to chocolate spot, ascochyta blight, stem nematodes, rust, *Orobanche crenata*, Bean Leaf Roll Virus (BLRV) and Bean Yellow Mosaic Virus (BYMV).

Disease	Sources
Chocolate spot	BPL 110, 112, 261, 266, 710, 1179, 1196, 1278, 1821; ILB 3025, 3026, 2282, 3033, 3034, 3036, 3106, 3107, 2302, 2320,; L82003, L82009
Ascochyta blight	BPL 74, 230, 365, 460, 465, 471, 472, 646, 818, 2485; ILB 752; L83118, L83120, L83124, L183125, L83127, L83129, L183136, L83142, L83149, L83151, L83155, L83156, L82001
Rust	BPL 7, 8, 260, 261, 263, 309, 409, 417, 427, 484, 490, 524, 533, 539; Sel.82 Lat. 15563-1, -2, -3, -4
Stem nematode	BPL 1, 10, 11, 12, 21, 23, 26, 27, 40, 63, 88, 183
Orobanche crenata	18009, 18025, 18035, 18054, 18105, LS 222, LS 225, 8/9-72, -85, -86, -128, -136, -137, -138, -139, -143, -152, -153
BLRV	BPL 756, 757, 758, 769
BYMV	BPL 1351, 1363, 1366, 1371

Table 13.4. Faba bean accessions with desirable values for various descriptors.

Descriptor/desired	Accessions (BPL)
Flowering (earliness)	180, 237, 247, 271, 280, 281, 286, 298, 548, 558, 561, 570, 589, 600, 609, 719, 726, 729, 1230
Lodging resistant	185, 218, 376, 385, 545, 643
Height of pods (tall)	39, 40, 42, 48, 50, 53, 58, 60, 65, 68, 69, 70, 71, 72, 76, 85, 828, 836, 844, 894, 969, 972, 1262, 1351, 2443
Pods per plant (high)	704, 787, 805, 978, 1703, 1732, 1739, 1758, 1766, 1827
Pod length (long)	250, 251, 255, 318, 339, 415, 416, 455, 632, 890, 1089, 1102, 1106, 1129, 1157. 1159, 1868, 1872, 1876
Seed yield (high)	22, 24, 46, 47, 52, 53, 65, 77, 133, 348, 379, 393, 442, 444, 451, 596, 725, 787, 936, 999, 1015, 1043, 1046, 1120, 1160
Autofertility (high)	17, 42, 43, 44, 50, 69, 70, 101, 296, 543, 555, 568, 569, 579, 580, 591, 614, 698, 703, 722, 1029, 1145, 1151, 1342, 1815
Protein content (high)	61, 171, 491, 495, 526, 527, 528, 603, 737, 976, 1014, 1386

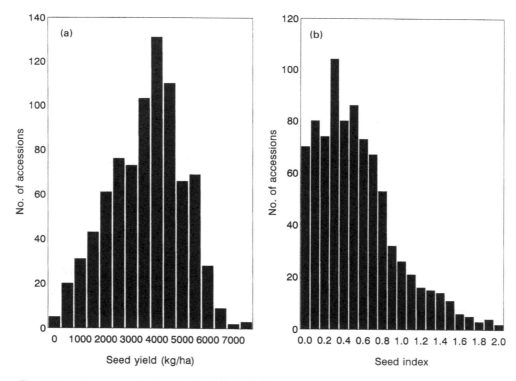

Fig. 13.2. Frequency distribution for (a) seed yield and (b) autofertility for 840 BPL faba bean accessions evaluated at Tel Hadya, Syria during 1985/86.

Table 13.5. Frequency distributions for 840 BPL faba bean accessions evaluated at Tel Hadya, Syria in 1985/86.

Descriptor/ Score		Frequency (%)	Descriptor/ Score		Frequency (%)
Growth habit	All accessions were indeterminate	100.0	Pod surface reflectance	Matte	97.0
Stem pigmentation	Absent	17.5		Glossy	2.4
	Weak	58.0		Mixed	0.6
	Intermediate	21.3	Pod colour	Light	35.7
	Strong	3.1		Dark	62.4
	Mixed	0.1		Mixed	1.9
Leaflet size	Small	37.6	Pod distribution	Uniform	93.1
	Medium small	24.9		Mainly basal	5.5
	Medium	26.2		Mainly terminal	1.4
	Medium large	8.9	Pod shattering	Non-shattering	90.4
	Large	2.4		Shattering	9.6
Stem colour	All accessions were light	100.0	Male fertility	All accessions were male fertile	100.0
Stipule spot pigmentation	Absent	2.0	Ground colour of testa	Dark brown	5.6
	Present	98.0		Light brown	91.2
Flower ground colour	White	99.8		Light green	0.1
	Dark brown	0.2		Dark green	0.2
Intensity of streaks on the standard petal	No streaks	0.4		Red	0.2
	Slight	54.3		Violet	2.0
	Moderate	33.8		White	0.1
	Intense	11.5		Grey	0.2
Wing petal colour	Uniformly white	0.5		Violet brown	0.2
	Spotted	99.5	Hilum colour	Black	93.1
Pod angle	Erect	42.9		Colourless	5.7
	Horizontal	40.2		Grey	0.1
	Pendant	16.9		Mixed	1.1
Pod shape	Subcylindrical	39.6	Seed shape	Flattened	28.3
	Flattened constricted	12.7		Angular	56.7
	Flattened non-constricted	47.6		Round	15.0

Table 13.6. Summary statistics for 840 BPL germplasm accessions evaluated at Tel Hadya, Syria during 1985/86.

Descriptor[†]	Check mean	Mean	Minimum	Maximum	CV (%)
DFLR (days)	98.5	101.4	91	127	7.3
DMAT (days)	185.3	186.5	173	200	2.4
PTHT (days)	91.8	93.4	52	152	13.9
HTFP (cm)	23.3	23.8	7.5	84.0	44.4
PPP	14.7	13.5	2.3	42.5	40.1
PL (cm)	9.4	7.6	2.0	14.8	23.2
HI (%)	39.8	32.0	5.0	63.7	24.0
HSW (g)	132.0	85.4	20.8	193.2	39.9
SYLD (kg/ha)	5420	3574	0	7510	40.3

[†] DFLR=days to flowering; DMAT=days to maturity; PTHT=plant height; HTFP=height to 1st pod; PPP=pods per plant; PL=pod length; HI=harvest index; HSW=100-seed weight; SYLD=seed yield.

Table 13.7. Summary statistics of autofertility indices and yield and its components for all tested BPL accessions.

Variable[†]	Mean	SE of mean	CV (%)	Minimum	Maximum	Range	df
SI	0.539	0.014	74.2	0.00	1.99	1.99	821
FPI	4.63	0.30	18.4	1.33	7.92	6.58	839
SYLD	3570	49.7	40.3	0	7510	7510	839
HSW	85.4	1.17	39.9	20.8	193.2	172.4	839
PPP	13.5	0.19	40.1	2.3	42.5	40.2	839
SPD	2.5	0.02	17.4	1.3	4.6	3.3	839

[†] SI=Seed index; FPI=flowers per inflorescence; SYLD=seed yield; HSW=100-seed weight; PPP=pods per plant; SPD=seeds per pod.

Table 13.8. Minimum, maximum and mean of seed index (autofertility) and standard deviations for BPLs according to country of origin[†].

Country	Number of BPLs	Seed index			
		Minimum	Maximum	Mean	SD
AFG	63	0.00	1.51	0.43	0.32
DZA	27	0.03	1.34	0.43	0.35
EGY	58	0.00	1.84	0.94	0.50
ESP	63	0.09	1.48	0.50	0.35
ETH	79	0.00	1.99	0.40	0.37
GBR	48	0.03	1.52	0.54	0.38
GRC	10	0.09	1.15	0.49	0.34
IRQ	55	0.00	1.92	0.73	0.51
JOR	18	0.00	1.42	0.70	0.52
JPN	10	0.26	0.98	0.61	0.26
LBN	29	0.03	0.93	0.46	0.27
MAR	12	0.15	1.44	0.59	0.41
SDN	16	0.04	1.96	0.79	0.57
SUN	13	0.00	0.78	0.26	0.26
SYR	27	0.00	1.87	0.71	0.41
TUN	28	0.16	1.43	0.50	0.30
TUR	84	0.00	1.46	0.54	0.30
YUG	12	0.06	1.33	0.64	0.43
Overall		0.00	1.99	0.57	0.39
F (Among/Within)				6.70**	

[†] Origin country codes as per ISO (International Standarization Organization) three-letter country codes.

Table 13.9. Minimum, maximum and mean of protein content (%) and standard deviations for BPLs according to country of origin[†].

Country	Number of BPLs	Protein (%)			
		Minimum	Maximum	Mean	SD
AFG	63	20.0	27.0	24.6	14
DZA	27	20.1	27.7	23.5	19
EGY	58	21.3	28.3	25.0	14
ESP	63	20.5	29.8	24.4	19
ETH	79	18.0	30.1	23.5	21
GBR	48	18.8	26.9	23.7	17
GRC	10	20.4	25.7	23.1	17
IRQ	55	21.2	28.8	24.6	15
JOR	18	19.5	25.8	23.0	16
JPN	10	22.0	28.7	26.6	20
LBN	29	21.8	27.0	24.3	12
MAR	12	20.2	26.5	23.1	19
SDN	16	22.8	27.5	25.3	14
SUN	13	19.3	27.8	23.7	28
SYR	27	20.1	26.1	23.2	15
TUN	28	20.5	27.2	24.2	15
TUR	84	19.7	28.5	23.6	16
YUG	12	19.5	25.6	22.5	19
Overall	652	18.0	30.1	24.1	17
F (Among/Within)				6.70**	

[†] Origin country codes as per ISO (International Standarization Organization) three-letter country codes.

Table 13.10. Distribution of the faba bean ILB and BPL collections from ICARDA for the period 1990-92.

Year	Country code[†]	No. of accessions	Year	Country code[†]	No. of accessions
1990	GBR	1	1993	AUS	516
	IND	20		DEU	120
	IRQ	49		ETH	822
	PRK	23		FRA	45
	RUS	2		GBR	1
	YEM	2		IRQ	48
	Subtotal	**97**		MAR	3
				POL	253
1992	BGD	20		USA	26
	CHN	11		**Subtotal**	**1834**
	DEU	222			
	EGY	92	1994	DZA	3
	ITA	22		EGY	38
	JOR	2		ETH	43
	MAR	9		JOR	2
	MEX	8		MAR	4
	USA	127		TUN	182
	Subtotal	**513**		**Subtotal**	**272**
				Total	**2716**

[†] Origin country codes as per ISO (International Standarization Organization) three-letter country codes.

Seed protein content varied between 18.0 and 31.0% with a mean of 24.0% (El-Sherbeeny and Robertson 1992). Accessions with a high protein content are listed in Table 13.4. There was moderate variation among countries for protein content (Table 13.9), with

Yugoslavia having the lowest and Japan the highest. Cooking time for faba bean has been determined in the world collections and showed a large variation.

Gaps in Germplasm Collections

One of the major gaps in the faba bean collections is the lack of passport data. Another is the relatively small number of accessions from China, the major producer of faba bean. The collection lacks adequate resistance sources to *Orobanche crenata* Forsk. The genetic base of resistance sources to chocolate spot also needs to be widened.

Use and Distribution

Since 1990, 2716 accessions of the faba bean ILB and BPL collections have been distributed from ICARDA (Table 13.10), most of them to countries in the developing world. Major uses of this material have been for screening for and receiving sources of resistance to rust, ascochyta blight, chocolate spot and *O. crenata*. Other requests have been made to broaden the genetic base of the crop in a particular country. Many faba bean breeding programmes built their foundation on ICARDA resources.

Seed Conservation

Faba bean has orthodox seeds that are easily maintained in the active collection at 0°C at a relative humidity of 16%. Seeds are dried to a seed moisture content of about 6% before storage. The major problem in conservation is the relatively large seed size, storage space and the low rates of multiplication and rejuvenation. There is a schedule of seed testing to ensure that only pathogen- and virus-free seed of faba bean are kept in the collection on a long-term basis. Also, during multiplication in the field, there is routine monitoring of the faba bean multiplication plots for viruses and pathogens. ICARDA has signed an agreement for the safety-duplication of its ILB and BPL collections with the Federal Institute of Agrobiology (FIA) in Vienna, Austria and about 1500 accessions have been duplicated.

Properties and Uses

Several excellent reviews describe composition of faba bean seed (Hill-Cottingham 1983; Halse 1994) and Table 13.11 summarizes the analyses. Amino acid composition of faba bean complements that of cereals and although it is deficient in essential amino acids it has adequate concentrations of arginine and leucine (Williams *et al.* 1994). Important antinutritional factors affect its utilization as a human food. The most important are the glucosides, convicine and vicine (Marquardt 1982) which cause a haemolytic anaemia that is often fatal when young males first eat faba bean. The cause of the disease has been traced to a deficiency in glucose-6-phosphate dehydrogenase (Mager *et al.* 1965). The levels of the glucosides are higher in the green seed and consumption of even small quantities can trigger an attack. Other antinutritional factors include L-Dopa, haemaglutinins, oligosaccharides, phytic acid, tannins, trypsin inhibitors and saponins (Williams *et al.* 1994).

The chief uses of faba bean for human consumption are summarized in Table 13.12. They are often consumed as a green salad, vegetable dish and dry seed. The dry seeds of faba bean are used whole in dishes such as *cous-cous* and *foul medames*, a mainstay of the diet in Egypt. In Ethiopia the dishes *kollo, kik, nirfro, gunkul, ashuk* and *endushdush* are made from faba bean by various methods including roasting, soaking and sprouting; seed is ground to a flour and used in such national dishes as *shiro* and *siljo*. In Asia whole seeds of faba bean are commonly used for snacks, either roasted or fried and are eaten like nuts. Dry seeds are also ground and flour used to make *falafel*, a doughnut-shaped dish which is deep-fried. In China faba bean seeds are important for processing and many products for food and industrial use are produced (Table 13.12). In Europe faba bean is used as a protein supplement in animal feed.

Table 13.11. Ranges for proximate composition (% of dry matter) for faba bean reported by various researchers.

Component	Amount (%)	Reference
Crude protein	18.0 - 31.0	El-Sherbeeny and Robertson 1992
	23 - 39	Halse 1994
	22 - 36	Bond 1977
	20.3 - 41.0	Chavan *et al.* 1989
Crude fibre	5.1 - 9.9	Frauen *et al.* 1984
	5.0 - 8.5	Chavan *et al.* 1989
	6.0 - 10.6	Eden 1968
	5.0 - 7.3	Ali *et al.* 1981
Carbohydrates	50.9 - 67.9	Chavan *et al.* 1989
	50.9 - 53.5	Ali *et al.* 1981
Mineral matter	2.7 - 3.7	Chavan *et al.* 1989
	3.1 - 4.8	Eden 1968
	2.7 - 3.0	Ali *et al.* 1981

Table 13.12. Common faba bean foods prepared from green pods, green seeds and dry seeds, including processed foods.

Type of material used	Type of food produced
Green pods	Sliced pods as green beans
Green seeds	Vegetable, salad, canned and frozen vegetable
Dry seeds	Foul medames, harira (soup), cous-cous, cooked beans, canned foul medames, fried snack (as nut type food), kik, kollo, gunkul, ashuk, endushdush
Ground dry seeds	Fafafel, shiro, siljo
Processed	Faba sauce (soya sauce substitute), noodles, vermicelli, protein additive, protein drinks, health foods
Industrial	Starch (papermaking, spinning, adhesive, weaving and medicine)

The type of cultivar desired is of paramount importance to a breeding programme and for faba bean includes: (1) landraces and mass selections from landraces, (2) open-pollinated populations and synthetics, (3) fully autogamous lines, (4) hybrids and (5) near-pure lines developed mostly by pedigree selection. All cultivars are open-pollinated, whether they are heterogeneous populations from mass selection or synthetics, or near-homozygous lines from pedigree programmes. Use of heterogeneous, heterozygous, open-pollinated cultivars, whether synthetics or products from mass selection, offers the exploitation of heterosis. If the attempt to convert faba bean to a truly self-pollinated crop (using the closed flower mutant and sources of autofertility available in the germplasm collection) is successful, then truly autogamous uniform cultivars could be developed. More importantly, breeding procedures would become more efficient.

Breeding

The breeding programme at ICARDA, initiated in 1977, stopped in 1991. A summary is given by Robertson *et al.* (1992). Germplasm collections are the main sources for resistances for diseases such as chocolate spot (Hanounik and Robertson 1987, 1988) and to *O. crenata* (Cubero and Martinez 1980). Most faba bean breeding programmes follow a pedigree breeding scheme with various degrees of pollination control. Pedigree selection is especially useful when selection is practised for such traits as disease and pest resistance, seed size, pod length and earliness. The ICARDA faba bean breeding programme made use of germplasm selected for disease resistances in the development of the BPL collection in a pedigree selection scheme which made use of an off-season nursery and alternate years of selection for disease resistance (Fig. 13.3). Selections for frost resistance and yield were made at Tel Hadya, Syria and for disease resistance at Lattakia,

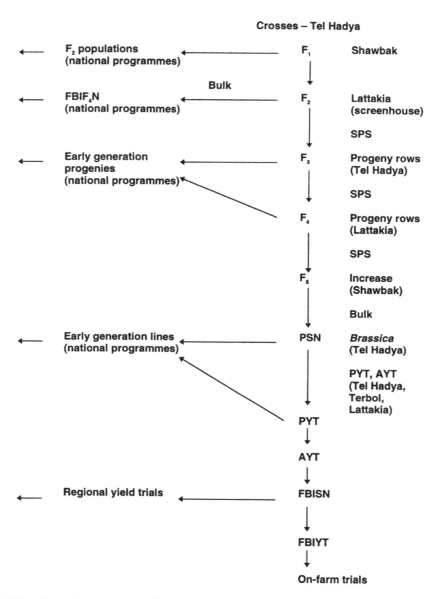

Fig. 13.3. Breeding scheme followed by the ICARDA breeding programme for disease resistance.

Syria. The off-season nursery was grown at Shawbak, Jordan. This scheme allowed the maximum use of disease-resistant sources from the germplasm collection for developing populations for use in national programmes for development of disease-resistant selections.

Prospects

Because of the considerable genetic variation in the *Vicia faba* germplasm collection, WANA breeding programmes are beginning to show interest in increasing areas of production. Exciting new plant types may alleviate some of the traditional problems of excessive vegetative growth with the accompanying large drop of flowers, young pods and lodging. The determinate genes and independent vascular supply types of faba bean are now available in Mediterranean-type germplasm through introgression with landraces from this region, offering the possibility for their use in the major production areas for faba bean. Disease-resistant varieties are becoming available from breeding programmes that have made use of sources from germplasm collections. Sources of resistance for two or more diseases are now becoming available for use in breeding programmes. Recent results in breeding for *O. crenata* resistance offer hope of control of this noxious weed. Newer biotechnological techniques offer the possibility of inserting herbicide resistance genes into faba bean with the possibility to control *O. crenata* by chemical means.

The genetic potential of faba bean is high, with yields of 5-6 t/ha reported. Using germplasm that offers solutions to many of the traditional problems of growing faba bean offers the promise of greatly increasing faba bean production by both increasing yield levels and alleviating the destabilizing effects of biotic and abiotic stress factors.

REFERENCES

Ali, A.E.M., G.E.E.A. Ahmed and E.K.B.E. Hardallou. 1981. Faba beans and their role in diets in Sudan. Pp. 317-318 *in* The Faba Bean (*Vicia faba* L.): A Basis for Improvement (P.D. Hebblethwaite, ed.). Butterworths, London, UK.

Bond, D.A. 1977. Breeding for zero-tannin and protein yield in field beans (*Vicia faba* L.). Pp. 348-360 *in* Protein Quality from Leguminous Crops. EUR 5686 EN.

Bond, D.A. and M. Pope. 1987. Proportion of cross-bred and selfed seed obtained from successive generations of winter bean (*Vicia faba* L.) crops. J. Agric. Sci., Camb. 108:103-108.

Bond, D.A. and M.H. Poulsen. 1983. Pollination. Pp. 23-76 *in* The Faba Bean (*Vicia faba* L.): A Basis for Improvement (P.D. Hebblethwaite, ed.). Butterworths, London, U.K.

Burton, G.W. 1970. Handling cross-pollinated germplasm efficiently. Crop Sci. 19:685-690.

Chavan, J.K., L.S. Kute and S.S. Kadam. 1989. Pp. 223-245 *in* CRC Handbook of World Legumes (D.K. Salunkhe and S.S. Kadam, eds.). CRC Press, Boca Raton, FL, USA.

Cubero, J.I. 1972. On the evolution of *Vicia faba* L. TAG 45:48-51.

Cubero, J.I. and A. Martinez. 1980. The genetics of the resistance of faba bean to *Orobanche crenata*. FABIS Newsl. 2:19-20.

Cubero, J.I. and M.J. Suso. 1981. Primitve and modern forms of *Vicia faba*. Kulturpflanze 29:137-145.

Eden, A. 1968. A survey of the analytical composition of field beans (*Vicia faba* L.). J. Agric Sci., Camb. 70:299-301.

El-Sherbeeny, M. and L.D. Robertson. 1992. Protein content variation in a pure line faba bean (*Vicia faba*) collection. J. Sci. Food Agric. 58:193-196.

Frauen, M., G. Robbelen and E. Ebmeyer. 1984. Quantitative measurement of quality determining constituents in seeds of different inbred lines from a world collection of *Vicia faba*. Pp. 279-296 *in* Vicia faba: Agronomy, Physiology and Breeding (P.D. Hebblethwaite, T.C.K. Dawkins, M.C. Heath and G. Lockwood, eds.). Marinus Nijhoff, The Hague, The Netherlands.

Halse, J.H. 1994. Nature, composition, and utilization of food legumes. Pp. 97 *in* Expanding the Production and Use of Cool Season Food Legumes (F.J. Muehlbauer and W.J. Kaiser, eds.). Proceedings of IFLRC II Conference, 12-16 April 1992, Cairo, Egypt. Kluwer, Dordrecht, The Netherlands.

Hanelt, P. 1972. Zur Geschichte des Anbaues von *Vicia faba* L. und ihrer verschiedenen Formen. Kulturpflanze 20:209-223.

Hanounik, S.B. and L.D. Robertson. 1987. New sources of resistance in *Vicia faba* to chocolate spot caused by *Botryis fabae*. Plant Dis. 72:696-698.

Hanounik, S.B. and L.D. Robertson. 1988. Resistance in *Vicia faba* germplasm to blight caused by *Ascochyta fabae*. Plant Dis. 73:202-205.

Hill-Cottingham, D.G. 1983. Chemical constituents and biochemistry. Pp. 159-180 *in* The Faba Bean (*Vicia faba* L.): A Basis for Improvement (P.D. Hebblethwaite, ed.). Butterworths, London, UK.

Kupicha, F.K. 1981. Pp. 377-381 *in* Advances in Legume Systematics Part 1 (R.M. Polhiss and P.H. Raven, eds.). Royal Botanic Gardens, Kew, UK.

Mager, J., G.N. Glaser, A. Razin, S. Bien and M. Noam. 1965. Metabolic effects of pyrimidines derived from faba bean glycosides on human erythrocytes deficient in glycose 6-phosphate dehydrogenase. Biochem. Biophys. Res. Comm. 20:235-240.

Marquardt, R.R. 1982. Favism. Pp. 343-353 *in* Faba bean Improvement: World Crops: Production, Utilization, Description, Vol. 6 (G.C. Hawtin and C. Webb, eds.). Martinus Nijhoff, The Hague, The Netherlands.

Maxted, N. 1993. A phenetic investigation of *Vicia* L. subgenus *Vicia* (Leguminosae-Vicieae). Bot. J. Linn. Soc. 111:155-182.

Muratova, V. 1931. Common beans, *Vicia faba* L. Suppl. Bull. Appl. Bot. Genet. Plant Breed. (Leningrad).

Robertson, L.D. 1985. Faba bean germplasm collection, maintenance, evaluation, and use. P. 15-21 *in* Faba Beans, Kabuli Chickpeas, and Lentils in the 1980s, Proceedings of an International Workshop (M.C. Saxena and S. Verma, eds.), 16-21 May, 1983, Aleppo, Syria. ICARDA, Aleppo, Syria.

Robertson, L.D. and M. El-Sherbeeny. 1988. Faba Bean Germplasm Catalog: Pure Line Collection. ICARDA, Aleppo, Syria.

Robertson, L.D. and M. El-Sherbeeny. 1991. Distribution of discretely scored descriptors in a pure line faba bean (*Vicia faba* L.) germplasm collection. Euphytica 57:83-92.

Robertson, L.D. and M.C. Saxena. 1993. Problems and prospects of stress resistance breeding in faba bean. P. 37-50 *in* Breeding for Stress Tolerance in Cool-Season Food Legumes (K.B. Singh and M.C. Saxena, eds.). John Wiley & Sons, Chichester, UK.

Robertson, L.D. and M.H. El-Sherbeeny. 1995. Autofertility in a pure line faba bean (*Vicia faba* L.) germplasm collection. Genet. Resour. and Crop Evol. 42:157-163.

Robertson, L.D., S.B. Hanounik, Z.A. Fatemi, R. Kallida, S.P.S. Beniwal and M.C. Saxena. 1992. Development and highlights of faba bean research in Morocco under INRA/ICARDA collaborative program. Al Awamia 78:83-112.

Schafer, H.I. 1973. Zur Taxonomie der Vicia narbonensis-Gruppe. Kulturpflanze 21:211-273.

Tao, Z.H. 1981. Faba bean production and research in China. FABIS Newsl. 3:24.

Williams, P.C., R.S. Bhatty, S.S. Deshpande, L.A. Hussein and G.P. Savage. 1994. Improving nutritional quality of cool season food legumes. Pp. 113-129 *in* Expanding the Production and Use of Cool Season Food Legumes (F.J. Muehlbauer and W.J. Kaiser, eds.). Proceedings of IFLRC II Conference, 12-16 April 1992, Cairo, Egypt. Kluwer, Dordrecht, The Netherlands.

Chapter 14

Soyabean

K. Dashiell and C. Fatokun

BOTANY AND DISTRIBUTION

Soyabean or soybean, *Glycine max* (L.) Merr. which also is synonymous with *Glycine soja* Sieb. & Zucc., *Glycine hispida* (Moench) Maxim. and *Soja max* (L.) Piper, belongs to the family Leguminosae, subfamily Papilionoideae, tribe Phaseoleae and genus *Glycine* Wild. The genus *Glycine* is divided into two subgenera: *Glycine* and *Soja*. While cultivated soyabean is in the subgenus *Soja* along with its annual wild relatives *G. soja* and *G. gracilis*, the subgenus *Glycine* contains nine wild perennial species (Juvik *et al.* 1985): *G. argyrea* Tindale, *G. canescens* F.J. Herm., *G. clandestina* Wendl., *G. cyrtoloba* Tindale, *G. falcata* Benth., *G. latifolia* (Benth.) Newell & Hymowitz, *G. latrobeana* (Meissn.) Benth., *G. tabacina* (Labill.) Benth. and *G. tomentella* Hayata. All of the nine perennial species are native to Australia with seven of them restricted to Australia.

Botany

The soyabean plant is an erect, bushy and hairy annual. It can grow to a height of 20-108 cm depending on the genotype and length of growing period before flowering because soyabeans are reproductively photosensitive. The time needed to attain maturity can range from 75 to 200 days and this period is of great importance in the adaptation of the crop to a particular latitude. All cultivars flower earlier when grown in environments with 14-16 hours of darkness (Purseglove 1977). Grey or tawny hairs are found on the stem, leaves, calyx and pods. Primitive cultivars tend to be prostrate and this is particularly so when plants are shaded. Improved soyabean varieties are generally determinate in growth. The stems have very few primary branches and no secondaries. In general, buds found in the primary leaf axils do not develop except when the shoot-tip is damaged.

Three types of stem growth occur in soyabean. In both determinate and semi-determinate, stem growth stops abruptly and the stem terminates in a long raceme. The third type is indeterminate where stem growth declines rather slowly, with the stem terminating in one or two pods or a small leaf. The root system is extensive and distributed mainly in the upper 20 cm of the soil. The radicle develops into the taproot, which is little different in thickness from its several branches, and can penetrate as deep as 2 m. When present, the root nodules are small, globose and sometimes may be irregular or lobed, especially when two or more infection areas develop close to one another and merge as they grow. Soyabean nodules are determinate because they do not have the extended terminal growth observed in some other legumes like *Medicago* and *Trifolium*. The mature structure of the nodules is retained until the 6th or 7th week after they start to senesce (Bergersen and Briggs 1958).

Four types of leaves are found on soyabean plants: first is a pair of seed leaves (cotyledons), followed by another pair of simple primary leaves, then the trifoliate and the prophylls. The trifoliate leaves are alternate in arrangement. The leaflets, with entire margins, vary in shape from ovate to lanceolate and also in size. Usually each leaf has three leaflets but occasionally four to seven may be found. Two small stipels subtend each terminal leaflet while the asymmetrical lateral leaflets have one each. The petiole is long, cylindrical and narrow, ending with a pulvinus at its base.

Origin, Domestication and Diffusion

The centre of diversity of a crop does not necessarily imply its centre of origin (Harlan 1971), but for soyabean China is both the centre of diversity and the centre of origin. Soyabean is believed to have been derived from *G. ussurienses* Regel & Maack, a slender, twining plant found growing wild throughout eastern Asia (Purseglove 1977). The wild, annual *Glycine soja* is, from most of the available evidence, the presumed ancestor of cultivated soyabean.

Soyabean was domesticated in China where, according to records, it has been in cultivation for over 3000 years. Since the process of domestication takes place over time it is suggested that for soyabean this may have occurred during the Shang Dynasty (ca. 1700-1100 BC) or even earlier (Hymowitz 1970). A semicultivated, weedy form of soyabean, *G. gracilis*, with an intermediate morphology between *G. max* and *G. soja*, is known only from Northeast China. These three species are also generally cross-compatible (Broich and Palmer 1980).

Soyabean is a very ancient crop of China where according to Hymowitz and Newell (1981) its primary gene centre is located. Records show that soyabean reached northeast, central and southern China and the Korean peninsula by the 1st century AD (Kwon 1972). Soyabean was introduced from China to Japan and southeast Asia, as well as south-central Asia where landraces have also developed. Movement of the crop to these places followed the Silk Route (Hymowitz and Newell 1981). Soyabean then spread to other parts of the world from this Asiatic region. It must have reached the Netherlands before 1737 as soyabean was described by Linnaeus. This implies that soyabean was one of the plants growing in the garden at Hartecamp (Hymowitz and Newell 1981). The Royal Botanic Garden in Kew, England recorded planting of soyabean for the first time in 1790 while in South America records show that the crop was present as early as 1882. According to Hymowitz and Newell (1981) soyabean was first mentioned in the American (USA) literature by Mease in 1804.

Because soyabeans are daylength-sensitive, they spread much easier between east and west than between north and south. Movement of the crop across latitudes was therefore rare (Hymowitz and Kaizuma 1981). On the basis of seed protein electrophoresis data, historical and agronomic literature, Hymowitz and Kaizuma suggested that the paths of soyabean movement from the eastern half of north China, the primary centres of diversity, were:

1. to the Asiatic regions of the former Soviet Union,
2. to northern China and Japan to southern Korea, and
3. to central China and Kyushu in Japan before moving northwards to Hokkaido.

There were further movements from:

4. coastal China to Taiwan,
5. central China to the northern half of the Indo-Pakistan subcontinent, and
6. Japan, south China and southeast China to central India.

The genotypes found in these places constitute the primary, secondary and tertiary genepools for soyabean.

Reproductive Biology

After a period of vegetative growth, the duration of which varies with genotype and the environment, plants reach the reproductive stage when the axillary buds develop into clusters of flowers containing 2-35 flowers each. In the determinate type, vegetative growth terminates at the appearance of the first flowers, whereas in the indeterminate types, flowering occurs at the 8th to 10th leaf stage. Three major genes are known to influence flowering and time to maturity in soyabean. On the plants the first flowers appear as from the 5th, 6th or even higher nodes on the main stem since the lower ones are vegetative. The bud in the axil of a trifoliate leaf that develops into an inflorescence has a stalk which is like a stem and possesses attributes of a stem by having epidermis, cortex and vascular tissue. The stalks expand following secondary growth from the vascular cambium. Flowers are borne on short axillary or terminal racemes on main stem and branches; they are small and white or lilac. Two bracteoles subtend each flower, of which the calyx is persistent and hairy. The five sepals are fused for most of their length with the two upper lobes behind the standard petal while the remaining three are below the keel. The standard petal is ovate and about 5 mm; the wings are narrow and obovate. The keel is shorter than the wings, its two petals fused along the base but free along the upper part.

Soyabean produces a large number of flowers, many of which abort. The proportion of flowers that produce mature pods ranges from 19 to 57% (Schaik and Probst 1958). Flower abortion can occur at several stages of its development, e.g. at flower bud initiation, at time of fertilization, or during any of the different stages of embryo development. Failure of fertilization is not a major cause of flower abscission in soyabean (Abernathy *et al.* 1977) as in most flowers that abscise, the ovules are fertilized and have developed proembryos. Pods which are retained on plants for more than 7 days from anthesis usually develop to maturity. Fertilized and retained ovules initiate cotyledons 1 week after anthesis with a mostly cellular endosperm. For another week the seed and embryo continue to grow while vascularization of the seed coat has taken place. There is a continued accumulation of dry matter and maximum seed weight is attained from 30 to 70 days after anthesis and, thereafter, losses in fresh weight of seeds and pods commence. The maximum seed growth rate is in the range of 5-15 mg seed^{-1} day^{-1} (Summerfield and Wien 1980). Depending on the variety and environment, seed maturity can take 50-80 days (Carlson and Lersten 1987).

GERMPLASM CONSERVATION AND USE

Presently, IITA maintains in its genebank a collection of about 2500 accessions of soyabean germplasm introduced from many parts of the world, for use in soyabean breeding and related research. The existence of a wide range of diversity in soyabean and its wild relatives for agronomically important traits has been reported by several workers. These traits range from simply inherited ones such as disease resistance to more complex ones such as maturity and adaptability. In eastern Asia, China and Japan different cropping systems involving soyabean have been adopted and as a consequence landraces have developed to suit these different systems. The soyabean cultivars in Japan have been classified into eight groups on the basis of days to flowering and days from flowering to maturity. These eight groups are identified as Ia, Ib, IIa, IIb, IIIb, IIIc, IVc and Vc. In China attempts also have been made to classify available germplasm into groups as has been done for USA and Japanese cultivars. No satisfactory grouping has emerged, mainly because in China the environment is quite variable owing to a wide range of latitude, altitude and cropping system including spring, summer, fall and winter plantings (Gai 1985).

The phenotypic frequencies of a few agronomic and morphologic attributes among native Chinese accessions are presented in Table 14.1. Throughout China genotypes characterized by determinate and indeterminate stem termination, erect and semi-erect growth habit, yellow seed coat, yellow cotyledon and broad leaf shape dominate their counterparts, respectively. Lines with purple and white flower colour were almost equally distributed as gray and tawny pubescence colour.

Genes conferring resistance to parasitic nematodes have been detected among Asian soyabean landraces. Parasitic nematodes adversely affect soyabean plants and it has been estimated in the USA that yield losses due to nematode infestations can be up to 10%. The most important of the nematodes are the soyabean cyst nematode, *Heterodera glycines* as well as some others in the *Meloidogyne* spp. There are also a few other

Table 14.1. Phenotypic frequencies (%) of agronomic and morphological characters of soyabean land cultivars in China (summarized from historical records; Gai 1985).

Character		Northern spring[†]	Northern summer[†]	Southern multiple	Total
% of strains		33.5	37.4	29.1	100.0
Planting type[‡]	SP	100.0	5.2	23.9	42.4
	SU	0	94.8	70.5	56.0
	FA	0	0	5.6	1.6
Stem termination[§]	IN	71.8	32.5	13.2	40.0
	SD	5.3	15.3	8.3	9.9
	DE	22.9	52.2	78.5	50.1
Growth type[¶]	ER	60.6	41.6	67.8	56.1
	SE	27.8	39.5	22.5	31.0
	SP	3.5	10.5	2.7	5.0
	PR	8.1	8.1	7.0	7.9
Seed coat colour[††]	YW	61.9	62.2	59.4	61.3
	GN	12.0	10.7	24.7	15.2
	BK	17.6	18.5	6.7	14.7
	B1	2.9	2.7	2.0	2.6
Cotyledon colour[‡‡]	YW	95.1	97.1	98.6	97.0
	GN	4.9	2.9	1.4	3.0
Flower colour[§§]	PU	47.9	49.0	67.6	54.1
	WH	52.1	51.0	32.4	45.9
Pubescence colour[¶¶]	GY	61.1	68.2	38.6	56.9
	TH	38.9	31.8	61.4	43.1
Leaf shape[*]	BD	93.9	96.1	95.6	95.4
	NW	6.1	3.9	4.4	4.6

[†]   Planting region/season.

[‡]   Planting type: SP - Spring, SU - Summer, FA - Fall (Autumn).

[§]   Stem termination: IN - Indeterminate, SD - Semi-determinate, DE - Determinate.

[¶]   Growth type: ER - Erect, SE - Semi-erect, SP - Semi-prostrate, PR - Prostrate.

[††] Seed coat colour: YW - Yellow, GN - Green, BW - Brown, BK - Black, B1 - Bicolour.

[‡‡] Cotyledon colour: YW - Yellow, GN - Green.

[§§] Flower colour: PU - Purple, WH - White.

[¶¶] Pubescence colour: GY - Grey, TW - Tawny.

[*]   Leaf shape: BD - Broad, NW - Narrow.

nematodes that could be of local importance. Some efforts have been expended toward breeding soyabean varieties with resistance to these nematodes in the USA. Following screening of a number of genotypes, a black-seeded cultivar (Peking), which was introduced into the USA in 1907 from China, was identified as resistant to two of the four races of the cyst nematode. Some other black-seeded soyabean types have also demonstrated reasonable levels of resistance to the nematode. Genetic studies indicate that three recessive gene pairs and a single dominant gene are responsible for resistance to cyst nematode. The dominant gene is, however, closely linked with the gene for coloured seed (Hague 1980).

The genetic base of most soyabean varieties grown in the USA is narrow since about 80% of their genepool can be traced to between seven and 10 introductions made from the same geographical area (Delannay *et al.* 1983). This situation could lead to catastrophes should an outbreak occur of any disease or pest for which no gene(s) for resistance is available among these varieties. There is therefore the need to broaden the base of genetic variability by exploiting the germplasm of the wild perennial relatives of cultivated soyabean.

The soyabean genetic collection has been divided into four categories, according to Palmer and Kilen (1987), as follows:

1. the Type Collection made up of strains that contain all published genes as well as those mutants or strains considered to have genetic potential by the Soybean Genetics Committee.
2. the Isoline Collection made up of near-isolines of single genes or combination of genes backcrossed to the adapted cultivars Clark, Harosoy and Lee.
3. the Linkage Collection containing all linkage combinations and various genetic recombinations.
4. the Cytological Collection made up of interchanges, inversions, deficiencies, trisomics and tetraploids.

Soyabean genotypes in the USA have been divided into 12 maturity groups and this is based on their adaptation to different latitudes. The genotypes making up maturity groups 00 through IV are adapted to the northern latitudes while strains in maturity groups V through X are adapted to shorter-day conditions. Over 70% of the germplasm, comprising forage crop, old US and Canadian cultivars, and plant introductions, maintained at Urbana and Stoneville (both in the USA), belong to the maturity groups 00 to V.

This classification of soyabean genotypes in the USA on the basis of maturity does not seem applicable most times to those grown in the tropics. A new classification that is based on photoperiodic response and maturity duration has been proposed for soyabeans grown in the tropics (Fig. 14.1).

Evaluation

Some morphological and biochemical attributes of soyabean plants have been studied in relation to their expression among genotypes of different geographical regions. Among these attributes are rhizobium response, root fluorescence, presence of flavonols in leaves and isozymes, among others.

Scientists at the International Institute of Tropical Agriculture (IITA) reported a soyabean germplasm line, TGm 685, from Indonesia with a high level of resistance to field weathering, a problem of soyabean in the humid tropics. In addition this line showed a very low level of anthracnose in the seed compared with seeds of some other lines that were susceptible to field weathering (Ndimande *et al.* 1981). Soyabean varieties showing resistance to field weathering have been developed and distributed to farmers in some parts of Nigeria. With the availability of these varieties farmers are now more favourably disposed to allow seeds to dry to a fairly low moisture level in the field before harvesting. Previously, viability of seeds of varieties grown by the farmers was very low when the seeds were left in the field to dry to low moisture content.

Germplasm lines (plant introductions) respond differenttly when inoculated with several fast-growing rhizobia strains. Fast-growing rhizobia strains collected from east-central China were found to inoculate and nodulate a *G. max* plant introduction (Peking) but caused the development of ineffective nodules with many USA improved varieties (Keyser *et al.* 1982). When Devine (1985) tested 285 plant introductions from Asia for their ability to nodulate with rhizobium strain USDA 205 he found that 56% in *G. max* did form nodules. Three dominant alleles designated as *Rj2*, *Rj3* and *Rj4* led to ineffective nodule response with some slow-growing rhizobium strains. Recently, IITA scientists have incorporated promiscuous nodulation into improved breeding lines.

The flavonol content in leaves is a heritable trait and alleles (*Fg1*, *fg2*, *Fg3* and *Fg4*) governing its inheritance are not random among accessions from different geographical regions. The allele *Fg1* occurred at a high frequency in *G. max* accessions from India while *Fg3* was found at a lower frequency than in accessions from China and Japan (Buttery and Buzzel 1973, 1976). They also found that allele *Fg1* was almost nonexistent in accessions of *G. soja* from Japan while on the other hand it occurred at high frequencies among accessions from China, Russia and Korea. Broich and Palmer (1981)

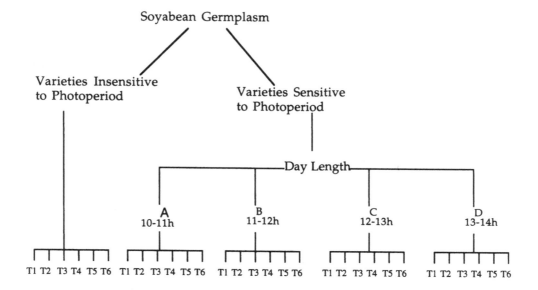

Days to Maturity Classification

T1 <70 days
T2 71 - 80 days
T3 81 - 90 days
T4 91 - 100 days
T5 101 - 120 days
T6 >120 days

Fig. 14.1. A soyabean classification system for use in the tropics (Shanmugasundaran 1976).

reported that allele *Fg3* was absent in 28 accessions of *G. soja* collected from Russia but was found in many of those obtained from China, Korea and Japan.

Genetic studies showed that root fluorescence is controlled by four independent alleles, three of which are recessive (*fr1*, *fr2* and *fr4*); the fourth is dominant (*Fr3*). Delannay and Palmer (1982) tested 572 accessions belonging to *G. max* for the presence or absence of root fluorescence and found that absence of fluorescence occurred at a higher frequency among European lines (19%) than among those of Asian origin (3.4%). Distribution of the different alleles also varied such that alleles *Fr3* and *fr4* were found only among accessions from Asia. According to Broich and Palmer (1981) negative root fluorescence is common in the wild in certain geographical areas but occurs rarely under cultivation. In their view the distinct geographical distribution of this trait in *G. soja* may be because the domestication process occurred in an area where negative root fluorescence was rare.

The presence, form and number of trichomes on soyabean plant parts are controlled by several known alleles. Broich and Palmer (1981) reported that among accessions evaluated, more than 95% of *G. soja* had sharp pubescence tip while less than 10% of *G. max* accessions showed this trait. The appressed form of pubescence is common in *G. soja* but not universal. It is also commonly found among accessions from Japan and Korea. Bernard (1975) reported that soyabean (*G. max*) cultivars characterized by appressed pubescence are heavily damaged and stunted by the potato leafhopper (*Empoasca fabae*) in the north-central USA. He belived that in Asia appressed pubescence may confer resistance to pod borer. Hence differences in selection pressures between natural and agricultural environments may be reponsible for the differences in the frequency of pubescence among genotypes in this subgenus.

Properties and Uses

When compared with other crops of world importance, soyabean can be described as neglected in terms of germplasm collecting for both close and distant relatives of the crop. Brown *et al.* (1985) have speculated that there is a risk of losing some wild soyabean relatives, especially those growing in the not easily accessible areas of southeast Asia. Concerted efforts aimed at collecting, characterizing and exploiting germplasm of more distant relatives of soyabean commenced in the 1970s (Marshall and Broue 1981).

Singh *et al.* (1993) stated that the wild relatives of soyabeans have not been sufficiently exploited for the improvement of the crop. The reason for this lack of exploitation of wild soyabean relatives has been attributed to the problems encountered while embarking on wide crosses. These problems include failure of hybrid embryos to develop and sterility of the interspecific hybrids. This trend is, however, changing with the availability of tissue culture techniques by which very young and potentially abortive interspecific hybrid embryos can now be successfully rescued and made to grow into plants.

The subgenus *Glycine* contains nine perennial species which are found mainly in Australia. Some of these are distributed over a wide range of environments. For example, *G. tomentella* is mainly found in the monsoonal tropical north, *G. canescens* in the central semi-arid region, *G. falcata* in the periodically flooded, heavy basaltic plains, *G. clandestina* in the subalpine woodlands to altitudes of 1450 m and *G. latrobeana* is found in the cooler southerly latitudes (42°S). The distribution of these perennial *Glycine* genotypes in such diverse environments is an indication of the potential existing for tolerance to adverse environments. This should be an advantage in the development of soyabean lines for adaptation to a wide range of environments (salinity, drought, etc.). In addition these perennial *Glycine* spp. show a wide range of diversity in their morphology. For example, some of them have a thick taproot system that penetrates deep into the soil, plant growth habit is variable, some produce chasmogamous and cleistogamous flowers and even set pods underground, as happens in *G. falcata*. These perennial *Glycine* spp. would therefore be excellent sources of genes for many desirable attributes that may not be readily encountered among the cultivated soyabean genotypes.

Leaf rust, caused by *Phakospora pachyrhizi*, is a disease that adversely affects soyabean productivity. Grain yield losses of up to 30% can be caused by this disease. Breeding soyabean lines with resistance to this disease has been chosen as an option for its control in the crop. Fortunately sources of resistance have been detected among the perennial *Glycine* spp. Genes for resistance to this disease have been found in the six perennial *Glycine* spp. that were tested against several races of the pathogen (Brown *et al.* 1985).

Breeding

Breeding soyabean varieties with resistance to pests and diseases received attention only after 1950. Some plant introductions were found to be particularly promising because of their resistance to some pests and diseases. Among these are PI 229358, PI 227687 and PI 171451 which were found to be extremely resistant to Mexican bean beetle. Following further field tests these lines showed multiple resistance to bean leaf beetle, corn earworm, striped blister leaf beetle, soyabean looper and the bollworm (Kilen *et al.* 1977). Yang (1980) has provided a list of varieties and accessions (PIs) with resistance to various soyabean diseases such as bacterial pustule, brown stem rot, downy mildew and mosaic virus.

A very extensive review of soyabean breeding methods is available (Fehr 1987). This report describes objectives of cultivar development, population development, inbred line development, inbred line evaluation, breeder seed production and commercial use of seed mixtures. There also have been several reports describing the soyabean breeding activities at IITA (Dashiell *et al.* 1985, 1987, 1991). These reports have described the importance of having soyabean germplasm available and then using it as a source for developing improved varieties. The procedures used to screen germplasm and breeding lines for seed longevity, promiscuous nodulation and disease and insect resistance are described. Presently, the goal of the soyabean breeding project at IITA is to develop improved soyabean varieties well adapted to the savannas of Tropical Africa. The specific objectives are to develop soyabean breeding lines with high and stable grain yield, resistance to the major diseases and insect pests, resistance to pod shattering, promiscuous nodulation, improved seed longevity and acceptable seed colour, oil and protein content. These objectives focus on developing soyabean varieties that are productive and easy to manage and market.

In 1995, IITA began to develop soyabean breeding lines that are improved for their contribution to the cropping system. In essence, we are developing soyabean varieties that contribute more to assisting a rotation crop such as maize. Our goal is to make the cropping system more productive and not just the soybean component of the system. One of the areas that we are focusing on is nitrogen fixation. In trials that were conducted in the savanna of Nigeria in 1994 and 1995, different varieties had 20-60% of their total nitrogen derived from the atmosphere (%NDFA) (Sanginga and Dashiell, unpublished data). Each of the varieties produced high grain yield and thus, when standard variety evaluation procedures are used, would probably qualify as a very good variety. However, with the additional information of %NDFA it is evident that some varieties (those with low %NDFA) are mining nitrogen from the soil and this may not be good for the crops on this same land during the next growing season. The varieties with relatively high %NDFA should be very good for the sustainability and productivity of the cropping systems.

In most areas of the world, the sole purpose of a soyabean crop is grain production. However, we believe that both grain and fodder production of soyabean will be important for the farming systems in Africa. In 1995, we had soyabean trials in the savanna of Nigeria. The dry fodder yield was 1500-4000 kg/ha for varieties that all had good grain yields (Dashiell and Sanginga, unpublished data). In these trials if only the grain yield had been measured, then all of the varieties would appear to be equally productive. However, with the additional information on fodder yield it is clear that there are big differences in the total market value of these varieties. Even in farming systems where soyabean fodder will not be fed to livestock, the high fodder yields will have value in the farming system as a mulch and as additional organic matter for soil improvement.

Prospects

The effort to develop soyabean varieties that contribute to making the farming system more productive has just begun. This is a radical change from previous efforts when the aims were primarily breeding for maximum soyabean grain yield. Of course, soyabean grain yield is still an important component of the productivity of the farming system. We have given two examples, of higher %NDFA and fodder yield, as traits that we are now breeding for. The soyabean germplasm collection will continue to be a very important source for the improvement of these and many other traits.

REFERENCES

Abernathy, R.H., R.G. Palmer, R. Shibles and J.C. Anderson. 1977. Histological observations on abscising and retained soybean flowers. Can. J. Plant Sci. 57:713-716.

Bergersen, F.J. and M.J. Briggs. 1958. Studies on the bacterial component of soybean root nodules: Cytology and organisation in the host tissue. J. Gen. Microbiol. 19:482-490.

Bernard, R.L. 1975. The inheritance of appressed pubescence. Soybean Genet. Newsl. 2:34-36.

Broich, S.L. and R.G. Palmer. 1980. Segregation patterns of some simply inherited traits in *Glycine max* x *G. soja* crosses. Soybean Genet. Newsl. 7:62-64.

Broich, S.L. and R.G. Palmer. 1981. Evolutionary studies of the soybean: the frequency and distribution of alleles among collections of *Glycine max* and *G. soja* of various origin. Euphytica 30:55-64.

Brown, A.H.D., J.E. Grant, J.J. Burdon, J.P. Grace and R. Pullen. 1985. Collection and utilisation of wild perennial *Glycine*. Pp. 345-352 in Proceedings World Soybean Research Conference III (R. Shibles, ed.). Westview Press, London.

Buttery, B.R. and R.I. Buzzel. 1973. Varietal differences in leaf flavonoids of soybean. Crop Sci. 13:103-106

Buttery, B.R. and R.I. Buzzel. 1976. Flavonol glycoside genes and photosynthesis in soybeans. Crop Sci. 16:547-550.

Carlson, J.B. and N.R. Lersten. 1987. Reproductive morphology. Pp. 95-134 in Soybeans: Improvement, Production and Uses, 2nd edition (J.R. Wilcox, ed.). American Society of Agronomy, Inc., Madison, WI, USA.

Dashiell, K.E. E.A. Kueneman, W.R. Root and S.R. Singh. 1985. Breeding tropical soybean for superior seed longevity and for nodulation with indigenous rhizobia. Pp. 133-139 in Soybean in Tropical and Subtropical Cropping Systems (S. Shanbmugasundaram and E.W. Sutzberger, eds.). Asian Vegetable Research and Development Center, Shanhua, Taiwan.

Dashiell, K.E. L.E.N. Jackai, G.L. Hartman, H.O. Ogundipe and B. Asafo-Adjei. 1991. Soybean germplasm diversity, uses and prospects for crop improvement in Africa. Pp. 203-212 in Crop Genetic Resources of Africa, Vol. II (N.Q. Ng, P. Perrino, F.Attere and H. Zedan, eds). Sayce Publishing, UK.

Dashiell, K.E., L.L. Bello and W.R. Root. 1987. Breeding soybeans for the tropics. Pp. 3-16 In Soybeans for the Tropics (S.R. Singh, K.O. Rachie and K.E. Dashiell, eds.). John Wiley & Sons Ltd., Chichester.

Delannay, X. and R.G. Palmer. 1982. Four genes controlling root fluorescence in soybean. Crop Sci. 22:278-281.

Delannay, X., D.M. Rodgers and R.G. Palmer. 1983. Relative genetic contributions among ancestral lines to North America soyabean cultivars. Crop Sci. 23:944-949.

Devine, T.E. 1985. Nodulation of plant introduction lines with the fast-growing rhizobia strain USDA 205. Crop Sci. 25:354 - 356

Fehr, W.R. 1987. Breeding methods for cultivar development. Pp. 249-293 In Soybeans: Improvement, Production and Uses, 2nd edition (J.R. Wilcox, ed.). American Society of Agronomy, Inc. Madison, WI, USA.

Gai, J. 1985. Utilizing the genetic diversity of annual soybean species. Pp. 353-360 in Proceedings World Soybean Research Conference III (R. Shibles, ed.). Westview Press, London.

Hague, N.G.M. 1980. Nematodes of legume crops. Pp. 199-205 in Advances in Legume Science (R.J. Summerfield and A.H. Bunting, eds.). Proceedings of International Legume Conference, Kew, 31 July - 4 August 1978. Vol. 1. Royal Botanic Gardens, Kew.

Harlan, J.R. 1971. Agricultural origins: Centers and noncenters. Science 174:468-474.

Hymowitz, T. 1970. On the domestication of the soybean. Econ. Bot. 24:408-421.

Hymowitz, T. and C.A. Newell. 1981. Taxonomy of the genus *Glycine*, domestication and uses of soybeans. Econ. Bot. 35:272-288.

Hymowitz, T. and N. Kaizuma. 1981. Soybean, seed protein electrophoresis profiles from 15 Asian countries or regions: hypotheses on paths of dissemination of soybeans from China. Econ. Bot. 35:10-23.

Juvik, G.A., R.L. Bernard and H.E. Kauffman. 1985. Directory of germplasm collections 1. II. Food legumes (Soyabean). IBPGR, Rome; INTSOY, Urbana-Champaign.

Keyser, H.H., B.B. Bohlool, T.S. Hu and D.F. Weber 1982. Fast-growing rhizobia isolated from root nodules of of soybeans. Science 215:1631-1632.

Kilen, T.C., J.H. Hatchett and E.E. Hartwig 1977. Evaluation of early generation soyabeans for resistance to soyabean looper. Crop Sci. 17:397-398.

Kwon, S.H. 1972. History and the land races of Korean soybean. SABRAO Newslett. 4:107-111.

Marshall, D.R. and P. Broue. 1981. The wild relatives of crop plants indigenous to Australia and their use in plant breeding. J. Aust. Inst. Agric. Sci. 47:149-154.

Ndimande, B.N., H.C. Wien and E.A. Kueneman. 1981. Soybean seed deterioration in the tropics. 1. The role of physiological factors and fungal pathogens. Field Crops Res. 4:113-121.

Palmer, R.G. and T.C. Kilen. 1987. Qualitative genetics and cytogenetics. Pp. 135-209 in Soybeans: Improvement, Production and Uses, 2nd edition (J.R. Wilcox, ed.). American Society of Agronomy, Madison, WI, USA.

Purseglove, J.W. 1977. Tropical Crops: Dicotyledons, pp. 264-273. Longman, London.

Schaik, P.H. van and A.H. Probst. 1958. Effects of some environmental factors on flower production and reproductive efficiency in soyabeans. Agron. J. 50:192-197.

Shanmugasundaram, S. 1976. Important considerations for the development of a soyabean classification system for the tropics. Pp. 63-64 and 191-196 *in* INTSOY Series no. 10 (R.M. Goodman, ed.). University of Illinois, USA.

Singh, R.J., K.P. Kollipara and T. Hymowitz. 1993. Backcross (BC2-BC4)-derived fertile plants from *Glycine max* and *G. tomentella* intersubgeneric hybrids. Crop Sci. 33:1002-1007.

Summerfield, R.J. and H.C. Wien. 1980. Effects of photoperiod and air temperature on growth and yield of economic legumes. Pp. 17-36 *in* Advances in Legume Science (R.J. Summerfield and A.H. Bunting, eds.). Proceedings of International Legume Conference, Kew, 31 July - 4 August 1978. Vol. 1. Royal Botanic Gardens, Kew.

Yang, C.Y. 1980. Developments in crop protection of *Glycine*. Pp. 323-335 *in* Advances in Legume Science (R.J. Summerfield and A.H. Bunting, eds.). Proceedings of International Legume Conference, Kew, 31 July - 4 August 1978. Vol. 1. Royal Botanic Gardens, Kew.

Chapter 15

Barley

J. Valkoun, S. Ceccarelli and J. Konopka

Barley is a cool-season crop in countries with a Mediterranean climate and is well adapted to stressful and extreme environments. Barley fields can be seen as high as 4800 m asl in the Himalayas, in latitudes over 60°N in Iceland and Scandinavia and in the rain-fed semi-arid regions of WANA with less than 250 mm annual rainfall. Barley is a principal food crop in highlands and marginal areas where other cereals will not grow, as well as animal feed and forage all over the world. It is also an important industrial crop, providing raw material for malt and beer production. Its straw is of better quality than that of wheat and is, therefore, a valuable complement of cattle and small ruminant diets. Barley is grown in a wide range of environments but nearly two-thirds of the world's production is grown in subhumid or semi-arid regions.

BOTANY AND DISTRIBUTION

Barley belongs to the tribe Triticeae of the grass family Poaceae together with other important cereals, wheat and rye. The main distinction from other members of the tribe is that each spike node bears three 1-flowered spikelets ('triplets') of which one, two or all three are fertile. The genus *Hordeum* includes about 30 species (Bothmer 1992a). According to the same author, the 45 taxa of the genus are mostly diploid ($2n=2x=14$ chromosomes, 28 taxa), but also tetraploid ($2n=4x=28$ chromosomes, 16 taxa) and hexaploid ($2n=6x=42$ chromosomes, 8 taxa) with a basic chromosome number $x=7$.

Botanical Names

Hordeum vulgare L., 1753, Sp. pl., p. 84; *H. aegiceras* Neux ex Royle, 1839, Illustr. Bot. Himal., p. 418; *H. sativum* Jessen, 1865, Deutschl. Gräser., p. 200; *H. eurasiaticum* Vav. et Bacht., 1948, Dokl. Akad. Nauk SSSR, 59, p. 974; *H. aethiopicum* Vav. et Bacht, *H. sinojaponicum* Vav. et Bacht., 1948, loc. cit., p. 975.

Vernacular Names

Barley (English), *orge* (French), *Gerste* (German), *cebada* (Spanish), *yachmen'* (Russian), *shai'r* (Arabic).

Morphology

Plants are usually up to 1 m tall; some landraces can be even taller, up to 1.5 m. Culms are mostly erect, up to 3 mm thick, with 3 to 4 nodes. Leaves are glabrous, rarely hairy, with long, prominent auricles surrounding the culm; however, ligule-less forms also exist. Leaf sheets are split almost to the base. Spikes with awns are 5-25 cm long; their colour ranges from white-yellowish over purplish to blackish, without the terminal fertile spikelet. There are 10-30 nodes, each bearing three one-flowered spikelets. All three spikelets are fertile in six-rowed forms, but only the central spikelet is fertile in two-rowed barley. In the latter form, the lateral spikelets are non-fertile with anthers (type *nutans*) or fully reduced (type *deficiens*). The central spikelet is sessile, with more or less flat narrow glumes; the lemma is glabrous or scabrid with five nerves, usually tightly covering the kernel (not in naked forms). The awn of the lemma is rough or glabrous, 3-18 cm long. Sometimes awns are absent or transformed into 'furcs'. Lateral spikelets are usually sessile, but may be pedicellate (3 mm long) when fertile. Glumes are 1-2 cm long, more or less flattened. Kernels in naked forms have a furrow on the dorsal side; their colour may be yellow, gray, green, violet or black. In contrast to other cereals, the aleurone layer has more than one row of cells (two to four). Plants are essentially autogamous, diploid, $2n=14$; tetraploids, $2n=28$, have been produced artificially. Both winter and spring forms exist.

Origin, Domestication and Diffusion

Archeological evidence from the Near East indicates that barley was first domesticated from its wild progenitor, *Hordeum vulgare* L. subsp. *spontaneum* (C. Koch) Thell., in the Fertile Crescent region more than 10 000 years ago. The earliest cultivated forms with non-brittle rachis were two-rowed hulled and also naked barley, which were found in the Prepottery Neolithic A layers in two sites dated 8000 BC. The six-rowed forms appeared some 1000-2000 years later. Barley and emmer wheat appear to be the first cultivated crops and together with two domesticated animals, sheep and goats, they formed a basis of the Neolithic farming system, which spread from the nuclear area to other parts of the Old World. With this Neolithic 'package', barley spread throughout Europe, around the Mediterranean, eastward to the Indus and southward to Ethiopia, but it did not reach China before the second millennium BC (Zohary and Hopf 1988; Harlan 1991). The archeological evidence is supported by the contemporary geographical distribution of subsp. *spontaneum*. The area of its massive stands in primary habitats includes or is located close to the Neolithic archeological sites of the Near East. However, the monophyletic two-rowed progenitor hypothesis has been questioned by some authors, who propose an alternative hypothesis that assumes six-rowed forms originated from an extinct six-rowed progenitor with a brittle rachis, similar to present *agriocrithon* forms. These are, however, secondary products of hybridization of subsp. *spontaneum* with six-rowed cultivated forms (Zohary 1959), not adapted to the survival in undisturbed habitats. Bothmer *et al.* (1991) provided a logical explanation of barley origin, arguing that the one-seeded diaspora in subsp. *spontaneum* is more functional for seed dispersal in the natural habitat than three seeds linked together. Therefore, the wild progenitor of all cultivated barley forms would be a two-rowed *spontaneum* type with a brittle rachis from which a diversity of tough-rachis forms developed through mutations and domestication. Traits like brittle/tough rachis, row number, seed cover or awn development are controlled by relatively few genes. Additional genetic diversity has been generated through repeated cycles of differentiation and hybridization (Harlan 1992). As there are no crossing barriers between cultivated barley and wild or weedy subsp. *spontaneum*, gene introgression in both directions is a common process at the field borders and/or within fields in West Asia and North Africa (WANA). Therefore, mutation, hybridization and gene recombination generated new genetic diversity including new brittle types. This type of barley evolution was first suggested by DeCandole in 1886 and later advocated by a number of other authors. There are, however, at least four other hypotheses on the origin of cultivated barley, postulated by Cberg, Takahashi, Bakhteev and Shao, respectively (Shao and Li 1987). Cberg suggested that *H. agriocrithon* was a common ancestor of both cultivated barley, *H. vulgare*, and two-rowed wild barley, *H. spontaneum*, while Takahasi favoured the diphyletic theory of barley origin. According to this, the two-rowed cultivated barley originated from *H. spontaneum* and the six-rowed cultivated form was derived from *H. agriocrithon*. Bakhteev identified the bottle-shaped wild barley form, *H. lagunculiforme* Bacht., to be the ancestor of all cultivated *H. vulgare* forms. However, Shao and Li (1987) questioned the latter hypothesis claiming that *H. lagunculiforme* was an unstable intermediary form in evolution from two-rowed to six-rowed wild barley. In their opinion, two-rowed wild barley was the oldest ancestor, which was the first stage of evolution of cultivated barley. The six-rowed wild barley, *H. agriocrithon*, was the second stage, while the cultivated form was the third and final stage of evolution.

Genetic Resources
Global holdings

The barley genepool, held in the *ex situ* genebank collections worldwide, is the second largest after wheat, with 486 724 accessions (FAO 1996). This represents 8.4% of global genetic resources holdings. As with other cereals, the major part of the barley genebank holdings are replications and the number of distinct accessions would be much smaller. An earlier estimate (Plucknett *et al.* 1987) indicates that only 20% of barley global collections may be distinct (unique) accessions. Chapman (1987) made an inventory of unique accessions of barley and concluded that both landraces and cultivars are represented in the global collections by the same number of unique accessions, i.e. 25 000. Genetic diversity of cultivated barley is relatively well represented in genebanks; it is believed that 85% of landraces have been collected. This high percentage is in contrast with 20% of wild *Hordeum* spp. (Plucknett *et al.* 1987).

Barley holdings in major genebanks are presented in Table 15.1. The highest proportion of landraces/obsolete cultivars is in PGRCE, Ethiopia (100%) and ICARDA (87%) collections. In spite of the impressive number of barley accessions in germplasm collections, only 10% are stored under long-term conditions, i.e. at temperatures below –10°C, and less than half of the total holdings are kept in medium-term cold stores (FAO 1996).

Genetic resources exchange among genebanks has generated many duplications or multiplications of original accessions. In order to rationalize and increase efficiency of national and international genetic resources programmes, attempts have been made to eliminate undesirable duplications from barley genebank holdings (Knüpffer 1988). The identification of tentative duplications usually starts with passport information screening of databases. However, it is advisable to verify the results of the database screening by morphological characterization and/or biochemical or molecular fingerprinting before duplicate accessions are eliminated from the genebank.

Recently, ICARDA used passport information in the European Barley Data Base (Knüpffer 1988) and GRIN (US Genetic Resources Information Network) for the identification of duplications between a new shipment received from IPK, Gatersleben, Germany and the barley germplasm collection held at the ICARDA Genetic Resources Unit (GRU). The results of the database screening were verified in the field by planting the 'duplicate' accessions side by side. Surprisingly, only 740 out of the total of 1100 'duplicates' were morphologically identical. Since all the accessions were landraces, the discrepancy between the database information and the seed sample may have resulted from selection of distinct morphotypes in originally heterogenous samples. This example clearly demonstrates that the 'rationalization' of germplasm collections has to be done with caution.

Table 15.1. The largest barley collections in the world.

No. of accessions	% of global holdings	LT[†] (%)	MT[‡] (%)	Institution	Country
41 360	8	0	100	PGRC	Canada
26 019	5	0	100	NSGC	USA
24 268[§]	5	80[§]	100[§]	ICARDA	Syria
23 766	5	0	100	IPSR	UK
18 210	4	0	100	CENARGEN	Brazil
17 768	4	0	98	VIR	Russia
16 351	3	100	0	NSSL	USA
12 648	3	0	100	PGRCE	Ethiopia
10 648	2	97	3	IPK	Germany
486 724	**100**	**10**	**42**	**Total global barley holdings**	

Source: FAO 1996.
[†] Long-term storage conditions.
[‡] Medium-term storage conditions.
[§] ICARDA database, 6 July 1996.

International Collaboration

Coordinated multilateral international collaboration started in 1981 when the International Board for Plant Genetic Resources (IBPGR) convened an *ad hoc* Working Group on Barley Genetic Resources to review the current status of collections and set up priorities for international action. In Europe, the coordination of genetic resources activities began in 1983 with the establishment of a Barley Working Group of the UNDP/IBPGR European Cooperative Programme for Conservation and Exchange of Genetic Resources. The European collaborative effort resulted in the development of a European Barley Core Collection (ECPGR 1983) and the European Barley Data Base (Knüpffer 1988). The regional effort was extended to the global level when an International Barley Genetic Resources Network (IBGRN) was established and an international Barley Core Collection (BCC) committee was formed (Knüpffer and van Hintum 1995).

In another initiative, IBPGR set up a world network of base collections. The IBPGR (now the International Plant Genetic Resources Institute, IPGRI) designated barley global base collections are at ICARDA, Aleppo, Syria and Plant Gene Resources of Canada (PGRC), Ottawa, Canada. Regional collections are held at the Nordic Gene Bank (NGB), Lund, Sweden for Europe, at the Plant Genetic Resources Centre (PGRC), Addis Ababa, Ethiopia for Africa, and at the National Institute for Agrobiological Research (NIAR), Tsukuba, Japan for Asia (IBPGR 1990). A new collection structure was proposed to increase the efficiency of the system (IBPGR 1992). In 1994, ICARDA and other CGIAR centres placed their germplasm collections under the auspices of the FAO. According to the agreement with the FAO, the germplasm is held in trust for the benefit of the global community.

Current Status of Barley Ex Situ Collections

The primary task of genetic resources programmes is to assemble genetically diverse collections with the traits required by breeders to develop cultivars for sustainable agriculture. An additional and increasingly important role of genebanks is to conserve the maximum genetic diversity of a crop genepool, particularly that in danger of genetic erosion.

Barley genepool

Harlan and de Wett (1971) classified the crop genepool into primary, secondary and tertiary pools, according to the relationships with the cultivated species. According to

Bothmer *et al.* (1991), the primary barley genepool has three major components: cultivars and breeding lines, landraces and the barley wild progenitor, *H. vulgare* subsp. *spontaneum*. As there are no crossability barriers within the primary genepool, gene transfer to adapted cultivars is feasible from three components. A single species, *H. bulbosum*, belongs to the secondary genepool. Crosses with cultivated barley mostly result in *bulbosum* chromosome elimination and production of *vulgare* haploids. This phenomenon has been widely exploited in the dihaploid method of barley breeding, which can shorten the breeding cycle and ensure complete homozygosity and, consequently, uniformity of the breeding products. Sometimes the *bulbosum* chromosomes are not eliminated and the embryo develops into a true hybrid. Chromosome meiotic pairing in such hybrids may be high but the fertility is extremely low, making gene transfer possible but very difficult. All other *Hordeum* species belong to the tertiary genepool. Strong crossability barriers and no or very low chromosome homology with *vulgare* chromosomes have prevented their contribution to barley breeding so far. This situation may change with new developments in biotechnology, and new, interesting methodologies could become available to barley breeders.

The barley wild progenitor, *H. vulgare* subsp. *spontaneum*, belongs to the primary genepool, there are no crossing barriers with cultivated barley and their chromosomes are fully homologous. As the species has evolved in the harsh and diverse environments in WANA, it is a valuable source of genes for stress tolerance and adaptation to marginal environments and low-input farming systems. Natural populations of wild barley in many parts of the Near East, a region of origin and primary genetic diversity, are continuously eroded by heavy overgrazing by small ruminants but wild barley can tolerate overgrazing better than wild wheat species.

Although the total number of wild barley accessions held in genebanks (16 100) is quite impressive, most of them (86%) are single-seed progenies from a very limited number of accessions from Palestine held at the Plant Breeding International (PBI), Cambridge, UK. Other collections represent diversity of specific areas and the area of distribution of the species is insufficiently represented (Bothmer 1992b).

To improve the geographical coverage of *spontaneum* collections, ICARDA has conducted a number of missions to countries of West and Central Asia and North Africa in which natural collections were sampled and introduced to ICARDA's genebank as new accessions (Tables 15.2 and 15.3).

Utilization of other wild *Hordeum* spp. in cultivated barley breeding has been very limited because of crossability problems, sterility of hybrids, resistance to chromosome doubling and linkage of desirable with undesirable traits (Fedak 1985). *Hordeum bulbosum* L. is the single species in the secondary genepool and its global holdings are approximately 1000 accessions. The other 30 species are rather distantly related to the cultigen and all belong to the tertiary genepool. It is estimated that 3200 accessions are held in 11 genebanks, but a large majority (95%) is concentrated in the Canadian-Scandinavian Collection. Attractiveness of this previously neglected part of the barley genepool will certainly increase with new advances in biotechnology.

Diversity in cultivated barley collections

Attempts have been made to analyze genetic diversity of barley global collections, using different means. Vavilov initiated the agro-ecological characterization of the global barley collection held at VIR (All-Union Institute of Plant Industry, now Vavilov All-Russian Scientific Institute of Plant Genetic Resources), St. Petersburg, Russia. This several-decade effort resulted in the identification of 37 agro-ecological groups or 'proles' (Kobyl'yansky and Lukyanova 1990):

Abyssinian (Ethiopian), Azerbaijano-Daghestanian, Anatolian, Apenninian, Apsheronian, Arabic (=proles jemenicum), Armenian, Armeno-Georgian, East-Siberian, Highland-Caucasian, Greek-Macedonian, Daghestanian, Far-East, Egyptian, West-

Table 15.2. Origin of *H. vulgare* subsp. *spontaneum* accessions held at ICARDA genebank.

Country	No. of accessions	Country	No. of accessions
Palestine	926	USSR	2
Syria	241	Cyprus	3
Jordan	146	Russia	3
Iran	71	Tajikistan	3
China	46	Azerbaijan	2
Lebanon	25	Uzbekistan	6
Turkmenistan	22	Pakistan	2
Iraq	28	Egypt	1
Libya	11	Kazakhstan	1
Turkey	48	Unknown	21
Afghanistan	7	**Total**	**1609**

Table 15.3. Barley germplasm collected by ICARDA.

Country	No. of accessions	
	Cultivated barley (subsp. *vulgare*)	Wild barley (subsp. *spontaneum*)
Morocco	459	–
Syria	220	113
Egypt	151	1
Jordan	107	139
Pakistan	102	2
Algeria	95	–
Ecuador	56	–
Iran	46	8
Lebanon	12	25
Turkmenistan	5	8
Tajikistan	5	1
Tunisia	3	–
Uzbekistan	1	2
Turkey	–	36
Iraq	–	8
Libya	–	7
Cyprus	–	3
Russia	–	2
Total	**1271**	**352**

European, West-European intensiv, West-Caucasian, West-Siberian, West-Ukranian, Indian, Iranian, Irano-Turkestanian, Chinese, Manchurian, Mongolo-Tibetian, Pamiro-Badakhshanian, Pyrenean, Coastal (USA), Northern (Scandinavia and north of the European Russia), North-African, North-European naked-grain, Cis-Caucasian, Syriaco-Palestinian, Steppic, Tibetian winter-type, Central-European and Japanic.

Takahashi (1987) identified two very different barley groups, the oriental and occidental, in studies of the world barley collection based on the geographical distribution of genes controlling morphological and physiological traits. The oriental group includes barley originating from Japan, Korea, China, the Hindukush region of Pakistan and Afghanistan and the highlands of Nepal, and the occidental barley comes from lowlands of Nepal, India, Turkey, Europe, the former USSR, North Africa and Ethiopia. Genetic variation of both the occidental and oriental type was found in barley from Southwest Asia, thus confirming its status as a centre of barley origin and primary diversity. Konishi (1988, 1995), studying geographical distribution of alleles of neutral complementary genes controlling brittle rachis, hybrid chlorosis and weakness and genetic variation among isozyme loci, confirmed the difference between East-Asian barley and barley of Western origin.

Germplam exchange between countries and the extensive utilisation of exotic germplasm in barley breeding resulted in the emergence of new centres of diversity (Peeters 1988). In his analysis of the global barley germplasm collection of the AFRC held at the Institute of Plant Science Research, Cambridge, UK comprising approximately 5000 accessions of worldwide origins and 12 qualitative and 18 quantitative traits, the highest mean genetic diversity was found in the US germplasm, followed by barley of Turkey, Japan, the USSR, China, West Germany and France. When the number of unique character combinations was considered, then the ranking was as follows: the USA, the USSR, West Germany, UK, Turkey, Japan, France, Afghanistan, Canada, Netherlands and Ethiopia. This study also indicates that erosion of genetic diversity in the centres of primary diversity may be accompanied by the opposite process in other parts of the world, i.e. increase in genetic diversity through germplasm introductions and the use of germplasm of worldwide origin held *ex situ* in genebanks. However, van Hintum (1994) questioned Peeters' conclusions because of a biased sampling procedure, low authenticity of the genebank material and no data on within-accession variation.

A total of 17 000 accessions of the USDA world barley germplasm collection were analyzed using five morphological descriptors (Tolbert *et al.* 1979). Mean diversity index values H', i.e. scaled Shannon-Weaver information index (H'), pooled over characters, showed wide variation. Romania, Hungary and the USSR were countries with the highest H' values, 0.66, 0.61 and 0.58, respectively. Again, as with the previous study, the validity of the results was disputed by van Hintum (1994).

Genetic diversity also has been studied in parts of the barley genepool, both in natural populations of the barley wild progenitor, *H. vulgares* subsp. *spontaneum*, and landraces of cultivated barley. Nevo (1992) found a clear geographical differentiation in 52 populations from three countries of West Asia. Out of 127 isoenzyme alleles in 27 loci, 65 occurred only in one country. In general, the alleles were not widespread but localized, having an 'island' structure of the genetic diversity. There were multilocus associations indicating the effect of natural selection. Landraces of cultivated barley have been analyzed on a regional or country basis in several studies including important countries from primary or secondary centres of genetic diversity, e.g. Ethiopia (Bekele 1983; Negassa 1985; Engels 1991), Iran (Brown and Munday 1982), Syria and Jordan (Ceccarelli *et al.* 1987; Weltzien 1989), Jordan (Jaradat 1989), the Himalayas (Murphy and Witcombe 1981; Konishi and Matsura 1991), Yemen and Nepal (Damania *et al.* 1985). These studies indicate that genetic diversity in barley landraces is distributed non-randomly and depends both on ecogeographic and ethnographic factors, and landraces are highly heterogeneous and a significant part of diversity is within landraces.

Results of six comparative studies of genetic diversity in cultivated barley (*H. vulgare* subsp. *vulgare*) and its wild progenitor were summarized by Brown (1992). Both allozyme and DNA data showed higher levels of diversity in wild barley, with one exception (Jana and Pietrzak 1988) which can be explained by a limited ecological niche of wild barley populations adjacent to cultivated fields. It was concluded that: (a) considerable molecular variation was present within landrace populations, (b) the barley species is more variable than other autogamous crops such as tomato and soybean, and (c) within bred cultivars the level of genetic diversity was much less. The predominantly autogamous breeding system in the wild and the cultivated subspecies of *H. vulgare* resulted in the following features of diversity:

1. Levels of diversity differ substantially from one population to another, reflecting the combined interaction of variation in whole-genome selection and repeated occurrence of bottleneck events.

2. Genetic variants at several loci are liable to be correlated in occurrence ('linkage

disequilibrium'). If an individual within a population carries a distinctive allele in one locus, then it is likely to do so at a second locus and a population that is polymorphic at one locus is likely to be polymorphic at a second.

3.   Variation shows geographical patterns both in levels and in allelic occurrence. In some loci it is adaptive, i.e. related to environmental variables.

The spontaneous (natural) genetic variation in barley germplasm collections has been found to be extensive. However, considerable new variation has been generated by artificial mutagenesis because barley has been a model for mutation research for 65 years. More mutants have been induced in barley than in any other flowering plant (Hockett and Nilan 1985). Although induced and spontaneous variants were mostly similar, induced mutations also produced forms of traits not found among the spontaneous variants. For example, waxy coating of the plant was induced in more than 1300 *eceriferum* mutants. Spontaneous variation for this trait is controlled by six loci, where analyses of induced variation revealed 77 loci, some having 15 to 20 alleles. Induced mutations extended genetic variability in traits as *erectoides*, anthocyanin synthesis, desynaptic chromosomes, mildew resistance and intermediate spike. At least 58 cultivars were released either through direct use of mutants or through hybridization of mutant lines (Hockett and Nilan 1985).

Diversity of barley germplasm collection held at ICARDA

The barley collection held at ICARDA is of relatively recent origin, because the centre was founded in 1976. In the first years of ICARDA's existence, barley germplasm was held at the Cereal Program as a working collection. In 1983, when the Genetic Resources Unit (GRU) was established, the barley genetic resources were transferred to the active collection and maintained under medium-term storage conditions. The base collection has been developed since 1989, when the long-term store facility became available. The present structure and genetic diversity of the barley collection resulted from ICARDA's acquisition and collecting strategy. From the very beginning the focus was on germplasm indigenous to the WANA region and landraces from other parts of the world. There were two main reasons for this strategy: (a) indigenous germplasm, i.e. the barley wild progenitor and cultivated barley landraces, is subjected to genetic erosion and (b) a valuable source of genes for adaptation and stress tolerance in breeding barley for low-input less-favourable environments exists in semi-arid regions of developing countries. Of the total 22 628 accessions, more than half (14 489) were received from the USDA Small Grain Collection, 2049 were obtained from Germplasm Resources Institute, CAAS, China and 1068 from the Institute of Plant Genetics and Crop Plant Research, Gatersleben, Germany. Twenty other donors contributed with smaller donations. The Cereal Improvement Program (now merged in the Germplasm Program) at ICARDA provided 2336 accessions, mostly breeding lines and named cultivars. Accessions collected by ICARDA in collaboration with national programmes are listed in Table 15.3. The collecting missions were undertaken to fill the gaps not only in the ICARDA barley germplasm collection but also in other global collections. Other major gaps were filled with germplasm received from IBPGR/IPGRI-supported missions to Bhutan, Libya, Morocco, Oman, Saudi Arabia and Yemen (399 accessions), and by donations from national programmes of Tunisia (490), Nepal (317), Morocco (258), Syria (209), Iraq (108) and Iran (100).

Cultivated barley germplasm held at ICARDA was characterized and evaluated for a number of descriptors. Characterization and preliminary evaluation data for more than 12 000 accessions were published in two catalogues (ICARDA 1986, 1988). The third catalogue with data on more than 9000 accessions is in press. Characterization data for growth class, kernel row number and kernel covering by country are shown in Table 15.4. The major part of this germplasm is spring barley (11 270 accessions), followed by facultative types (5607) and winter barley (4141). Six-rowed barley with

14 545 (70%) accessions is more frequent than two-rowed barley (6312 accessions, i.e. 30%, including the types with rudimentary sterile florets). The proportion of two-rowed barley is higher than in the USDA collection, where 23% was reported by Tolbert *et al.* (1979) and lower than the 37% found in European genebanks (Knüpffer *et al.* 1987). Barley with covered (hulled) kernels prevails (17 978 accessions), naked barley is less frequent (2274). The percentage of naked barley (11%) is similar to that in the USDA collection (14%). Frequency distributions for other descriptors are summarized in Table 15.5.

Characterization data were used for estimating phenotypic diversity for individual characters in 21 countries well represented in the ICARDA barley germplasm collection. The Shannon-Weaver information index was used for the measurement and comparisons of phenotypic diversity. The index was calculated as: $H_s = -3_i[p_i \ln (p_i)]$, where p_i is the proportion of the total number entries in the ith category. Relative indices H_{SR} were calculated according to Andrivon and de Vallavieille-Pope (1995), i.e. each H_s value was divided by its maximum value ($\ln N$), where N is the maximum number of the character categories. The overall country diversity $\overline{H}_{SR}$ was calculated as the arithmetic mean of the character H_{SR} values. World collection diversity was estimated from pooling over countries. The diversity estimates by character and country and the mean country diversity are shown in Table 15.6.

The highest diversity in the world estimates was for rachilla hair (1.00), growth class (0.92) and growth habit (0.87), while the lowest diversity was in hoodedness/awnedness (0.17) and stem colour (0.26). The mean diversity in the ICARDA world collection (0.65) was higher than in any individual country, as well as higher than in the USDA barley collection (0.57) when calculated by the same method from the data in Tolbert *et al.* (1979). However, their data include only five characters, while 12 characters were used in our estimates of diversity in the ICARDA barley collection. The diversity estimates from ICARDA characterisation data were substantially higher than the estimates by Tolbert *et al.* (1979) for all developing countries except Pakistan, e.g. Turkey (0.57 vs. 0.43), Syria (0.56 vs. 0.43), Ethiopia (0.55 vs. 0.48), Iran (0.55 vs. 0.46), Afghanistan (0.55 vs. 0.46), China (0.55 vs. 0.36), India (0.53 vs. 0.43), Morocco (0.50 vs. 0.28), Jordan (0.45 vs. 0.31) and Tunisia (0.34 vs. 0.13). Our data indicate that barley germplasm from countries of the primary and secondary centres of diversity such as Turkey, Syria, Iran, Afghanistan, Ethiopia, China, India and Morocco possesses high phenotypic diversity, as well as that from developed countries which introduced exotic germplasm and utilized it extensively in their breeding programmes (USA, Japan, Russia and Germany).

Peeters (1988) estimated overall country diversity in the barley collection of the Agriculture and Food Research Council (AFRC), UK using the Shannon-Weaver index to describe variation for 18 qualitative traits and standard deviation to assess the variation in 12 quantitative traits. The overall country means over all traits were calculated for both parameters and these were subsequently used to calculate average country ranks. We employed the same method for the calculation of average country ranks using 12 categorical characters (Table 15.7). The first eight countries displaying the highest diversity in the Peeters study were also included in our diversity estimates. Thus, it is interesting to compare the ranks in the two studies. In general, our ranking of countries from West and Central Asia and Africa is much higher (Table 15.8). This, again, indicates that ICARDA's systematic focus on the WANA region countries in the germplasm collection and acquisition resulted in higher genetic diversity conserved *ex situ* than in the USDA or AFRC collections.

Table 15.4. Geographical distribution of traits in ICARDA barley collection.

Origin	Total samples	Growth class			Kernel row		Kernel	
		Winter	Facult.	Spring	six	two	Naked	Covered
China	3039	174	1946	772	2488	346	1027	1808
Ethiopia	2692	99	324	2216	1433	1043	130	2436
USA	2244	1154	491	547	1726	437	79	2041
Turkey	1393	491	351	534	653	723	8	1358
Morocco	743	44	94	507	527	117	12	628
Switzerland	689	49	102	518	81	583	28	634
Tunisia	602	40	134	417	575	16	-	588
Colombia	569	5	43	517	554	10	-	564
Nepal	494	20	19	444	478	3	194	266
Germany	492	83	115	288	125	353	16	457
Syria	443	4	10	100	43	242	2	278
India	434	48	41	338	398	23	81	316
Iran	428	63	97	265	238	186	13	261
Yugoslavia	423	273	69	81	322	101	1	419
Greece	339	42	140	155	308	28	6	333
Ukraine	327	91	85	144	133	181	8	299
Afghanistan	285	47	59	171	223	46	25	238
Unknown	284	59	69	110	150	78	11	210
Japan	262	84	65	112	223	35	71	176
Russia	259	158	41	51	193	52	5	232
Pakistan	248	4	54	180	236	2	112	53
Austria	238	20	127	59	46	159	3	60
Spain	226	121	85	19	186	38	-	225
Libya	202	-	4	197	183	18	12	187
Korea	196	94	68	32	192	-	41	143
Egypt	196	9	3	135	140	7	-	146
Canada	184	51	33	94	144	31	3	168
Iraq	175	11	103	45	90	70	1	71
South Africa	146	5	8	132	20	124	-	144
UK	140	56	23	56	54	78	2	125
Jordan	137	2	8	114	36	91	-	121
Algeria	127	21	9	2	31	1	-	32
Poland	117	47	38	30	30	83	1	107
USSR	114	56	32	24	75	35	7	100
Peru	110	9	28	71	103	5	72	36
Yemen	106	-	5	90	16	89	4	91
Czechoslo-vakia	98	25	41	30	4	92	-	72
Georgia	94	19	17	57	50	43	-	93
Hungary	84	20	27	15	16	46	-	62
Australia	84	27	20	32	52	25	2	71
France	72	36	17	17	39	30	2	67
Sweden	68	23	17	25	25	38	1	61
Azerbaijan	58	49	7	1	55	1	-	57
Finland	53	2	13	38	19	34	2	50
Oman	44	-	-	-	-	-	-	-
Denmark	43	13	12	16	8	33	-	41
Bulgaria	38	30	2	6	34	4	1	37
Netherlands	37	20	8	6	16	17	-	33
Romania	36	19	5	12	24	11	1	32
Turkmenistan	34	12	14	7	26	7	3	30
Mongolia	33	3	1	29	24	9	18	15
Bhutan	30	8	1	3	12	-	2	10
Italy	29	3	14	10	19	8	6	21
Albania	28	5	10	13	19	9	-	28
Portugal	18	7	7	4	16	2	1	17
Mexico	18	2	4	12	16	2	-	18
Uzbekistan	16	7	5	2	11	3	-	12
Saudi Arabia	16	-	1	15	3	13	-	16
Lebanon	16	1	4	11	13	3	-	13
Argentina	16	4	3	8	9	6	1	14
Armenia	15	9	3	3	9	6	-	15
Bolivia	14	2	4	4	9	1	1	9
Chile	9	7	2	-	7	2	1	8
Tajikistan	8	-	3	5	4	4	1	7
Sudan	8	2	4	1	7	-	2	4

Origin	Total samples	Growth class			Kernel row		Kernel	
		Winter	Facult.	Spring	six	two	Naked	Covered
Venezuela	7	1	4	2	7	-	-	6
Cyprus	7	-	-	1	1	-	-	1
Norway	6	2	-	3	3	1	-	4
Belgium	6	5	1	-	6	-	-	6
Uruguay	5	2	1	2	3	2	-	5
Palestine	5	-	2	-	2	-	-	2
Ecuador	4	1	1	1	3	-	-	3
Belarus	4	-	1	3	-	4	-	4
United Arab Emirates	4	2	2	-	3	1	-	4
Zimbabwe	3	-	2	1	3	-	-	3
Lithuania	3	1	1	1	-	3	-	3
Brazil	3	-	3	-	1	2	-	1
Paraguay	2	2	-	-	2	-	-	2
Latvia	2	-	1	1	-	2	-	2
Kyrgyzstan	2	-	1	-	-	1	-	1
Guatemala	2	-	2	-	1	1	1	-
Estonia	2	-	-	2	1	1	-	2
Maldives	1	1	-	-	1	-	-	1
Kazakhstan	1	1	-	-	-	1	-	1
Ireland	1	-	1	-	-	1	-	1
Greenland	1	-	-	1	1	-	-	1
Central Afr. Rep.	1	1	-	-	1	-	1	-
ICARDA lines	2336	233	395	1303	1505	421	260	1666
Grand total	**22628**	**4141**	**5607**	**11270**	**14545**	**6312**	**2274**	**17978**

GERMPLASM DISTRIBUTION AND USE

The demand for cultivated barley germplasm held at GRU/ICARDA has been high in the last 7 years. From 1990 to 1996, the GRU distributed 27 256 seed samples on request from its barley active collection. Of these, 14 966 barley germplasm samples went to users in 30 countries and 12 290 samples were provided to barley breeders and scientists at ICARDA. Thus, on average, 3890 barley germplasm samples were distributed to users every year. As our data for 1992-94 indicate, the number of distinct accessions is around 60% of the total samples distributed. Therefore, we estimate that some 10% of the total 22 628 cultivated barley holdings are annually distributed to users at ICARDA and worldwide. The germplasm provided to the ICARDA Cereal Improvement Program contributed to the development of barley improved germplasm which was released as 80 new varieties in 32 countries (ICARDA 1995).

Properties and Uses

Barley grain is used for different purposes – as feed for animals, malt and food for human consumption. Archeological and historical evidence show that barley has been used in human food several thousand years ago. Barley preceded wheat as food grain in ancient Egypt and Nubia and it was eaten by Roman gladiators known as 'Hordearii' (Bhatty 1992). At the beginning of this century, it was still an important component of rural people's diet in Scandinavian countries. However, as living standards increased, barley was replaced by wheat and rice because of the superior quality of wheat gluten, which is important for baked products, and the better organoleptic quality of rice. Nevertheless, it is still an important food grain in developing countries. The consumption per person and year in the 1986-89 period was 68.3 kg in Morocco, 19.0 kg in Ethiopia, 18.1 kg in Algeria, 15.4 kg in Afghanistan, 11.5 kg in Iraq, 10.6 kg in Tunisia, 8.9 kg in Libya and 7.1 kg in Iran (Bhatty 1992). In the highlands of the Andes and Himalayas, barley is an essential food grain. About 7% of the world total production (11 million t) is used in beer and malt production (Munck 1992). The majority of the world's barley grain production is utilized for animal feed. It is a significant source of

Table 15.5. Frequency distribution of categories in 13 characters.

Category	Code	No. of access.	Description
Awn roughness	ARG	1 180	smooth
		19 468	rough
Spike density	SDE	2 074	lax (rachis internode >4 mm)
		16 964	intermediate
		1 918	dense
Growth habit	GHA	2 548	erect
		11 235	intermediate
		5 333	prostrate
Hoodedness/awnedness	H_A	168	sessile hoods
		213	elevated hoods
		129	awnless or awnleted (<2 cm)
		20 366	awned, on all six rows for six-rowed forms
		147	awned on central rows for six-rowed forms only
Lemma colour	LCO	14 343	white/brown
		3 257	purple or black
		1 931	other
Awn colour	ACO	4 390	white
		8 371	yellow
		2 904	brown
		738	reddish/brown
		465	black
Glume colour	GCO	3 531	white
		6 175	yellow
		1 165	brown
		1 061	black
		794	other
Grain colour (pericarp)	KCO	1 491	white
		3 149	blue
		1 030	black
		6 949	other
Rachilla hair length	RHL	7 181	short
		5 731	long
Stem colour	SCO	607	green
		187	purple
		12 156	other
Frost damage	FRD	675	1 damaged completely (1-9 scale)
		399	3 very poor
		1 292	5 fair
		3 070	7 good
		173	9 no damage
Resistance to lodging	LOD	1 920	1 excellent (1-9 scale)
		11 454	3 good
		2 125	5 fair
		1 775	7 poor
		1 151	9 very poor
Powdery mildew resistance	PM	14	1 resistant (1-9 scale)
		1 862	3 moderate resistant
		2 030	5 moderate
		1 003	7 moderate susceptible
		174	9 very susceptible

carbohydrates for different animals: swine, small ruminants and poultry. Protein quality (high lysine content) is important in protein-deficient regions of the world.

A barley plant produces at least as much straw as grain, therefore its global production is estimated at 180 million t. It has different uses depending on region or country. It is utilized as fertilizer, raw material for industry, fuel and animal feed. For example, straw is a valuable feed for small ruminants in WANA, which, in dry years, may have a higher price than barley grain. Barley landraces from those countries have soft and highly palatable straw; this trait is rare or lacking in the modern lodging-resistant varieties. Breeding for straw quality has been neglected by barley breeders; therefore ICARDA included high straw quality in its breeding objectives.

Breeding

Modern barley breeding started at the end of the last century and on several occasions barley research and breeding 'pioneered' the way for other crops. During a relatively short history the breeders employed mutation breeding, hybrid breeding and population improvement programmes involving composite crosses and evolutionary breeding. Male-sterile facilitated recurrent selection was used in breeding for disease resistance and interspecific and intergeneric crosses were used for the same purpose. Recently, haploid breeding based on the bulbosum system or anther culture became part of routine programmes focused on accelerated development of homozygous lines from segregating populations. Innovative biotechnological methods, including transformation, provide new possibilities, unthinkable some 30 years ago.

Formal, or institutional, breeding has been highly efficient in improving barley yield levels, malting quality and disease resistance. However, its efficiency has remained largely confined to favourable environments, or to environments which could be made favourable by adding fertilizer and irrigation, and by chemical control of weeds, pests and diseases.

ICARDA, one of the centres of the CGIAR system, gradually recognized the limitations of the formal breeding in barley improvement for stress-affected low-input farming systems in the semi-arid regions of WANA and other parts of the world. Resource-poor farmers, who practise approximately 60% of global agriculture and produce 15-20% of the world food (Francis 1986), have not known the benefit of the Green Revolution. Some 1.4 billion people are dependent on agriculture practised in stressful environments (Pimbert 1994).

To better serve those farmers who were neglected in the first Green Revolution, ICARDA breeders and researchers developed a new breeding strategy based on selection of segregating materials in target environments with farmers' participation. Whenever possible, the selection is done on farmers' fields, using their level of inputs, their technology and their experience and preferences.

Typical characteristics of formal breeding programmes in several crops are:

1. they generally produce genetically uniform cultivars (pure lines, clones, hybrids),
2. they are largely conducted either in good environments or in well-managed experiment stations where growing conditions are optimum or near-optimum,
3. in most grain crops selection is almost exclusively for grain yield and disease resistance,
4. they promote cultivars which can be grown over large areas (widely adapted in a geographical sense), and
5. they do not involve the clients (the farmers) in any of the steps which will eventually lead to new cultivars, except perhaps in the final field testing of a few promising lines.

Table 15.6. Estimates of H_{SR} for 21 countries and 12 characters and mean diversity $\overline{H}_{SR}$ over all characters.

Origin	N'(MAX)	ACO²	ARG	GCL	GCO	GHA	H/A	KCO	KCV	LCO	RHL	RNO	SCO	Mean
Switzerland	669	0.67	0.04	0.62	0.42	0.69	0.03	0.36	0.25	0.02	0.96	0.37	0.04	0.37
France	70	0.69	0.26	0.94	0.55	0.86	0.00	0.62	0.19	0.00	0.97	0.62	0.07	0.48
Spain	229	0.38	0.28	0.84	0.51	0.53	0.00	0.76	0.00	0.49	0.84	0.41	0.00	0.42
Greece	339	0.50	0.11	0.89	0.42	0.80	0.00	0.53	0.05	0.81	0.55	0.26	0.03	0.41
Yugoslavia	423	0.68	0.15	0.81	0.63	0.83	0.00	0.69	0.02	0.60	0.79	0.50	0.04	0.48
Hungary	62	0.60	0.12	0.97	0.55	0.66	0.00	0.68	0.00	0.61	0.97	0.59	0.00	0.48
Germany	486	0.66	0.30	0.87	0.57	0.65	0.04	0.63	0.21	0.55	0.95	0.62	0.16	0.52
Russia	277	0.77	0.21	0.88	0.69	0.80	0.06	0.81	0.14	0.43	1.00	0.54	0.04	0.53
United States	2168	0.66	0.73	0.93	0.71	0.88	0.30	0.83	0.23	0.17	0.92	0.50	0.06	0.58
India	424	0.60	0.13	0.59	0.68	0.95	0.14	0.68	0.73	0.66	0.85	0.21	0.09	0.53
Japan	258	0.71	0.04	0.98	0.71	0.95	0.11	0.70	0.87	0.47	0.71	0.38	0.00	0.55
China	2892	0.35	0.10	0.72	0.67	0.68	0.14	0.64	0.94	0.86	1.00	0.35	0.11	0.55
Pakistan	238	0.40	0.00	0.56	0.18	0.83	0.00	0.37	0.91	0.16	0.70	0.04	0.58	0.40
Afghanistan	279	0.65	0.23	0.84	0.61	0.96	0.03	0.66	0.43	0.71	0.97	0.42	0.03	0.55
Iran	425	0.57	0.29	0.83	0.61	0.93	0.00	0.79	0.26	0.97	0.77	0.62	0.00	0.55
Turkey	1373	0.67	0.47	0.99	0.63	0.92	0.02	0.74	0.03	0.64	1.00	0.63	0.07	0.57
Syria	285	0.70	0.99	0.41	0.70	0.59	0.01	0.87	0.06	0.58	0.56	0.38	0.89	0.56
Jordan	128	0.59	0.20	0.29	0.76	0.42	0.00	0.65	0.00	0.62	0.84	0.53	0.52	0.45
Ethiopia	2640	0.92	0.05	0.48	0.88	0.80	0.00	0.84	0.29	0.56	0.84	0.85	0.10	0.55
Tunisia	591	0.39	0.10	0.69	0.39	0.45	0.00	0.32	0.00	0.44	0.50	0.12	0.65	0.34
Morocco	643	0.69	0.10	0.58	0.54	0.63	0.30	0.53	0.13	0.33	0.94	0.53	0.75	0.50
World	18995	0.77	0.30	0.92	0.81	0.87	0.17	0.82	0.49	0.73	1.00	0.65	0.26	0.65

Table 15.7. Countries sorted by decreasing mean diversity ($\bar{H}_{SR}$) of their germplasm across all characters[1].

Country	$N^2_{(MAX)}$	Percentage evaluated[3]	$\bar{H}_{SR}$	Average rank[4]
United States	2168	98	0.58	7.7
Turkey	1373	91	0.57	6.7
Syria	285	68	0.56	10.0
Ethiopia	2640	95	0.55	8.9
Japan	258	90	0.55	9.3
Iran	425	66	0.55	9.7
Afghanistan	279	83	0.55	9.7
China	2892	68	0.55	9.9
Russia	277	93	0.53	8.1
India	424	86	0.53	9.4
Germany	486	88	0.52	9.5
Morocco	643	97	0.50	10.8
France	70	97	0.48	10.2
Yugoslavia (former)	423	86	0.48	10.8
Hungary	62	81	0.48	12.0
Jordan	128	86	0.45	12.7
Spain	229	70	0.42	11.4
Greece	339	82	0.41	14.3
Pakistan	238	76	0.40	15.4
Switzerland	669	98	0.37	14.0
Tunisia	591	92	0.34	17.0

[1] For observations taken on 12 categorical characters
[2] For explanation see Table 15.3.
[3] Per entry and across all 12 individual characters
[4] Over 21 countries (for calculation method see text)

Table 15.8. Ranking of overall country diversity in barley collections.

Country	ICARDA	Peeters (1988)
Turkey	1	2
USA	2	1
Russia (former USSR)	3	4
Ethiopia	4	18
Japan	5	3
India	6	11
Germany	7	6
Afghanistan	8-9	16
Iran	8-9	21
China	10	5
Syria	11	-
France	12	7
Yugoslavia	13-14	8
Morocco	13-14	34

Assumptions of formal breeding programmes are that:
1. selection must be conducted under good growing conditions where heritability is higher, and therefore response to selection is also higher,
2. yield increases can only be obtained through replacement of locally adapted landraces (Brush 1991) which are low yielding and disease susceptible,
3. breeders know better than farmers the characteristics of a successful cultivar,

and

4. when farmers do not adopt improved cultivars it is because of ineffective extension and/or inefficient or insufficient seed production capabilities: the hypothesis that the breeder might have bred the wrong varieties is rarely considered.

Because of the success that breeding has had in good environments, these characteristics and assumptions are not questioned even when the objective is to improve yield and yield stability for poor farmers in stressful environments. The implicit assumption is that what has worked well in favourable conditions must also be appropriate to unfavourable conditions, and very little attention has been given to developing new breeding strategies for less favourable environments.

In the last few years there has been mounting evidence that these assumptions are not valid, and that the special problems of unfavourable environments and their farming systems must be addressed in different ways. Unfavourable environments are defined as those where crop yields are commonly low owing to the concomitant effects of several abiotic and biotic stresses. The semi-arid areas of Syria, where barley-livestock is the predominant farming system, are a good example of such environments where not only low annual rainfall, but also rainfall distribution, low winter temperatures, high temperatures and hot winds from anthesis to grain filling are important abiotic stresses. The frequency, timing, intensity and duration of each of these stresses, as well as their specific combinations, vary from year to year. However, low yields of barley are common, crop failures occur one year out of ten, and yields of 3 t/ha or more are expected less than 15% of the time. By contrast, in relatively favourable environments yields of 1 t/ha or less have a frequency of about 10%, and yields above 2.5 t/ha occur more than 40% of the time.

Because of the probability of low yields and crop failures in unfavourable environments, the use of inputs such as fertilizers, pesticides and weed control is uneconomical and risky for resource-poor farmers. Therefore, the adoption of 'improved agronomic practices' has been limited, and one economic solution to increase crop yields in unfavourable environments can be through breeding. Many of the environments where 'improved technologies' in general, and 'improved cultivars' in particular, have had little or marginal impact have some characteristics in common with those described for the semi-arid areas of Syria. These are the unpredictability and variability of climatic conditions, the consequent high probability of crop failures which discourages the use of inputs.

Resource-poor farmers in many regions of the word practising agriculture in these or similar situations have adopted a strategy based on both intraspecific and interspecific diversity (Martin and Adams 1987). Different crops are grown in the same field at the same time (interspecific diversity), and the cultivars of the different crops are frequently genetically heterogeneous (intraspecific diversity). A second level of intraspecific diversity is obtained by growing, at the same time and in the same field, different cultivars of the same crops (Haugerud and Collinson 1990). The type of diversity which prevails in different areas depends on both climatic and socioeconomic conditions and farmers' response to these. In the dry areas of WANA, for example, barley is often the only feasible rain-fed crop, and the cultivars which are grown at present, and which have been grown for centuries, are genetically heterogeneous (Ceccarelli *et al.* 1987; Weltzien and Fischbeck 1990).

This diversity which is typical of resource-poor farming is in marked contrast with the uniformity pursued by formal breeding and production practices in most crops grown in favourable environments, and is one of the causes for a different mechanism of seed supply. While in high-input agriculture served by formal breeding, the seed market is the main source of seed supply, particularly for grain crops, in resource-poor agriculture the seed is usually produced on the farm, after some form of selection done by the farmer, or it is purchased from neighbouring farmers (Almekinders *et al.* 1994). Formal breeding thus not only tries to replace diversity with uniformity, but also tries to reach farmers with the seed of new cultivars through mechanisms and institutions which are not familiar, are not efficient and often are not trusted by resource-poor farmers.

Breeding for unfavourable environments based on selection (not merely testing) in the target environments is undoubtedly more complex than selection for favourable environments, largely because of the year-to-year variation. Procedures and methodologies developed for favourable environments need to be modified. The methodology for barley enhancement at ICARDA is the following:

1. Breeding material (including parental material and segregating populations) is evaluated in the target environments using farmers' agronomic practices, including rotations. In the driest site (long-term average rainfall of 233 mm) this means no use of fertilizers, pesticides and weed control. Farmers' fields are inspected one or two cropping seasons earlier and those where the farmer's crop is sufficiently uniform are selected as 'experiment sites'. Concurrently, the material is evaluated at the main experiment station (long-term average rainfall of 373 mm) with a level of inputs commonly used in moderately favourable areas. In all the experiment sites the material is evaluated strictly under rain-fed conditions.

2. Experimental designs have evolved from the randomized block design to the lattice design (introduced in 1984), to α-lattice design (introduced in 1993). This has progressively improved our control of environmental variability.

3. Segregating populations are evaluated as bulks for 3 years, taking advantage of the large year-to-year variation in total rainfall, rainfall distribution and temperature patterns. Each year bulks yielding less than the check are discarded. Individual plant selection is done only within the selected bulks.

4. Selection is done for high grain yield at each of the experiment sites, regardless of the performance in other experiment sites. This promotes breeding material with specific adaptation.

5. In addition to grain yield, traits used as selection criteria are plant height, tillering, straw softness and disease resistance in the two driest sites; earliness, lodging and disease resistance in the wettest sites.

Using this methodology for barley breeding at ICARDA, direct selection in unfavourable environments revealed that locally adapted landraces could be a useful source of breeding material that would have been missed had the evaluation taken place only in high-yielding environments. The presence of useful diversity within landraces has been documented in many crops. The diversity within barley landraces collected in Syria and Jordan has been documented by Ceccarelli *et al.* (1987, 1995), van Leur *et al.* (1989), Weltzien (1988, 1989) and Weltzien and Fischbeck (1990).

International breeding programmes aim to assist national programmes to increase agricultural production by developing superior cultivars. This is traditionally done through very large breeding programmes which develop fixed or semi-fixed lines with an average good performance across many environments (often well-managed experiment stations). The interaction between international and national programmes has been largely a one-way, "top-down" process (Simmonds and Talbot 1992) where international programmes develop germplasm, distribute it as 'international nurseries', and national programmes test and eventually release it as varieties. This has commonly excluded the use of locally adapted germplasm, which often performs poorly in favourable conditions such as those of experiment stations, and encouraged its displacement.

The adoption of a positive interpretation of genotype by environment (GE) interaction by international breeding programmes has been advocated as a way to address the need of small, resource-poor farmers, who have been bypassed by the Green Revolution (Stroup *et al.* 1993). To exploit specific adaptation fully and make positive use of GE interactions, an international breeding programme should devolve most of the selection work to national programmes by gradually replacing the traditional international nurseries with earlier generation material. Early distribution of breeding material reduces the danger of useful lines being discarded because of their relatively poor performance at some test sites.

In 1991 ICARDA's barley breeding programme started a gradual process of devolution of selection work to the four Maghreb countries (Ceccarelli *et al.* 1994). When fully implemented, national programmes in north Africa will receive from ICARDA's barley breeding programme only targeted F_2 segregating populations (based on crosses designed by national programmes), and yield trials consisting of lines derived from these F_2 populations selected in-country. Selection between F_2 populations will be in the different agro-ecological environments within each country under conditions as similar to farmers' fields as possible. Lines selected from superior F_2 populations will be advanced at ICARDA and then yield-tested in different locations within each country.

However, decentralization to national programmes of the selection component of an international breeding programme will not respond to the needs of resource-poor farmers if it is only a decentralization from one experiment station to another. This can be solved by what may be considered the most extreme decentralization and possibly the most effective way of exploiting specific adaptation, i.e. farmers' participation in selection under their own conditions.

Indeed, there is evidence that when farmers are involved in the selection process, their selection criteria may be very different from those of the breeder (Hardon and de Boef 1993; Sperling *et al.* 1993). There is also evidence that, when breeders and farmers select in the same environment, farmers' selection can be effective, implying that farmers possess considerable knowledge which is almost totally neglected by formal plant breeding programmes.

A typical example of different selection criteria between farmers and breeders can be found in barley used as animal feed. Breeders often use grain yield as the sole selection criterion which usually brings with it high harvest index and lodging resistance. However, in unfavourable environments lodging is often not a problem because of moisture stress, and farmers are interested not only in grain yield, but also in forage yield and in the palatability of both grain and straw.

Farmers' participation in the ICARDA barley breeding programme to date has been informal and consisted of discussions during field visits and occasional inspection and selection by farmers of breeding lines. The most significant contribution of this informal participation has been the incorporation by the breeders of plant height under drought and softness of the straw as selection criteria in breeding barley for dry areas. A crop which remains tall even in very dry years is important to farmers because it reduces their dependence on costly hand-harvesting, while soft straw is considered important in relation to palatability. It is obvious that these two characteristics represent a drastic departure from the typical selection criteria used in breeding high-yielding cereal crops – short plants with stiff straw and high harvest index. It is also obvious that cultivars possessing the two characteristics considered important by farmers in dry areas will not be suited for cultivation in high-yielding environments because of their lodging susceptibility – a further indication of the importance of specific adaptation. Lines extracted from Syrian barley landraces show little variation in straw softness but considerable variation, both between and within collection sites, for plant height. The most promising avenue to improve plant height under drought is offered by the use of the wild progenitor of cultivated barley, still widely distributed along the Fertile Crescent where, particularly in the driest areas, it can be easily identified at a distance because of its tallness. For example, while the mean plant height of 1532 breeding lines tested in 1994/95 at a site which received only 222 mm rainfall was 23.5 cm, the shortest lines were only 12.5 cm tall, and the most widely cultivated landrace (Arabi Aswad) was about 25 cm, some of the lines derived from crosses with *H. spontaneum* were taller than 40 cm. They were also significantly taller than Zanbaka, a pure line selected from A. Aswad and already grown by some farmers because of its plant height.

A formal plant breeding programme could combine the concept of a positive use of GE interaction with the utilization of farmers' knowledge by evaluating a wider range of germplasm under farmers' field conditions and in conjunction with farmers. In those communities where extension services and conventional seed production systems are not able to reach resource-poor farmers, and farmers traditionally use their own seed from one cropping season to another, this will provide a direct link between formal plant breeding and farmers. The benefit to the farmers will be direct access to improved germplasm. The benefit to all the community will be the maintenance of genetic diversity within a crop because different farmers are likely to select different materials. Eventually, the benefit to formal breeding programmes could be a higher efficiency by using farmers' selection criteria.

The use of high-input selection environments in a market-driven agriculture has been largely responsible for the trend of modern plant breeding toward narrowing the genetic base of our crops accompanied by a trend towards homogeneity: one clone, one pure line, one hybrid (Simmonds 1983). Uniformity and broad adaptation are very useful attributes to accommodate large-scale centralized seed production (Davis 1990). Although the merits of genetic uniformity have been questioned in developed countries (Wolfe 1991), it is still very popular in breeding programmes and seed production systems of developing countries at both the national and international levels. This is in contrast with the genetic diversity that characterizes agriculture in marginal environments. Genetically heterogeneous landraces are still the backbone of agricultural systems in many developing countries, mainly in marginal environments where their replacement by modern, genetically uniform varieties bred for favourable environments has proved to be a difficult task at the levels of inputs farmers can afford.

Breeding for specific adaptation to unfavourable environments implies a re-evaluation of the important role that genetic resources such as landraces can play because they possess adaptive features to these environments. This is the first consequence on biodiversity of breeding for specific adaptation.

A second consequence of exploiting specific adaptation on biodiversity is that the number of varieties (not necessarily homogeneous) of a given crop grown at any time will be large. The benefits of maintaining genetic diversity within a crop over large areas have been discussed extensively in the literature in relation to resistance to pests and diseases and do not need further justification. A major constraint of breeding for specific adaptation is the problem of how to distribute many varieties among farmers. However, the distribution of specifically adapted varieties to resource-poor farmers does not have to follow the conventional release-seed production-seed certification schemes used in developed countries. Indeed, there are examples of successful distribution and adoption of varieties through non-market methods (Grisley 1993).

Prospects and Limitations

Barley has retained or even strengthened its position in the WANA region, where the area harvested increased by 23.9% between 1979-81 and 1992-94. This trend will probably continue, because barley is planted in degraded rangelands to increase their productivity and provide feed for flocks of small ruminants. In general, it can be envisaged that the area of barley will be increasing in some developing countries, because barley can adapt better to unfavourable conditions and low-input farming systems than wheat. It will be vitally important for resource-poor farmers. The shift in the importance of barley among the regions and growers will have implications for breeding objectives, strategy and methods, as described previously. Stress-tolerant and disease-resistant germplasm will be essential for the low-input farming system.

The decline of barley growing in Europe and North America will be difficult to stop, unless alternative, economically viable markets are found. One possible chance might be if barley is rediscovered as a food grain in the developed countries. New barley cultivars with thin or no hulls, waxy starch, high lysine and high β-glucan contents make the crop suitable for use in many food products (Bhatty 1992).

Barley has had a long history of cultivation, as well as of breeding. The extent of its cultivation in the developed countries depends on agricultural policies and market conditions. In many developing countries, there is no alternative to barley in stress-affected marginal environments. The limitation may be that most of the breeding effort has been spent on breeding for high-input barley production systems in developed countries, on malting barley, in particular, and breeding for agroclimatic conditions and farming systems in developing countries has been neglected. As the area of barley will be increasing in semi-arid regions of developing countries, in contrast to diminishing hectarage in Europe and possibly North America, priority has to be given to the development of improved germplasm appropriate for the low-input farming systems prevailing in those areas. The indigenous germplasm, landraces and wild relatives, held *ex situ* in genebanks, will be essential for the success in this endeavour.

REFERENCES

Almekinders, C.J.M., N.P. Louwaars and G.H. de Bruijn. 1994. Local seed systems and their importance for an improved seed supply in developing countries. Euphytica 78:207-216.

Andrivon, D. and C. de Vallavieillle-Pope. 1995. Race diversity and complexity in selected populations of fungal biotrophic pathogens of cereals. Phytopathology 85:897-905.

Bekele, E. 1983. Allozyme genotypic composition and genetic distance betwen the Ethiopian landrace populations of barley. Hereditas 98:259-267.

Bhatty, R.S. 1992. Dietary and nutritional aspects of barley in human food. Pp. 913-923. *in* Barley Genetics VI. (L. Munck, ed.). Munksgaard International Publishers, Copenhagen, Denmark.

Bothmer, R. von, N. Jacobsen, C. Baden, R.B. Jorgensen and I. Linde-Laursen. 1991. An Ecogeographical Study of the Genus *Hordeum*. Systematic and Ecogeographic Studies on Crop Genepools No. 7. Internationational Board for Plant Genetic Resources, Rome, Italy.

Bothmer, R. von. 1992a. The genepool of barley and preservation of wild species of *Hordeum*. Pp. 32-35 *in* International Crop Network Series No. 9. International Board for Plant Genetic Resources, Rome, Italy.

Bothmer, R. von. 1992b. The wild species of *Hordeum*: Relationships and potential use for improvement of cultivated barley. Pp. 3-18 *in* Barley: Genetics, Biochemistry, Molecular biology and Biotechnology (P. Shewry, ed.). Biotechnology in Agriculture No. 5, CAB International, Wallingford, UK.

Brown, A.H.D. 1992. Genetic variation and resources in cultivated barley and wild *Hordeum*. Pp. 669-682 *in* Barley Genetics VI (L. Munck, ed.). Munksgaard International Publishers, Copenhagen, Denmark.

Brown, A.H.D. and J. Munday. 1982. Population-genetic structure and optimal sampling of landraces of barley from Iran. Genetica 58:85-96.

Brush, S.B. 1991. A farmer-based approach to conserving crop germplasm. Econ. Bot. 45:153-161.

Ceccarelli, S, W. Erskine, S. Grando and J. Hamblin. 1994. Genotype x environment interaction and international breeding programmes. Exp. Agric. 30:177-187.

Ceccarelli, S., S. Grando and J.A.G. van Leur. 1987. Genetic diversity in barley landraces from Syria and Jordan. Euphytica 36:389-405.

Ceccarelli, S., S. Grando and J.A.G. van Leur. 1995. Understanding landraces: The Fertile Crescent's barley provides lesson to plant breeders. Diversity 11:112-113.

Chapman, C.G.D. 1987. Barley genetic resources, the status of collecting and conservation. Pp. 43-49 *in* Barley Genetics V, Proceedings of the Fifth International Barley Gemetics Symposium, Okayama,

Japan.

Damania, A.B., M.T. Jackson and E. Porceddu. 1985. Variation in wheat and barley landraces from Nepal and the Yemen Arab Republic. Z. Pflanzenzüchtg. 94:13-24.

Davis, J. 1990. Breeding for intercrops-with special attention to beans for intercropping with maize. *In* Research Methods for Cereal/Legume Intercropping (S.R. Waddington, A.F.E. Palmer and O.T. Edge, eds.). Proceedings of a Workshop on Research Methods for Cereal Legume Intercropping in Eastern and Southern Africa, Lilongwe, Malawi, 23-27 January 1988.

ECPGR. 1983. Report of a Working Group on Barley. UNDP/IBPGR European Cooperative Programme for Conservation and Exchange of Crop Genetic Resources. International Board for Plant Genetic Resources, Rome, Italy.

Engels, J.M.M. 1991. A diversity study in Ethiopian barley. Pp. 131-139 *in* Plant Genetic Resources of Ethiopia (J.M.M. Engels, J.G. Hawkes and M. Worede, eds.). Cambridge University Press, Cambridge, UK.

FAO. 1996. The State of the World's Plant Genetic Resources for Food and Agriculture. FAO, Rome, Italy.

Fedak, C. 1985. Wide crosses in *Hordeum*. Pp. 155-186 *in* Barley (D.C. Rasmusson, ed.). Agronomy Monograph No. 26. American Society for Agronomy, Madison, WI, USA.

Francis, C.A. 1986. Multiple Cropping Systems. Macmillan, New York.

Grisley, W. 1993. Seed for bean production in Sub-Saharan Africa, issues, problems, and possible solutions. Agric. Syst. 43:19-33.

Hardon, J.J. and W.S. de Boef. 1993. Linking farmers and breeders in local crop development. Pp. 64-71 *in* Cultivating Knowledge. Genetic diversity, farmer experimentation and crop research (W. de Boef, K. Amanor, K. Wellard and A. Bebbington, eds.). Int. Techn. Publ. Ltd.

Harlan, J.R. 1991. On the origin of barley: a second look. Barley Genetics II:45-50.

Harlan, J.R. 1992. Crops & Man. Second edition. American Society of Agronomy, Crop Science Society of America, Madison, WI, USA.

Harlan, J.R. and J.M.J. de Wett. 1971. Towards a rational classification of cultivated plants. Taxon 20:509-517.

Haugerud, A. and M.P. Collinson. 1990. Plants, genes and people: improving the relevance of ant breeding in Africa. Exp. Agric. 26:341-362.

Hockett, E.A. and R.A. Nilan. 1985. Genetics. Pp. 187-230 *in* Barley (D.C. Rasmusson, ed.). Agronomy Monograph No. 26, American Society for Agronomy, Madison, WI, USA.

IBPGR. 1990. Annual report 1989. International Board for Plant Genetic Resources, Rome, Italy.

IBPGR. 1992. Barley Genetic Resources. International Crop Network Series No. 9. International Board for Plant Genetic Resources, Rome, Italy.

ICARDA. 1986. Barley Germplasm Catalog I. The International Center for Agricultural Research in the Dry Areas (ICARDA), Aleppo, Syria.

ICARDA. 1988. Barley Germplasm Catalog Part II. The International Center for Agricultural Research in the Dry Areas (ICARDA), Aleppo, Syria.

ICARDA. 1995. ICARDA Annual report 1995. The International Center for Agricultural Research in the Dry Areas (ICARDA), Aleppo, Syria.

Jana, S. and L.N. Pietrzak. 1988. Comparative assessment of genetic diversity in wild and primitive cultivated barley in a center of diversity. Genetics 119:981-990.

Jaradat, A.A. 1989. Diversity within and between populations of two sympatrically distributed *Hordeum* species in Jordan. Theor. Appl. Genet. 78:653-656.

Knüpffer, H. 1988. The European Barley Database of the ECP/GR: an introduction. Kulturpflanze 36:135-162.

Knüpffer, H. and Th. J.L. van Hintum. 1995. The barley core collection: an international effort. Pp. 171-178 *in* Core Collections of Plant Genetic Resources (T. Hodgkin, A.H.D. Brown, Th.J.L. van Hintum and E.A.V. Morales, eds.). John Wiley & Sons, Chichester, UK.

Knüpffer, H., Ch.O. Lehmann and F. Scholz. 1987. Barley genetic resources in European genebanks: The European barley database. Pp. 75-82 *in* Barley Genetics V, Proceedings of the Fifth International Barley Genetics Symposium, Okayama, Japan..

Kobyl'yansky, V.D. and M.V. Lukyanova (eds.). 1990. Flora of cultivated plants. Volume II, Part 2 Barley [in Russian]. Agropromizdat, Leningrad, USSR.

Konishi, T. 1988. Genetic differentiation and geographical distribution of barley. Pp. 237-243 *in* Crop Genetic Resources of East Asia. Proceedings of the International Workshop on Crop Genetic Resources in East Asia, IBPGR, Rome, Italy..

Konishi, T. 1995. Geographical diversity of isozyme genotypes in barley. Kyushu University Press, Fukuoka, Japan.

Konishi, T. and S. Matsura. 1991. Geographic differentiation of isozyme genotypes of Himalayan barley

(*Hordeum vulgare*). Genome 34:704-709.

Martin, G.B. and M.W. Adams. 1987. Landraces of *Phaseolus vulgaris* (Fabaceae) in Northern Malawi. II. generation and maintenance of variability. Econ. Bot. 41:204-215.

Munck, L. 1992. The contribution of barley to agriculture today and in the future. Pp. 1099-1109 *in* Barley Genetics VI (L. Munck, ed.). Munksgaard International Publishers, Copenhagen, Denmark.

Murphy, P.J. and J.R. Witcombe. 1981. Variation in Himalayan barley and the concept of centers of diversity. Pp. 26-36 *in* Barley Genetics IV (M.J.C. Asher, R.P. Ellis and A.M. Hayter, eds.). Edinburgh, UK.

Negassa, M. 1985. Patterns of phenotypic diversity inan Ethiopean barley collection, and the Arussi-Bale Highland as a centre of origin of barley. Hereditas 102:139-150.

Nevo, E. 1992. Origin, evolution, population genetics and resources for breeding of wild barley, *Hordeum spontaneum*, in the Fertile Crescent. Pp. 19-43 *in* Barley: Genetics, Biochemistry, Molecular Biology and Biotechnology (P. Sewry, ed.). Biotechnology in Agriculture No. 5, CAB International, Wallingford, UK.

Peeters, J.P. 1988. The emergence of new centres of diversity: evidence from barley. Theor. Appl. Genet. 76:17-24.

Pimbert, M.P. 1994. The need for another research paradigm. Seedling, July 1994: 20-25.

Plucknett, D.L., N.J.H. Smith, J.T. Williams and N. Murthi-Anishetty. 1987. Genebanks and the World's Food. Princeton University Press, Princeton, NJ, USA.

Shao, Q. and A. Li. 1987. Unity of genetic population for wild barley and cultivated barley in Himalaya area. Pp. 35-41 *in* Barley Genetics V, Proceedings of the Fifth International Barley Genetics Symposium, Okayama, Japan.

Simmonds, N.W. 1983. Plant breeding, the state of the art. Pp. 5-25 *in* Genetic Engineering of Plants. An Agricultural Perspective (T. Kosuge, C.P. Meredith and A. Hollaender, eds.). Plenum Press, New York.

Simmonds, N.W. and M. Talbot. 1992. Analysis of on-farm rice yield data from India. Exp. Agric. 28:325-329.

Sperling, L., M.E. Loevinsohn and B. Ntabomvura. 1993. Rethinking the farmer's role in plant breeding: local bean experts and on-station selection in Rwanda. Exp. Agric. 29:509-519.

Stroup, W.W., P.E. Hildebrand and C.A. Francis. 1993. Farmer participation for more effective research in sustainable agriculture. Pp. 153-186 *in* Technologies for Sustainable Agriculture in the Tropics (J. Ragland and R. Lal, eds.). ASA Spec. Publ. 56, ASA, CSSA and SSSA, Madison, WI, USA.

Takahashi, R. 1987. Genetic features of East Asian barleys. Pp. 7-20 *in* Barley Genetics V, Okayama, Japan.

Tolbert, D.M., C.O. Qualset, S.K. Jain and J.C. Craddock. 1979. A diversity analysis of a world collection of barley. Crop Sci. 19:789-794.

van Hintum, Th.J.L. 1994. Drowning in the Genepool, managing genetic diversity in genebank collections. Thesis, Swedish Academy of Agricultural Sciences, Department of Plant Breeding Research, Sval'v, Sweden.

van Leur, J.A.G., S. Ceccarelli and S. Grando. 1989. Diversity for disease resistance in barley landraces from Syria and Jordan. Plant Breed. 103:324-335.

Weltzien, E. 1988. Evaluation of barley (*Hordeum vulgare* L.) landraces populations originating from different growing regions in the Near East. Plant Breed. 101:95-106.

Weltzien, E. 1989. Differentiation among barley landrace populations from the Near East. Euphytica 43:29-39.

Weltzien, E. and G. Fischbeck. 1990. Performance and variability of local barley landraces in near-eastern environments. Plant Breed. 104:58-67.

Wolfe, M.S. 1991. Barley diseases: maintaining the value of our varieties. Barley Genetics VI:1055-1067.

Zohary, D. 1959. Is *Hordeum agriocrithon* Cberg the ancestor of six-rowed barley? Evolution 13:279-280.

Zohary, D. and M. Hopf. 1988. Domestication of Plants in the Old World. Clarendon Press, Oxford, UK.

Chapter 16

A. Maize

S. Taba

World maize production in 1992-94 oscillated between 470 and 569 million t produced on about 127-132 million ha. The average yields were in the range of 3.7-4.3 t/ha, which are the highest among the most important world cereal food crops: wheat, rice and maize (FAO 1995). Its production, however, is not evenly distributed. More than 40% of world maize production comes from the United States. Maize is widely adapted between 55°N and S latitudes (Guidry 1964) and at altitudes from sea level to 3600 m in cool tropical highlands of the Andes. Adapted maize germplasm is cultivated in tropical lowlands, tropical and subtropical mid-altitudes, temperate and cool tropical highland climates.

Hybrid maize was first introduced in the United States before World War II and further development of single-cross hybrids from the 1960s in most temperate maize-growing countries has been a significant factor in increasing maize production (Hallauer *et al.* 1988). Hybrid maize technology is being employed for the other maize types grown in tropical, mid-altitude and highland maize production regions where, however, an important part of maize production is the use of landraces (traditional local maize varieties) and improved varieties for food preparations of preferred grain texture and colour.

BOTANY AND DISTRIBUTION

Cultivated maize is *Zea mays* L. (Species Plantarum 971. 1753) or *Zea mays* L. subsp. *mays* Iltis (Iltis and Doebley 1980), and two of its relatives, *Tripsacum* and Teosinte, are described later in this volume. It is also called corn in English, *maïs* in French, *mais* in German/Italian, *maíz* in Spanish, *milho* in Portuguese, *yùmí* in Chinese, *khao phot* in Thai, *jagung* in Indonesia, *tomorokoshi* in Japanese, *bokolo* in Ethiopia, *chimanga* in Malawi and *zorrat* in Persian, to name a few.

Origin, Distribution and Diffusion

Corn was domesticated well before 4000 BC in Tehuacán, Puebla State, Mexico (MacNeish 1985). In Mesoamerica maize became the dietary staple by about 1500 BC (Goodman 1988a). The oldest known archaeological samples of maize were found in caves near Tehuacán (Mangelsdorf *et al.* 1964, 1967). Original dating based on analysis of surrounding strata placed their origin as far back as 5000 BC, but recent direct analyses of the samples using accelerator mass spectrometry (AMS) suggest a more likely date of around 3600 BC (dendro-calibrated in calendric years; Long *et al.* 1989). This early maize had very small, fragile, eight-rowed ears, but there is a debate about whether it was wild or domesticated (Mangelsdorf 1974; Bird 1980, 1984; Wilkes 1989; Benz 1994) and concerning its subsequent evolution. According to Bird (1980), a small-eared early domesticate reached Central America within 1000 years of domestication and there hybridized with the teosinte *Zea luxurians*. Then a variable set of more productive types, including a lineage that led to Olotillo, was spread back to Mexico. By 2000 years ago, a new complex of archaeological types, precursors to Nal-Tel and Chapalote, became abundant in Mexico (Mangelsdorf 1974; Benz 1994). Not only did teosinte introgression cause the explosive evolution of maize, but hybridization between maize races containing varying amounts of teosinte germplasm and various teosinte races has continued to contribute to the development of new maize races (Wellhausen *et al.* 1952; Wilkes 1977, 1979, 1989). Kato (1984), postulating the origin of maize from teosinte, suggested that genetic introgression between maize and teosinte could have taken place in the early domestication periods, but has been limited since by the rapid development of genetic barriers.

Central American maize could have reached South America at about 2000 BC, where agricultural, ceramic-using cultures had already developed (Grobman *et al.* 1961; Bird 1980, pers. comm.). However, Pearsall (1994) claims the possible introduction of maize to northern South America from lower Central America around 5000 BC, based on maize pollen remains and phytoliths in archaeological strata at various sites in Colombia and Ecuador. In North America, maize reached the southwest before 1000 BC (Adams 1994) and appeared as a food in the New England and eastern New York areas around 1000 AD (Bendremer and Dewar 1994). Archaeological remains from the West Indies site of En Bas Saline, northern Haiti, comprise pop and floury corn types and date to as early as about 1250 AD (Newsom and Deagan 1994).

Columbus found maize in Cuba and introduced it in Europe (Mangelsdorf 1974). Subsequent introductions of early flint types from Canada and the northern USA in the 18th century and of Corn Belt temperate dent maize in the late 19th and early 20th centuries added to already adapted genepools (Brandolini 1968; Trifunovic 1978). In the 16th century maize spread into Asia via the Mediterranean trade route, the Atlantic and Indian Ocean sea routes, and Magellan's voyage in the Pacific (the Philippines and eastern Indonesia) (Brandolini 1970). West Indian maize races were brought to Shikoku, Japan, in about 1580 by Portuguese sailors (Suto and Yoshida 1956). Maize entered Africa from Spain and Italy. Lowland tropical races from Brazil, the Guyanas and the Paraná basin were also introduced to West African coastal regions by traders. Dutch settlers in southern Africa brought flint floury types early in the 17th century and new introductions came from the southern USA and northern Mexico to southern and eastern Africa in the late 19th and early 20th centuries (Brandolini 1970). Southern dent and northern flint from the USA were introduced by European settlers in the Southern Cone of South America in the 19th century (Timothy *et al.* 1961; Paterniani and Goodman 1977).

Reproductive Biology

Maize is an annual crop and has a wide span of crop duration. Tropical highland maize matures in 8-10 months. Early tropical lowland maize can be harvested in 80-90 days, depending on the season. Maize varieties or landraces differ in the numbers of days to pollen shed and to silk emergence from the ear shoot, in numbers of grain on an ear, ear shape, grain type and plant/ear height. These agronomic traits are related to the reproductive capacity of the maize plant. Therefore, seed regeneration and multiplication of bank accessions need the knowledge of where they are best adapted to produce the seed.

The ear shoot arises from an internode near the centre of the culm (stem) which terminates in the tassel. The ear bears female flowers, each with a fertile and an aborted floret. Each functional female floret has a single ovary, which has an elongated style, or silk. The tassel consists of central and lateral branches which are covered with numerous paired spikelets (pedicellate, sessile). Each spikelet produces three anthers that shed pollen. Fertilization takes place when pollen is transferred to silk either by airborne or by artificially directed pollination. The silk is receptive to pollen along its entire length. In 12-24 hours after pollination, a pollen tube reaches the embryo sac to cause double fertilization (Poethig 1982). The monoecious reproductive system of maize facilitates the formation of progenies by the breeders and geneticists for experiments and crop improvement.

For multiplication and regeneration of the bank accessions, controlled pollination is usually necessary. It is a labour-intensive and expensive operation. Each ear shoot is covered with a glassine bag to avoid contamination before silk emergence, and a day before pollination, a tassel bag is placed on the flowering tassel to collect pollen. When pollen sheds in the morning it is collected in the tassel bag and poured on to silks which have emerged from the ear shoot.

Hybrid seed production is also carefully controlled to maintain a high degree of genetic purity (Wych 1988). Maize breeders produce half-sib or full-sib progenies and make selections within and among them to develop inbred lines, hybrids, synthetics and varieties.

GERMPLASM CONSERVATION AND USE

Systematic collecting, characterization and utilization of tropical maize landraces began in Mexico in the early 1940s, as part of cooperative breeding research between the Rockefeller Foundation and the Mexican Ministry of Agriculture, and similar efforts later by the NAS-NRC. They revealed enormous variation in landraces (Wellhausen *et al.* 1952, 1957; Brown 1953; Suto and Yoshida 1956; Hatheway 1957; Roberts *et al.* 1957; Brieger *et al.* 1958; Timothy *et al.* 1961, 1963; Grobman *et al.* 1961; Grant *et al.* 1963; Brandolini 1968; Mochizuki 1968; Costa-Rodrigues 1971; Paterniani and Goodman 1977; Wellhausen 1988; Avila and Brandolini 1990). The first catalogue of maize genetic resources (NAS-NRC 1954-55) deals with original strains of maize in the Americas. These are partly preserved in national germplasm banks and largely at CIMMYT (Taba 1994a). Updated lists of current collections from Latin America, including CIMMYT holdings, were published on CD-ROM by the Latin American Maize Project (LAMP) in 1992 and updated in 1995.

Acquisitions from previously non-collected areas by national maize or genetic resources programmes, in collaboration with IBPGR/IPGRI (Reid and Konopka 1988), augmented the number of Latin American maize accessions in the 1970s and 1980s. The last decade has seen extensive collaboration through the LAMP (Salhuana 1988; Eberhart *et al.* 1995) and a project coordinated by CIMMYT whereby 13 national banks in the Americas are regenerating some 7000 endangered accessions of maize landraces they hold.

According to a 1992 survey, open-pollinated landraces still account for some 42% of the developing country maize area (CIMMYT 1994). Original strains of local landraces, if not properly preserved *in situ* and *ex situ*, will be lost. In addition to landraces, current and obsolete elite germplasm (inbred lines, synthetic varieties, populations, etc.) from breeding programmes should be preserved in national and, in some cases, international germplasm banks. Enhanced materials from breeding programmes can be accessioned in germplasm banks by registering the germplasm in the journal *Crop Science* or with the American Society of Agronomy.

Maize by nature is an outbreeding crop from which inbreds and hybrids have been created (East 1908; Shull 1909; Gracen 1986; Russell 1991; Bjarnason 1994; Vasal and McLean 1994). For open-pollinated varieties, landraces, synthetics and populations, CIMMYT has used outbreeding strategies based on population genetics for preservation and regeneration (Breese 1989; Crossa 1989; NRC 1993; Crossa *et al.* 1994). Maintenance of inbred lines requires inbreeding or within-line sibbing, seed conservation approaches that resemble those used for self-pollinating crops such as rice and wheat (NRC 1993). Currently CIMMYT maize accessions total more than 14 700, and new introductions are constantly being added. Passport data on CIMMYT maize germplasm bank accessions have been compiled and published (CIMMYT 1988; LAMP 1992). To safeguard these collections, duplicate samples of about four-fifths of the accessions are kept at the US National Seed Storage Laboratory (NSSL) in Fort Collins, Colorado, USA. Base collection seed at CIMMYT is kept in sealed containers at subzero temperatures and low humidity and remains viable for 50-100 years. Seed in the active collection is kept at just above freezing (0-2°C) and constitutes the 'working' bank from which seed requests are filled. The current bank storage facility was built in 1972. Construction of a new bank facility began in 1995, with funding from the government of Japan. A significant function of a maize germplasm bank is to replenish seed samples. An optimum sample size for regenerating non-inbred accessions is determined by the gene frequencies of rare alleles present in the accession (Breese 1989; Crossa 1989; NRC 1993). For landrace collections or other panmictic maize populations, 100 or more ears are produced per accession, to include genotypes or alleles that occur at frequencies of 5% (Crossa 1989). In addition, depending on the contribution of gametes of individual plants to the next generation, the genetic structure of the accession can change according to the effective population size (Crossa *et al.* 1994). Population genetics has been used to derive practical procedures for regenerating germplasm bank accessions (Hallauer and Miranda 1981; Crossa *et al.* 1994).

In international meetings at CIMMYT in 1988, leading authorities noted that unique landrace samples were in dire need of regeneration. With funding from USAID through Project Noah and NSSL, some 7000 endangered accessions were scheduled for regeneration (Listman 1994). For crop germplasm collections, Brown (1989a, 1989b) used statistical theory on neutral alleles to suggest that a subset containing 10% of the original accessions could represent over 70% of the genetic variation in the collection. For example, the estimated number of maize landrace accessions in germplasm banks throughout Latin America and the Caribbean is some 26 000 (CIMMYT 1994), so a sampling of 2600 accessions, if properly chosen, could fairly represent the genetic diversity of maize landraces in the region. To arrive at this 10% subset, the CIMMYT maize germplasm bank has developed stratifying methods that help avoid duplication and ensure a representative sampling.

Anderson and Cutler (1942) used short, descriptive race names to group maize accessions that possess recognizable, common features. Since then, these common features have been accepted as characteristics of particular races by maize germplasm collectors and germplasm bank managers (NAS-NRC 1954-55; Gutierrez 1974; Bird 1982). LAMP evaluation trials have grouped the accessions based on the region within each homologous area (LAMP 1992; Eberhart *et al.* 1995). Thus, the trials often consisted of several taxonomic maize races. In field evaluation, Ward's Minimum Variance Cluster Analysis is used at CIMMYT to group accessions based on multiple traits considered stable over environments (Goodman and Paterniani 1969; Sanchez *et al.* 1993).

Mexican and Caribbean races were the first materials evaluated by CIMMYT, and the resulting datasets have been used to establish breeder-targeted core subsets of the Tuxpeño and Cónico complexes (Silva 1992; Crossa *et al.* 1994; Taba *et al.* 1994). From 848 Tuxpeño accessions and accession composites from the bank passport database, 175 were selected, based on lodging and adaptation in multilocation trials in 1988-90, and evaluated at two sites in Mexico to obtain characterization data for cluster and principal component analyses. Figures 16A.1 and 16A.2 show the results of principal component analysis and cluster analysis on 80 accessions of a core subset chosen from homogenous substrata created by the cutting point applied to a dendrogram from classification analysis (cluster analysis) on 175 accessions.

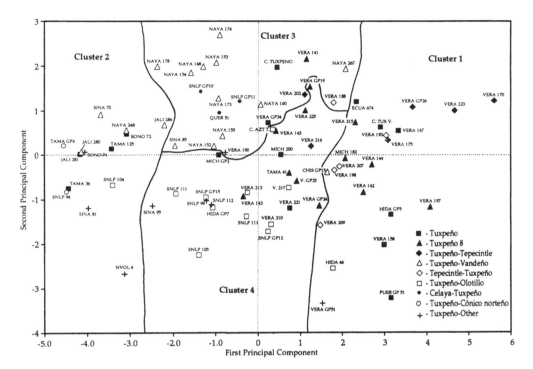

Fig. 16A.1. Principal component analysis (PCA) of diversity for 80-accessions of the Tuxeño maize race complex, using the means of trial data from two sites for eleven morphological/agronomic traits. The accessions are identified with the collection numbers or names to show collection sites and with the symbols for race classification. Solid lines delimit the four main clusters.

It has been a close to a half century since the great diversity of Mexican races of maize was systematically sampled and studied (Wellhausen *et al.* 1952). Samples from the original collections were widely distributed, and seed exchange among the collaborating institutions has continued in recent times (Reid and Konopka 1988; Taba 1994b). Work under LAMP has only accomplished part of this objective (LAMP 1992; Eberhart *et al.* 1995). Given the continued importance of certain landraces to food production in the region (CIMMYT 1994), future efforts must entail both *ex situ* conservation of the original collections and *in situ* preservation of locally important maize races.

To obtain current information on the use of landraces in Latin America, in spring 1995 a questionnaire was sent to managers of the 13 germplasm banks participating in the cooperative Latin American maize regeneration project. The results showed that many landraces are cultivated, and that the same taxonomic races are often found in different countries, but they may be specially adapted to local growing conditions or farmer requirements (Appendix I-III). Special food preparations are the motive behind the continued use of certain landraces, as has been reported before (Goodman and Bird 1977). Such locally adapted varieties may have acquired special properties from farmer selection over the many generations they have been grown and can be targets of *in situ* (on-farm) conservation, as defined in Article 2 of the Convention on Biological Diversity that followed the United Nations Conference on Environment and Development (UNCED) in 1992 (Krattiger *et al.* 1994).

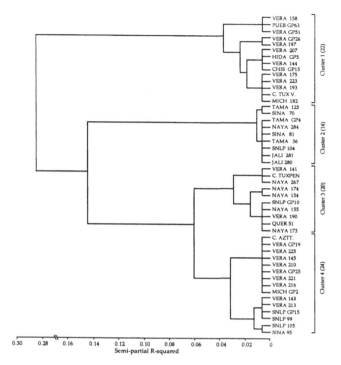

Fig. 16A.2. Dendrogram of an 80-accession subset of the Tuxpeño maize race complex. The number of accessions per cluster is given in parentheses. The accessions representing each cluster are listed.

The sampling strategy for collecting germplasm and choosing core subsets of bank collections is designed to cover rare and widespread, as well as common and localized, classes of alleles (Marshall and Brown 1975; NRC 1993). Some races are common in a given country, which facilitates their conservation on-farm. Once such a landrace has been monitored *in situ*, a core subset can be developed; enhancement of the race can begin as part of forming the subset (Crossa *et al.* 1994; Taba 1994b).

A landrace's hallmark is that it maintains the genetic properties that meet farmers' needs (Ortega-Paczka 1973; Brush 1991). About 10% of Mexican maize races are utilized in breeding programmes; the rest are condemned to extinction in the long run unless they are also improved (Marquez-Sanchez 1993). Soleri and Smith (1995) found that both genetic shift and genetic drift had occurred in populations of Hopi maize conserved *ex situ*, compared with those conserved *in situ*. Geneflow from improved germplasm to landraces can be achieved without diminishing biodiversity. Marquez-Sanchez (1993) reported a method of improving maize landraces in Mexico via limited backcrossing. Crossa *et al.* (1990) reported that narrow-ear races and their derivatives, such as Tabloncillo, Chapalote, Jala and Harinoso de Ocho, had better combining ability than other landraces studied. A core subset of a race usually contains several subracial diversity groupings (Taba *et al.* 1994; Silva 1992). Selected geneflow, even into individual subracial groups, adds useful biodiversity to a race, assists farmers and thus contributes to the likelihood of the race's conservation *in situ*. Soleri and Smith (1995) suggested that a central goal of *in situ* conservation is to maintain both the genetic diversity and the population structure that farmers can exploit through local adaptation. Some 250 Latin American maize races could benefit from enhancement (Goodman and Brown 1988) and some are already in the process (Marquez-Sanchez 1993).

In other parts of the world where landraces are utilized extensively, they can be improved in a similar manner (Brown and Robinson 1992). Modern biotechnology should play an important role in characterizing landraces and developing a methodology to help them 'evolve' in their habitats. Such work could begin with a locally important race – for example, race Bolita, in Mexico. Efforts are underway in the USA to broaden the germplasm base of Corn Belt dent, through the introgression of tropical germplasm (Hallauer 1978; Goodman 1988b, 1992; Troyer 1990; Tiffany *et al.* 1992; Michilini and Hallauer 1993).

Properties and Uses

An early classification of maize was by its kernel types: dent, flint, floury, sugary, pop. Recently, morocho grain type was added to the grain types in the CIMMYT germplasm bank database. It has been included in the flint or semi-floury category in the early classification. It is related to floury backgound as it has a soft floury texture inside the grain, surrounded by the hard flint texture, large grain size cultivated in the Andean region (Taba 1995). The current accession characterization database indicates that the CIMMYT maize bank has the following proportions of each grain type: dent (61.6%), flint (27.7%), floury (9.5%), sugary (0.6%), pop (0.3%) and morocho (0.17%) in a total of 14 700 accessions.

In the United States specialty corn has been developed by breeding, such as high oil corn, waxy corn, high amylose, high starch, popcorn, sweet corn and pipe corn (Alexander 1988). High-quality protein maize (QPM) utilizes opaque–2 mutant gene (NRC 1987). Traditional food uses of maize in Latin America are listed in Appendix II. Whole kernel is composed (% dry basis) of starch (67.8-74.0%), fat (3.9-5.8%), protein (8.1-11.5%), ash (1.37-1.5%), sugar (1.61-2.22%) in dent corn hybrids grown in the US and the endosperm contains starch (86.4-88.9%), fat (0.7-1.0%), protein (6.9-10.4%), ash (0.2-0.5%) and sugar (0.5-0.8%) (Alexander 1988).

Maize is used mainly for animal feed in developed-country economies. Maize products of fermentation and distilling industry are alcohols and whisky, the dry-milling industry produces grits, meal and flour, and the wet-milling industry produces corn starch, oil and syrups. Many other food and feed products and industrial derivatives are produced from maize (Watson 1988). About 10% of maize grown in the US is used for silage (Agricultural Statistics 1990).

Breeding

Elite germplasm sources identified since the initial collections in the Americas have been incorporated into breeding composites, groups, genepools and populations by the CIMMYT Maize Program and national maize breeding programmes worldwide (CIMMYT 1982, 1992; Gracen 1986; Pandey and Gardner 1992; Gerdes *et al.* 1993; Bjarnason 1994; Vasal and McLean 1994). Landrace accessions were first characterized and superior accessions then selected for direct release as varieties such as V520 (San Luis Potosí 20), V520c (Capitaine) and Rocamex V-7 (Hidalgo 7) (Wellhausen 1950). Other breeding populations for various maize types have been improved (CIMMYT 1974, 1982; Pandey and Gardner 1992; Eagles and Lothrop 1994; Taba 1995). Inbred lines and hybrids have been developed from improved germplasm at CIMMYT (Bjarnason 1994; Vasal and McLean 1994). Unimproved landraces, such as tropical floury and morocho maize, are still grown locally by many subsistence farmers (CIMMYT 1994).

Germplasm bank collections can provide genetic diversity for use by breeders concerned with genetic vulnerability and seeking unique genetic variation. Doebley *et al.* (1985) reported high levels of isozymic variation in 34 maize races from Mexico. Seventy-two percent of this variation resided within accessions, 27% among accessions, and for the 13-enzyme system encoded by 23 loci, an average of 7.09 alleles per locus was recorded, which indicated a level of variation comparable to that of the maize wild relative, teosinte (Doebley *et al.* 1984). Plant breeding tends to narrow the genetic diversity of a crop (Goodman 1988b, 1990). Germplasm development strategies should include the introgression of genetic variation from landrace cultivars or older varieties, as attempted in LAMP (Eberhart *et al.* 1995).

The use of maize landraces at CIMMYT is exemplified by breeding work with the race Tuxpeño and related Mexican dent racial complexes. Breeding favours a reduced plant architecture for greater per-plant production efficiency, and excess plant and ear height can be reduced without significant change to other agronomic traits (Johnson *et al.* 1986; Russell 1991). Russell (1985) has shown simple correlation coefficients for 18 plant, ear and grain traits with grain yield for 28 maize cultivars of four open-pollinated varieties and four single-cross hybrids representing each 10-year period from 1930 to 1980 in the USA. At CIMMYT, beginning in the 1970s, researchers have conducted extensive work to reduce plant height in Tuxpeño Crema I (Johnson *et al.* 1986). Table 16A.1 shows results of selection cycles 0, 6, 9, 12 and 15 for evaluations of plant height, grain yield, harvest index, ears per plant, lodging and other traits at three locations over 2 years in Mexico. Judged from the accessions included, Tuxpeño Crema I probably contains Tuxpeño karyotypic groups of the races Tuxpeño, Vandeño and Celaya (Kato 1984; Bretting and Goodman 1989) and minor introgressions of Cónico Norteño and Eto Blanco. Tuxpeño Crema I includes a variation of Coahuila which is shorter and relatively earlier maturing than the race from Veracruz, with a possible introgression of Cónico Norteño (Taba *et al.* 1994). Michoacán and Colima accessions include Celaya and Vandeño. Studies of chromosome knob patterns of 57 Tuxpeño accessions (Kato 1988) placed the materials in four groups corresponding to north-central Mexico. The similar ear anatomy of Mexican dent Tuxpeño, Vandeño and Celaya races was noted by Benz (1986), who placed them with Tepecintle in the 'unaffiliated' category of the Mexican narrow ear complex.

Tuxpeño populations have been widely distributed in 43 countries in tropical Asia, Africa, and Latin America (CIMMYT 1986). CIMMYT researchers collaborated extensively with scientists from the International Institute of Tropical Agriculture (IITA), Nigeria, in the early 1980s to develop populations of maize based on La Posta (Tuxpeño race) that would possess resistance to streak virus, an important maize disease in sub-Saharan Africa (Tang and Bjarnason 1993). Their efforts were largely successful, except for the fact that Tuxpeño is susceptible to *Striga hermonthica*, a parasitic flowering plant whose economic importance in maize-cropping areas of sub-Saharan Africa has greatly increased in recent years (Kim 1994). The CIMMYT maize physiology unit has also developed drought-tolerant maize populations based on Tuxpeño Crema I cycle 11 (Edmeades *et al.* 1992; Bolaños and Edmeades 1993; Bolaños *et al.* 1993) using reduced anthesis-silking interval as a phenotypic marker for increased yield under drought. Drought-tolerant genotypes did not show deeper rooting, which could be desirable for water uptake under severe drought. Improved cycles of selection (C_6 and C_8) under water stress at the CIMMYT experiment station at Tlaltizapán outyielded non-tolerant cultivars and were more stable over a range of environments, including irrigation, rain-fed conditions and limited irrigation (Byrne *et al.* 1995).

Between 1966 and 1990, national programme cooperators released 147 varieties and hybrids developed from CIMMYT's and IITA's mainly Tuxpeño populations, and releases cover approximately 3.8 million hectares worldwide (CIMMYT 1992). Tuxpeño germplasm figures in the pedigrees of many CIMMYT genepools and populations (CIMMYT 1982). Tuxpeño accessions are also included in Thai Composite 1 (later called Suwan 1), a downy mildew resistant variety developed by Thai researchers and

Table 16A.1. Results of modified full-sib recurrent selection for reduced plant height in Tuxpeño Crema I (Johnson *et al.* 1986)[†].

Cycle	Plant height (cm)	Grain yield (Mg/ha)	Harvest index	Ears/plant	Lodging (%)	Days to silking
0	282	3.17	0.30	0.70	43	73.4
6	219	4.29	0.40	0.87	12	67.1
9	211	4.48	0.40	0.90	14	66.7
12	202	4.93	0.41	0.93	9	65.7
15	179	5.40	0.45	0.98	5	61.6
LSD (*P*=0.05)	22	0.30	0.04	0.12	–	0.6
Change per cycle (%)	−2.39**	4.43**	3.10**	2.50**	6.70**	1.0**

[†] Data from trials grown at or near optimum plant density at two or three locations in Mexico, 1978-79.
** Significant at *P*=0.01. Percent change is based on cycle 0.

subsequently used in breeding research worldwide (Sriwatanapongse *et al.* 1993). Some modifier genes have been shown to reduce protein quality, compared with that of soft endosperm opaque-2 (QPM with 8-9% protein, 2.6-2.9% lysine and 0.68-0.78% tryptophan in protein) (Villegas *et al.* 1992). Galinat (1995) discussed the importance of adaptive gene complexes in breeding for biotechnology-derived quality traits such as high lysine, high oil, sweetness, etc., drawing on the natural diversity of maize. QPM modifiers come from diverse germplasm sources used at CIMMYT, and QPM is used in China, South America, a few countries in Africa and the USA (Mertz 1992). Research on QPM at CIMMYT resulted in many quality protein pools, populations and inbreds (Bjarnason and Vasal 1992; Pixley and Bjarnason 1993).

REFERENCES

Adams, K.R. 1994. A regional synthesis of *Zea mays* in the prehistoric American southwest. Pp. 273-302 *in* Corn and Culture in the Prehistoric New World (S. Johannessen and C.A. Hastorf, eds.). Westview Press Inc., Boulder, Colorado.

Agricultural Statistics. 1990. U.S. Government Printing Office, Washington, DC.

Alexander, D. E. 1988. Breeding special nutritional and industrial types. Pp. 869-880 *in* Corn and Corn Improvement (G.F. Sprague and J.W. Dudley, eds.). ASA Monograph 18. ASA, Madison, WI, USA.

Anderson, E. and H.C. Cutler. 1942. Races of *Zea mays*: I. Their recognition and classification. Ann. Mo. Bot. Gard. 29: 69-88.

Avila, G. and A.G. Brandolini. 1990. I Mais Boliviani, p. 99. Istituto Agronomico Per L'oltremare, Firenze, Italia.

Bendremer, J.C.M. and R.E. Dewar. 1994. The advent of prehistoric maize in New England. Pp. 369-394 *in* Corn and Culture in the Prehistoric New World (S. Johannessen and C.A. Hastorf, eds.). Westview Press Inc., Boulder, Colorado.

Benz, B.F. 1986. Racial systematics and the evolution of Mexican maize. Pp. 121-136 *in* Studies in the Neolithic and Urban Revolutions, The V. Gordon Childe Colloquium, Mexico (L. Manzanilla, ed.). BAR International Series 349, Oxford.

Benz, B.F. 1994. Reconstructing the racial phylogeny of Mexican maize: Where do we stand? Pp. 157-179 *in* Corn and Culture in the Prehistoric New World (S. Johannessen and C.A. Hastorf, eds.). Westview Press Inc., Boulder, Colorado.

Bird, R. Mck. 1980. Maize evolution from 500 B.C. to the present. Biotropica 12(1):30-41.

Bird, R. Mck. 1982. Maize and teosinte germplasm banks. Pp. 351-355 *in* Maize for Biological Research (W.F. Sheridan, ed.). Plant Molecular Biology Association, Charlottesville, Virginia.

Bird, R. Mck. 1984. South American maize in Central America? pp. 39-65 *in* Pre-Colombian Plant Migrations (D. Stone, ed.). Peabody Museum, Harvard University, Cambridge, MA, USA.

Bjarnason, M. and S.K. Vasal. 1992. Breeding of quality protein maize (QPM). Plant Breed. Rev. 9:181-216.

Bjarnason, M. (ed.) 1994. The Subtropical, Midaltitude, and Highland Maize Subprogram. Maize Program Special Report. CIMMYT, Mexico, D.F.

Bolaños, J. and G.O. Edmeades. 1993. Eight cycles of selection for drought tolerance in lowland tropical maize. I. Responses in grain yield, biomass, and radiation utilization. Field Crops Res. 31:233-252.

Bolaños, J., G.O. Edmeades and L. Martinez. 1993. Eight cycles of selection for drought tolerance in lowland tropical maize. III. Responses in drought-adaptive physiological and morphological traits. Field Crops Res. 31:269-286.

Brandolini, A. 1968. European races of corn. Ann. Corn Sorghum Res. Conf. Proc. 24:36-48.

Brandolini, A. 1970. Maize. Pp. 273-309 *in* Genetic Resources in Plants: Their Exploration and Conservation (O.H. Frankel and E. Bennett, eds.). F.A. Davis Co., Philadelphia.

Breese, E.L. 1989. Regeneration and Multiplication of Germplasm Resources in Seed Genebanks: The Scientific Background. IBPGR, Rome.

Bretting, P.K. and M.M. Goodman. 1989. Karyotypic variation in Mesoamerican races of maize and its systematic significance. Econ. Bot. 43(1):107-124.

Brieger, F.G., J.T.A. Gurgel, E. Paterniani, A. Blumenschein and M.R. Alleoni. 1958. Races of maize in Brazil and other eastern South American countries. Publication 593. NAS-NRC, Washington, DC.

Brown, A.D.H. 1989a. The case for core collections. Pp. 136-156 *in* The Use of Plant Genetic Resources (A.D.H. Brown, O.H. Frankel, D.R. Marshall and J.T. Williams, eds.). Cambridge University Press, New York.

Brown, A.D.H. 1989b. Core collections: A practical approach to genetic resources management. Genome 31(21):818-824.

Brown, W.L. 1953. Maize of the West Indies. Trop. Agric. 30:141-170.

Brown, W.L. and H.F. Robinson. 1992. The status, evolutionary significance, and history of Eastern Cherokee maize. Maydica 37:29-39.

Brush, S.B. 1991. A farmer -based approach to conserving crop germplasm. Econ. Bot. 45(2):153-165.

Byrne, P.F., J. Bolaños, G.O. Edmeades and D.L. Eaton. 1995. Gains from selection under drought versus multilocation testing in related tropical maize populations. Crop Sci. 35:63-69.

CIMMYT. 1974. Proceedings of worldwide maize improvement in the 70's and the role for CIMMYT. Maize Program internal document.

CIMMYT. 1982. CIMMYT's maize program: An overview. CIMMYT. El Batan, Mexico. Maize Program internal document.

CIMMYT. 1988. Maize Germplasm Bank Inquiry System. Mexico, D.F.

CIMMYT. 1992. 1991-92 CIMMYT World Maize Facts and Trends: Maize Research Investment and Impacts in Developing Countries. Mexico, D.F.

CIMMYT. 1994. CIMMYT 1993/94 World Maize Facts and Trends. Maize Seed Industries Revisited: Emerging Roles of the Public and Private Sectors. Mexico, D.F.

CIMMYT Maize Staff. 1986. Improving on excellence: Achievements in breeding with the maize race Tuxpeño. CIMMYT, Mexico, D.F.

Costa-Rodrigues, L. 1971. Races of maize in Portugal. Agron. Lusit. 31:239-284.

Crossa, J. 1989. Methodologies for estimating the sample size required for genetic conservation of outbreeding crops. Theor. Appl. Genet. 77:153-161.

Crossa, J., S. Taba and E.J. Wellhausen. 1990. Heterotic patterns among Mexican races of maize. Crop Sci. 30(6):1182-1190.

Crossa, J., S. Taba, S.A. Eberhart and P. Bretting. 1994. Practical considerations for maintaining germplasm in maize. Theor. Appl. Genet. 89:89-95.

Doebley, J.F., M.M.Goodman and C.W. Stuber. 1984. Isozyme variation in *Zea* (Gramineae). Syst. Bot. 9:203-218.

Doebley, J.F., M.M. Goodman and C.W. Stuber. 1985. Isozyme variation in the races of maize from Mexico. Am. J. Bot. 72 (5):629-635.

Eagles, H.A. and J.E. Lothrop. 1994. Highland maize from Central Mexico: Its origin, characteristics, and use in breeding programs. Crop Sci. 34:11-19.

East, E.M. 1908. Inbreeding in corn. Rep. Conn. Agric. Exper. Stat. 1907, pp. 419-428.

Eberhart, S.A., W. Salhuana, R. Sevill and S. Taba. 1995. Principles for tropical maize breeding. Maydica 40:339-355..

Edmeades, G.O., J. Bolaños and H.R. Lafitte. 1992. Progress in breeding for drought tolerance in maize. *In* Proc. 47th Ann. Corn Sorgh. Ind. Res. Conference (D. Wilkinson, ed.). American Seed Trade Association (ASTA), Washington, DC.

FAO. 1995. FAO yearbook: production. Vol. 48. FAO statistics series no. 125. Rome.

Galinat, W.C. 1995. El Origen del Maíz: El Grano de la Humanidad. The origin of Maize: Grain of Humanity. Econ. Bot. 49 (1):3-12.

Gerdes, J.T., C.F. Behr, J.G. Coors and W.F. Tracy. 1993. Compilation of North American Maize Breeding Germplasm. Crop Science Society of America, Inc., Madison, WI, USA.

Goodman, M.M. 1988a. The history and evolution of maize. CRC Critical Rev. in Plant Sci. 7(3):197-220.

Goodman, M.M. 1988b. US maize germplasm: Origins, limitations, and alternatives. Pp. 130-148 *in* Recent Advances in The Conservation and Utilization of Genetic Resources: Proceedings of the Global Maize Germplasm Workshop. CIMMYT, Mexico, D. F.

Goodman, M.M. 1990. Genetic and germplasm stocks worth conserving. J. Hered. 81(1):11-16.

Goodman, M.M. 1992. Choosing and using tropical corn germplasm. *In* Proc. Annu. Corn Sorghum Ind. Res. Conf. 47:47-64. ASTA, Washington, DC.

Goodman, M.M. and W.L. Brown. 1988. Races of Corn. Pp. 33-79 *in* Corn and Corn Improvement (G.F. Sprague and J.W. Dudley, eds.). ASA monograph 18. ASA, Madison, Wisconsin.

Goodman, M.M. and R. Mck. Bird. 1977. The races of maize IV: Tentative grouping of 219 Latin American races. Econ. Bot. 31:204-221.

Goodman, M.M. and E. Paterniani. 1969. The races of Maize. III. Choices of appropriate characters for racial classification. Econ. Bot. 23:265-273.

Gracen, V.E. 1986. Sources of temperate maize germplasm and potential usefulness in tropical and subtropical environments. Adv. Agron. 39:127-172.

Grant, U.J., W.H. Hatheway, D.H.Timothy, C. Cassalett D., and L.M. Roberts. 1963. Races of maize in Venezuela. NAS-NRC Publication 1136. Washington DC.

Grobman, A., W. Salhuana and R. Sevilla, with P.C. Mangelsdorf. 1961. Races of maize in Peru. NAS-NRC Publication 915. Washington DC.

Guidry, N.P. 1964. A graphic summary of world agriculture. USDA Misc. Pub. 705. U.S. Govt. Print. Office, Washington, DC.

Gutierrez G., M. 1974. Maize germplasm preservation and utilization at CIMMYT. Pp. 1-21 *in* proceedings of worldwide maize improvement in the 70's and the role for CIMMYT. CIMMYT, Mexico, D.F.

Hallauer, A.R., W.A. Russell and K.R. Lamkey. 1988. Corn breeding. Pp. 463-564 *in* Corn and Corn Improvement (G.F. Sprague and J.W. Dudley, eds.). ASA Monograph 18. ASA, Madison Wisconsin, USA.

Hallauer, A.R. 1978. Potential of exotic germplasm for maize improvement. Pp. 229-247 *in* Maize Breeding and Genetics (D.B. Walden, ed.). John Wiley & Sons, New York.

Hallauer, A.R. and J.B. Miranda Filho. 1981. Quantitative genetics in maize breeding. Iowa State University Press, Ames, Iowa.

Hatheway, W.H. 1957. Races of maize in Cuba. NSA-NRC Publication 453. Washington, DC.

Iltis, H.H. and J.F. Doebley. 1980. Taxonomy of *Zea* (Gramineae). II. Subspecific categories in the *Zea mays* complex and a generic synopsis. Am. J. Bot. 67 (6):994-1004.

Johnson, E.C., K.S. Ficher, G.O. Edmeades and A.F.E. Palmer. 1986. Recurrent selection for reduced plant height in lowland tropical maize. Crop Sci. 26:253-260.

Kato Y., T.A. 1984. Chromosome morphology and the origin of maize and its races. Evol. Biol. 17:219-253.

Kato Y., T.A. 1988. Cytological classification of maize race populations and its potential use. Pp. 106-117 *in* Recent Advances in The Conservation and Utilization of Genetic Resources: Proceedings of the Global Maize Germplasm Workshop. CIMMYT, Mexico, D.F.

Kim, S.K. 1994. Genetics of maize tolerance of *Striga hermonthica*. Crop Sci. 34:900-907.

Krattiger, A.F., J.A. McNeely, W.H. Lesser, K.R. Miller, Y. St. Hill and R. Senanayake (eds.) 1994. Widening Perspectives on Biodiversity. IUCN, Gland, Switzerland & International Academy of the Environment, Geneva, Switzerland.

LAMP. 1992. Data of the Latin American Maize Project. CD-ROM.

Listman, G.M. 1994. Rescue of Latin American maize progresses lays groundwork for future collaborative missions. Diversity 10(4):26-27.

Long, A., B.F. Benz, D.J. Donahue, A.J.T. Jull and L.J. Toolin. 1989. First direct AMS dates on early maize from Tehuacán, Mexico. Radiocarbon 31(3):1035-1040.

MacNeish, R.S. 1985. The archaeological record on the problem of the domestication of corn. Maydica 30:171-178.

Mangelsdorf, P.C. 1974. Corn: Its Origin, Evolution, and Improvement. Belknap Press, Harvard University Press, Cambridge, Massachusetts.

Mangelsdorf, P.C., R.S. MacNeish and W.C. Galinat. 1964. Domestication of corn. Science 143:538-545.

Mangelsdorf, P.C., R.S. MacNeish and W.C. Galinat. 1967. Prehistoric wild and cultivated maize. Pp. 178-200 *in* The Prehistory of the Tehuacán Valley I: Environment and Subsistence (D.S. Byers, ed.). University of Texas Press, Austin.

Marquez-Sanchez, F. 1993. Mejoramiento genético de maices criollos mediante retrocruza limitada. Pp. 16-19 *in* El maíz en la década de los 90, Memorias del simposio. SARH, Jalisco, Mexico.

Marshall, D.R. and A.H.D. Brown. 1975. Optimum sampling strategies in genetic conservation. Pp. 53-80 *in* Crop Genetic Resources for Today and Tomorrow (O.H. Frankel and J.G. Hawkes, eds.). Cambridge University Press.

Mertz, E.T. 1992. Discovery of high lysine, high tryptophan cereals. Pp. 1-8 *in* Quality Protein Maize (E.T. Mertz, ed.). The American Association of Cereal Chemists, St. Paul, Minnesota.

Michilini, L.A. and A.R. Hallauer. 1993. Evaluation of exotic and adapted maize (*Zea mays* L.) germplasm crosses. Maydica 38:275-282.

Mochizuki, N. 1968. Classification of local strains of maize in Japan and selection of breeding materials by application of principal component analysis. Pp. 173-178 *in* Symposium on Maize Production in

Southeast Asia. Agriculture, Forestry, and Fisheries Research Council, Ministry of Agriculture and Forestry, Tokyo.

NAS-NRC (National Academy of Science-National Research Council). 1954-55. Collections of original strains of corn I, II. Report of the Committee on Preservation of Indigenous Strains of Maize. Washington, DC.

Newsom, L.A. and K.A. Deagan. 1994. *Zea mays* in the West Indies: The archaeological and early historic record. Pp. 203-217 *in* Corn and Culture in the Prehistoric New World (S. Johannessen and C.A. Hastorf, eds.). University of Minnesota Publications in Anthropology no. 5. Westview Press, Boulder, Colorado.

NRC. 1987. Quality protein maize. NRC, National Acad. Press, Washington DC.

NRC. 1993. Managing Global Genetic Resources: Agricultural Crop Issues and Policies, pp.131-172. Washington, DC.

Ortega-Paczka, R.A. 1973. Variación en maíz y cambios socioeconómicos en Chiapas, México, 1946-1971. Master's thesis, Colegio de Postgraduados, ENA, Chapingo, Mexico.

Pandey, S., and C.O. Gardner. 1992. Recurrent selection for population, variety, and hybrid improvement in tropical maize. Adv. Agron. 48:2-79.

Paterniani, E., and M.M. Goodman. 1977. Races of maize in Brazil and adjacent areas. CIMMYT, Mexico, D.F.

Pearsall, D.M. 1994. Issues in the analysis and interpretation of archaeological maize in South America. Pp. 245-272 *in* Corn and Culture in the Prehistoric New World (S. Johannessen and C.A. Hastorf, eds.). Westview Press, Boulder, Colorado.

Pixley, K.V. and M.S. Bjarnason. 1993. Combining ability for yield and protein quality among modified-endosperm opaque-2 tropical maize inbreds. Crop Sci. 33:1229-1234.

Poethig, R.S. 1982. Maize-the plant and its parts. Pp. 9-18 *in* Maize for Biological Research (W.F. Sheridan, ed.). University of North Dakota Press, Grand Forks, ND, USA.

Reid, R. and J. Konopka. 1988. IBPGR's role in collecting and conserving maize germplasm. Pp. 9-16 In Recent Advances in The Conservation and Utilization of Genetic Resources: Proceedings of the Global Maize Germplasm Workshop, CIMMYT, Mexico, D.F.

Roberts, L.M., U.J. Grant, R. Ramirez E., W.H. Hatheway and D.L. Smith, with P.C. Mangelsdorf. 1957. Races of maize in Colombia. NAS-NRC publication 510. Washington, DC.

Russell, W.A. 1985. Evaluations for plant, ear, and grain traits of maize cultivars representing seven eras of breeding. Maydica 30:85-96.

Russell, W.A. 1991. Genetic improvement of maize yields. Adv. Agron. 46:245-298.

Salhuana, W. 1988. Seed increase and germplasm evaluation. Pp. 29-38 *in* Recent Advances in The Conservation and Utilization of Genetic Resources: Proceedings of the Global Maize Germplasm Workshop. CIMMYT, Mexico, D.F.

Sanchez G., J.J., M.M. Goodman and J.O. Rawlings. 1993. Appropriate characters for racial classification in maize. Econ. Bot. 47:44-59.

Shull, G. H. 1909. A pure-line method in corn breeding. Rep. Am. Breed. Assoc. 5:51-59.

Silva C., E.G. 1992. Estudio agronómico y taxonómico de la raza de maíz "Cónico", su colección central y perspectiva de uso en mejoramiento. Master's thesis. Colegio de Postgraduados, Montecillo, Mexico.

Soleri, D. and S.E. Smith. 1995. Morphological and phenological comparisons of two Hopi maize varieties conserved *in situ* and *ex situ*. Econ. Bot. 49:56-77.

Sriwatanapongse, S., S. Jinahyon and S.K. Vasal. 1993. Suwan-1: Maize from Thailand to the world. CIMMYT, Mexico, D.F.

Suto, T. and Y. Yoshida. 1956. Characteristics of the oriental maize. Pp. 375-530 *in* Land and Crops of Nepal Himalaya (H. Kihara, ed.). Vol. 2. Kyoto: Fauna and Flora Res. Soc., Kyoto University.

Taba, S. (ed.) 1994a. The CIMMYT Maize Germplasm Bank: Genetic Resource Preservation, Regeneration, Maintenance, and Use. Maize Program Special Report. CIMMYT, Mexico, D.F.

Taba, S. 1994b. The future: Needs and activities. Pp. 52-58 *in* The CIMMYT Maize Germplasm Bank: Genetic Resource Preservation, Regeneration, Maintenance, and Use. Maize Program Special Report. CIMMYT, Mexico, D.F.

Taba, S. 1995. Use of exotic germplasm and long-term gene pool development for Andean highland maize improvement. CIMMYT Maize Program, internal document.

Taba, S., F. Pineda E. and J. Crossa. 1994. Forming core subsets from the Tuxpeño race complex. Pp. 60-81 *in* The CIMMYT Maize Germplasm Bank: Genetic Resource Preservation, Regeneration, Maintenance, and Use. Maize Program Special Report,. CIMMYT, Mexico, D.F.

Tang, C.Y. and M.S. Bjarnason. 1993. Two approaches for the development of maize germplasm resistant to maize streak virus. Maydica 38:301-307.

Tiffany, G.D., O.C. Howie Smith, W. Salhuana and D. Misevic. 1992. Considerations in selection of progeny from elite and broad based germplasm. *In* Proc. Annu. Corn Sorghum Ind. Res. Conf. 47:164-176. ASTA, Washington, DC.

Timothy, D.H., B. Pena V. and R. Ramirez E. with W.L. Brown and E. Anderson. 1961. Races of maize in Chile. NAS-NRC Publication 847. Washington, DC.

Timothy, D.H., W.H. Hatheway, U.J. Grant, M. Torregroza C., D. Sarria V. and D. Varela A. 1963. Races of maize in Ecuador. NAS-NRC Publication 975. Washington, DC.

Trifunovic, V. 1978. Maize production and maize breeding in Europe. Pp. 41-56 *in* Maize Breeding and Genetics (D.B. Walden, ed.). Wiley & Sons, New York.

Troyer, A.F. 1990. A retrospective view of corn genetic resources. J. Hered. 81:17-24.

Vasal, S.K. and S. McLean (eds.) 1994. The Lowland Tropical Maize Subprogram. Maize Program Special Report. CIMMYT, Mexico, D.F.

Villegas, E., S.K. Vasal and M. Bjarnason. 1992. Quality protein maize - What is it and how was it developed? Pp. 27-48 *in* Quality Protein Maize (E.T. Mertz, ed.). The American Association of Cereal Chemists, St. Paul, Minnesota.

Watson, S.A. 1988. Corn marketing, processing, and utilization. Pp. 881-940 *in* Agricultural Statistics. 1990 U.S. Govt. Print. Office, Washington, DC.

Wellhausen, E.J. 1950. El programa de mejoramiento del maíz en Mexico. Pp. 119-149, presentation *in* La primera asamblea Latinoamericana de fitogenetistas. Oficinas Estudios Especiales. S.A.G., Mexico.

Wellhausen, E.J. 1988. The indigenous maize germplasm complexes of Mexico: Twenty-five years of experience and accomplishments in their identification, evaluation, and utilization. Pp. 17-28 *in* Recent Advances in The Conservation and Utilization of Genetic Resources: Proceedings of the Global Maize Germplasm Workshop, CIMMYT, Mexico, D.F.

Wellhausen, E.J., A. Fuentes O. and A.H. Corzo with P.C. Mangelsdorf. 1957. Races of maize in Central America. Publication 511. NAS-NRC, Washington, DC.

Wellhausen, E.J., L.M. Roberts and E. Hernandez X. in collaboration with P.C. Mangelsdorf. 1952. Races of Maize in Mexico. The Bassey Institute, Harvard University, Cambridge, Massachusetts.

Wilkes, H.G. 1977. Hybridization of maize and teosinte in Mexico and Guatemala and improvement of maize. Econ. Bot. 31:254-293.

Wilkes, H.G. 1979. Mexico and Central America as a center for the origin of maize. Crop Improv. (India) 6:1-18.

Wilkes, H.G. 1989. Maize: Domestication, racial evolution, and spread. Pp. 441-455 *in* Foraging and Farming (D.R. Harris and G.C. Hillman, eds.). Unwin Hyman, London.

Wych, R.D. 1988. Production of hybrid seed corn. Pp. 565-607 *in* Corn and Corn Improvement (G.F. Sprague and J.W. Dudley, eds.). ASA Monograph 18. ASA, Madison Wisconsin, USA.

Appendix I. Extant maize landraces and their distribution.

Country	No. of races	Data source
Argentina	46	Ing. Lucio Roberto Solari, INTA, Pergamino, Argentina
Bolivia	46	Dr. Gonzalo Avila, Pairumani, Cochabamba, Bolivia
Chile	23	Ing Orlando Paratori, INIA, Santiago, Chile
Colombia	16	Dr. Carlos Diaz, ICA, Medellín, Colombia
Cuba	7	Ing. Cecilio Marcos Torres, Institute L. Dimitrova, El Tomeguin, Cuba
Ecuador	23	Ing. Edison Silva, INIAP, Quito, Ecuador
Mexico	48	Dr. Francisco Cardenas, INIFAP, Mexico
Peru	52	Ing. Ricardo Sevilla, PCIM, Lima, Peru
Paraguay	7	Ing. Manuel Santiago Paniagua, CRIA, Paraguay
Venezuela	19	Ing. Agr. Elena Mazzani, Arnoldo Bejarano and Victor Segovia, FONAIAP, Venezuela

Appendix II. Maize landraces cultivated for specific uses.

No. of races	Primary use	Secondary use	Tertiary use	Country
16	Grain (13), polenta, humita, locro, choclo, flour, semolas, mote, chilcan, ulpada, misko pitapi, pochoclo, pororo, masa, bread, caldo majao, anchi de semola, cooked flour, frangollo, mazamorra, glucose, starch, alcohol, chichoca, biscochos, alfajor, turron, tamal, yupe, guaycha	Feed grain, forage, Paraguayan soup. baipai, yapora, bori, calapi, sanco, chicha, tulpo, pire, locro chilcan, ulpada, soup, choclo, tijtinchas, flour, tostadas, chilcon, biscochos, pire, chicha morada, starch, alcohol	Tortas, tortillas, croquettes, feed grain, ulpada, soup, misko, pitapi, bread, pastel, choclos, chicha, mazamorra with cheese	Argentina
42	Grain (all), choclo, soup, chicha, choclo-tamal, reposteria, tostado (parched grains), tostado (popcorn), mote	Burritas, tamales, soup, mote, choclo, chicha, choclo-tamal, tostado, api (mazamorra), pito	Mote, soup	Bolivia
13	Grain (3), choclo, chicken feed, chuchoca, harina tostada, maiz reventon	Chuchoca, chicken feed, humitas, guisos	Humitas, guisos	Chile
10	Grain (5), choclos, drink with chocolate	Arepas, forage, feed, grain		Colombia
7	Green ears	Grain (all)		Cuba
9	Grain (7), toasted grain, choclo, mote, soups, drinks, popcorn, atole	Choclo, grano tostado, mote, flour	Mote, flour, grano tostado, chicha	Ecuador
12	Grain (all), bread, toasted grain, elotes, pozole, totopos, tamales	Elotes, pozole, memelas, tlacoyos, huacholes, toasted grain, popcorn, tortillas	Gordas de maiz, maiz crudo reposado, ponteduro con panela	Mexico
10	Grain (all), mote, cancha, choclo, chicha	Choclo, chicha, mote	–	Peru
7	Grain (all), human food	Animal food	–	Paraguay

Definitions: alfajor=cake; arepas=large, thick tortilla; atole=sweet drink of corn flour with milk; biscocho=sweet roll; cancha=toasted grain; chicha=corn beer; chichoca=cooked grits; choclo=green ear; chuchoca=cooked grits; elotes=green ears; grano tostado=toasted grain; guisos=any main dish from maize; harina cocida=dough; harina tostada=floury toasted; humita=sweet dough cooked; humitas=sweet masa cooked; maíz reventon=popcorn; mazamorra=candy; memelas=large, flat fried corn cakes; mote=hominy; palomitas=popcorn; polenta=cooked flour; ponteduro con panela=candy; pozole=whole grains in meat broth; reventado=popcorn; tamal/tamales=steamed dough filled with meat and/or sauce; tlacoyos=fried cakes with beans inside; tostada=toasted tortilla; totopos=a type of small tortilla; turron=candy.

Appendix III. Maize landraces, in order of priority, used in breeding

Country	No. of races
Argentina	12
Bolivia	14
Chile	4
Colombia	4
Cuba	7
Ecuador	6
Mexico	11
Peru	10
Venezuela	1

B. *Tripsacum*

J. Berthaud, Y. Savidan, M. Barré and O. Leblanc

Although they are very different morphologically, the perennial genus *Tripsacum* is genetically related to *Zea*, annual and perennial, and hybrids have been produced from crosses between them (Mangelsdorf and Reeves 1931; Harlan and de Wet 1977; Leblanc *et al.* 1995b). Plants can be recovered from maize x *Tripsacum* hybrids through a series of backcrosses, suggesting the feasibility of gene exchange between the two genera. Galinat (1977) has shown that introgression between maize and *Tripsacum* is also possible.

BOTANY AND DISTRIBUTION
The taxonomy in this chapter is based on Cutler and Anderson (1941), Randolph (1970) and de Wet *et al.* (1976, 1981, 1982, 1983). *Tripsacum* and *Zea* belong to the subtribe Tripsacinae of the Andropogonae of the Poaceae family. Two sections are distinguished by the presence (*Fasciculata*) or absence (*Tripsacum*) of a pedicel on a male spikelet of the pair. The basic chromosome number is $x=18$, whereas in *Zea* $x=10$. Several ploidy levels have been observed in this genus, including some we have found that have not been reported in the literature (Table 16B.1).

Tripsacum plants are perennial. They do not easily allow definite species identification because variation for many characters does not permit a clear-cut determination. We have developed the following key for species from Mexico based on observations of materials in our field genebank. The values proposed for the number of racemes on a terminal (RAC1) and secondary inflorescences (RAC2) are the average of observed values. Abbreviations of species are listed in Table 16B.2.

- **TBV**. Vegetative traits: basal sheaths in a fan-like position; strong brace roots; pilosity on basal sheaths (mostly on mid-rib). Inflorescences: RAC1=4, RAC2=3; flowering mostly synchronous on terminal and secondary inflorescences; erect or slightly bent rachis; sessile paired spikelets.
- **TDH**. Vegetative traits: basal sheaths with variable pilosity, often hairs near the ligule; sheaths not fan-like. Inflorescences: RAC1=3, RAC2=1; infrequent ramifications on secondary inflorescences; erect rachis; sessile paired spikelets.
- **TDM**. Vegetative traits: strong stems; basal sheaths pilose to glabrous; sheaths not fan-like. Inflorescences: RAC1=12, RAC2=5; rachis erect to pendulous, often bent; one spikelet sessile, the other sessile or on a short pedicel.
- **TIT**. Vegetative traits: almost no hairs on basal sheaths; intense tillering, sometime from the main stems. Inflorescences: RAC1=4, RAC2=2; erect, becoming pendulous over the flowering period; one sessile and one shortly pedicellate spikelet.

- **TJL**. Vegetative traits: almost no hairs on basal sheaths; high stems. Inflorescences: RAC1=8, RAC2=3; pendulous; long male glumes; one sessile and one shortly-pedicellate spikelet.
- **TLC**. Vegetative traits: basal sheaths with hairs; leaves very often blue. Inflorescences: RAC1=6, RAC2=3; erect to pendulous; paired spikelets, often one sessile and one pedicellate.
- **TLT**. Vegetative traits: strong decumbent stems of indeterminate growth; no hairs on basal sheaths; not all stems produce inflorescences during flowering. Inflorescences: RAC1=RAC2=4; erect rachis for the diploid form and almost pendulous for the triploid; short, purple male glumes; sessile spikelets for the diploid form and sessile and pedicellate paired spikelets for the triploid.
- **TLX** and **TMZ**. Vegetative traits: basal sheaths with long stinging hairs; a few strong stems per plant; large leaves. Inflorescences: RAC1=27, RAC2=13; pendulous.
- **TMN**. Vegetative traits: comparable to TIT, no hairs on sheaths, small plant. Inflorescences: RAC1=RAC2=1; sessile paired spikelets.
- **TPL**. Vegetative traits: basal sheaths with long, non-stinging hairs, also often hairs on terminal sheaths. Inflorescences: RAC1=15, RAC2=8; bent-to-pendulous; short-to-long pedicel of spikelet.
- **TZP**. Vegetative traits: Small plant, glabrous basal sheaths, very thin stems. Flexuous leaves, curly and reddish-brown when old. Inflorescences: RAC1=RAC2=1. Erect inflorescences, sessile paired spikelets.

We have not proposed a key for the South American species because morphological variation is minimal in our field genebank samples.

Origin and Centre of Diversity

Mexico and Guatemala are the centres of diversity for the genus. *Tripsacum* species are widely distributed in the Americas and most numerous in Mexico (de Wet *et al.* 1983; personal observations in herbaria). In South America, *Tripsacum* species have been collected at 250-650 m asl for TAA and TAH, 500-1700 m asl for TMR and 600-1400 m asl for TCD. *Tripsacum* in Central America is found from sea level to 2600 m asl. TDH is found from 100 to 2600 m asl, and the TDH population at the highest elevation is a diploid. The general distribution of *Tripsacum* species appears to be structured more in relation to climate than to altitude. TLX (and TMZ) and TLT are clearly adapted to the humid tropics. It is difficult to find a relationship between altitude and ploidy level. The lack of relationship between altitude and ploidy level could be best explained by the existence of polyploid series in most *Tripsacum* species. Near Tuxtla Gutierrez, Chiapas, in southeast Mexico, at altitudes ranging from 750 to 1200 m asl, it is possible to find TIT (tetraploid and pentaploid), TJL (tetraploid), TLX (diploid) and TMN (diploid). In the northern state of Sinaloa, Mexico, diploid, triploid and tetraploid forms of TPL grow together in the same population or as neighbouring populations.

Tripsacum species extend from longitude 42°N to 24°S. The published information on general distribution is not always accurate. TBV was first described as endemic in Valle de Bravo, state of Mexico (de Wet *et al.* 1976), but in fact it is as widespread as TDH. Further, TZP was reported as coming from Cañón del Zopilote, Guerrero and other parts of Mexico (Randolph 1970), but is actually limited to the Cañón; the specimens reported from other locations, though resembling this species, belong to TDH. Finally, TLT was reported from only Belize, Guatemala, Costa Rica and Honduras (Pohl 1980; de Wet *et al.* 1982), but we collected samples from wild diploid and triploid TLT populations in Mexico.

Table 16B.1.   Distribution of ploidy levels in *Tripsacum* species collected from wild populations in Mexico, 1989-92.

Species or hybrids	No. of populations	2x	3x	4x	5-6x
			No. of populations with:		
BV	36	1	1[t]	35	–
BVLC	2	–	–	2	–
DH	32	4[t]	1[t]	29	2[t]
DHBV	2	–	–	2	–
DHIT	8	–	–	8	–
DM	20	–	1[t]	20	2[t]
DMBV	2	–	–	2	–
DMLT	1	–	–	1	–
DMPL	2	–	–	2	–
IT	13	–	2[t]	10	2[t]
JL	2	–	–	2	–
LC	9	–	–	9	–
LCPL	1	–	–	1	–
LT	9	7	2[t]	0[‡]	–
LX	4	4	–	–	–
MN	1	1	–	–	–
MZ	13	5	4[t]	5	–
MZPL	2	–	–	2	–
PL	10	4[t]	3[t]	5	1[t]
ZP	5	5	1[t]	1[t]	–
Total	**174**	**31**	**15**	**136**	**7**
%		16.4[§]	7.9	72	3.7

[t] New ploidy levels compared with literature.
[‡] Ploidy level found in Randolph (1970), but the specimen had mistakenly been described as atypical *T. latifolium*.
[§] Percentages are taken from the total population/species/ploidy combinations. In some populations there are several ploidy levels and sum of % greater than 100.

Table 16B.2. Abbreviations of *Tripsacum* species.

TAD	=*T. andersonii* Gray	TIT	=*T. intermedium* de Wet and Harlan
TAA	=*T. australe* var. *australe* Cutler and Anderson	TJL	=*T. jalapense* de Wet and Brink
TAH	=*T. australe* var. *hirsutum* de Wet and Timothy	TLC	=*T. lanceolatum* Ruprecht ex Fournier
TBV	=*T. bravum* Gray	TLT	=*T. latifolium* Hitchc.
TCD	=*T. cundinamarce* de Wet and Timothy	TLX	=*T. laxum* Nash
TDD	=*T. dactyloides* (L.) L.	TMZ	=*T. maizar* Hernandez and Randolph
TDH	=*T. dactyloides* var. *hispidum* (Hitchc.) de Wet and Harlan	TMN	=*T. manisuroides* de Wet and Harlan
TDM	=*T. dactyloides* var. *mexicanum* de Wet and Harlan	TPR	=*T. peruvianum* de Wet and Timothy
TMR	=*T. dactyloides* var. *meridionale* de Wet and Timothy	TPL	=*T. pilosum* Scribner and Merrill
TFL	=*T. floridanum* Porter ex Vasey	TZP	=*T. zopilotense* Hernandez and Randolph

The various *Tripsacum* species in Mexico can be placed in three groups based on distribution, although these groups do not present clear-cut boundaries:

1. **Northwestern group**. This group comprises various forms of TLC found in the Durango region and Sierra Madre Occidental. TLC has been encountered in these

mountains and in the USA in the state of Arizona. All plants checked for chromosome number are tetraploid ($2n=72$).

2. **Southern group**. This includes TIT, TJL, TLT, TLX, TMZ and TMN. According to de Wet *et al.* (1982), TJL, TLT and TLX are found in Guatemala. There are also specimens of TLT in Belize and Honduras. This group covers not only southern Mexico from Guerrero to Oaxaca, Veracruz and Chiapas, but also Central America. It involves species with ploidy levels from diploid to pentaploid. However, TMN and TLX have only diploid specimens; TLT, diploid and triploid.

3. **Central group**. This group includes TBV, TDM, TPL, TZP and some forms of TLX, TMZ. These species are found from Jalisco to Guerrero, Michoacán and Mexico states. One species (TDH, related to the TDD found in the eastern USA) extends to the northeast through the Sierra Madre Oriental. The group is very diverse, comprising species such as TZP, which are endemic in specific locations, and broadly distributed species such as TBV and TDH. Except for TDM, species of this group exhibit a range of ploidy levels (see description of polyploid series, below).

These groups were formed using the geographical origins of *Tripsacum*, but also illustrate the history of *Tripsacum* species. Species from the southern group would be more related to the South American group, which are diploid, with the exception of TPR (de Wet *et al.* 1981). South American species are distributed from Venezuela to the south of Brazil. Samples in the CIMMYT collection come from Venezuela, Colombia and Peru.

Reproductive Biology

If one considers each species/ploidy/population combination as a unit, in the CIMMYT collection 16.6% of the plants are diploid, 8% are triploid, 71.5% tetraploid and 3.8% are pentaploid and hexaploid. Most *Tripsacum* species exhibit a polyploid series (Table 16B.1) from diploid to tetraploid and in some cases pentaploid and hexaploid. The series does not exhibit drastic morphological variations, suggesting that polyploidization is occuring within the species and not only through interspecific hybridization between a diploid and a polyploid species. Our observations and those of Leblanc *et al.* (1995a) found that polyploid plants from 48 populations tested were apomictic. *Tripsacum* is then a genus with the same reproductive organization as other agamic complexes: diploid plants reproduce sexually while polyploids rely on apomixis.

Observations on triploids have shown these plants to be apomictic, and male and female fertile (Moreno Perez 1994). They could act as genetic bridges between diploid and tetraploid forms. Analyses of diversity using molecular markers (Barre *et al.*, unpublished data) showed geneflow across species and ploidy levels to be quite free. Apomixis has occasionally been described as an evolutionary dead end. From analyses of *Tripsacum*, however, it can be considered as favouring hybridization and geneflow, through the fixation and propagation of hybrid forms.

Tripsacum andersonii, guatemala grass, has 64 chromosomes (Levings *et al.* 1976). As of this writing, the species is thought to be the result of a hybridization event between *Zea* (10 chr) and *Tripsacum* (54 chr=$3x$). On the basis of morphological similarities between the two species, de Wet *et al.* (1983) proposed *T. latifolium* ($2n=36$) as the putative *Tripsacum* parent. Studies by Talbert *et al.* (1990) have shown that the *Zea* genome is from *Zea luxurians* and the *Tripsacum* genome is not *T. latifolium* ($2x$), but could be *T. maizar* or *T. laxum*. By analyzing at CIMMYT the diversity of the material surveyed in Mexico, two types were found in Mexican *T. latifolium* accessions, both with the same gross morphology but one diploid (as is normal in the species) with paired sessile spikelets, and the other triploid with paired spikelets, one sessile and one shortly pedicellate. We therefore propose that the triploid *T. latifolium* is a hybrid between a diploid *T. latifolium* and another *Tripsacum* species of the *Fasciculata* section (to explain the pedicellate spikelets).

Results of molecular marker analysis at CIMMYT suggest that TLT (3x) originated in the hybridization event: *T. latifolium* (2x) x *T. maizar* (2x) → *T. latifolium* (3x=54 chr), with an unreduced gamete from one of the parents. A second hybridization event led to the creation of *T. andersonii*: *T. latifolium* (3x=54) x *Zea luxurians* (2n=20) → *T. andersonii* (54 + 10 chr). The first event must have occurred several times, as fingerprinting (isozymes) indicates that the two triploid *T. latifolium* populations are different. The second event was probably unique, as the more than 20 different accessions of *T. andersonii* from several South American countries show exactly the same morphology, isozyme pattern and DNA fingerprint. *T. andersonii* is the only example of a natural *Zea* x *Tripsacum* hybrid.

RFLP fingerprinting shows that polyploid plants are vegetatively propagated by apomixis in wild populations, but that apomixis does not limit genotypic diversity (Table 16B.3). Still, some clones are represented by more plants than others. Populations comprising several species and ploidy levels have been found in the wild, but many have few (10-100) plants representing only a few clones.

GERMPLASM CONSERVATION AND USE

Despite its widespread distribution in the Americas, *Tripsacum* is not a very common plant. Populations of various size exist, but their colonizing ability is inferior to that of plants such as *Panicum maximum*. Our surveys in Mexico have shown that *Tripsacum* species are also widely distributed, and we have located more than 150 different populations. Some plants have been transferred to the field bank, but these populations continue to exist *in situ*, permitting normal geneflow and the processes that led to the current natural diversity. *Tripsacum* populations are very sensitive to grazing. Land-tenure modifications that change cattle-raising practices could endanger certain populations, but *Tripsacum* species should at most be considered vulnerable, not endangered. During 1989-92, we collected samples from 158 *Tripsacum* populations in Mexico, the centre of diversity for this genus. We obtained 2500 accessions as cuttings, some 1000 of which have been established in a field collection on the CIMMYT experiment station at Tlaltizapán, state of Morelos, Mexico (940 m asl, 18°N latitude).

It appears necessary to establish a living collection of *Tripsacum* in an appropriate subtropical environment to ensure the preservation of most of the species. Studying genotypic diversity through RFLP fingerprinting and ploidy levels through chromosome counts allowed us to identify duplicate accessions. We verified that in wild populations of *Tripsacum*, apomictic plants are always polyploid, an important step toward propagating the collection through seeds, both for conservation and distribution. One option is vegetative propagation (cuttings), which was the method used to establish our base collection. However, the disadvantages of using cuttings for distribution are well known, so we would consider apomictic propagation a feasible alternative, especially since apomixis exists in most populations of our base collection. Before conserving seeds as accessions, their viability should be assessed through germination tests. It should also be noted, though, that dormancy is a common phenomenon in *Tripsacum* seeds. We eliminated the dormancy problem through embryo rescue and culture on an N6 medium.

Diploid plants do not exhibit apomixis; thus, all diploid plants are different from one another, being reproduced through recombination and open-pollination (a fact confirmed by molecular marker studies). Depending on the pollen sources, seed from diploid plants may be diploid or triploid, and distribution of germplasm identical to that collected is only possible through vegetative propagation or time-consuming controlled pollinations within populations. If some genetic variation is permissible,

Table 16B.3. Distribution of *Tripsacum* genotypes in the wild population 'La Toma'.

Species and types	No. of chromosomes	No. of plants	Species and types	No. of chromosomes	No. of plants
BV1	72	33	DM12	54	1
BV2	72	3	DM13	54	2
DM1	72	4	DM14	72	1
DM2	72	2	DM15	72	1
DM3	72	2	DM16	72	1
DM4	72	27	DM17	54	1
DM5	54	5	DM18	54	1
DM6	90	1	DM19	54	1
DM7	108	1	DM20	72	1
DM8	72	1	DM21	72	1
DM9	72	1	DM22	72	1
DM10	72	1	DM23	72	1

then seeds constitute a viable means of distribution, but progenies must be checked to cull morphological and cytological off-types. The most common *Tripsacum* off-types are produced as $2n+n$ seedlings (i.e. an unreduced female gamete fertilized by a normal male gamete). This 'ploidy building' mechanism was observed in most progenies from the live collection and also in progenies from seeds collected directly in the wild. Observed frequencies of these off-types vary from 3 to 35%, depending on the genotypes.

Properties and Uses

In the USA, *T. dactyloides* is cultivated in forage breeding programmes at the USDA research station, Woodward, Oklahoma, and at Iowa State University. *Tripsacum andersonii* is also used as a forage plant. The spread of this crop is related to its use as a fodder for Guinea pigs by indigenous people in South America (Hernandez X. 1970). Now that a large collection of wild *Tripsacum* populations is available, it may be worthwhile to find new opportunities for *Tripsacum* as a forage crop in the tropics.

Breeding

Gene transfer to maize is possible. The introgression route has been reviewed by Harlan and de Wet (1977). A transfer programme for apomixis is currently underway at CIMMYT, applying new molecular tools. In addition to the potential benefits for farmers in developing countries, the transfer of apomixis from *Tripsacum* to maize constitutes an interesting case study that shows the way for transferring other useful genes. As one example, IITA (D. Berner, pers. comm.) and the University of Bristol, UK (A. Lane, pers. comm.) have identified *Tripsacum* plants that possess resistance to *Striga hermonthica*, a parasitic flowering plant which causes significant economic damage to maize in sub-Saharan Africa and to which no known source of resistance exists in maize.

REFERENCES

Cutler, H.C. and E. Anderson. 1941. A preliminary survey of the genus *Tripsacum*. Ann. Miss. Bot. Gard. 28:249-269.

de Wet, J.M.J., D.E. Brink and C.E. Cohen. 1983. Systematics of *Tripsacum* section *Fasciculata* (Gramineae). Am. J. Bot. 70:1139-1146.

de Wet, J.M.J., J.R. Gray and J.R. Harlan. 1976. Systematics of *Tripsacum* (Gramineae). Phytologia 33:203-227.

de Wet, J.M.J., J.R. Harlan and D.E. Brink. 1982. Systematics of *Tripsacum dactyloides*. Am. J. Bot. 69:1251-1257.

de Wet, J.M.J., D.H. Timothy, K.W. Hilu and G.B. Fletcher. 1981. Systematics of South American *Tripsacum* (Gramineae). Am. J. Bot. 68:269-276.

Galinat, W.C. 1977. The origin of corn. *In* Corn and Corn Improvement (G.F. Sprague, ed.). Am. Soc. Agron., Agron. series, vol. 18:1-47.

Harlan, J.R. and J.M.J. de Wet. 1977. Pathways of genetic transfer from *Tripsacum* to *Zea mays*. Proc. Nat. Acad. Sci. USA 74:3494-3497.

Hernandez X., E. 1970. Apuntes sobre la exploración etnobotánica y su metodología. Colegio de Postgraduados. Escuela, Nat. Agron. Chapingo, Mexico.

Leblanc, O., D. Grimanelli, D. Gonzalez de Leon and Y. Savidan. 1995a. Detection of the apomixis mode of reproduction in maize-*Tripsacum* hybrids using maize RFLP markers. Theor. Appl. Genet. 90:1198-1203.

Leblanc, O., M.D. Peel, J.G. Carman and Y. Savidan. 1995b. Megasporogenesis and megagametogenesis in several *Tripsacum* species (Poaceae). Am. J. Bot. 82:57-63.

Levings, C.S., D.H. Timothy and W.W.L. Hu. 1976. Cytological characteristics and nuclear DNA buoyant densities of corn, teosinte, *Tripsacum* and corn-*Tripsacum* hybrids. Crop Sci. 16:63-66.

Mangelsdorf, P.C. and R.G. Reeves. 1931. Hybridization of maize, *Tripsacum*, and *Euchlaena*. J. Heredity 22:339-343.

Moreno Perez, E. 1994. Estudios citológicos en poblaciones naturales de *Tripsacum*. Master's Thesis 67, Colegio de Postgraduados, Montecillos, state of Mexico.

Pohl, R.W. 1980. Family #15, Gramineae, *Tripsacum*. Pp. 574-577 in *Flora Costaricensis* (W. Burger, ed.). Fieldiana Botany, New series, No 4., Field Museum Natural History.

Randolph, L.F. 1970. Variation among *Tripsacum* populations of Mexico and Guatemala. Brittonia 22:305-337.

Talbert, L.E., S. Larson and J.F. Doebley. 1990. The ancestry of *Tripsacum andersonii*. Pp. 100-103 *in* Proceedings II Int. Symposium on Chromosome Engineering in Plants (G. Kimber, ed.), August 13-15.

C. Teosinte

S. Taba

Teosinte is a wild relative and the closest relative of maize. It thrives in wild or cultivated fields by dispersing seed or, in the case of perennial forms, through rhizomes. The plant is adapted to mid- and high-elevation regions in Mexico and Central America that are seasonally dry and receive summer rains (Wilkes 1967). Only in the last 100 years has teosinte become known to the rest of the world.

BOTANY AND DISTRIBUTION
Teosinte was first classified as *Euchlaena mexicana*, a botanical grass, by Schrader in 1832, on the basis of a sample of seed sent from Mexico to Germany by a mine engineer. The common name teosinte derives from the Nahuatl[1] *teocintle* associated with the annual population (*Zea luxurians*, race Guatemala) that grows wild in southeastern Guatemala (Wilkes 1967).

Origin, Domestication and Diffusion
Archeological specimens of teosinte are scarce in Mexico, but the various local names for the plant suggest a long association between man and teosinte in the maize-oriented cultures of Mesoamerica. Seed of this population was sent to France in 1869, whence it was distributed worldwide as a potential fodder crop in the late 19th century. The teosinte later increased in tropical Florida, USA and distributed widely as 'Florida teosinte' was descended from the sample sent to France.

A general map of teosinte distribution was published by Wellhausen *et al.* (1952). Wilkes (1967) reported in detail on the distribution of annual teosinte in Mexico and Central America. The discovery of the perennial diploid teosinte *Zea diploperennis* in 1978 (Iltis *et al.* 1979), as well as the rediscovery of the perennial tetraploid teosinte *Zea perennis* by Guzman (1978), spurred much research and additional exploration and collecting of teosinte in Mexico. Later Sanchez G. and Ordaz S. (1987) provided new descriptions of teosinte's distribution in Mexico and Sanchez G. *et al.* (1995) updated that information and characterized Mexican teosinte populations using morpho-agronomic and chromosome knob data. Wilkes (1988, 1993) and CIMMYT (1986) added a report on the status of populations *in situ*. A new taxonomic classification of teosinte which integrated it and maize in the genera *Zea* was summarized by Doebley (1990a, 1990b). For the purposes of conservation and utilization, natural variations in teosinte can be characterized either through descriptions of ecogeographic races of populations *in situ* (Wilkes 1967) or taxonomic classifications (Doebley 1990a, 1990b) (Table 16C.1).

---

[1] Language of the Aztec Indians.

Table 16C.1. Taxonomy of *Zea*.

Wilkes (1967) revised	Iltis and Doebley (1980) Doebley (1990a, 1990b) revised
Section *Euchlaena* (Schrader) Kuntze	Section *Luxuriantes* Doebley and Iltis
	Zea diploperennis Iltis, Doebley and Guzman
Zea perennis (Hitchc.) Reeves and Mangelsdorf	*Zea perennis* (Hitchc.) Reeves and Mangelsdorf
Zea mexicana (Schrader) Kuntze	*Zea luxurians* (Durieu) Bird
Race Guatemala	
	Section *Zea*
	Zea mays L. subsp. *mexicana* (Schrader) Iltis
Race Chalco	Race Chalco
Race Central Plateau	Race Central Plateau
Race Nobogame	Race Nobogame
Race Durango*	Race Durango[†]
Race Balsas	subsp. *parviglumis* Iltis and Doebley
	Within-race variations[‡]
Race Huehuetenango	subsp. *huehuetenangensis* (Iltis and Doebley) Doebley
Section *Zea*	
Zea mays L.	subsp. *mays* (L.) Iltis

[†] Race Durango was included previously in race Central Plateau (Wilkes 1967). After recollecting of the population it was recognized as race Durango (CIMMYT 1986; Sanchez G. *et al.* 1995).
[‡] Within-race variations have been shown in race Balsas (Doebley 1990a, 1990b; Razo L. 1989; Sanchez G. *et al.* 1995) to form possible ecogeographic races or groups (see text).

Although able to cross-fertilize with maize (*Zea mays* L.), teosinte is morphologically and genetically distinct from that crop. It has solitary female spikelets (compared with paired ones in maize) that are two-ranked (versus many-ranked in maize), a shattering rachis (i.e. the cob breaks into pieces, whereas in maize it does not) and fruitcase-enclosed kernels (maize kernels have no hard covering) (Galinat 1992). Numerous studies on teosinte's taxonomy (Bird 1978; Iltis *et al.* 1979; Doebley and Iltis 1980; Iltis and Doebley 1980; Doebley 1990a, 1990b), cytology (Kato Y. 1984; Kato Y. and Lopez R. 1990; Sanchez G. *et al.* 1995), genetics (Mazzoti and Velasquez 1962; Beadle 1980; Doebley 1984; Galinat 1985, 1988, 1992, 1995; Mangelsdorf 1986; Allen *et al.* 1989; Kermicle and Allen 1990), ecology, geography and taxonomy (Wilkes 1967, 1977, 1986, 1988, 1993; Orozco and Cervantes S. 1986; Benz 1988; Benz *et al.* 1990; Sanchez G. *et al.* 1995), as well as molecular analyses (Smith *et al.* 1984, 1985; Doebley *et al.* 1987a, 1987b; Doebley 1990a, 1990b) and recent reviews (Wilkes 1986; Benz 1987; Iltis 1987; Goodman 1988; Galinat 1992) have contributed much to knowledge on the possible role of teosinte in the domestication and evolution of maize and on the natural diversity of teosinte.

Collections of teosinte populations have been made at more than 100 sites in Mexico by genetic resource experts of the Mexican National Institute of Forestry, Agriculture, and Livestock Research (INIFAP) and other collaborators in Mexico. Sanchez G. (pers. comm.) estimates that another 20% uncollected populations might exist in Mexico. New teosinte sites have been discovered by scientists acting on information from local agronomists who had sighted what they thought to be teosinte. Figure 16C.1 depicts the geographic distribution of races of teosinte as identified in Table 16C.1. Teosinte in Honduras is considered extinct in the wild (Wilkes 1967). A teosinte collection of *Zea luxurians* from Rancho Apacunca, Department of Chinandega, Nicaragua (collection number: Iltis, Medina and Castrillo 30919 from the University of Wisconsin) in 1991 has been increased at CIMMYT. Given its use there as a fodder crop, however, it may have been introduced from southeastern Guatemala to the collecting site in recent times, in response to the local need for such a crop.

Teosinte Races in Mexico and Guatemala

Race Guatemala

This race has been distributed worldwide as a fodder crop. It is known as *teocintle* or *milpa silvestre* and is found in small, isolated populations in broad valleys and hills at 900-1200 m asl in the Departments of Jutiapa, Jalapa and Chiquimula, southeastern Guatemala. Wilkes (1993) reported a roughly three-quarter reduction in teosinte sites, number of populations and population sizes from 1962-63 to 1991. A Florida collection of the race exists.

At higher latitudes the plant produces abundant tillers, which parallels the tillering pattern in *Tripsacum* (Wilkes 1988). Male tassel branches are few (1-10) and extend from the short central branching axis; they are stiff, straight and erect, and the central spike has the same morphology as the lateral branches (Doebley and Iltis 1980; Wilkes 1967). The large terminal chromosome knobs are more numerous than for any other teosinte and suggest a distant relation to the annual Mexican teosintes (Kato Y. 1976, 1984; Kato Y. and Lopez R. 1990; Smith *et al.* 1984). Race Guatemala requires an artificially shortened daylength to flower in temperate environments.

Race Huehuetenango

This race grows at 500-1650 m asl in northwestern Guatemala, bordering on Mexico, where it is known as *salic* or *milpa de rayo*. Wilkes (1993) reported population sizes that were only about 10% of those observed in 1963. The central spike is somewhat stiffer and stronger than that in section *Zea* and more densely set with spikelets than lateral spikes (Wilkes 1967; Doebley and Iltis 1980). The chromosome knob pattern is similar to that of *Zea diploperennis*; the presence of many terminal knobs (Kato Y. and Lopez R. 1990) seems to indicate a close relationship with races of the section

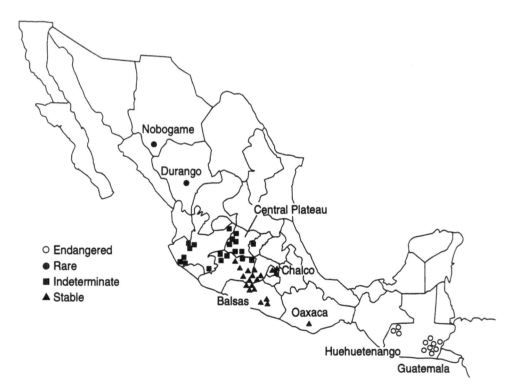

Fig. 16C.1. Current status of teosinte populations in Mexico and Guatemala.

Luxuriantes. Tassel morphology, data on cytoplasmic cpDNA and plant and seed morphology support its placement in the phylogenetic/taxonomic section *Zea* as subsp. *huehuetenangensis* (Doebley 1990a, 1990b).

Race Balsas and its variations

This race grows in the seasonally dry, thorn-scrub vegetation of the mountains of the Balsas River basin, south-central Mexico. Balsas teosinte is known as *maíz de pajaro* or *maíz de huiscatote* in the Balsas area and *atzitzinte* in the Chilpancingo area of Guerrero State. The variation in this subrace is represented by collections from Mazatlán, a site south of Chilpancingo in central Guerrero State (Wilkes 1967, 1977; Sanchez G. *et al.* 1995; Orozco and Cervantes S. 1986; Iltis 1987; Doebley 1984). Collections from Mazatlán possess a trait, resulting from farming practices in the area, which allows its seed to lie dormant for as long as a year (Wilkes 1977).

Another distinct group comes from elevations of 1200-1700 m asl within 18-19°N in the states of Guerrero, Oaxaca (at the recently collected site of San Cristobal Honduras), Mexico and Michoacán, bordering the Rio Balsas basin. Teloloapán, Arcelia, Huetamo and Valle de Bravo are the representative sites for this central Balsas teosinte. According to Sanchez G. *et al.* (1995), it differs slightly in maturity and plant type from teosinte of central Guerrero and also differs clearly from race Central Plateau. Another ecotype separated from the above two groups in Jalisco has earlier maturity, relatively large seed and some tolerance to leaf rust (*Puccinia sorghi* Schw.). Still other Balsas-type teosintes grow in Nayarit State at 20°N and about 900 m asl and at Amatlán, near Tepoztlán, Morelos. Morpho-agronomic and chromosome knob variations (Sanchez G. *et al.* 1995; Orozco and Servantes S. 1986) seem to coincide largely with isozyme variations (Doebley *et al.* 1984; Smith *et al.* 1985) within race Balsas. Balsas teosinte, especially accessions from Valle de Bravo, state of Mexico and central Michoacán, has the greatest similarity to maize of any teosinte. Doebley (1990a, 1990b) suggested *Zea mays* subsp. *parviglumis* should be divided into races Jalisco, Central Balsas and Central Guerrero, based on numerical analysis of isozyme variations. However, much more variation appears to exist even within these subdivisions (Razo L. 1989; Sanchez G. *et al.* 1995; Orozco and Cervantes S. 1986). Maize chloroplast DNA patterns were similar to those of Balsas and Huehuetenango teosinte (Timothy *et al.* 1979; Doebley *et al.* 1987b; Doebley 1990a, 1990b). Good compatibility between race Balsas and maize has been reported (Kermicle and Allen 1990; Goodman *et al.* 1983) and Iltis (1987) and Doebley (1990b) have even suggested that maize was domesticated in the Balsas basin.

Race Central Plateau

Collecting sites range from around 1700 to 2150 m asl in the states of Michoacán, Guanajuato and Jalisco. The common name is *maíz de coyote* in Michoacán and Guanajuato. The southernmost sites are in Michoacan at 19°N, extending to 20°N in Guanajuato. One population is located at 21°N, 1950 m asl in Jalisco (Sanchez G. *et al.* 1995). Populations of Opopeo at 2320 m asl and Ciudad Hidalgo at 2040 m asl in Michoacán are considered very similar to race Chalco, based on cluster analysis using morpho-agronomic traits (Sanchez G. *et al.* 1995), despite the fact that these collecting sites are in the general distribution area of race Central Plateau. In the same study, chromosome knob frequencies used for the cluster analysis separated race Central Plateau very well from the other races studied. The population of Degollado, Jalisco, 1650 m asl, was classified as race Balsas by Sanchez G. and Ordaz S. (1987), but isozymic and morphological analyses (Doebley *et al.* 1984) suggest that it belongs to race Central Plateau. Modern agriculture is apparently taking its toll on the distribution of Central Plateau teosinte. Teosinte was disappearing at Copandaro, Michoacán, in 1993. Both plant and seed of the weedy form (adapted to cultivated maize fields) are larger than those of the wild form found on limestone outcroppings along the field margins, when grown together in a uniform field and closer to race Chalco in plant and seed type (Wilkes 1967, 1977).

Race Durango

This race is closely associated with race Central Plateau in many characteristics. It is locally known as *maicillo*. There are three known sites in the Valley of Guidiana, northeast of Durango City, from which collections were made along a river bank and irrigation canals: Francisco Villa, Puente Dalila-Hacienda de Dolores and Puente Gavilan (Sanchez G. *et al.* 1995). First sighted early this century, Durango teosinte subsequently remained hidden from the view of teosinte hunters until fairly recently and was recognized as a separate race after its rediscovery (CIMMYT 1986; Sanchez G. *et al.* 1995). It is vulnerable to loss from local agricultural practices such as maize and sorghum farming and could disappear at any time. Nonetheless, it has been able to hold its own under similar conditions for the last 50 years (Wilkes, pers. comm.).

Race Chalco

Known by the local name of *acece* in Chalco, valley of Mexico, this race also grows in Toluca, state of Mexico (Wilkes and Taba 1993) and at Ciudad Serdán, in the San Juan Atenco municipality and Tlachichuca, in the San Salvador El Seco municipality, both of Puebla State (Sanchez G., pers. comm.). Like maize production in the area, Chalco teosinte is seriously affected by the rapid expansion of nearby Mexico City (Listman 1994). Chalco is resistant to leaf rust (Wilkes 1967; Doebley 1984; Sanchez G. *et al.* 1995). It has a cross-incompatibility factor which functions like *Ga1-s* from popcorn against normal dent maize with *ga1*, designated as *Ga1-s*:Chalco and which has stronger expression in heterozygotes than *Ga1-s* (Kermicle and Allen 1990). This factor possibly acts as an isolating mechanism between maize and teosinte; even so, F_1 hybrids are frequent (Wilkes, pers. comm.).

Race Nobogame

This race is adapted to a valley in the Sierra Madre Occidental at 26°N, 1750-1920 m asl, the northernmost site for teosinte in Mexico. The tassel has a few, short lateral branches that are widely spaced on the main axis, with a central spike and tassel glumes as large as those of races Central Plateau and Chalco (Wilkes 1967; Iltis and Doebley 1980; Sanchez G. *et al.* 1995). Race Nobogame crosses freely with dent maize (Kermicle and Allen 1990).

Perennial Teosintes

There are two forms: a tetraploid (*Zea perennis*; $2n=40$) and a diploid (*Zea diploperennis*; $2n=20$) which has the same number of chromosomes as the annual teosintes described above. Both perennial teosintes are late maturing with slow seedling growth, wide and short leaves, large tassel glumes very much like those of race Guatemala and tolerance to leaf rust (Sanchez G. *et al.* 1995). According to Doebley and Iltis (1980), *Z. diploperennis* has a robust plant type and rhizomes with internodes shorter than *Z. perennis*, often forming tuber-like short shoots. *Z. perennis* plants are more slender. Norton *et al.* (1985) reported that *Zea diploperennis* supported the fewest *Pratylenchus scribner* and *Helicotylenchus pseudorosbustus* nematodes in field tests among teosinte and maize geneotypes. The tetraploid was first reported early this century at Ciudad Guzman, Jalisco (Collins 1921) and was rediscovered by Guzmán in 1978 at Los Depósitos (1650 m asl) and at Piedra Ancha (2100 m asl) in Ciudad Guzman on the northern slope of the Nevado de Colima at 19°N, Jalisco. *Zea diploperennis* was first discovered at the east end of the Sierra de Manantlán, Jalisco, at La Ventana, head of the San Miguel valley, 19°31'N, 2250 m asl and subsequently at Las Joyas (1800 m asl) and Manantlán, 1350 m asl (Iltis *et al.* 1979). Population San Miguel is grown in rotation with maize for use as a forage crop. Population Las Joyas grows in an area where slash-and-burn farming is practised and grazing occurs. Teosinte was apparently introduced at the Manantlán sites in recent times (Benz *et al.* 1990). The Sierra de Manantlán Biosphere Reserve was created in 1987 for *in situ* conservation of perennial teosinte and other endemic species (Sanchez-Velasquez 1991). Isozymic variations (Doebley *et al.* 1984; Doebley 1990a, 1990b) and chloroplast cpDNA restriction site analysis (Doebley 1990a, 1990b) differentiated diploid and tetraploid perennial teosintes in a hierarchical classification system, although *Z. perennis* is considered an autotetraploid derived from *Z. diploperennis* (Galinat and Pasupuleti 1982). Eubanks (1995) has obtained fertile hybrid plants with paired kernel rows from crosses between *Tripsacum dactyloides* and *Z. diploperennis*, and suggests that such a cross may have played a role in the origin and evolution of maize.

GERMPLASM CONSERVATION AND USE

Staff of the INIFAP maize germplasm bank, Mexico, have been active in collecting and preserving teosinte (Sanchez G. and Ordaz 1987). In the mid-1980s, CIMMYT, INIFAP and ICTA began coordinating efforts to monitor teosinte populations *in situ* (CIMMYT 1986; Wilkes 1993; Sanchez G. *et al.* 1995). Monitoring visits have been made in recent times to the following sites:

- The regions of Jutiapa, Jalapa and Huehuetenango[2]
- Central Guerrero, central Balsas and San Cristobal Honduras in San Pedro Juchatengo Province, Oaxaca
- Uriangato-Moroleón in southern Guanajuato and Copandaro-Penjamillo-Cuitzeo in north central Michoacán
- Francisco Villa and Puente Gavilán in the valley of Guadiana, in Durango
- Guadalupe y Calvo, Nabogame, in Chihuahua
- Chalco, Los Reyes and Texcoco in the valley of Mexico
- Toluca in the valley of Toluca.

Monitoring trips serve to (1) collect a representative sampling with sufficient seed for both preservation and utilization, and (2) determine the current status of the population. Future monitoring trips in Mexico will cover additional sites in the states of Jalisco, Puebla, where a new collection was recently made (Sanchez G. pers. comm.) and Chiapas, where a teosinte population is reported to exist near Villa Flores in the Freylesca region. Most teosintes except Balsas are considered 'vulnerable', according to the terms of the Species Survival Commission of the International Union for Conservation of Nature and Natural Resources (IUCN) in Switzerland (CIMMYT 1986; Wilkes 1988). Grazing often affects populations of races Guatemala, Huehuetenango, part of Balsas and Central Plateau, and Nobogame. The races Nobogame and Durango are considered rare, as they are scarce enough that they can be eliminated easily. However, with recent collections in Toluca state of Mexico and in Puebla State, race Chalco can be considered stable, despite such cases as the Chalco population at Los Reyes that survives in a single field on the urban outskirts of Mexico City (Listman 1994). Monitoring trips have confirmed no immediate threat of extinction for other populations, despite occasionally significant reductions in their size (races Guatemala and Huehuetenango, for example, have shrunk to 25 and 10%, respectively, of their size in 1963 and are considered endangered). However, *in situ* monitoring should be intensified, with the possibility of organizing *in situ* conservation of the endangered Guatemalan populations (Wilkes 1993). Races Central Plateau and Durango are in modern agricultural regions where land use and cropping patterns will greatly affect their survival.

---

[2] These locations are in Guatemala; the rest are in Mexico.

It is problematic for maize germplasm banks to conduct regular seed increase programmes for teosinte, given its ability to contaminate experimental plots of maize and to outcross with maize or other teosinte accessions. Seed increase must be done in isolation, using open-pollination among more than 100 plants, if possible. Each year a few accessions are regenerated at CIMMYT in isolation from experimental maize plots. Ideally, a permanent seed-increase plot for teosinte managed by germplasm bank personnel is needed. The CIMMYT Maize Germplasm Bank preserves samples collected in the 1960s and some recently collected during the monitoring trips, but needs additional samples from currently known locations in Mexico and Guatemala. Teosinte accessions are preserved in medium- and long-term storage, once seed has been cleaned and dried at 23-25% RH.

Properties and Uses

Teosinte has been used as a fodder crop and for studies on maize evolution. So far the plant has been of little use for maize improvement or hybrid development (Goodman 1988). However, results of experiments using teosinte germplasm to increase yield in maize hybrid combinations were encouraging (Cohen and Galinat 1984). Viral resistances were reported in perennial teosintes (Nault *et al.* 1982), but their use in maize improvement has been limited, because maize lines were found that confer similar resistance (Louie *et al.* 1990). Some teosinte races show cross-incompatibility with normal maize genotypes, making it difficult to introgress their germplasm into maize. *Zea diploperennis* seems to have a barrier to fertilization with maize (Sanchez G., pers. comm.). On the other hand, such races as Balsas and Nobogame cross quite easily with maize. Maize races Camelia Vicuña and Arrocillo Amarillo were cross-compatible with Guerrero teosinte (Castro G. 1970). Kermicle and Allen (1990) reported that race Central Plateau has a dominant barrier complex on chromosome four different from the incompatibility factor in race Chalco. In the same study, races Nobogame (*Z. mays* subsp. *mexicana*) and Balsas (*Z. mays* subsp. *parviglumis*) accepted pollen from dent maize and set seeds, but *Zea luxurians* only partially set seeds and subsp. *huehuetenangensis* was incompatible through the first backcross generation. Allen *et al.* (1989) developed cytolines having different teosinte cytoplasm and the maize genome of inbred W23. They found a teosinte-cytoplasm-associated miniature trait (TCM) expressed by section *Luxuriantes* and which is countervailed by a dominant nuclear gene, denoted *Rcm1* (rectifier), present in many maize inbreds. Balsas teosinte may have the greatest genetic diversity of any teosinte, making it a logical candidate for use in introgressive hybridization with maize. Teosinte may contribute to maize improvement in the future in the same manner as maize landraces have: by increasing genetic diversity in the crop.

Acknowledgements

The author gratefully acknowledges the valuable suggestions of Garrison Wilkes, Robert McK. Bird, Takeo Angel Kato Yamakake and José de Jesus Sánchez González and G. Michael Listman on this section.

REFERENCES

Allen, J.O., G.K. Emenhiser and J.L. Kermicle. 1989. Miniature kernel and plant: Interaction between teosinte cytoplasmic genomes and maize nuclear genomes. Maydica 34:277-290.

Beadle, G.W. 1980. The ancestry of corn. Sci. Am. 242(1):112-119.

Benz, B.F. 1987. Racial systematics and the evolution of Mexican maize. Pp. 121-136 *in* Studies in the Neolithic and Urban Revolutions: The V. Gordon Childe Colloquium, Mexico 1986 (L. Manzanilla, ed.). BAR International Series 349, Oxford, England.

Benz, B.F. 1988. *In situ* conservation of the genus *Zea* in the Sierra de Manantlán Biosphere Reserve. Pp. 59-69 *in* Recent Advances in the Conservation and Utilization of Genetic Resources: Proceedings of the Global Maize Germplasm Workshop. CIMMYT, Mexico, D.F.

Benz, B.F., L.R. Sanchez-Velasquez and F.J. Santana Michel. 1990. Ecology and Ethnobotany of *Zea diploperennis*: Preliminary investigations. Maydica 35:85-98.

Bird, R. Mck. 1978. A name change for Central American teosinte. Taxon. 27:361-363.

Castro G., M. 1970. Frequencies of maize by teosinte crosses in a simulation of a natural association. Maize Gen. Coop. Newsl. 44:21-24.

CIMMYT. 1986. CIMMYT Research Highlights, 1985. Mexico, D.F.

Cohen, J.I. and W.C. Galinat. 1984. Potential use of alien germplasm for maize improvement. Crop. Sci. 24:1011-1015.

Collins, G.N. 1921. Teosinte in Mexico. J. Hered. 12:339-350.

Doebley, J.F. 1984. Maize introgression into teosinte - A reappraisal. Ann. Mo. Bot. Gard. 71:1100-1113.

Doebley, J.F. 1990a. Molecular systematics of *Zea* (Gramineae). 1990. Maydica 35:143-150.

Doebley, J.F. 1990b. Molecular evidence and evolution of maize. Econ. Bot. 44, Suppl. 3:6-27.

Doebley, J.F. and H.H. Iltis. 1980. Taxonomy of *Zea* (Gramineae). I. A subgeneric classification with key to taxa. Am. J. Bot. 67:982-993.

Doebley, J.F., M.M. Goodman and C.W. Stuber. 1984. Isoenzymatic variation in *Zea* (Gramineae). Syst. Bot. 9:203-218.

Doebley, J.F., M.M. Goodman and C.W. Stuber. 1987a. Patterns of variation between maize and Mexican annual teosinte. Econ. Bot. 41(2):234-246.

Doebley, J.F., W. Renfroe and A. Blanton. 1987b. Restriction site variation in the *Zea* chloroplast genome. Genetics 117:139-147.

Eubanks, M. 1995. A cross between two maize relatives: *Tripsacum dactyloides* and *Zea diploperennis* (Poaceae). Econ. Bot. 49(2):172-182.

Galinat, W.C. 1985. The missing links beween teosinte and maize: A review. Maydica 30:137-160.

Galinat, W.C. 1988. The origin of corn. Agronomy 18:1-31.

Galinat, W.C. 1992. Evolution of corn. Adv. Agron. 47:203-229.

Galinat, W.C. 1995. El Origen del maiz: El grano de la humanidad. The origin of maize: Grain of humanity. Econ. Bot. 49(1):3-12.

Galinat, W.C. and C.V. Pasupuleti. 1982. *Zea diploperennis*: II. A review on its significance and potential value for maize improvement. Maydica 27:213-220.

Goodman, M.M. 1988. The history and evolution of maize. CRC Crit. Rev. Plant Sci. 7(3):197-220.

Goodman, M.M., J.S.C. Smith, J.F. Doebley and C.W. Stuber. 1983. Races of teosinte show differential crossability with maize when maize is used as the female parent. Maize. Gen. Coop. Newsl. 57:130-131.

Guzman, R. 1978. Redescubrimiento de *Zea perennis* (Gramineae). Phytologia 38:17-27.

Iltis, H.H. 1987. Maize evolution and agricultural origins. Pp. 195-213 *in* Grass Systematics and Evolution (T. Soderstrom, K. Hilu, C. Campbell and M. Barkworth, eds.). Smithsonian Institute Press, Washington, DC.

Iltis, H.H. and J.F. Doebley. 1980. Taxonomy of *Zea* (Gramineae) II. Subspecific categories in the *Zea mays* complex and a generic synopsis. Am. J. Bot. 67:994-1004.

Iltis, H.H., J.F. Doebley, R. Guzman and B. Pazy. 1979. *Zea diploperennis* (Gramineae): A new teosinte from Mexico. Science 203:186-188.

Kato Y., T.A. 1976. Cytological studies of maize. Massachusetts Agricultural Experiment Station Research Bulletin. No. 635.

Kato Y., T.A. 1984. Chromosome morphology and the origin of maize and its races. Evol. Biol. 17:219-253.

Kato Y., T.A. and A. Lopez R. 1990. Chromosome knobs of the perennial teosintes. Maydica 35:125-141.

Kermicle, J.O. and J.O. Allen. 1990. Cross-incompatibility between maize and teosinte. Maydica 35:399-408.

Listman, G.M. 1994. Los Reyes teosinte: Going, going...gone! Tracking teosinte outside of Mexico City. Diversity 10(1):34-36.

Louie, R., J.K. Knoke and W.R. Findley. 1990. Elite maize germplasm: Reactions to Maize Dwarf Mosaic and Maize Chorotic Dwarf Viruses. Crop Sci. 30:1210-1215.

Mangelsdorf, P.C. 1986. The origin of corn. Sci. Am. 255(2):72-78.

Mazzoti, L.B. and R. Velasquez. 1962. Interacciones nucleo-citoplásmicas. Revista de la Facultad de Agronomía, Universidad Nacional de la Plata, Tomo XXXVIII.

Nault, L.R., D.T. Gordon, U.D. Domesteegtand and H.H. Iltis. 1982. Response of annual and perennial teosintes (*Zea*) to six maize viruses. Plant Dis. 66(1):61-62.

Norton, D.C., J. Edwards and P.N. Hinz. 1985. Nematode populations in maize and related species. Maydica 30:67-74.

Orozco, J.L. and T. Cervantes S. 1986. Relación entre poblaciones de teocintle anual Mexicano (*Zea mexicana* [Schrader] Kuntze). Agrociencia 64:215-235.

Razo L., A. 1989. Clasificación de razas de maíz y teocintle anual de México, según información de nudos cromosómicos. Facultad de Ciencias, Universidad Nacional Autónoma de México, México, D.F.

Sanchez G., J.J. and L. Ordaz S. 1987. El teocintle in México: Distribución actual de poblaciones. IBPGR, Systematic and ecogeographic studies on crop genepools no. 2.

Sanchez G., J.J., T.A. Kato Y., M. Aguilar S., J.M. Hernandez C. and A. Lopez R. 1995. Systematic and Ecogeographic Studies on Crop Genepools: Distribución y Caracterización del teocintle. Mexican National Institute of Forestry, Agriculture and Livestock Research (INIFAP) and Colegio de Postgraduados, Montecillo, Mexico, Mexico, D.F.

Sanchez-Velasquez, L.R. 1991. *Zea diploperennis*: Mejoramiento genético del maíz, ecología y la conservación de recursos naturales. Tiempos de ciencia 24 (julio-septiembre): 1-8. University of Guadalajara, Jalisco, Mexico.

Smith, J.S.C., M.M. Goodman and C.W. Stuber. 1984. Variation within teosinte. III. Numerical analysis of allozyme data. Econ. Bot. 38(1):97-113.

Smith, J.S.C., M.M. Goodman and C.W. Stuber. 1985. Relationships between maize and teosinte of Mexico and Guatemala: Numerical analysis of allozyme data. Econ. Bot. 39:12-24.

Timothy, D.H., C.S. Levings III, D.R. Pring, M.F. Conde and J.L. Kermicle. 1979. Organelle DNA variation and systematic relationships in the genus *Zea*: teosinte. Proc. Natl. Acad. Sci. USA 76:4220-4224.

Wellhausen, E.J., L.M. Roberts, E. Hernandez-Xolocotzi and P.C. Mangelsdorf. 1952. Races of Maize in Mexico. Cambridge, Massachusetts: The Bussey Institute, Harvard University.

Wilkes, H.G. 1967. Teosinte: The Closest Relative of Maize. Cambridge, Massachusetts: The Bussey Institute, Harvard University.

Wilkes, H.G. 1977. Hybridization of maize and teosinte in Mexico and Guatemala and the improvement of maize. Econ. Bot. 31:254-293.

Wilkes, H.G. 1986. Maize: Domestication, racial evolution and spread. *In* Plant Domestication and Early Agriculture. XI. World Archaeological Congress. Allen Unwin, London.

Wilkes, H.G. 1988. Teosinte and the other wild relatives of maize. Pp. 70-80 *in* Recent Advances in the Conservation and Utilization of Genetic Resources. Proceedings of the Global Maize Germplasm Workshop. CIMMYT, Mexico, D.F.

Wilkes, H.G. 1993. Conservation of maize crop relatives in Guatemala. Pp. 75-88 *in* Perspectives on Biodiversity: Case Studies of Genetic Resource Conservation and Development. AAAS Publication 93-10S. AAAS, Washington, DC.

Wilkes, H.G. and S. Taba. 1993. Teosinte in the valley of Toluca, Mexico. Maize Gen. Coop. Newsl. 67:21.

Chapter 17

Pearl Millet[1]

K.N. Rai, S. Appa Rao and K.N. Reddy

Pearl millet is a coarse grain cereal annually grown on the Indian subcontinent and in Africa, primarily for its grain and secondarily for its stover that is variously used as fuel, fencing and roofing material (Rachie and Majmudar 1980). In the USA, Australia and South Africa it is grown as a warm-season forage crop.

India accounts for 10.5 million ha out of the total area of 11.2 million ha annually planted to pearl millet on the Indian subcontinent. In western Africa, 17 countries grow pearl millet on 10.8 million ha, but the five major countries accounting for 90% of the total area are Niger (3.5 million ha), Nigeria (3.2 million ha), Burkina Faso (1.1 million ha), Mali (1.0 million ha) and Senegal (0.9 million ha).

BOTANY AND DISTRIBUTION

Pennisetum, the largest genera in the tribe Paniceae, consists of more than 140 species (Clayton 1972) and is divided into five sections: *Gymnothrix, Eu-Pennisetum, Penicillaria, Heterostachya* and *Brevivalvula*. The genus includes both annuals and perennials, sexual and asexual reproduction, and apomictic species. Species have large variation in chromosome number, $2n=10$ to $2n=72$, in multiples of 5, 7, 8 and 9 (Table 17.1). The lowest chromosome number ($2n=2x=10$) occurs in *P. ramosum* (Hochst.) Schmeinf. Those with $x=7$ chromosomes include cultivated pearl millet, its wild and weedy subspecies, *P. schweinfurthii* Pilger ($2n=2x=14$) and *P. purpureum* Schumach. ($2n=4x=28$). *Pennisetum massaicum* Stapf ($2n=16$ and 32) is the only known species with $x=8$. All other species have $x=9$. Pearl millet belongs to the section *Penicillaria*. Previously, the most widely accepted name for pearl millet was *Pennisetum typhoides* (Burm.) Stapf & Hubb. It was also referred to as *P. typhoideum* L. C. Rich. and *P. americanum* (L.) Leeke. Currently, the most accepted and correct name for pearl millet is *P. glaucum* (L.) R. Br. (de Wet 1987). Most of its wild relatives are found in Africa (Hanna 1987), but they also have been collected from Israel, Japan, India, Australia, New Zealand, Polynesia, Mexico and Central and South America.

Pennisetum germplasm has been classified into primary, secondary and tertiary genepools. The primary genepool includes all those taxa that can easily cross with the cultivated pearl millet, including all varieties of the cultivated *P. glaucum*, and those in its wild progenitor [*P. glaucum* subsp. *violaceum* (=*monodii* Maire)] and weedy form (*P. glaucum* subsp. *stenostachyum* Kloyzcsh ex. A. Br. and Bouche). The secondary genepool includes those species that also cross easily with the cultivated types, but do not produce fertile hybrids. Elephant or napiergrass (*P. purpureum*), a rhizomatous perennial, belongs to this genepool. It is an allotetraploid ($2n=2x=28$) with A'A'BB genomic

[1] Submitted as Journal Article no. 1897 by ICRISAT.

Table 17.1. Somatic chromosome number of some wild *Pennisetum* species, and number of accessions assembled at ICRISAT Asia Center, Patancheru, India.

Pennisetum species	Chromosome no.[†] (2*n*=)	Genepool	Accessions[‡] (no.)	Countries (no.)
glaucum				
subsp. *monodii* (=violaceum)	14	1	382	13
purpureum	28	2	16	4
ramosum	10	3	6	3
schweinfurthii	14	3	5	2
massaicum	16, 32	3	3	2
alopecuroides	18	3	1	1
atrichum	36	3	–	–
bambusiforme	36	3	–	–
basedowii	54	3	–	–
cattabasis	18	3	–	–
cenchroides	36	3	5	3
clandestinum	36	3	1	–
distachyum	36	3	–	–
divisum	36	3	2	1
flaccidum	18, 36	3	5	1
frutescens	36	3	–	–
hohenackeri	18	3	5	1
hordeoides	18	3	1	–
lanatum	18	3	1	–
latifolium	36	3	–	–
macrostachyum	54	3	1	–
macrourum	36	3	1	–
nervosum	36	3	–	–
nodiflorum	18	3	–	–
notarisii	36, 54	3	–	–
orientale	18, 36, 54	3	20	1
pedicellatum	36, 54	3	132	8
polystachyon	36, 54	3	79	12
pseudotrilicoides	18	3	–	–
schimpeii	18	3	–	–
setaceum	27	3	2	2
setosum	54	3	–	–
squamulatum	54	3	1	–
subangustum	36, 54	3	–	–
tempisquense	72	3	–	–
thunbergii	18	3	3	1
trisetum	36	3	–	–
villosum	18, 36, 54	3	1	–
Unknown species	–	–	10	–

[†] Jauhar 1981; Hanna 1987.

[‡] Does not include *P. glaucum* subsp. *stenostachyum*.

constitution in which the A′ genome is homologous or at least homeologous to the A genome of the primary genepool. Its hybrids with cultivated pearl millet have *n*=21 and are sterile. The tertiary genepool includes the remainder of the species, which either do not cross with cultivated pearl millet or require special techniques to cross.

On the basis of seed shape, Brunken *et al.* (1977) classified the world collection into four races: typhoides, nigritarum, globosum and leonis. Pearl millet is known by different names in different parts of the world. In Europe, it is known as bulrush millet and cattail millet. In India, it is known as *bajra, bajri, sajja, sajje, cumbu, ganti,* etc. In western Africa, its various types are called *souna, sanio, soumna, nara, nyali, dauro, gero* and *maiwa*; in southern Africa, *babala, nyauti, mausa, mahangu* and *munga*; and in Arab Africa and the Arabian Peninsula, *dokhn*.

Origin, Domestication and Diffusion

The greatest morphological diversity in pearl millet occurs in western Africa, south of the Sahara desert and north of the forest zone. Its likely wild progenitor (*monodii*) also occurs in the drier, northern part of this zone. Therefore, Harlan (1971) suggested that pearl millet has a diffused belt of origin, stretching from Senegal to western Sudan. The last wet phase in the Sahara Desert occurred 3000-5000 BC (Mann *et al.* 1983). As the most recent dry phase began and the rainfall patterns shifted from winter to summer, wheat and barley became unadapted (Clark 1962), and the inhabitants of the region were forced to experiment with local grasses (de Wet 1977). Pearl millet probably came from experimental crops that were domesticated 3000 years ago along lake edges in what is now the Sahara. Portères (1976) suggested that pearl millet is a product of multiple domestications: post-domestication differentiation arising from adaptation to varied agro-ecological environments, geneflow from the wild progenitor(s), and local selection by farmers for diverse traits and needs. On the basis of isozyme analysis of the cultivated pearl millet and its wild progenitor from western Africa, Tostain and Marchais (1989) provide evidence to support the theory of multiple domestications along the Sahelian distribution belt of the wild progenitor.

The western African origin of pearl millet is well established. Northern Senegal-Mali and southern Mauritania appear to be its original home 4000 years ago. About 3000 years ago, it reached eastern Africa, spread to India and reached southern Africa 2000 years ago.

Reproductive Biology

Pearl millet inflorescence takes 5-10 weeks from sowing to emerge through the flag-leaf sheath, depending on the genotype and growing conditions (mainly temperature and daylength). Pearl millet is protogynous, with stigma emergence preceding anther emergence. Stigmas emerge when the spikelets are mature, irrespective of the degree of inflorescence emergence (Rangaswami Ayyangar *et al.* 1933). Under warm weather conditions in Tifton, Georgia, USA, the highest number of anthers exserted around sunrise (Burton 1983). Post-harvest dormancy has been reported in pearl millet (Burton 1983). A large percentage of viable seeds remain dormant for 14 days after harvesting when grain moisture content is substantially low. However, pearl millet seeds have been reported to germinate even in the inflorescence of the standing crop under wet conditions.

The total growth and development of a pearl millet plant can be divided into three phases: GS_1 or vegetative phase (sowing to panicle initiation), GS_2 or reproductive phase (panicle initiation to flowering) and GS_3 or grain-filling phase (flowering to physiological maturity). The number of tillers initiated per plant under normal and extended daylength has been found to be the same, but those developing to maturity were considerably reduced under extended daylength conditions (Carberry and Campbell 1985). A small increase in GS_1 duration results in a significant increase in leaf area and total dry weight at flowering but not in productive tillers or grain yield (Alagarswamy and Bidinger 1985; Carberry and Campbell 1985). This could be related to continued stem growth throughout the GS_3 phase, in direct competition with the panicle growth (Craufurd and Bidinger 1988).

GERMPLASM CONSERVATION AND USE

By the end of 1995, ICRISAT had assembled 20 503 accessions of cultivated pearl millet from 47 countries and 688 accessions of its wild relatives from 23 countries (Table 17.2). This represents the largest collection of pearl millet assembled and conserved anywhere in the world. Among the wild species, *monodii* constitutes the bulk of germplasm, followed by *pedicellatum* Trin. and *polystachyon* (L.) Schult. (Table 17.1).

Table 17.2. Geographical sources and number of pearl millet germplasm assembled at ICRISAT Asia Center, Patancheru, India (as on 31 Dec 1995).

Country of origin	Cultivated	Wild relatives *monodii*	Others	Total
Africa				
Algeria	5	–	–	5
Benin	46	–	–	46
Botswana	82	–	–	82
Burkina Faso	863	4	1	868
Cameroon	916	3	79	998
Cape Verde Islands	2	–	–	2
Central African Rep.	146	–	10	156
Chad	101	32	3	136
Congo	8	–	–	8
Ethiopia	2	–	–	2
Gambia	15	–	–	15
Ghana	283	–	–	283
Kenya	98	–	–	98
Lesotho	–	–	4	4
Malawi	300	–	12	312
Mali	1069	87	22	1178
Mauritania	6	24	7	37
Morocco	4	–	–	4
Mozambique	31	–	2	33
Namibia	1118	–	8	1126
Niger	1092	139	39	1270
Nigeria	1906	8	3	1917
Senegal	403	12	–	415
Sierra Leone	59	–	1	60
Somalia	4	–	–	4
South Africa	162	–	–	162
Sudan	587	16	11	614
Tanzania	483	–	20	503
Togo	515	–	–	515
Tunisia	6	–	–	6
Uganda	123	–	1	124
Zaire	11	–	–	11
Zambia	154	–	3	157
Zimbabwe	1382	2	2	1386
Asia				
India	7655	2	118	7775
Korea (South)	1	–	–	1
Lebanon	108	–	–	108
Myanmar	10	–	–	10
Pakistan	160	–	–	160
Former USSR	16	–	–	16
Sri Lanka	–	–	2	2
Turkey	2	–	–	2
Yemen	290	–	–	290
Europe				
Germany	3	–	–	3
UK	31	–	1	32
Americas				
Brazil	2	–	–	2
Mexico	10	–	–	10
USA	225	5	5	235
Oceania				
Australia	8	–	–	8
Total	**20503**	**334**	**354**	**21191**

Several *Pennisetum* species have recently been collected from high elevations in the western Himalayas (Appa Rao *et al.* 1992). Sizeable collections are maintained at two other locations: ORSTOM, Bondy, France, and the Institut Sénégalais de Recherches Agricoles (ISRA), Bambey, Senegal. ORSTOM maintains over 2700 accessions, comprising cultivated pearl millet and its wild/weedy relatives. Other locations with small holdings include different locations in India and USA. About 500 accessions are conserved in Canada, 32 in Italy, 291 in Malawi, 36 in Pakistan, 32 in Tunisia and 24 in Zambia. The recently established Southern African Development Community (SADC) Plant Genetic Resources Center (SPGRC) near Lusaka, Zambia, has begun conserving germplasm from the SADC countries.

The pearl millet germplasm assembled at ICRISAT has been evaluated in batches of 1000-2000 accessions every year during 1974-93 rainy (June-Oct) and post-rainy (Nov-Apr) seasons in Alfisols at ICRISAT Asia Center (IAC), Patancheru (18°N). Various morphological traits related to yield, adaptation and consumers' preference as well as yield potential were recorded according to the descriptors for pearl millet (IBPGR and ICRISAT 1993). Although impressive, the collection represents the genetic resources of several countries poorly or not at all (Genetic Resources Division, ICRISAT 1995).

Cluster bagging is used to maintain pearl millet germplasm at IAC (Appa Rao 1980). In this method, about 120 plants of an accession are grown in a row. Five adjoining plants are used to form a cluster, leading to 24 clusters for an accession. One panicle of each plant within a cluster is enclosed with one large Kraft paper bag at the beginning of the panicle emergence. At harvest, an approximately equal quantity of seed of each panicle bagged in an accession is bulked to reconstitute the accession. Sibbing occurs only within a cluster, and there is also some degree of selfing in those panicles that flower first and last. Since it would be difficult to know what percentage of selfing occurs in different accessions, Burton (1985) suggested a selfed bulk method for maintaining pearl millet germplasm. At IAC, landraces are maintained by cluster bagging and inbred lines and genetic stocks by selfing. Four trait-specific genepools with respect to large panicle size, high tillering, large grain size and early maturity, have been developed at IAC (Table 17.3). A short-term storage room of 680 m^3, maintained at 18°C temperature and 30% RH, is used for temporary holding of seeds while they are dried and prepared for subsequent transfer to medium- and long-term storage facilities. Germplasm accessions of pearl millet and its wild relatives assembled at ICRISAT are conserved under appropriate medium-term storage conditions that meet international standards recommended by IPGRI (IBPGR 1985) to ensure maximum survival of conserved seed. One room with a capacity of 125 m^3, maintained at 4°C temperature and 20% RH, holds the active collection and conserves the seeds for as long as 15-20 years. Here seeds dried at 7-9% moisture content are stored in aluminium cans with screw caps that have rubber gaskets. The base collection is conserved in a room of 130 m^3 capacity, maintained at –20°C. It is expected to conserve seeds for more than 50 years (Mengesha *et al.* 1989). In this collection, seeds are dried to 4-6% moisture content and stored in hermetically sealed laminated aluminium foil packets to extend their viability. Good-quality seed is produced during the post-rainy season and harvested at least a week after physiological maturity as such seeds have been found to be the best in terms of viability retention during conservation (Rao *et al.* 1991).

A pearl millet landrace is highly heterogeneous because the species is allogamous. Nevertheless, the greatest variability is reflected in the phenotypic diversity among the landraces. An example of the strikingly large variability among landraces is given in Table 17.4. Large phenotypic diversity in pearl millet germplasm has been observed for several characters. Two such characters are panicle size and panicle shape (Fig. 17.1). Grain shape is another distinctive character. It is so unique to a particular landrace that it was used as the basis to classify pearl millet into different races (Brunken *et al.* 1977).

Table 17.3. Composition of four pearl millet genepools developed at ICRISAT Asia Center, Patancheru, India.

Genepool	Random matings (no.)	Accessions	
		Number	Origin
Early genepool (EGP)	6	1143	24 countries
High-tillering genepool (HTGP)	4	1093	28 countries
Large-grain genepool (LGGP)	4	887	19 countries
Large-panicle genepool (LPGP)	4	804	22 countries

Table 17.4. Diversity for quantitative characters in the world collection of pearl millet germplasm, evaluated at ICRISAT Asia Center, Patancheru, India.

Character[†]	Accessions		Range	
	No.	Mean±SE	Minim. (IP no.)	Maxim. (IP no.)
Days to flowering-R	16 259	75.4±0.18	33 (IP 4021)	159 (IP 11945)
Days to flowering-PR	16 124	71.2±0.09	32 (IP 7846)	138 (IP 7373)
Plant height-R (cm)	16 228	246.3±0.51	30 (IP 10402)	480 (IP 15537)
Plant height-PR (cm)	16 125	160.4±0.30	25 (IP 10401)	425 (IP 13016)
Total tillers-R	16 187	2.7±0.01	1 (IP 3407)	35 (IP 3110)
Productive tillers-R	16 115	2.1±0.01	1 (IP 3035)	19 (IP 3110)
Panicle exsertion-R (cm)	15 718	3.6±0.05	−45 (IP 9208)	29 (IP 4278)
Panicle length-R (cm)	16 209	28.1±0.09	5 (IP 12649)	120 (IP 12370)
Panicle length-PR (cm)	16 123	25.6±0.09	4 (IP 15625)	125 (IP 10379)
Panicle thickness-R (mm)	16 210	24.0±0.03	8 (IP 8128)	58 (IP 18583)
Panicle thickness-PR (mm)	16 125	23.3±0.04	9 (IP 10402)	61 (IP 14070)
1000-grain mass-PR (g)	16 408	8.6±0.01	1.5 (IP 15352)	21.3 (IP 11407)

[†] R=rainy season, PR=post-rainy season.

Useful variability in pearl millet germplasm has been found for several other characters. For instance, four sweet-stalk accessions have been identified that have 13.5-19.4% soluble sugar in stalks at flowering (similar to a normal control) and 8.5-11.9% at maturity (more than twice the normal control) (Appa Rao et al. 1982). Genetically enhanced lines from the germplasm have been produced that may have significant values in a breeding programme. For instance, some of these lines have high protein content (19.8% protein) (Singh et al. 1987) while others have high levels of seedling thermotolerance (Peacock et al. 1993). Many lines with high levels of resistance to downy mildew (Sclerospora graminicola (Sacc.) J. Schröt.), rust (Puccinia substriata Ell. & Barth var. indica Ramachar & Cumm.) (Singh 1990; Singh et al. 1990a, 1990b), smut (Tolyposporium penicillariae Bref.) (Thakur et al. 1992) and ergot (Claviceps fusiformis Loveless) (Thakur et al. 1993) also have been developed from accessions in this collection. Genes for photoperiod insensitivity (Singh et al. 1994) and brown midrib (Gupta 1995), several genes for dwarfism (Appa Rao et al. 1986) and leaf glossiness (Appa Rao et al. 1987), sources of cytoplasmic-nuclear male sterility (Appa Rao et al. 1989; Rai 1995), and useful marker traits of applied value (Appa Rao et al. 1988; Rai et al. 1995) also have been identified in the germplasm. Improved germplasm of pearl millet produced at ICRISAT includes about 50 composites of diverse origin and varying morphological characteristics (Rai and Anand Kumar 1994) and about 2500 inbred lines. It is likely that an intensive search in the germplasm combined with a system of multi-stage, selfed-progeny testing and recombination, as followed for the development of ergot resistance (Thakur et al. 1982), may lead to the production of stocks with enhanced levels of resistance to Striga, stem borer, head miner and drought.

The most conspicuous geographical pattern of diversity on both the Indian Subcontinent and in Africa relates to maturity. For instance, in western Africa, there is a north-south gradient, with the early maturing types (70-90 days) in the Sahelian zone in the north and the late-maturing types (120-180 days) in the Sudanian zone in the south. There is also a medium-maturing type (90-120 days) (Anand Kumar and Appa Rao 1987) that is grown in the southern part of the Sahelian zone and northern part of the Sudanian zone. The oasis millets of western and northern Africa and the desert types from Rajasthan and Gujarat states of India have extra-early maturity (60-65 days). The flowering of most of these types is least responsive to extended daylength. Other landraces, including Djane of Hoggar province of Algeria, Faya and Ligui of Chad, and Massue landraces from Mauritania also represent oasis pearl millet (Bono 1973; Anand Kumar and Appa Rao 1987). The greatest diversity in pearl millet lies in western Africa. A diversity analysis of pearl millet from western Africa showed that landraces from Niger and Senegal, characterized by early maturity and long panicle size (30-150 cm) were not distinct (Bilquez and Le Conte 1969; Zongo *et al.* 1988). Another study showed that regional patterns of variation were more marked in Mali than in Senegal (Marchais 1982). Two distinct groups were noticed: the Niafunke Timbouctou region and the Dogoun plateau region. Wilson *et al.* (1990) classified pearl millet landraces from central Burkina Faso into 10 clusters, which could be further put into three major geographical groups. Bono (1973) analyzed 11 characters of pearl millet from six countries of western Africa. Accessions from Senegal and Niger grouped together. The second group consisted of accessions from Mali, Ivory Coast, Mauritania and Burkina Faso. Within Niger, Ankoutess formed one cluster while Zongo, Haini-Kirei and Matam Hatchi formed another cluster. Geographical sources more likely to provide promising pearl millet germplasm for various characters related to yield, farmers' preference and disease resistance are given in Table 17.5.

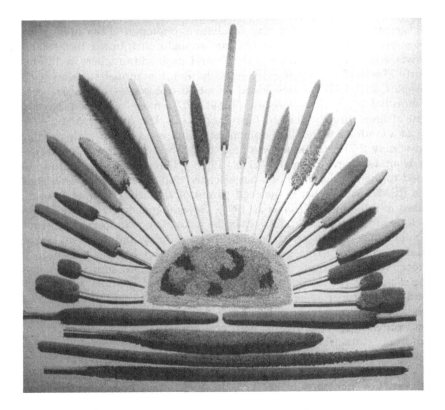

Fig. 17.1. Diversity for panicle size and shape in pearl millet.

Table 17.5. Geographical distribution of promising germplasm sources for various characters in pearl millet.

Character	Country of origin
Earliness	India, Pakistan, Botswana, Lebanon, Togo, Ghana, USA
Short height	Pakistan, Lebanon, Sierra Leone, USA
High productive tillering	Sierra Leone, Kenya, Benin, Uganda, Yemen, India
Long panicle	Niger, Nigeria, Namibia, South Africa, Tanzania, Zimbabwe
Thick panicle	Togo, Namibia, Ghana, Tanzania, Zimbabwe, Botswana
Large grain	Togo, Ghana, Benin, Burkina Faso, Yemen
Good exsertion	Togo, Ghana, India, Pakistan
Panicle shape	
Cylindrical	Malawi, Zambia, Mali, Central African Republic, Sudan
Candle	Sierra Leone, Pakistan, Namibia, Lebanon, Benin, Uganda, USA
Grain colour	
White	Lebanon, Cameroon, Ghana, Central African Republic
Yellow	Sudan, Burkina Faso, Central African Republic, Sierra Leone
Grain shape	
Globular	Benin, Burkina Faso, Togo, Sudan, Cameroon, Central African Rep.
Obovate	Malawi, Lebanon, Kenya, South Africa
Disease resistance	
Downy mildew	Nigeria, Niger, Mali, Burkina Faso, Senegal, Togo, Cameroon
Smut	Cameroon, Mali, Nigeria, Senegal
Ergot	India, Nigeria, Togo, Uganda

Breeding

Although the germplasm utilization story of pearl millet is similar to that of other crops in that only a small fraction has been touched, extensive use has been made of the cultivated germplasm, especially the facultative photoperiod-sensitive type. The Genetic Resources Division (GRD) at IAC has annually distributed 1000-9000 samples in most years since 1975 to researchers at ICRISAT itself and to others in different parts of the world (Table 17.6). During 3 years (1987-89), the annual supplies were about 12 700 samples. ICRISAT has developed about 50 composites of diverse origin and varying morphological characteristics (Table 17.7), most of which have a large component of lines directly derived from landraces. The most dramatic example of direct use of a landrace is the development of a large-seeded and high-yielding open-pollinated variety (ICTP 8203) that was bred at IAC by selection within a large-seeded accession from northern Togo (Rai *et al.* 1990). Direct selection within the same landrace led to the development of a large-seeded and downy mildew resistant male-sterile line (ICMA 88004) that is the seed parent of an early maturing hybrid (ICMH 356) released in India in 1993 (Rai *et al.* 1995). Several genetically enhanced lines have been developed from germplasm sources in the facultative photoperiod-sensitive group of pearl millet that confers resistance to downy mildew and rust, smut, ergot and seedling-heat tolerance. This group of germplasm also has been used to develop genetic stocks with novel traits (Mengesha *et al.* 1990) and new sources of male sterility (Appa Rao *et al.* 1989; Rai 1995).

The obligate photoperiod-sensitive group of pearl millet germplasm has been used less in breeding. A recent evaluation of this group of materials at IAC, however, shows that it may provide good sources of hard and vitreous grains, large seed size in the background of high-tillering plants with thin panicles, and new sources of high levels of resistance to downy mildew and rust. The utilization of this group of germplasm in ICRISAT's African regional programme has led recently to the release of two open-pollinated varieties (Lohani *et al.* 1995). Wild species have been least used because of the large amount of genetic variability available in pearl millet landraces and breeding materials. However, there have been a few targeted attempts and the results have been

Table 17.6. Dissemination of pearl millet germplasm[†] by ICRISAT Asia Center, Patancheru, India, 1974-95.

Year	ICRISAT	India	Asia and Oceania	Africa	Europe	Americas	Total
1974	–	75	–	–	–	–	75
1975	–	1179	–	–	–	–	1 179
1976	–	2327	–	–	–	–	2 327
1977	403	1518	–	154	4	12	2 091
1978	1079	2846	–	53	4	25	4 007
1979	4385	3141	23	1428	105	256	9 338
1980	2909	1359	172	68	33	80	4 621
1981	2466	732	18	2200	–	235	5 651
1982	1846	726	285	537	141	61	3 596
1983	854	580	267	96	144	7	1 948
1984	3317	186	22	3325	53	251	7 154
1985	2348	420	184	1555	225	75	4 807
1986	1349	5080	358	2108	88	90	9 073
1987	1672	8047	184	2643	–	182	12 728
1988	1420	8734	125	2063	215	220	12 777
1989	779	7902	75	3902	20	14	12 692
1990	814	2864	67	415	4	171	4 335
1991	1903	2366	–	890	174	–	5 333
1992	248	2820	7	43	6	–	3 124
1993	1498	5749	60	51	18	5	7 381
1994	684	1336	22	26	50	225	2 343
1995	2196	589	–	222	76	73	3 156
Total	32170	60576	1869	21779	1360	1982	119 736

[†] Number of germplasm samples distributed; excludes breeding lines, trials and nurseries supplied by the Genetic Enhancement Division of ICRISAT.

encouraging. For instance, the use of *P. glaucum* subsp. *monodii* led to the identification of new sources of cytoplasmic-nuclear male sterility (CMS) (Marchais and Pernes 1985; Hanna 1989). Hanna (1992) has summarized the use of this wild relative for new sources of resistance to leaf diseases, that of *P. purpureum* for sources of forage traits, stiff stalk and restorer genes of the A_1 CMS system, and that of *P. squamulatum* Fresen for apomictic gene(s).

Recurrent selection and pedigree breeding are extensively used in the periodic improvement of pearl millet. A multilocation-multiyear study of various recurrent selection cycle bulks of four composites showed genetic gains of 3.6-4.9% per cycle in three composites without any adverse changes in plant height and maturity (Rattunde and Witcombe 1993). About 50 composites of diverse origin and varying plant architecture (Table 17.7) have been developed/assembled by ICRISAT and subjected to varying cycles of recurrent selection (Rai and Anand Kumar 1994).

A moderate level of backcross breeding continues to be an integral part of the programme. Following a sidecar method with limited backcrossing, seven tall composites have been converted to dwarf types (Rai 1990). The Nigerian dwarf composite (NCD_2) derived from this programme has been found to be a useful source of seedling thermotolerance (Peacock *et al.* 1993). A recent development in breeding approach that appears to be picking up momentum is farmers' participation in the selection of potential cultivars at the early stages of their development. The extent of their participation may go as deep as the choice of parental lines for making crosses and even doing mass selection on-farm (Weltzien *et al.* 1995).

Both open-pollinated varieties (OPVs) and hybrids have been successfully produced and adopted for large-scale cultivation in India. WC-C75 is the first ICRISAT-bred open-pollinated variety, which was released in India in 1982 (Andrews *et al.* 1985).

Table 17.7.   Major features of some pearl millet composites developed/assembled at ICRISAT Asia Center, Patancheru, India.

Composite	Origin	Major features
I. ICRISAT Asia Center Composites		
African Population 88 (AfPop 88)	ICRISAT	Very long duration, tall height, long panicles, photoperiod-sensitive
African Population 90 (AfPop 90)	ICRISAT	Long duration, very long panicles, very vigorous
Bold Seeded Early Composite (BSEC)	ICRISAT	Short duration, medium height, large seed size
Dwarf Composite (D2C)	ICRISAT	Dwarf height, medium duration
Early B-Composite (EBC)	ICRISAT	Short duration, medium height, large seed size, maintainer of A_1 CMS
Early Composite (EC)	ICRISAT	Short duration, medium height, good tillering
Early Composite II (EC II)	ICRISAT	Mid-short duration, medium height
Early Composite 87 (EC 87)	ICRISAT	Short duration, medium height, large seed size, moderate tillering
Early Composite 89 (EC 89)	ICRISAT	Short duration, medium height, large seed size, moderate tillering
Early Composite 91 (EC 91)	ICRISAT	Short duration, medium height, long panicles, moderate tillering, smut resistance
Early Rajasthan Population 91 (ERajPop 91)	ICRISAT	Very short duration, high tillering, good adaptation to low-yield environments
Early Smut Resistant Composite II (ESRC II)	ICRISAT	Short duration, medium height, long panicles, smut resistance
Ergot Resistant Composite (ERC)	ICRISAT	Very long duration, tall height, long panicles, ergot resistance
Ex-Bornu (EB)	Kano, Nigeria	Long duration, tall height, long panicles
Extra-Early B-Composite (EEBC)	ICRISAT	Very short duration, medium height, large seed size, maintainer of A_m CMS
High Growth Rate Population (HiGroP)	ICRISAT	Long duration, medium height, high vegetative growth rate
High Head Volume B-Composite (HHVBC)	ICRISAT	Long duration, dwarf height, large seed size, maintainer of A_m CMS (20-30% plants maintainer of A_1 CMS)
High-Tillering B-Composite (HTBC)	ICRISAT	Medium duration, medium height, good tillering, maintainer of A_m CMS (20-30% plants maintainer of A_1 CMS)
High-Tillering Population 88 (HiTiP 88)	ICRISAT	High tillering, medium duration, compact thin panicles
High-Tillering Population 89 (HiTiP 89)	ICRISAT	High tillering, medium duration, compact thin panicles
ICRISAT Restorer Composite II (ICRC II)	ICRISAT	Medium duration, medium height, medium panicles, restorer for A_1 CMS
Intervarietal Composite (IVC)	ICRISAT	Medium duration, tall height, long panicles
Large Grain Population (LaGraP)	ICRISAT	Large grain size, medium height, mid-short duration
Medium Composite (MC)	ICRISAT	Medium duration, tall height, long panicles
Medium Composite 88 (MC 88)	ICRISAT	Medium duration, medium height, large seed size, moderate tillering
Medium Composite 91 (MC 91)	ICRISAT	Medium duration, medium height, large seed size
New Elite Composite (NELC)	ICRISAT	Medium duration, tall height, long panicles, medium-large seed size
Nigerian Composite (NC)	Samaru, Nigeria	Long duration, tall height, long panicles
Senegal Population (SenPop)	ICRISAT	Long duration, tall height, long panicles, high vegetative growth rate

Composite	Origin	Major features
Smut Resistant B-Composite (SRBC)	ICRISAT	Medium duration, medium height, smut resistance, maintainer of A, CMS
Smut Resistant Composite (SRC)	ICRISAT	Long duration, tall height, long panicles, smut resistance
Smut Resistant Composite II (SRC II)	ICRISAT	Medium duration, tall height, long panicles, smut resistance
Super Serere Composite (SSC)	ICRISAT	Medium duration, tall height, long panicles, large seed size
Western Rajasthan Population 88 (WRajPop 88)	ICRISAT	High basal and nodal tillering, small grain size, mid-short duration
World Composite (WC)	Samaru, Nigeria	Medium duration, tall height, long panicles
II. ICRISAT Sahelian Center Composites		
Early-Maturing Composite (EMC)	ISC[†]	Short duration, medium height, medium-long panicles, grey seeds
Intervarietal Composite (GRGB)	ISC	Medium duration, medium-tall height, medium-long panicles, grey and white seeds
INRAN/ICRISAT Intervarietal Composite-2 (INMG-2)	INRAN[‡]/ ICRISAT	Medium duration, tall height, medium-long panicles
INRAN/ICRISAT Intervarietal Composite-3 (INMG-3)	INRAN/ ICRISAT	Medium duration, medium-tall height, medium-long panicles
ISC Intervarietal Composite (ISC-851)	ISC	Medium duration, tall height, long panicles
Long Head Gene Pool (LHGP)	ISC	Medium duration, tall height, long panicles
Medium-Maturing Composite (MMC)	ISC	Medium duration, tall height, medium-long panicles
III. SADC/ICRISAT Composites		
Namibian Composite-90 (NC-90)	SADC[§]/ ICRISAT	Short duration, medium height, good tillering
SADC Bold Grain Composite (SDBGC)	SADC/ ICRISAT	Bold seeds, short duration, medium height
SADC Bristled Composite (SDBC)	SADC/ ICRISAT	
SADC Dwarf Composite (SDDC)	SADC/ ICRISAT	Dwarf height, medium to long duration, good tillering
SADC Early Composite (SDEC)	SADC/ ICRISAT	Short duration, medium to tall height, medium to long panicles, good tillering
SADC Late-Maturing Composite (SDLMC)	SADC/ ICRISAT	Long duration, tall height, long panicles
SADC Medium-Maturing Composite (SDMMC)	SADC/ ICRISAT	Medium duration, tall height, long panicles
SADC White Grain Composite (SDWGC)	SADC/ ICRISAT	White-seeded, early to medium duration, medium to tall height, medium to long panicles, good tillering
Tanzania SADC Late-Maturing Composite (TSLMC)	SADC/ ICRISAT	Very late and photoperiod-sensitive, tall height, long panicles

[†] ICRISAT Sahelian Center (Niger).
[‡] Institut national de recherches agronomiques du Niger (Niger).
[§] Southern African Development Community (Botswana).

About a dozen OPVs have been developed by ICRISAT (or by ICRISAT in collaboration with NARS) that have varying levels of adoption in India and other countries in western and southern Africa (Rai and Anand Kumar 1994).

Properties and Uses
Pearl millet grains have about 67% starch, 11% protein and 5% lipids (Table 17.8). Singh *et al.* (1987) reported 14.4-19.8% stable protein content in some of the high protein

lines compared with 9.2-10.5% in two commercial open-pollinated varieties.

Pearl millet is used as a summer annual crop for pasture in the USA, South America and Australia, and as fodder in Korea. Preliminary studies indicate that its forage is more succulent and has higher crude protein than sorghum and maize, with the other chemical constituents being comparable (Singh *et al.* 1974). Andrews and Kumar (1992) have summarized food uses of pearl millet grain. Blending of up to 20% pearl millet flour with wheat flour has been shown to be acceptable to consumers in Senegal and Sudan (Perten 1983). Fine milling of pearl millet flour, however, will be necessary for this purpose to avoid a gritty texture in the bread.

Pearl millet grain is also used to produce beer in Africa, and grain can be used in weaning food, baby food, snacks and bakery products. A review of results from feed experiments, comparing pearl millet with maize and sorghum (Bramel-Cox *et al.* 1995), indicates that the feed value of pearl millet is at least equivalent to maize and generally superior to sorghum because of its high protein content and quality, protein efficiency ratio and metabolizable energy.

Prospects

The potential to use germplasm for making further genetic advances is high and good opportunities exist for making pearl millet an important component of sustainable farming systems and diversified food uses. In the drylands of the Indian subcontinent and African regions, pearl millet will continue to be an important coarse grain cereal crop. Development of input infrastructures that can make cultivation of other crops more economical may lead to a decline in pearl millet area, but that may be a short-term scenario. As irrigation water resources are depleted, short-duration, drought-tolerant and water-use-efficient crops like pearl millet offer attractive options for sustainable agriculture. Drought tolerance of pearl millet has evolved over several millennia and further genetic improvement of this character remains the greatest challenge. A review of results related to improved management practices for higher and stable yields in Sahelian western Africa leads to the conclusion that inherent low soil fertility and limited, untimely cultural operations are often more limiting to production than drought (Fussell *et al.* 1987). Higher soil fertility and improved cultural practices improve the water-use efficiency of pearl millet.

To halt the encroachment of marginal dryland and environmental degradation in Africa, and reduce the pressure on land on the Indian subcontinent, future pearl millet production increases will have to come from productivity enhancement. Large-scale cultivation of improved cultivars, especially single-cross hybrids, would result in a downy mildew challenge, as observed in India (Dave 1987). Numerous resistance sources and screening/breeding techniques should enable effective control of this disease through host-plant resistance. New approaches will be required to assess whether adequate resistance levels for *Striga*, stem borer and head miner are available in the germplasm. In the short term, integrated management appears to be the best option for these three biotic production constraints. Pearl millet can become an important component of intensive agriculture in both traditional and non-traditional environments. It is being evaluated for its potential in the double-cropping system after wheat in the central Midwest of the USA (Andrews *et al.* 1993). With the availability of major genes and other sources responsible for photoperiod-insensitive early flowering (33-38 days), it is possible to breed extra-short-duration cultivars (60-65 days) with moderate grain yield potential (3 t/ha) that can be grown as a catch crop.

Production needs to be increased to feed a growing human population in the traditional pearl millet growing areas of Asia and Africa. Pearl millet is nutritionally superior to sorghum and maize and comparable to wheat and rice. Advances in grain processing and improved keeping quality of flour will substantially increase its utilization in traditional and non-traditional food preparations. Considering the

Table 17.8. Chemical composition of pearl millet grain.

Chemical constituent	No. of genotypes	Mean	Range
Major constituent (%)			
Starch	44	66.7	62.8-70.5
Amylose	44	25.9	21.9-28.8
Soluble sugars	36	2.1	1.4-2.6
Reducing sugars	136	0.17	0.1-0.26
Crude fibre	36	1.3	1.1-1.8
Protein	20 704	10.6	5.8-20.9
Lipids	36	5.1	4.1-6.4
Ash	36	1.9	1.1-2.5
Minerals and trace element (mg per 100 g of grain)			
Potassium	20	370	294-460
Phosphorus	20	260	185-363
Magnesium	20	106	46-128
Calcium	20	38	13-52
Iron	20	16.9	4.0-58.1
Zinc	20	4	1.0-6.6
Copper	20	7.9	0.6-21.2
Manganese	20	1.5	0.2-1.8

Source: Jambunathan and Subramanian 1988.

relative feed value of pearl millet compared with that of maize and sorghum for cattle, swine and poultry, and the increasing demand for livestock products with the rising incomes, pearl millet as a feed grain has potential for international trade.

Limitations

Pearl millet is relatively better adapted than maize (*Zea mays* L.) and sorghum (*Sorghum bicolor* (L.) Moench) to marginal agro-ecological environments with a multitude of abiotic stress factors. These factors, along with biotic constraints, negligible use of purchased management inputs and lack of national pricing policies, contribute to a perpetual low grain yield of pearl millet, around 600 kg/ha in the traditional farming system.

Pearl millet is a C_4 species with a very high photosynthetic efficiency and biomass production potential. Apart from numerous production constraints, the plant architecture of traditional cultivars with a harvest index (HI) of 15-20% is a major reason for its low productivity. Pearl millet is endowed with enormous genetic variability for both grain and forage yield components, aspects of plant architecture that may be useful to breed improved cultivars with >40% HI and traits related to adaptation to biotic and abiotic stress factors, and better nutrition and consumers' preference.

REFERENCES

Alagarswamy, G. and F.R. Bidinger. 1985. The influence of extended vegetative development and d_2 dwarfing gene in increasing grain number per panicle and grain yield in pearl millet. Field Crops Res. 11:265-269.

Anand Kumar, K. and S. Appa Rao. 1987. Diversity and utilization of pearl millet germplasm. Pp. 69-82 *in* Proceedings of the International Pearl Millet Workshop, 7-11 Apr 1986, ICRISAT Center, India (J.R. Witcombe and S.R. Beckerman, eds.). ICRISAT, Patancheru, India.

Andrews, D.J. and K.A. Kumar. 1992. Pearl millet for food, feed, and forage. Adv. Agron. 48:89-139.

Andrews, D.J., S.C. Gupta and P. Singh. 1985. Registration of WC-C75 pearl millet. Crop Sci. 25:199-200.

Andrews, D.J., J.F. Rajewski and K.A. Kumar. 1993. Pearl millet: a new feed grain crop. Pp. 198-208 *in* New Crops (J. Janick and J. Simons, eds.). John Wiley & Sons, New York, USA.

Appa Rao, S. 1980. Progress and problems of pearl millet germplasm maintenance. Pp. 279-282 *in*

Trends in Genetical Research on *Pennisetum* (V.P. Gupta and J.L. Minocha, eds.). Punjab Agricultural University, Ludhiana, India.

Appa Rao, S., P.N. Mathur and M.H. Mengesha. 1992. Collection of germplasm from western Himalayas. Indian J. Plant Genet. Resour. 5:1-6.

Appa Rao, S., M.H. Mengesha and C. Rajagopal Reddy. 1986. New sources of dwarfing genes in pearl millet (*Pennisetum americanum*). Theor. Appl. Genet. 73:170-174.

Appa Rao, S., M.H. Mengesha and C. Rajagopal Reddy. 1987. Glossy genes in pearl millet. J. Hered. 78:333-335.

Appa Rao, S., M.H. Mengesha and C. Rajagopal Reddy. 1988. Inheritance and linkage relationships of qualitative characters in pearl millet (*Pennisetum glaucum*). Indian J. Agric. Sci. 58:840-843.

Appa Rao, S., M.H. Mengesha and C. Rajagopal Reddy. 1989. Development of cytoplasmic male-sterile lines of pearl millet from Ghana and Botswana germplasm. Pp. 817-823 *in* Perspectives in Cytology and Genetics (G.K. Manna and U. Sinha, eds.). All India Congress of Cytology and Genetics, Kalyani, India.

Appa Rao, S., M.H. Mengesha and V. Subramanian. 1982. Collection and preliminary evaluation of sweet-stalk pearl millet (*Pennisetum*). Econ. Bot. 36:286-290.

Bilquez, A.F. and J. Le Conte. 1969. Relations entre mils sauvages et mils cultivés: étude de l'hybride *Pennisetum typhoides* Stapf & Hubb. x *Pennisetum violaceum* L. (Rich). Agron. Trop. 24:249-257.

Bono, M. 1973. Contribution a la morpho-systématique des *Pennisetum* annuels cultivés pour leur grain en Afrique occidentale francophone. Agron. Trop. 28:229-355.

Bramel-Cox, P.J., K. Anand Kumar, J.D. Hancock and D.J. Andrews. 1995. Sorghum and millets for forage and feed. Pp 325-363 *in* Sorghum and Millets: Chemistry and Technology (D.A.V. Dendy, ed.). American Association of Cereals Chemists, St. Paul, Minnesota, USA.

Brunken, J., J.M.J. de Wet and J.R. Harlan. 1977. The morphology and domestication of pearl millet. Econ. Bot. 31:163-174.

Burton, G.W. 1983. Breeding pearl millet. Plant Breed. Rev. 1:162-182.

Burton, G.W. 1985. Collection, evaluation and storage of pearl millet germplasm. Field Crops Res. 11:123-129.

Carberry P.S. and L.C. Campbell. 1985. The growth and development of pearl millet as affected by photoperiod. Field Crops Res. 11:207-217.

Clark, J.D. 1962. The spread of food production in sub-Saharan Africa. J. Afr. Hist. 3:211-228.

Clayton, W.D. 1972. Gramineae. Pp. 349-512 *in* Flora of West Tropical Africa (2nd edn.), Vol.III. Part 2 (J. Hutchinson and J.M. Dalziel, eds. and F.N. Hepper, rev.). Crown Agents for Oversea Governments and Administrations, London, UK.

Craufurd, P.O. and F.R. Bidinger. 1988. Effects of the duration of the vegetative phase on shoot growth, development and yield in pearl millet (*Pennisetum americanum* (L.) Leeke. J. Exp. Bot. 39:124-139.

Dave, H.R. 1987. Pearl millet hybrids. Pp. 121-126 *in* Proceedings of the International Pearl Millet Workshop, 7-11 Apr 1986, ICRISAT Center, India (J.R. Witcombe and S.R. Beckerman, eds.). ICRISAT Patancheru, India.

de Wet, J.M.J. 1977. Domestication of African cereals. Afr. Econ. Hist. 3:15-32.

de Wet, J.M.J. 1987. Pearl millet (*Pennisetum glaucum*) in Africa and India. Pp. 3-4 *in* Proceedings of the International Pearl Millet Workshop, 7-11 Apr 1986, ICRISAT Center, India (J.R. Witcombe and S.R. Beckerman, eds.). ICRISAT, Patancheru, India.

Fussell, L.K., P.G. Serafini, A. Bationo and M.C. Klaij. 1987. Management practices to increase yield and yield stability of pearl millet in Africa. Pp. 255-268 *in* Proceedings of the International Pearl Millet Workshop, 7-11 Apr 1986, ICRISAT Center, India (J.R. Witcombe and S.R. Beckerman, eds.). ICRISAT, Patancheru, India.

Genetic Resources Division, ICRISAT, 1995. Annual Report 1994. Patancheru 502 324, Andhra Pradesh, India: Genetic Resources Division, International Crops Research Institute for the Semi-Arid Tropics. (Semi-formal publication).

Gupta, S.C. 1995. Inheritance and allelic study of brown midrib trait in pearl millet. J. Hered. 86:301-303.

Hanna, W.W. 1987. Utilization of wild relatives of pearl millet. Pp. 33-42 *in* Proceedings of the International Pearl Millet Workshop, 7-11 Apr 1986, ICRISAT Center, India (J.R. Witcombe and S.R. Beckerman, eds.). ICRISAT, Patancheru, India.

Hanna, W.W. 1989. Characteristics and stability of a new cytoplasmic-nuclear male-sterile source in pearl millet. Crop Sci. 29:1457-1459.

Hanna, W.W. 1992. Utilization of germplasm from wild species. Pp. 251-257 *in* Desertified Grasslands: Their Biology and Management. Academic Press, London, UK.

Harlan, J.R. 1971. Agricultural origins: centers and non-centers. Science 14:468-474.

IBPGR and ICRISAT. 1993. Descriptors for pearl millet (*Pennisetum glaucum* (L.) R. Br.). IBPGR, Rome,

Italy and ICRISAT, Patancheru, India.

IBPGR. 1985. Documentation of Genetic Resources: Information Handling Systems for Genebank Management (J. Konopka and J. Hanson, eds.). IBPGR, Rome, Italy.

Jambunathan, R. and V. Subramanian. 1988. Grain quality and utilization of sorghum and pearl millet. Pp. 133-139 *in* Biotechnology in Tropical Crop Improvement: Proceedings of the International Biotechnology Workshop, 12-15 Jan 1987, ICRISAT Center, India. ICRISAT, Patancheru, India.

Jauhar, P.P. 1981. Cytogenetics and Breeding of Pearl Millet and Related Species. Alan R. Liss, Inc., New York.

Lohani, S.N., P. Sereme and R. Zangre. 1995. Registration of IKMP1 and IKMP2 pearl millet. Crop Sci. 35:590-591.

Mann, J.A., C.T. Kimber and F.R. Miller, 1983. The origin and early cultivation of sorghum in Africa. Tex. Agric. Exp. St. Bull. 1454.

Marchais, L. 1982. La diversité phénotypique des mils penicillaires cultivés au Sénégal et au Mali. Agron. Trop. 37:68-80.

Marchais, L. and J. Pernes. 1985. Genetic divergence between wild and cultivated pearl millet (*Pennisetum typhoides*). I. Male sterility. Z. Pflanzenzücht. 95:103-112.

Mengesha, M.H., S. Appa Rao and C. Rajagopal Reddy. 1990. Characteristics and potential uses of some genetic stocks of pearl millet in the world collection. Indian J. Plant Genet. Resour. 3:1-7.

Mengesha, M.H., P.P. Khanna, K.P.S. Chandel and N.K. Rao. 1989. Conservation of world germplasm collections of ICRISAT mandate crops. Pp. 65-69 *in* Collaboration on Genetic Resources: Summary Proceedings of a Joint ICRISAT/NBPGR (ICAR) Workshop on Germplasm Exploration and Evaluation in India, 14-15 Nov 1988, ICRISAT Center, India. ICRISAT, Patancheru, India.

Peacock, J.M., P. Soman, R. Jayachandran, A.U. Rani, C.J. Howarth and A. Thomas. 1993. Effects of high soil surface temperature on seedling survival in pearl millet. Exp. Agric. 29:215-225.

Perten, H. 1983. Practical experience in processing and use of millet and sorghum in Senegal and Sudan. Cereal Foods World 28:680-683.

Portères, R. 1976. African cereals: *Eleusine*, fonio, black fonio, teff, *Brachiaria*, *Paspalum*, *Pennisetum* and African rice. Pp. 409-452 *in* Origins of African Plant Domestication (J.R. Harlan, J.M.J. de Wet and A.B.L. Stemler, eds.). Mouton, The Hague, Netherlands.

Rachie, K.O. and J.V. Majmudar. 1980. Pearl Millet. Pennsylvania University Press, University Park, USA.

Rai, K.N. 1990. Development of high-yielding dwarf composites of pearl millet. Crop Improv. 17:96-103.

Rai, K.N. 1995. A new cytoplasmic-nuclear male sterility system in pearl millet. Plant Breeding 114:445-447.

Rai, K.N. and K. Anand Kumar. 1994. Pearl millet improvement at ICRISAT - an update. Int. Sorghum and Millets Newsl. 35:1-29.

Rai, K.N., A.S. Rao and C.T. Hash. 1995. Registration of pearl millet parental lines ICMA 88004 and ICMB 88004. Crop Sci. 35:1242.

Rai, K.N., K. Anand Kumar, D.J. Andrews, A.S. Rao, A.G.B. Raj and J.R. Witcombe. 1990. Registration of ICTP 8203 pearl millet. Crop Sci. 30:959.

Rangaswami Ayyangar, G.N., C. Vijiaraghavan and V.G. Pillai. 1933. Studies on *Pennisetum typhoides* (Rich.). The pearl millet. Part I. Anthesis. Indian J. Agric. Sci. 3:688-694.

Rao, N.K., S. Appa Rao, M.H. Mengesha and R.H. Ellis. 1991. Longevity of pearl millet (*Pennisetum glaucum*) seeds harvested at different stages of maturity. Ann. Appl. Biol. 119:97-103.

Rattunde, H.F. and J.R. Witcombe. 1993. Recurrent selection for increased grain yield and resistance to downy mildew in pearl millet. Plant Breeding 110:63-72.

Singh, J., T.R. Raju, L.L. Relwani, A. Kumar and A.K. Mehta. 1974. Forage yields and chemical composition of different strains of pearl millet. Indian J. Agric. Res. 8:179-184.

Singh, P., U. Singh, B.O. Eggum, K.A. Kumar and D.J. Andrews. 1987. Nutritional evaluation of high protein genotypes of pearl millet (*Pennisetum americanum*) (L.) Leeke). J. Sci. Food Agric. 38:41-48.

Singh, S.D. 1990. Sources of resistance to downy mildew and rust in pearl millet. Plant Dis. 74:871-874.

Singh, S.D., G. Alagarswamy, B.S. Talukdar and C.T. Hash. 1994. Registration of ICML 22 photoperiod insensitive, downy mildew resistant pearl millet germplasm. Crop Sci. 34:1421.

Singh, S.D., S.B. King and P. Malla Reddy. 1990a. Registration of five pearl millet germplasm sources with stable resistance to downy mildew. Crop Sci. 30:1164.

Singh, S.D., S.B. King and P. Malla Reddy. 1990b. Registration of five pearl millet germplasm sources with stable resistance to rust. Crop Sci. 30:1165.

Thakur, R.P., K.N. Rai, S.B. King and V.P. Rao. 1993. Identification and utilization of ergot resistance in pearl millet. Research Bulletin no. 17. ICRISAT, Patancheru, India.

Thakur, R.P., S.B. King, K.N. Rai and V.P. Rao. 1992. Identification and utilization of smut resistance in pearl millet. Research Bulletin no. 16. ICRISAT, Patancheru, India.

Thakur, R.P., R.J. Williams and V.P. Rao. 1982. Development of ergot resistance in pearl millet. Phytopathology 72:406-408.

Tostain, S. and L. Marchais. 1989. Enzyme diversity in pearl millet (*Pennisetum glaucum*). 2. Africa and India. Theor. Appl. Genet. 77:634-640.

Weltzien, R.E., M.L. Whitaker and M.M. Anders. 1995. Farmer participation in pearl millet breeding for marginal environments. Presented at the IDRC/IPGRI/FAO (ICCPPGR) Workshop on Participatory Breeding Approaches, 26-28 June 1995, Wageningen, Netherlands.

Wilson, J.P., G.W. Burton, J.D. Zongo and I.O. Dicko. 1990. Diversity among pearl millet landraces collected in central Burkina Faso. Crop Sci. 30:40-43.

Zongo, J.D., M.C. Sedogo, P. Sereme and G.R. Zangre. 1988. Synthèse des prospections du mil [*Pennisetum glaucum* (L.) R. Br.] au Burkina Faso. Pp. 121-131 *in* Proceedings of Regional Pearl Millet Improvement Workshop (L.K. Fussell and J. Werder, eds.). Institute for Agricultural Research, Samaru, Nigeria and ICRISAT Sahelian Center, Niamey, Niger.

Chapter 18

Small Millets[1]

K.E. Prasada Rao and J.M.J. de Wet

Small millets are small-grained cereals mainly grown in arid, semi-arid or montane zones as rain-fed crops under marginal and submarginal conditions of soil fertility and moisture. Small millets are important to global agriculture and are major cereal crops, grown in fairly large areas of South Asia, China, the former USSR and Africa. They are also found in areas of the United States and Europe on a limited scale. Although precise estimates on their area and production are not available, these crops may occupy between 18 and 20 million ha, producing 15-18 million tonnes of grain. The region-wise distribution of area is 6.3 million ha in South Asia, 5 million ha in China, 4 million ha in USSR and 3 million ha in Africa.

Finger millet is the principal small millet species grown in South Asia, followed by kodo millet, foxtail millet, little millet, proso millet and barnyard millet, in that order. Foxtail millet and proso millet are important in China and the latter is grown extensively in southwestern USSR. In Africa, finger millet, teff and fonio have local importance (Riley 1988).

The average global productivity of small millets is almost 1 t/ha. There has been a trend in the last two decades to replace these crops with major cereals like maize and wheat, which has been a factor in the reduction of area under these crops. Presently, small millets are cultivated in areas where they produce a more dependable harvest than any other crop. This has been largely responsible for their continued presence and cultivation in many parts of the world. There is now an increasing realization of this fact, and a greater awareness that these crops merit more research and development.

BOTANY AND DISTRIBUTION

All the crop species of small millets (also called minor millets) belong to the family Poaceae (Gramineae). Of them, the genera *Setaria, Panicum, Paspalum, Echinochloa* and *Digitaria* are classified under the tribe Paniceae of subfamily Panicoideae, and the other two genera *Eleusine* and *Eragrostis* under the tribe Eragrostideae of subfamily Chloridoideae (Prasada Rao and Mengesha 1988). Doggett (1989) and this chapter include the following crop species under small millets: Finger millet, *Eleusine coracana* (L.) Gaertn.; Foxtail millet, *Setaria italica* (L.) Beauv.; Proso millet, *Panicum miliaceum* L.; Little millet, *Panicum sumatrense* Roth. ex Roem. & Schult.; Barnyard millet, *Echinochloa crus-galli* (L.) Beauv.; *Echinochloa colona* (L.) Link; Kodo millet, *Paspalum scrobiculatum* L.; Teff, *Eragrostis tef* (Zucc.) Trotter; Fonio millet, *Digitaria exilis* Stapf and *Digitaria iburua* Stapf.

---

[1] Submitted as Journal Article no. 1894 by ICRISAT.

Finger millet is an old tropical cereal widely grown in eastern Africa, southern Africa and South Asia. Foxtail, proso and barnyard millets have been important, especially in Asia. Kodo and little millets continue to be important in Asia in times of famine.

Origin, Domestication and Diffusion

The most important plant of this group is finger millet, which was domesticated in Africa, probably in the Ethiopian region. It was introduced to India perhaps more than 3000 years ago (Doggett 1989). It is a tropical crop, grown from sea level to 3000 m asl. It is the most widely grown small millet in India and Africa, and can be very productive. In Africa, finger millet is grown abundantly in the Rift Valley region: Uganda, Kenya, Tanzania, Zaire, Rwanda, Burundi, Malawi, Zambia and Mozambique. Significant amounts are also grown in Zimbabwe. Cultivated finger millet is *Eleusine coracana* (L.) Gaertn. subsp. *coracana*. The closest wild relative is *Eleusine coracana* subsp. *africana* (Kenn.-O'Byrne) Hilu & de Wet. Wild finger millet (subsp. *africana*) is native to Africa but was introduced as a weed to the warmer parts of Asia and America. Derivatives of hybrids between subsp. *coracana* and subsp. *africana* are companion weeds of the crop in Africa. *Eleusine indica* (L.) Gaertn. is a diploid ($2n=18$) and *Eleusine coracana* subsp. *africana* and subsp. *coracana* are tetraploid ($2n=36$). The weedy *E. indica* is genetically isolated from finger millet (de Wet 1995).

Eleusine coracana and *E. africana* have a basic chromosome number of nine ($2n=4x=36$), regularly show the formation of 18 bivalents during meiosis, and apparently are allotetraploid in origin (Hiremath and Chennaveeraiah 1982). Cultivated finger millets are divided into races and subraces on the basis of inflorescence morphology (Table 18.1).

Eleusine coracana is predominantly self-fertilized. The subspecies *africana*, however, does cross occasionally with subsp. *coracana* to produce fully fertile hybrids. Derivatives of such crosses are aggressive colonizers and are grouped under the race spontanea (de Wet *et al.* 1984b).

Members of the race compacta are commonly referred to as cockscomb finger millets in both Africa and India. The race plana is characterized by large spikelets (8-15 mm long) that are arranged in two more or less even rows along the rachis, giving the inflorescence branch a flat ribbon-like appearance. The race vulgaris is the common finger millet of Africa and Asia (Fig. 18.1a). Four subraces are recognized on the basis of inflorescence morphology.

Foxtail millet (*Setaria italica*) is grown as a cereal in southern Europe and in temperate, subtropical and tropical Asia. Its closest wild relative is *S. italica* subsp. *viridis* ($2n=18$), the spontaneous green foxtail, extensively variable, widely distributed in temperate Eurasia, and extensively naturalized as a weed in the temperate parts of the New World. Foxtail millet was domesticated in the highlands of central China; remains of cultivated foxtail are known from the Yang-Shao culture dating back some 5000 years. Comparative morphology suggests that foxtail millet spread to Europe and India as a cereal soon after its domestication (Prasada Rao *et al.* 1987). Foxtail millet was widely cultivated across southern Eurasia until the early 20th century. It remains an important cereal in India and China (Naciri and Belliard 1987). China is the principal producer of foxtail millet today. Comparative morphological study of the accessions resulted in the recognition of the races and subraces presented in Table 18.2.

Setaria pumila (Poir.) Roem. & Schult (yellow foxtail, $2n=36$) is occasionally sown across southern India. Cultivated kinds differ from wild *S. pumila*, which naturally colonizes cultivated fields, primarily by efficient natural seed dispersal. Large inflorescences, colonizing ability of grains and ease of harvesting make this species a favourite wild cereal across southern Asia (Bor 1960).

Table 18.1. Races and subraces of finger millet (*Eleusine coracana*).

Species	Subspecies	Race	Subrace
E. coracana	africana	africana	–
	africana	spontanea	–
	coracana	elongata	laxa
	coracana	elongata	reclusa
	coracana	elongata	sparsa
	coracana	plana	seriata
	coracana	plana	confundere
	coracana	plana	grandigluma
	coracana	compacta	–
	coracana	vulgaris	liliacea
	coracana	vulgaris	stellata
	coracana	vulgaris	incurvata
	coracana	vulgaris	digitata

Source: Prasada Rao *et al.* 1993.

Table 18.2. Races and subraces of foxtail millet (*Setaria italica*).

Species	Subspecies	Race	Subrace
S. pumila	–	–	–
S. italica	viridis	–	–
	italica	moharia	aristata
	italica	moharia	fusiformis
	italica	moharia	glabra
	italica	maxima	compacta
	italica	maxima	spongiosa
	italica	maxima	assamense
	italica	indica	erecta
	italica	indica	glabra
	italica	indica	nana
	italica	indica	profusa

Source: Prasada Rao *et al.* 1993.

Study of accessions of *Setaria italica* subsp. *italica* ($2n=18$) suggests three morphologically distinct races (Prasada Rao *et al.* 1987): race moharia is centred in Europe and southwest Asia, maxima in former Transcaucasian Russia and the Far East, and indica in India and the rest of southern Asia. Grammie (1911) suggested, on the basis of comparative morphology, that Indian cultivars were derived from primitive Manchurian foxtail millets. Cultivars of the race moharia often resemble members of wild subsp. *viridis* (green foxtail) in phenotype, except that they have lost the ability for natural seed dispersal.

Extensively variable, the race maxima is characterized by spikelets closely arranged on elongated lateral branches, giving the inflorescence a lobed appearance (Fig. 18.1b). The race was introduced into the USA, where it is grown for bird feed, and also occurs in Nepal and Assam, along the southern foothills of the Himalayas, and in Georgia, former Russia (Prasada Rao *et al.* 1987).

The race indica is cultivated in many parts of India and is extensively variable. It was probably derived from a combination of moharia cultivars from southwest Asia and maxima cultivars from China. Plants in which the bristles bear a spikelet at the tip occur across the range of foxtail millet cultivation in India (Prasada Rao *et al.* 1987).

The botanical name of proso millet (*Panicum miliaceum*) is derived from the Latin word *millium* which means millet (Rangaswamy Ayyangar 1938). The chromosome numbers have been reported to be $2n=36, 72$ in some of the Indian races (Bor 1960). In the former USSR, this millet is known by the name *proso*. The common English names of this millet are common millet, hog millet and broom corn millet. The progenitor of proso millet is native to Manchuria. The species was introduced into Europe as a cereal

Fig. 18.1.   (a) Finger millet, mostly grown in India and Africa (race vulgaris, subrace digitata); (b) foxtail millet (race maxima); (c) little millet, mostly grown in the hilly Eastern Ghats region of India (race robusta); (d) barnyard millet, grown in India (race robusta); (e) Kodo millet, cultivated in India (race irregularis).

at least 3000 years ago. The bran or seed coat is always creamy white (Martin and Leonard 1959). The comparative morphological study of the accessions suggests the recognition of five cultivated races (Table 18.3).

Little millet (*Panicum sumatrense*) is widely cultivated as a cereal across India, Nepal and western Myanmar. The chromosome number of *P. sumatrense* (previously named *P. miliare*) has been reported to be 2n=36 (Bor 1960). It is particularly grown in the Eastern Ghats of India, where it forms an important part of tribal agriculture. The species is divided into subsp. *sumatrense*, cultivated little millet and subsp. *psilopodium* (Trin.) de Wet comb. nov., its wild progenitor. These two subspecies cross where they are sympatric to produce fertile hybrids, derivatives of which are often weedy in little millet fields (de Wet *et al.* 1984a).

Panicum sumatrense subsp. *sumatrense* includes a morphologically variable cultivated complex that is widely cultivated in Sri Lanka, India, Nepal and Myanmar. The comparative morphological study of the accessions resulted in the recognition of the races and subraces given in Table 18.4. The race robusta includes plants that are erect, or produce erect flowering culms from a short, geniculate base. Flowering culms are 120-190 cm tall and robust. Terminal inflorescences are 20-46 cm long, open or compact and strongly branched (Fig. 18.1c).

Two species of *Echinochloa* are grown as cereals. *Echinochloa crus-galli* is native to temperate Eurasia and was domesticated in Japan around 4000 years ago. *Echinochloa colona* (L.) Link is widely distributed in the, tropics and subtropics of the Old World. It was domesticated in India. The species is morphologically allied to *E. crus-galli*, but hybrids between them are sterile, although both the species are hexaploids (2n=54). *Echinochloa colona* differs consistently from *E. crus-galli* in that it has smaller spikelets with membraneous rather than chartaceous glumes. Hybrids between wild and cultivated taxa of *E. colona* and between those of *E. crus-galli* are fertile. Cultivated *E. colona* is variable. It is grown as a cereal crop in India (Kashmir) and Sikkim (de Wet *et al.* 1983). The comparative morphological study of the accessions resulted in the recognition of the species, subspecies and races given in Table 18.5. Barnyard millet remains an important cereal only in the tropics and subtropics of India. Four morphological races are recognized, although these do not have geographical, ecological or ethnological unity. The race laxa is confined to Sikkim where races robusta (Fig. 18.1d), intermedia and stolonifera are also grown.

Kodo millet (*Paspalum scrobiculatum*) is widely distributed in damp habitats across the Old World tropics. The chromosome number of this species has been reported to be 2n=40 and n=20 (Hiremath and Dandin 1975). It is harvested as a wild cereal in West Africa and in India. The species was domesticated in India around 3000 years ago. Raceme morphology allows for the recognition of three racial complexes (Table 18.6). The most common kodo millet is characterized by racemes with the spikelets arranged in two rows on one side of a flattened rachis (regularis). In most fields of kodo millet, plants with irregularly arranged spikelets also occur. Rarely are these aberrant kinds grown as pure stands. Two kinds of aberrations occur. In one kind, the spikelets are arranged along the rachis in 2-4 irregular rows (irregularis) (Fig. 18.1e). In the other, the lower part of each raceme is characterized by irregularly arranged spikelets, while spikelet arrangement becomes more regularly two-rowed in the upper part of the raceme (variabilis).

Teff (*Eragrostis tef*) is an important cereal crop cultivated in Ethiopia. It is extensively cultivated in Ethiopia for grain, and as a fodder crop in Australia and South Africa. The wild ancestor of teff has not been determined with certainty, but it probably is *E. pilosa* (L.) Beauv. (de Wet 1995). Although the exact date of its domestication is not known, there is no doubt that teff is an ancient crop of Ethiopia (Costanza 1974; Tadessa 1975). It is believed that teff domestication first took place somewhere in the northern highlands of Ethiopia. Teff is an annual grass forming

Table 18.3. Races of cultivated proso millet (*Panicum miliaceum*).

Species	Subspecies	Race
P. miliaceum	miliaceum	miliaceum
	miliaceum	patentissimum
	miliaceum	contractum
	miliaceum	compactum
	miliaceum	ovatum

Source: Prasada Rao *et al.* 1993.

Table 18.4. Races and subraces of cultivated little millet (*Panicum sumatrense*).

Species	Subspecies	Race	Subrace
P. sumatrense	psilopodium	–	–
	sumatrense	nana	laxa
	sumatrense	nana	erecta
	sumatrense	robusta	laxa
	sumatrense	robusta	compacta

Source: Prasada Rao *et al.* 1993.

Table 18.5. Species, subspecies and races of barnyard millet (*Echinochloa*).

Species	Subspecies	Races
E. colona	colona	–
	frumentacea	stolonifera
	frumentacea	intermedia
	frumentacea	robusta
	frumentacea	laxa
E. crus-galli	crus-galli	crus-galli
	crus-galli	macrocarpa
	crus-galli	utilis
	crus-galli	intermedia

Source: Prasada Rao *et al.* 1993.

Table 18.6. Races of kodo millet (*Paspalum scrobiculatum*).

Species	Races
P. scrobiculatum	regularis
	irregularis
	variabilis

Source: Prasada Rao *et al.* 1993.

scanty tufts. The chromosome number is $2n=40$. The flowers are cleistogamic, thus self-pollinated. Double fertilization is typically the means of reproduction (Mengesha and Guard 1966). The fruit is a caryopsis (Tadessa 1975). Tadessa (1975) enumerated 35 cultivars in *E. tef* and described them botanically after conducting field and laboratory studies of plants grown at the College of Agriculture, Alemaya, Ethiopia.

Several species of *Digitaria* are harvested as wild cereals in Africa (Busson 1965; Jardin 1967). Two species, *Digitaria exilis* (fonio) and *Digitaria iburua* (black fonio), are cultivated in West Africa (Portères 1976; de Wet 1977). Fonio millet (*Digitaria exilis* Stapf), also known as hungry rice, is grown as a cereal crop throughout the savanna zone of West Africa. In parts of Guinea and Nigeria, it is the staple crop (Doggett 1989). Fonio is largely cultivated from Senegal to Chad, as are pearl millet (*Pennisetum glaucum* (L.) R. Br.) and sorghum (*Sorghum bicolor* (L.) Moench). The primary centre of origin and evolution of fonio is the upper river valleys of The Gambia, Casamance, Senegal and Niger rivers in the Fouta-Djalon Plateau (Portères 1951, 1955; Bezpaly 1984). According to Portères (1955, 1976) and Purseglove (1972), the cultivation of fonio in West Africa goes back to the Neolithic period. This cereal is among the oldest indigenous crops. Its origins date back to 5000 BC (Ndoye and Nwasike 1993). Fonio

(*D. exilis*) grains are tiny; 1000-grain mass is about 0.5-0.6 g (Ndoye and Nwasike 1993). Black fonio (*D. iburua*) is known only as a cultivated crop (Stapf 1915). The species is grown by the Hausa of Nigeria between Jos and Zaria. It is also sporadically cultivated around Zinder in Niger, Azaguie in the Ivory Coast, Kande and Atalote in Togo, and between Birni and Natitingou in Benin (former Dahomey) (Portères 1955; Clayton 1972). Black fonio is often grown between rows of sorghum or pearl millet, and frequently as a mixture with *D. exilis* (true fonio). It is a more aggressive species than *D. exilis*, and often provides a crop when true fonio fails as a result of drought. Grains of black fonio are cooked like rice, or as stews.

GERMPLASM CONSERVATION AND USE

A total of 9094 accessions of small millets were assembled from 46 countries at ICRISAT (Table 18.7). The following list gives the locations of other accessions.

Teff	Plant Genetic Resource Centre, Ethiopia.
Fonio	ICRISAT Sahelian Center, Niamey, Niger
Finger millet	ICRISAT Asia Center (IAC) (Patancheru, India); Southern African Development Community (SADC) and ICRISAT-Southern and Eastern Africa (Bulawayo Zimbabwe and Nairobi, Kenya)
Foxtail, proso, barnyard, kodo and little millets	ICRISAT Asia Center (IAC), Patancheru, India

At ICRISAT Asia Center, Patancheru, India (IAC) all small millet germplasm accessions are sown for seed increase normally in the rainy season to meet seed requests from institutions from various countries. Since all six small millet crop species are reported as mostly self-pollinated, they are maintained without pollination control. IBPGR/IPGRI has assigned special responsibility to ICRISAT by providing funds to characterize the available germplasm.

A total of 4984 accessions have been characterized for important morpho-agronomic descriptors (Table 18.8). Parameters for some of the important morphological characters are given in Table 18.9. All the assembled germplasm of small millets, a total of 9094 accessions, is conserved at IAC under medium-term storage conditions of 4°C and 20% relative humidity.

So far, 42 130 seed samples have been distributed from the ICRISAT genebank (Table 18.10).

The Plant Genetic Resource Center (PGRC), Ethiopia, is holding about 2650 accessions of teff (Kebebew 1988). The most important development since 1986 is the completion of the work on the characterization of 2225 germplasm lines. The minimum and maximum values for some of these characters are given in Table 18.11.

Fonio has received very little attention from research organizations at either the regional level or the international level. In the past, small programmes were carried out in Nigeria and Mali. ORSTOM, France has 315 accessions of which 62 are from Togo, 100 from Mali and 21 from Burkina Faso (Froment and Renard 1996).

Properties and Uses

Small millets are considered coarse grains and are consumed where other food grains are expensive or difficult to grow. Compared with other cereals the information available on small-millet processing for food and industrial uses is very limited (Malleshi 1989). Small millets are used in India mainly to prepare *roti* (unleavened pancakes), *mudde* or dumpling, and thin porridge (*ambali*). Except for *roti* and *mudde*, debranned millets are cooked similar to rice. *Upma* (coarse-ground and boiled), *dosai*

Table 18.7. Geographic origin of the small millet germplasm collection at ICRISAT Asia Center as of December 1995.

Country	Finger	Foxtail	Proso	Barnyard	Kodo	Little	Total
Africa							
Burundi	15	–	–	–	–	–	15
Cameroon	8	8	–	–	–	–	16
Egypt	–	–	–	1	–	–	1
Ethiopia	27	1	–	–	–	–	28
Kenya	440	8	1	–	–	–	449
Malawi	248	–	1	2	–	–	251
Mozambique	1	–	–	–	–	–	1
Nigeria	19	–	–	–	–	–	19
Senegal	5	–	–	–	–	–	5
South Africa	–	4	–	–	–	–	4
Sudan	–	–	–	2	–	–	2
Tanzania	39	–	–	–	–	–	39
Uganda	918	1	–	–	–	–	919
Zambia	125	–	–	–	–	–	125
Zimbabwe	941	–	–	–	–	–	941
Asia							
Afghanistan	–	20	17	–	–	–	37
Bangladesh	–	–	2	–	–	–	2
China	–	60	1	–	–	–	61
India	1315	931	106	419	542	443	3756
Iran	–	4	9	–	–	–	13
Iraq	–	–	2	–	–	–	2
Japan	–	–	1	164	–	–	165
Korea	–	52	73	–	–	–	125
Lebanon	–	33	–	–	–	–	33
Maldives	3	–	–	–	–	–	3
Myanmar	–	6	–	–	–	2	8
Nepal	740	22	6	–	–	–	768
Pakistan	1	29	34	11	–	–	75
Philippines	–	–	–	98	111	–	209
Sri Lanka	18	14	2	–	2	–	36
Syria	–	119	290	2	–	–	411
Taiwan	–	28	–	–	–	–	28
Turkey	–	23	48	–	–	–	71
Former USSR	–	67	129	16	–	–	212
Europe							
Hungary	–	10	9	–	–	–	19
Italy	7	–	–	–	–	–	7
Spain	–	1	1	–	–	–	2
Switzerland	–	1	–	–	–	–	1
UK	28	20	4	4	–	–	56
Former Yugoslavia	–	–	1	–	–	–	1
Germany	1	–	12	–	–	–	13
Americas							
Argentina	–	–	1	–	–	–	1
Canada	–	–	1	–	–	–	1
Mexico	–	2	13	–	–	–	15
USA	7	38	65	–	–	–	110
Oceania							
Australia	–	–	2	–	–	–	2
Origin not known	55	9	–	–	–	–	64
Total	**4953**	**1495**	**831**	**715**	**655**	**445**	**9094**

Table 18.8. Small millet germplasm evaluated at ICRISAT Asia Center, India.

Crop	Suggested access. no.	Countries (no.)	No. accessions Assembled	Evaluated	Desc- riptors
Finger millet (*Eleusine coracana*)	IE	20	4953	1948	32
Foxtail millet (*Setaria italica*)	ISe	24	1495	1195	34
Proso millet (*Panicum miliaceum*)	IPm	26	831	736	34
Little millet (*Panicum sumatrense*)	IPmr	1	445	302	32
Barnyard millet (*Echinochloa* spp.)	IEc	8	715	516	31
Kodo millet (*Paspalum scrobiculatum*)	IPs	2	655	287	26

(fermented batter, cooked as thin pancakes) and *idli* (steamed) are popular south Indian preparations that are sometimes prepared from small millets. Some desserts made from foxtail millet are tastier and have a far better texture than those made from rice and wheat. Flakes can be prepared from pearled small millets. Because of their small size, quick hydration and soft texture, these flakes require less energy to produce than do flakes from rice or maize. Malting of finger millet is a traditional process in India, mostly used in infant foods and in milk-thickener formulations, conventionally called *ragi malt* (Malleshi and Hadimani 1993). Traditional opaque beer is a commercial product in several finger millet-producing countries in Africa (Gomez 1993).

Foxtail millet is native to China and is cultivated mainly for food; the straw is used as feed. The grain is used to prepare medicinal foods (Jiaju 1989). Proso millet, one of the most popular cereals in northern China, commands the same price as wheat. It is generally dehulled, and then steamed or boiled in a manner similar to rice. It is also sometimes sweetened with sugar and eaten as a dessert or flaked (Hulse *et al.* 1980). The other small millets (little, kodo and barnyard millets) are usually cooked after dehulling (Pushpamma 1989). Barnyard millet is grown as a forage crop in the USA, where it has been reported to produce eight harvests per year (Hulse *et al.* 1980). Teff flour is most widely used for making a pancake called *injera*. It is also used to make porridge and alcoholic drinks called *tella* and *katikalla*. The straw is used to reinforce mud-plastered walls, and as feed for cattle (Ketema 1989). Fonio is used in West Africa to make porridge, or added to other cereals as meal. It can be used as fodder, and the stems as roofing material (Spore 1995). In some countries in West Africa, fonio is still the cereal reserved for special occasions. In Nigeria and Togo it is used to make non-alcoholic drinks like *burujuto* or *pito* (Ndoye and Nwasike 1993).

Breeding

ICRISAT has designated finger millet (*E. coracana*) as an additional millet for special study in the highlands of eastern and southern Africa, where it is particularly important as a food crop. Projects have been initiated by ICRISAT-Southern and Eastern Africa at its regional centres at Nairobi (Kenya) and Bulawayo (Zimbabwe), with the objective of cooperating with national programmes to identify and improve finger millet genotypes with high yield potentials. The research priority will be to breed cultivars resistant to blast disease caused by *Pyricularia* (ICRISAT 1991). Two improved finger millet varieties, Lima (a selection from IE 2929) and FMV 2 (a selection from IE 2247), have been released for cultivation in Zambia and Zimbabwe, respectively (SADC/ICRISAT Southern Africa Programs 1993). Some breeding work on small millets is being done by the national agricultural research systems (NARS) in several countries.

Table 18.9. Parameters for some important quantitative characters of small millets, evaluated at ICRISAT Asia Center, Patancheru, India.

Crop	Days to 50% flowering	Plant height (cm)	Basal tillers (no.)	Flag leaf length (mm)	Flag leaf width (mm)	Sheath length (mm)	Peduncle length (mm)	Exsertion (mm)	Inflorescence length (mm)
Finger millet									
Mean	79.10	102.90	6.56	351.00	13.38	105.60	219.80	112.90	100.60
SD	10.42	15.54	3.94	70.70	3.78	20.08	40.92	40.89	32.96
Range	54-120	45-165	1-70	100-700	5-20	20-280	18-380	10-800	10-320
Foxtail millet									
Mean	52.41	105.80	7.75	286.20	20.56	136.20	299.50	165.30	156.40
SD	10.31	31.18	6.65	82.33	6.24	32.31	59.73	52.56	55.67
Range	32-135	20-175	1-80	30-520	5-40	50-260	0-500	10-360	10-320
Proso millet									
Mean	33.87	60.74	–	231.40	20.41	82.81	181.80	99.70	195.70
SD	2.89	17.83	–	48.64	6.86	17.10	64.63	60.80	60.16
Range	28-50	23-133	–	100-380	10-30	30-170	50-400	-50-320	22-400
Little millet									
Mean	54.50	116.10	19.06	256.80	12.23	107.80	144.70	37.42	298.20
SD	12.11	25.59	7.99	76.28	4.12	22.65	41.56	43.06	65.95
Range	39.98	60.240	5-46	120-560	5-30	50-180	60-280	-80-280	27-500
Barnyard millet									
Mean	44.68	74.38	–	191.90	17.99	87.94	165.20	78.64	137.20
SD	10.43	24.18	–	74.85	7.46	21.51	63.37	66.60	54.73
Range	37-90	52-242	–	100-420	5-40	40-260	95-520	-50-800	100-280
Kodo millet									
Mean	87.75	53.19	23.56	249.60	9.86	153.70	–	–	58.29
SD	8.19	10.37	7.68	64.88	1.95	22.95	–	–	19.56
Range	65-103	30-90	10-48	24-440	5-15	80-220	–	–	20-120

Table 18.10. Distribution of small millet germplasm from the ICRISAT genebank, 1978-95.

Distributed to:	No. of accessions	Distributed to:	No. of accessions
Africa		**Asia (cont.)**	
Benin	6	Nepal	1316
Burkina Faso	37	Pakistan	114
Burundi	25	Saudi Arabia	60
Djibouti	10	Sri Lanka	220
Ethiopia	749	Taiwan	320
Ghana	4	Thailand	92
Kenya	4366	Former USSR	206
Mali	89	**Oceania**	
Mozambique	30	Australia	50
Niger	21	**Europe**	
Nigeria	444	Czechoslovakia	17
Rwanda	125	Denmark	30
Senegal	29	France	125
Somalia	10	Germany	18
Sudan	33	Italy	34
Uganda	800	Romania	22
Zambia	2007	UK	191
Zimbabwe	1198	Former Yugoslavia	1970
Asia		**Americas**	
Bangladesh	522	Argentina	25
Bhutan	17	Brazil	180
China	515	Canada	61
India	24382	Mexico	289
Indonesia	48	Uruguay	100
Korea	392	USA	674
Malaysia	127	Venezuela	20
Myanmar	10	**Total**	**42130**

Table 18.11. Some agronomic and morphological characters of teff, evaluated at the Plant Genetic Resource Centre, Ethiopia.

Characters	Minimum	Maximum
Days to germination	4	12
Days to germination to heading	26	54
Days to heading to maturity	29	76
Days to germination to maturity	62	123
Culm thickness (mm)	1.5	4
Culm length (cm)	11	82
Peduncle length (cm)	7	42
Panicle length (cm)	14	65
Total plant height (cm)	31	55
Grain yield per panicle (g)	0.4	3
Grain yield per plant (g)	4	22
Straw yield per plant (g)	20	90
Shoot biomass yield per plant (g)	26	105
Leaf area (flag leaf) (cm^2)	5	22
Harvest index	7	38

Source: Ketema 1993.

In India, the All India Coordinated Millets Improvement Project started its work on finger millet at five centres: Bangalore, Coimbatore, Majhera, Sangla and Vizianagaram. Work on the other millets was started with help from the International Development Research Centre (IDRC); the sites chosen were Dindari (Madhya Pradesh) for kodo millet, Semiliguda (Orissa) for little millet, Dholi (Bihar) for proso millet, Nandyal (Andhra Pradesh) for foxtail millet and Almora (Uttar Pradesh) for barnyard millet

(Doggett 1993). Research on finger millet improvement in Africa has been done in Zambia, Kenya, Uganda, Tanzania, Malawi and Ethiopia. Genetic improvement has been largely limited to selection in germplasm lines except at Serere, Uganda, where some breeding work is in progress, using a male-sterile line introduced some 15 years ago (Doggett 1989).

There has been good progress in China in breeding foxtail millet; over the years, more than 100 improved varieties have been released in different provinces. The breeding approaches adopted include systematic selection, hybridization and radiation-induced mutations. The Hebei Baxia Agricultural Institute has developed two good hybrids, which are cultivated on more than 2000 ha (Jiaju and Yuzhi 1993). Breeding work on proso millet in the Prvolzhye province in Russia is aimed at high productivity, smut resistance and improved grain quality. A step-by-step procedure is used, involving complex crosses combining many parents. Saratovskoye 6 and Saratovskoye 8 are some of the recent releases that meet these requirements (Ilyn *et al.* 1993). Extensive research on teff is conducted in Ethiopia (and very little in other countries), where it is the staple food. Tareke (1975) reported the first successful intraspecific crosses, and this signalled a new era in teff breeding research (Ketema 1989).

The importance of research in fonio improvement has been recognized in Mali, where a programme for the promotion of Indigenous Cereals of the Sahel has been initiated in Bamako. A project for domesticating fonio is being undertaken at the Agricultural Research Station at Farako-Ba in western Burkina Faso, with assistance from the International Atomic Energy Agency (IAEA) based in Vienna, Austria. Although the project has only just begun, it is already showing great promise (Spore 1995).

Prospects

Different species of small millets possess many unique traits that should make them an important component of improved agricultural systems (Riley 1988). For example, they are generally fast-maturing and could fit into more intensive cropping systems. A short-duration millet could be used as a catch or relay crop in association with other, slower-maturing crops. Rapid seed multiplication is generally fairly simple and seed costs are low. The small seeds generally store well for long periods, ensuring a continued food supply during the dry season or when there is a crop failure. The seeds often require less cooking or preparation time than those of other cereals. Millets can be processed in a number of ways for use in traditional and novel preparations.

Some millets perform well in difficult or marginal farming situations. Teff, for example, is tolerant of waterlogged and acidic soils. Proso millet can tolerate both drought and salinity. Foxtail millet is adapted to low-fertility soils. Because many millets mature quickly, they can escape such stress conditions as drought. Many varieties of millets have excellent nutritional properties, with high levels of such essential minerals as iron and calcium. Finger millet, especially, is known to be particularly important in the diets of people involved in hard manual work. Millets are generally highly valued for their fodder. A new Indian variety of foxtail millet, SLA 326, is extremely popular with farmers in Andhra Pradesh, India. In addition to its high grain yield, the straw is highly palatable as livestock feed. For these farmers, the economic value of foxtail millet straw is almost equal to the value of the grain. Recent work at the Dryland Agricultural Centre in Bangalore, India, has shown that little millet and barnyard millet produced more forage yield per day under dry conditions than any other forage crop tested.

With the development of a dehiscent fonio variety, post-harvest treatment became much easier. With a combination of higher yields, improved processing and already proven market demand, fonio could have a more significant future than might once have been expected (Spore 1995). Small millets, currently assigned only a subsistence role, could be grown as surplus crops for sale and processing. They could have application in a range of foods: composite flour, bread, biscuits, pastas, instant *uji* and other 'instant' food items for sale in urban areas (Dendy 1993).

REFERENCES

Bezpaly, I. 1984. Les plantes cultivées en Afrique occidentale. Pp. 84-87 *in* Ouvrage sous la direction de Oustimenko. Bakoumovski. Editions MIR, Moscow.

Bor, N.L. 1960. Grasses of Burma, Ceylon, India and Pakistan. Pergamon, London, UK.

Busson, F. 1965. Plantes Alimentaires de l'Ouest Africain. L'imprimerie Leconte, Marseilles, France.

Clayton, W.D. 1972. Gramineae. Pp. 349-512 *in* Flora of West Tropical Africa (2nd edn.), Vol. III. Part 2 (J. Hutchinson and J.M. Dalziel, eds. and F.N. Hepper, rev.). Crown Agents for Oversea Governments and Administration, London, UK.

Costanza, S.H. 1974. Literature and Numerical Taxonomy of Teff (*Eragrostis tef*). MSc Thesis, B.S. Cornell University, Illinois, USA.

de Wet, J.M.J. 1977. Domestication of African cereals. Afr. Econ. Hist. 3:15-32.

de Wet, J.M.J. 1995. Minor cereals - various genera (Gramineae). Pp. 202-208 *in* Evolution of Crop Plants (J. Smartt and N.W. Simmonds, eds.). Longman Scientific and Technical, Harlow, UK.

de Wet, J.M.J., K.E. Prasada Rao and D.E. Brink. 1984a. Systematics and domestication of *Panicum sumatrense* (Gramineae). J. Agric. Trad. Bot. Appl. 30:159-168.

de Wet, J.M.J., K.E. Prasada Rao, D.E. Brink and M.H. Mengesha. 1984b. Systematics and evolution of *Eleusine coracana* (Gramineae). Am. J. Bot. 71:550-557.

de Wet, J.M.J., K.E. Prasada Rao, M.H. Mengesha and D.E. Brink. 1983. Domestication of sawa millet, *Echinochloa colona*. Econ. Bot. 37:283-291.

Dendy, D.A.V. 1993. Opportunities for non-traditional uses of the minor millets. Pp. 259-270 *in* Advances in Small Millets (K.W. Riley, S.C. Gupta, A. Seetharam and J.N. Mushonga, eds.). Oxford & IBH Publishing Co, New Delhi, India.

Doggett, H. 1989. Small millets - a selective overview. Pp. 3-18 *in* Small Millets in Global Agriculture (A. Seetharam, K.W. Riley and G. Harinarayana, eds.). Oxford & IBH Publishing Co, New Delhi, India.

Doggett, H. 1993. Introduction. Pp. 3-8 *in* Advances in Small Millets (K.W. Riley, S.C. Gupta, A. Seetharam and J.N. Mushonga, eds.). Oxford & IBH Publishing Co, New Delhi, India.

Froment, D. and C. Renard. 1996. Fonio *Digitaria exilis* Stapf. *In* Crop Production in Tropical Africa. Belgian Agency for Cooperation and Development, Brussels, Belgium (in press).

Gomez, M.I. 1993. Preliminary studies on grain quality evaluation for finger millet as a food and beverage use in the southern African region. Pp. 289-296 *in* Advances in Small Millets (K.W. Riley, S.C. Gupta, A. Seetharam and J.N. Mushonga, eds.). Oxford & IBH Publishing Co, New Delhi, India.

Grammie, G.A. 1911. Millets of the genus *Setaria* in the Bombay Presidency and Sind. Mem. Dep. Agric. India Bot. Ser. 4:1-8.

Hiremath, S.C. and M.S. Chennaveeraiah. 1982. Cytogenetical studies in wild and cultivated species of *Eleusine* (Gramineae). Caryologia 35:57-69.

Hiremath, S.C and S.B. Dandin. 1975. Cytology of *Paspalum scrobiculatum* Linn. Curr. Sci. 44:20-21.

Hulse, J.H., M.E. Laing and E.O. Pearson. 1980. Sorghum and the Millets: Their Composition and Nutritive Value. Academic Press, London, UK.

ICRISAT. 1991. ICRISAT Report 1990. ICRISAT, Patancheru, India.

Ilyn, V.A., E.N. Zolotukhin, I.P. Ungenfukht, N.P. Tikhonov and B.K. Markin. 1993. Importance, cultivation and breeding of proso millet in Povolzhye province in Russia. Pp. 109-116 *in* Advances in Small Millets (K.W. Riley, S.C. Gupta, A. Seetharam and J.N. Mushonga, eds.). Oxford & IBH Publishing Co, New Delhi, India.

Jardin, C. 1967. List of Foods Used in Africa. FAO, Rome, Italy.

Jiaju, C. 1989. Utilization of small millets in China. Pp. 347-350 *in* Small Millets in Global Agriculture (A. Seetharam, K.W. Riley and G. Harinarayana, eds.). Oxford & IBH Publishing Co, New Delhi, India.

Jiaju, C. and Q. Yuzhi. 1993. Recent developments in foxtail millet cultivation and research in China. Pp. 101-107 *in* Advances in Small Millets (K.W. Riley, S.C. Gupta, A. Seetharam and J.N. Mushonga, eds.). Oxford & IBH Publishing Co, New Delhi, India.

Kebebew, F. 1988. The activities of the Plant Genetic Resource Centre/Ethiopia (PGRC/E) on teff and minor millets. Pp. 21-22 *in* Small Millets - Recommendations for a Network. Manuscript Report IDRC-MR 171e. International Development Research Centre, Ottawa, Canada.

Ketema, S. 1989. Cropping systems, production technology, pests, diseases, utilization and forage use of millets with special emphasis on teff in Ethiopia. Pp. 309-314 *in* Small Millets in Global Agriculture (A. Seetharam, K.W. Riley and G. Harinarayana, eds.). Oxford & IBH Publishing Co, New Delhi, India.

Keteme, S. 1993. Teff crop improvement, nutrition and utilization. Pp. 61-65 *in* Advances in Small Millets (K.W. Riley, S.C. Gupta, A. Seetharam and J.N. Mushonga, eds.). Oxford & IBH Publishing Co, New Delhi, India.

Jardin, C. 1967. List of Foods Used in Africa. FAO, Rome, Italy.

Malleshi, N.G. 1989. Processing of small millets for food and industrial uses. Pp. 325-339 *in* Small Millets in Global Agriculture (A. Seetharam, K.W. Riley and G. Harinarayana, eds.). Oxford & IBH Publishing Co, New Delhi, India.

Malleshi, N.G. and N.A. Hadimani. 1993. Nutritional and technological characteristics of small millets and preparation of value-added products from them. Pp. 271-287 *in* Advances in Small Millets (K.W. Riley, S.C. Gupta, A. Seetharam and J.N. Mushonga, eds.). Oxford & IBH Publishing Co, New Delhi, India.

Martin, J.H. and W.H. Leonard. 1959. Principles of Field Crop Production. Macmillan, New York, USA.

Mengesha, M.H. and A.T. Guard. 1966. Development of the embryo of teff, *Eragrostis tef.* Can. J. Bot. 44: 1071-1075.

Naciri, Y. and J. Belliard. 1987. Le millet *Setaria italia* - une plante à redécouvrir. J. Agric. Trad. Bot. Appl. 36:65-87.

Ndoye, M. and C.C. Nwasike. 1993. Fonio millet (*Digitaria exilis* Stapf) in West Africa. Pp. 85-94 *in* Advances in Small Millets (K.W. Riley, S.C. Gupta, A. Seetharam and J.N. Mushonga, eds.). Oxford & IBH Publishing Co, New Delhi, India.

Portères, R. 1951. Géographie alimentaire - berceaux agricoles et migrations des plantes cultivées en Afrique intertropicale. C.R. Somm. Séances Soc. Biogéogr. 239:16-21.

Portères, R. 1955. Les céréales mineures du genre *Digitaria* en Afrique et Europe. J. Agric. Tradit. Bot. Appl. 2:349-675.

Portères, R. 1976. African cereals: *Eleusine*, fonio, black fonio, teff, *Brachiaria, Paspalum, Pennisetum* and African rice. Pp. 409-452 *in* Origins of African Plant Domestication. (J.R. Harlan, J.M.J. de Wet and A.B. Stemler, eds.). Mouton, The Hague, Netherlands.

Prasada Rao, K.E. and M.H. Mengesha. 1988. Minor millets germplasm resources at ICRISAT. Pp. 38-45 *in* Small Millets - Recommendations for a Network. Manuscript Report IDRC-MR 171e. International Development Research Centre, Ottawa, Canada.

Prasada Rao, K.E., J.M.J. de Wet, D.E. Brink and M.H. Mengesha. 1987. Infraspecific variation and systematics of cultivated *Setaria italia*, foxtail millet (Poaceae). Econ. Bot. 41:108-116.

Prasada Rao, K.E., J.M.J. de Wet, V. Gopal Reddy and M.H. Mengesha. 1993. Diversity in the small millets collection at ICRISAT. Pp. 331-346 *in* Advances in Small Millets (K.W. Riley, S.C. Gupta, A. Seetharam and J.N. Mushonga, eds.). Oxford & IBH Publishing Co, New Delhi, India.

Purseglove, J.W. 1972. Tropical Crops: Monocotyledons. Longman, London, UK.

Pushpamma, P. 1989. Utilization of small millets in Andhra Pradesh (India). Pp. 321-324 *in* Small Millets in Global Agriculture (A. Seetharam, K.W. Riley and G. Harinarayana, eds.). Oxford & IBH Publishing Co, New Delhi, India.

Rangaswamy Ayyangar, G.N. 1938. Studies in the millet *Panicum miliaceum* Linn. Madras Agric. J. 26:196-206.

Riley, K. 1988. Small millets with big potential - introductory remarks. Pp. 2-5 *in* Small Millets - Recommendations for a Network. Manuscript Report IDRC-MR 171e. International Development Research Centre, Ottawa, Canada.

SADC/ICRISAT Southern Africa Programs. 1993. Annual Report 1992. SADC ICRISAT Sorghum and Millet Improvement Program, Bulawayo, Zimbabwe.

Spore, C. 1995. A small cereal with big promise. Bull. Tech. Cent. Agric. Rural Cooperation (Wageningen) 55:5.

Stapf, O. 1915. Iburu and Fundi, two cereals of Upper Guinea. Kew Bull. 1915:318-386.

Tadessa, E. 1975. Teff (*Eragrostis tef*) Cultivars: Morphology and Classification: Part II. Debre Zeit Junior College and Agricultural Experiment Station Bulletin Number 66. Alemaya University of Agriculture, Dire Dawa, Ethiopia.

Tareke, B. 1975. A breakthrough in teff breeding technique. FAO Inf. Bull. Cereal Improv. and Prod. 12:11-13.

Chapter 19

Rice

M.T. Jackson, G.C. Loresto, S. Appa Rao, M. Jones, E.P. Guimaraes and N.Q. Ng

Rice feeds half the world's people, mainly in Asia. Their food security and crop biodiversity depend upon continued access to seed developed from thousands of locally adapted varieties of *Oryza sativa* and *O. glaberrima* that Asian and African farmers have grown for generations, the more than 20 species of wild rice native to Asia, Africa, Latin America and Oceania, and the related genera in the tribe Oryzeae. Worldwide, about 80 million ha of rice are grown under irrigated conditions, the most important rice production system, with average yields of 3-9 t/ha. Athough four CGIAR centres (IRRI, WARDA, CIAT and IITA) hold and use rice germplasm, only IRRI has a global mandate to conserve and improve germplasm. Other centres have regional or continental mandates in Africa and Latin America.

The aggregate population of the less-developed countries grew from 2.3 billion in 1965 to 4.1 billion in 1991. Asia accounted for 59% of the global population, about 92% of the world's rice production and 90% of global rice consumption. Bangladesh, China, India, Indonesia, Myanmar, Thailand and Vietnam are the world's largest rice producers, accounting for about 78% of world production (IRRI 1995). Even with rice providing 35-80% of total calories consumed in Asia and with a slowing of growth in total planted area, production has so far kept up with demand. The world's annual rough rice production, however, will have to increase by almost 70% over the next 35 years to keep up with population growth and income-induced demand for food. The urban poor in Asia spend a large part of their income on rice and the consumption of rice will continue to increase as incomes and urbanization increase.

BOTANY AND DISTRIBUTION

Besides domesticated *Oryza*, other wild *Oryza* and related genera of the tribe Oryzeae are distributed throughout the tropics (Tables 19.1 and 19.2). The basic chromosome number is $n=12$. *Oryza* species are generally grouped into four complexes of closely related species (Chang and Vaughan 1991; Vaughan 1989, 1994). Species of the *O. ridleyi* complex inhabit lowland swamp forests, and species of the *O. meyeriana* complex are found in upland hillside forests. The *O. officinalis* complex consists of diploid and tetraploid species found throughout the tropics. The *O. sativa* complex consists of the wild and weedy relatives of the two rice cultigens as well as the cultigens themselves. The wild relatives of *O. glaberrima* in Africa consist of the perennial rhizomatous species *O. longistaminata*, which grows throughout sub-Saharan Africa, and annual and weedy relatives are found primarily in West Africa. Among the wild relatives of *O. sativa*, the perennial *O. rufipogon* is widely distributed over South and Southeast Asia, southeast China and Oceania. Other forms are found in South America, usually in deepwater swamps (Chang 1976a).

Table 19.1. Taxa in the genus *Oryza*: the species complexes and genome groups (adapted from Chang and Vaughan 1991 and Vaughan 1994).

Species complex	Taxon	Genome group	Distribution
O. sativa	*O. glaberrima*	A^gA^g	West Africa
	O. barthii	A^gA^g	Sub-Saharan Africa
	O. longistaminata	A^lA^l	Sub-Saharan Africa, Madagascar
	O. sativa	AA	Worldwide
	O. nivara	AA	Tropical and subtropical Asia
	O. rufipogon	AA	Tropical and subtropical Asia
	O. meridionalis	AA	Northern Australia
	O. glumaepatula	$A^{gl}A^{gl}$	South and Central America
O. ridleyi	*O. longiglumis*	Tetraploid	Indonesia (Irian Jaya), Papua New Guinea
	O. ridleyi	Tetraploid	Southeast Asia, Papua New Guinea
O. meyeriana	*O. granulata*	Diploid	South and Southeast Asia
	O. meyeriana	Diploid	Southeast Asia
O. officinalis	*O. officinalis*	CC	Tropical and subtropical Asia
	O. minuta	BBCC	Philippines, Papua New Guinea
	O. eichingeri	CC	Sri Lanka, Sub-Saharan Africa
	O. rhizomatis	CC	Sri Lanka
	O. punctata	BBCC, BB	Sub-Saharan Africa, Madagascar
	O. latifolia	CCDD	South and Central America
	O. alta	CCDD	South and Central America
	O. grandiglumis	CCDD	South America
	O. australiensis	EE	Northern Australia
Species not yet assigned to a complex	*O. schlechteri*	Diploid	Indonesia (Irian Jaya), Papua New Guinea
	O. brachyantha	FF	Sub-Saharan Africa

Table 19.2. Genera in the tribe Oryzeae (adapted from Chang and Vaughan 1991).

Genus	No. of species	Distribution	Environment
Oryza	22	Pan-tropical	tropical
Leersia	17	Worldwide	temperate / tropical
Chikusiochloa	3	China, Japan	temperate
Hygroryza	1	Asia	temperate / tropical
Porteresia	1	South Asia	tropical
Zizania	3	Europe, Asia, North America	temperate / tropical
Luziola	11	North and South America	temperate / tropical
Zizaniopsis	5	North and South America	temperate / tropical
Rhynchoryza	1	South America	temperate
Maltebrunia	5	Tropical and southern Africa	tropical
Prosphytochloa	1	Southern Africa	temperate
Potamophila	1	Australia	temperate / tropical

Origin, Domestication and Diffusion

The centre of origin of rice and the exact time and place of its first development may never be known. The most convincing archaeological evidence for domestication, dated to 4000 BC, was discovered in Thailand (Solheim 1972). In the early Neolithic era, rice was grown in forest clearings under a system of shifting cultivation; puddling the soil and transplanting seedlings were likely refined in China (Chang 1976b, 1976c). In Southeast Asia rice originally was produced under dryland conditions in the uplands and only recently came to occupy the vast river deltas. Diffusion has carried rice to every continent except Antarctica.

Wetland rice cultivation came to the Philippines during the second millennium BC and Deutero-Malays carried the practice to Indonesia about 1500 BC. The crop was introduced to Japan no later than 100 BC (Chang 1976a, 1976c). It reached India in early times and by 1000 BC was a major crop in Sri Lanka. Rice was introduced about 344-324 BC to Greece and the Mediterranean by returning members of Alexander the Great's expedition to India and spread gradually throughout southern Europe and to a few locations in North Africa (Huke and Huke 1990). Rice cultivation was introduced to the New World by early European settlers (Grist 1986). Early in the 18th century rice spread to Louisiana, but not until the 20th century did it reach California's Sacramento Valley (Adair *et al.* 1975). The California introduction corresponded with that of the first successful crop in Australia's New South Wales (Huke and Huke 1990).

The primary centre of diversity for *O. glaberrima* is in the swampy basin of the upper Niger River (Chang 1976a). Two secondary centres lie to the southwest near the Guinean coast. In West Africa, *O. glaberrima* is a dominant crop grown in the flooded areas of the Niger and Sokoto River basins (Chang 1985). Ecological diversification in *O. sativa*, which involved cycles of hybridization, differentiation and selection, was enhanced when ancestral forms of the cultigen were carried by farmers and traders to higher latitudes, higher elevations, dryland sites, seasonably deepwater areas and tidal swamps (Chang 1985). Two major ecogeographic races differentiated as a result of isolation and selection: indica, adapted to the tropics, and japonica, adapted to the temperate regions and tropical uplands. Selections made to suit cultural and socioreligious traditions added diversity, especially grain size, shape, colour and endosperm properties (Chang 1985). Today, thousands of rice varieties are grown in more than 100 countries.

Reproductive Biology

The morphology of rice is divided into the seedling, vegetative organs and reproductive organs (Chang and Bardenas 1965; Vergara 1991). Growth duration is 3-6 months and potential grain yield is primarily determined before heading. The life history of rice has three growth phases: vegetative, reproductive and ripening (Fig. 19.1). A 120-day variety, when planted in a tropical environment, spends about 60 days in the vegetative phase, 30 days in the reproductive phase and 30 days in the ripening phase (Yoshida 1981; Vergara 1991). Heading is considered a synonym for anthesis in rice. It takes 10-14 days for a rice crop to complete heading because there is variation in panicle exsertion among tillers of the same plant and among plants in the same field (Yoshida 1981). Agronomically, heading is usually defined as the time when 50% of the panicles have exserted. The length of ripening varies among varieties from about 15 to 40 days. Ripening is also affected by temperature with ranges from about 30 days in the tropics to 65 days in cool, temperate regions such as Hokkaido, Japan and Yanko, NSW, Australia (Yoshida 1981; Vergara 1991).

GERMPLASM CONSERVATION AND USE

The full spectrum of germplasm in the genus *Oryza* consists of the following:
- Wild *Oryza* species, which occur throughout the tropics, and related genera, which occur worldwide in both temperate and tropical regions.
- Natural hybrids between the cultigens and wild relatives and primitive cultivars of the cultigen in areas of rice diversity.
- Commercial varieties, obsolete varieties, minor varieties and special-purpose types in the centres of cultivation.
- Pure line or inbred selections of farmers' varieties, elite lines of hybrid origin, F_1 hybrids, breeding materials, mutants, polyploids, aneuploids, intergeneric and interspecific hybrids, composites, cytoplasmic sources from breeding programmes and, more recently, transgenic lines produced through genetic engineering.

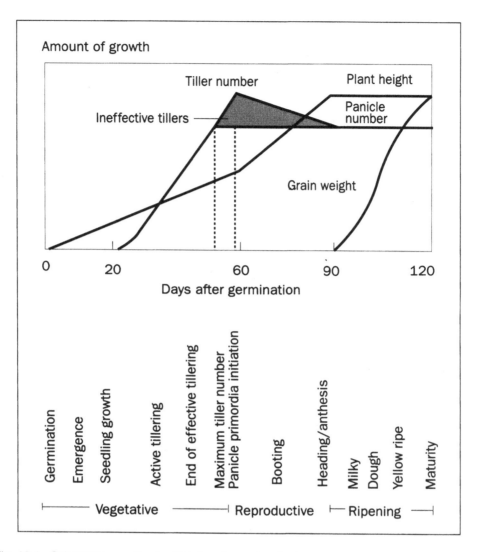

Fig. 19.1. Schematic growth of a 120-day rice variety in the tropics.

Conservation *ex situ* is a safe and efficient way to conserve rice genetic resources and to make the germplasm readily available to breeders and other researchers (Ford-Lloyd and Jackson 1986). Orthodox rice seed can be dried to a moisture content of ±6% and stored at subzero temperatures to keep them viable for decades and longer.

More than 219 000 accessions of cultivated and wild rices are stored in genebanks in more than 40 countries (Bettencourt and Konopka 1990). In Latin America and Europe, the accessions of *O. sativa* represent germplasm that has been introduced or developed as part of national rice breeding efforts. Accessions of *O. sativa* and *O. glaberrima* in Asia and Africa mostly represent indigenous landrace varieties that comprise the important germplasm heritage of these countries. The widespread distribution of wild rices throughout the tropics of Asia, Africa, South America and the Caribbean is reflected in their representation in national germplasm collections. The number of rice germplasm accessions does not necessarily reflect the genetic diversity of the rice crop because of the considerable exchange and duplication of genetic materials among genebanks.

Almost 100 000 samples of cultivated and wild rices are held in India, at the National Bureau of Plant Genetic Resources, and in the People's Republic of China, at the National Genebank of the Institute of Crop Germplasm Resources, Chinese Academy of Agricultural Sciences. Important collections are maintained in many other Asian countries, including the Philippines, Thailand and Indonesia. Japan holds about 12 000 accessions at the National Institute of Agrobiological Resources, in Tsukuba, which has a particular responsibility for the temperate-adapted japonica rices.

The most geographically diverse collections of rice germplasm are held in trust in the genebanks of two CGIAR centres, IRRI and IITA (Table 19.3). Together these institutes carry the responsibility for the long-term preservation of more than 86 100 samples of *O. sativa*, 3750 samples of *O. glaberrima* and 2900 samples of the 20 wild species in the genus *Oryza* – a total of almost 93 000 accessions. Add to these the several thousand accessions that WARDA maintains in working collections, many of them derived from its own breeding programmes, and the figure for rice germplasm in the CGIAR centres approaches 100 000 samples. About 34% of the germplasm conserved at IITA is a duplicate of accessions in the collection at IRRI.

Germplasm Collecting and Acquisition

Germplasm collecting has traditionally been a collaborative activity between the CGIAR centres and national programmes, and in Africa the effort has also involved other organizations such as ORSTOM, IRAT, IDESSA and IPGRI. Between 1972 and 1993, IRRI scientists have participated in 84 collecting missions in Bangladesh, Bhutan, India, Nepal and Sri Lanka in South Asia; in Cambodia, Indonesia, Lao PDR, Malaysia, Myanmar, Papua New Guinea, Philippines, Thailand and Vietnam in Southeast Asia; and in Botswana, Zambia and Madagascar in Africa. These missions resulted in the collection of more than 11 900 cultivated rice samples and 1908 wild species samples.

Beginning in 1978, WARDA scientists at the Mangrove Swamp Rice Program in Rokupr, Sierra Leone, collected traditional mangrove swamp rice varieties. By 1986, the collection stood at 754 accessions of mainly *O. sativa* and a few *O. glaberrima* from the Gambia, Guinea, Guinea Bissau, Nigeria, Senegal and Sierra Leone. The WARDA Continuum Program at Bouaké in Côte d'Ivoire also collected and received upland rice varieties from NARS between 1985 and 1993. Many accessions were also received from IRAT, IITA and ORSTOM. This collection now encompasses more than 4800 samples of *O. sativa*, more than 1200 samples of *O. glaberrima* and some wild species. Samples have been collected from most of the major West Africa rice-growing countries, including Benin, Burkina Faso, Cameroon, Chad, Côte d'Ivoire, The Gambia, Guinea, Guinea Bissau, Liberia, Mali, Niger, Nigeria, Senegal, Sierra Leone and Togo. Between 1976 and 1990, IITA carried out 60 exploration missions in more than 30 countries in Africa and collected more than 6000 cultivated and wild rice samples.

Between 1977 and 1985 IRAT and ORSTOM collected samples around Lake Chad, the Senegal River basin, Mali, Côte d'Ivoire, Guinea and Guinea Bissau. Their collection consists of three main species: *O. glaberrima* (222), *O. breviligulata* (=*O. barthii*) (19) and *O. longistaminata* (53), the perennial allogamous wild species (Bezançon *et al.* 1984; De Kochko 1985; Charrier and Hamon 1991).

Seed Conservation Facilities

The genebank at IRRI has operated since 1977 and after extensive renovation in 1993-94 was renamed the International Rice Genebank (IRG). It operates in accordance with the internationally accepted standards adopted by the FAO Commission on Plant Genetic Resources.

Table 19.3. The number of germplasm accessions of cultivated and wild rice species at IRRI and IITA in December 1995 (*O. sat.=Oryza sativa; O. glab.=Oryza glaberrima*).

Country	IRRI			IITA		
	O. sat.	*O. glab.*	Wild	*O. sat.*	*O. glab.*	Wild
Afghanistan	69			2		
Argentina	73		1	1		
Australia	94		66	16		
Austria	2					
Bangladesh	5462		77	23		
Belize	3					
Benin	4		2	75	16	
Bhutan	229					
Bolivia	11			1		
Botswana			2	7		11
Brazil	849		38	143		
Brunei	159		13			
Bulgaria	27					
Burkina Faso	457	52	2	861	180	1
Burundi	36					
Cambodia	1398		84			
Cameroon	66	27	49	84	69	1
Canada						
Central African Rep.	9			61	1	7
Chad	32	21	56	43	22	12
Chile	7					
Colombia	160		3	10		
Congo	1	1		1		
Costa Rica	12		7	1		
Côte d'Ivoire	841	10	11	79	209	5
Cuba	161		4	1		
Dominican Republic	10					
Ecuador	45			2		
Egypt	62	1		130	2	
El Salvador	37			24		
Ethiopia	6		12			
Fiji	26					
Former Yugoslavia	2					
France	192					
French Guiana			1			
Gabon	1					
Gambia	180	4	6	337	47	1
Ghana	198	15	6	163	44	
Greece	3					
Guatemala	13		8			
Guinea-Bissau	52	12	1	70	15	
Guinea-Conakry	549	110	14	484	77	3
Guyana	80		1			
Haiti	47					
Honduras	1					
Hong Kong	10					
Hungary	64					
India	14948		635	157		
Indonesia	8454		96	34		
Iraq	15					
Islamic Republic of Iran	195					
Italy	170					
Jamaica	6			4		
Japan	1103			16		
Kenya	253		10			
Republic of Korea	1098			104		

Country	IRRI			IITA		
	O. sat.	*O. glab.*	Wild	*O. sat.*	*O. glab.*	Wild
Korea DPR	7					
Lao PDR	1396		25	10		
Liberia	1413	416	1	1021	627	
Madagascar	1000		3	743		
Malawi	11			280		
Malaysia	2741		56	21		
Mali	46	271	121	53	136	8
Mauritania	2		1	3		
Mauritius	1					
Mexico	119		3	4		
Micronesia	7					
Morocco	2					
Mozambique	54					
Myanmar	1727		142	9		
Nepal	1472		16			
Netherlands	14					
New Zealand	1					
Nicaragua			1			
Niger	3	1	12	13	35	5
Nigeria	323	176	35	1092	1076	68
Pakistan	1073					
Panama	11		1	3		
Papua New Guinea	12		46	2		
Paraguay			1			
People's Rep. of China	7774		58	5		
Peru	65			16		
Philippines	4454		115	178		
Poland	9					
Portugal	69					
Puerto Rico	40			14		
Romania	32					
Russia & CIS	337			14		
Rwanda	2					
Saudi Arabia	1					
Senegal	552	85	18	548	81	11
Sierra Leone	775	24	18	358	28	1
Solomon Islands	2					
Spain	42					
Sri Lanka	2010		104	11		
Sudan	15	2	6			
Surinam	96		7	17		
Taiwan	1681		79	12		
Tanzania	125	2	37	328	4	6
Thailand	5084		522	21		
Togo	6			5	15	
Tunisia	1					
Turkey	53					
Uganda			15			
United Kingdom	8					
United States	1106			40		
Uruguay	5					
Venezuela	27		3	2		
Vietnam	1588		49			
Zaire	56		2	28		
Zambia	30		7	484	18	
Zimbabwe	93			44	2	
(Unknown)	1059	25	56	330	104	24
Total	**76614**	**1255**	**2765**	**9343**	**2808**	**164**

The IRG facilities include:
- An Active Collection for medium-term storage and distribution of samples maintained at a temperature of +4°C, 927 m³, capacity for about 110 000 accessions of ±500 g each.
- A Base Collection for long-term (50->100 yr) conservation maintained at a temperature of −20°C, 164 m³, capacity for 108 000 accessions, each with two aluminium cans, ±60 g each.
- Two screenhouses with a combined area of >4000 m². One is used for the cultivation of low viability or low seed stock accessions of cultivated rice. The other is used exclusively for the cultivation of the wild rices, in pots or special seedbeds.
- A seed drying room at 15°C and 15% RH, where seeds equilibrate to ±6% moisture content.
- A seed testing and germplasm characterization laboratory.
- A data management laboratory, with four computer work stations connected to the IRRI local area network.
- A conservation support laboratory for tissue culture of low viability and seed stock accessions, for cytological and biosystematic studies of the collection.
- A molecular marker laboratory, for studies of isozymes, random amplified polymorphic DNA (RAPD) and other markers.
- Access to >10 ha of field space on the IRRI Central Research Farm - upland site, with assured irrigation facilities for the multiplication and rejuvenation of germplasm and also field characterization.

The IITA also has both active and base collection seed storage facilities for rice germplasm conservation. The germplasm maintained by WARDA, however, including breeding lines, is currently stored in three working collections at each of its research sites: the Continuum Program at Bouaké, Côte d'Ivoire; the Irrigated Sahel Program at St. Louis in Senegal, and at WARDA's lowland breeding unit at IITA.

Rice can be grown almost throughout the year at the IRRI Central Research Farm, which is located at 121°15'E longitude and 14°13'N latitude. Germplasm is multiplied or rejuvenated for long-term conservation from November to May. Studies are conducted to determine optimal conditions for seed quality and longevity. In recent research at Los Baños, Kameswara Rao and Jackson (1996a, 1996b) identified the environmental factors that affect seed quality and therefore potential longevity of storage. Extending the work of Ellis *et al.* (1993) conducted under controlled conditions at the University of Reading in the UK, the field research in the Philippines included more rice varieties and a range of environmental conditions. Changes in seed quality during the ripening stage were studied in 16 rice cultivars, representing the three main types of *O. sativa* – indica, japonica and javanica – and one cultivar of *O. glaberrima*, grown during the 1992-93 dry season (November-May) at Los Baños, Philippines (Kameswara Rao and Jackson 1996a). Kameswara Rao and Jackson (1996b) also studied changes in seed quality during development and maturation in three japonica cultivars and one indica cultivar planted on three different dates in October and November 1993 and early January 1994. The changes in germplasm multiplication and rejuvenation that have been introduced at IRRI, coupled with the post-harvest handling of the seeds, have significantly enhanced the quality of rice germplasm stored in the IRG. If seeds do not germinate after 7 days under optimal conditions, their viability is determined using the topographical tetrazolium test (Ellis *et al.* 1985). Dehulling often improves germination, but studies indicate that different species respond to different temperature regimes (F.C. de Guzman, pers. comm.). For export of germplasm to some countries, a hot water treatment of seeds at 57°C for 15 minutes is a seed health requirement to control the seedborne nematode *Aphelenchoides besseyi*. Checking seed health before conservation has permitted the preparation of 10-g packs of seeds in the Active

Collection, ready for distribution. In the case of the wild species, only 10-20 seeds are distributed per sample.

In West Africa, seed is produced at each of WARDA's main research stations during the post-rainy season. The standing crop is inspected by experts and only disease- and insect-free seed is harvested. The moisture content is brought down to 6-8% before storage in the cold room. At each station, a 500-2000 g seed sample of each accession is kept in appropriate containers in air-conditioned rooms with temperatures ranging from 18 to 20°C and 20-30% RH. IITA maintains a collection of over 12 500 accessions, which include some accessions from WARDA and IRRI. Rice germplasm maintained at IITA was multiplied at rice paddies on site at IITA's station in Ibadan. During the growing season, rice plants were inspected by plant quarantine officers. Seeds harvested from healthy plants were certified for distribution.

For the active collection, up to about 500 g of seeds of each accession are stored in a plastic screw-top jar in a cold store conditioned at 5°C and 30% RH. Requests for germplasm samples from rice researchers are taken from this collection. For the base collection, about 150 g of seeds of each accession with seed germination rate known to be greater than 85% is dried to 5% moisture content in a drying cabinet or room conditioned at less than 10% RH and 20°C, and then sealed into aluminium cans or aluminium foil envelopes and stored in a cold room at –20°C.

At IRRI, the wild species are grown in pots in a restricted-access screenhouse. All wild species accessions are grown under a quarantine agreement with the Philippine Bureau of Plant Industry. Perennial species are maintained as living plants when seeds are difficult to produce. Since the panicles of wild rices shatter at maturity, they are bagged with nylon nets after pollination and mature seeds accumulating in the nets are collected for conservation. The seeds are dried to 6% moisture content, and small samples are stored in the Active and Base Collections.

For many years, the National Seed Storage Laboratory (NSSL) at Fort Collins, Colorado, USA, has provided a 'black box' duplicate safety storage for the germplasm from the IRG. In 1993 IRRI signed a formal agreement with USDA-ARS under which this 'black box' facility has operated. Sealed boxes of rice accessions in aluminium foil packs (20 g per accession) are placed in the –20°C vaults at NSSL, and remain sealed. The IITA, WARDA and IRRI also duplicate a proportion of the rice germplasm they hold in trust among the three institutes.

For several countries, including Sri Lanka, Cambodia and the Philippines, the germplasm conserved in the IRG represents a more or less complete duplicate of national rice collections. For other countries, such as India and the People's Republic of China, only a proportion of national collections are duplicated at IRRI. The IRG has provided an important safety net for several national conservation efforts, when for one reason or another, national collections were lost or where a genebank has not been established within a country. In this respect the germplasm conserved at IRRI has had a significant impact on national conservation efforts and whole collections have been restored to the country of origin. A notable example is Cambodia, where the cultivation of deepwater rices was actively discouraged during the period of political and civil strife of the 1970s. As a consequence, these rices were abandoned and lost in several provinces. Fortunately, earlier germplasm had been collected and conserved at IRRI and when the political climate changed, it was possible to return samples of each accession to Cambodia. In the cases of the Philippines and Sri Lanka, duplicate storage of national germplasm was an essential step during the development of a germplasm conservation infrastructure within each country. Once a medium-term genebank had been constructed in 1992 at the Philippine Rice Research Institute (PhilRice) located in Nueva Ecija Province on the island of Luzon, IRRI was requested to make available a complete set of Philippine accessions that now form the basis of the national rice germplasm collection. In Sri Lanka, a modern genebank was opened in the late 1980s at

Table 19.4. Germplasm restored to donors from the IRRI genebank.

Country	Year	No. of samples
Cambodia	1981-1989	524
India	1986, 1988, 1993, 1994	7111
Indonesia	1994	416
Mexico	1995	110
Nepal	1981	537
Pakistan	1982, 1984, 1985, 1994	3678
Philippines	1987, 1988	1973
Senegal	1988	517
Sri Lanka	1989, 1991	1950
Thailand	1994	392

Peradeniya, near Kandy, and local germplasm was restored to that country from accessions conserved at IRRI.

In 1994, IRRI received a request from the National Bureau of Plant Genetic Resources (NBPGR) in India for samples of the Assam Rice Collection which had been sent to IRRI for duplicate storage in the 1960s. Within a few months of receipt of this request, the IRG was able to return more than 5000 samples of this valuable germplasm to India, where it once again forms part of that country's germplasm heritage. In the same year, the IRG also restored rice germplasm to Thailand and Indonesia. In 1995, the Pakistan Agricultural Research Council requested the restoration of Pakistani accessions stored in the IRG. A list of all germplasm restoration is given in Table 19.4.

Germplasm Characterization and Evaluation
The IRRI and IITA germplasm has been characterized using 44 morphological and agronomic characters and according to the IRRI-IBPGR descriptors for *O. sativa*, published in 1980. Scientists have screened thousands of accessions and identified those having resistance to biotic stresses (Tables 19.5 and 19.6). Resistance to some stresses like grassy stunt virus was not found in *O. sativa*, but its identification in one accession (IRGC 101508) of *O. nivara* from India and its use in rice breeding led to the release of IR36 that at one time was the most widely cultivated variety of any cereal, occupying more than 11 million ha in its heyday (Swaminathan 1982). Based on extensive evaluation and analysis of resistance genes, researchers found distinct 'hot spots' for different pests and diseases (Table 19.7). Resistance to bacterial blight was found in Bangladesh and the Philippines; for blast in Vietnam, Lao PDR, Myanmar, Thailand; and brown planthopper in Sri Lanka. Vaughan (1991) also has indicated that the frequency of resistance genes to major pests and diseases is not uniformly spread throughout Asia. These areas deserve further collecting to acquire more sources and possibly new sources in a different genetic background.

Characterization of rice germplasm at WARDA's main research station at M'be, near Bouaké in Côte d'Ivoire, showed very wide variation in important morphological and agronomic traits within the traditional landraces of *O. sativa* and *O. glaberrima* as well as within some wild rice populations (Jones *et al.* 1993). In their evaluation studies, germplasm with rapid and vigorous vegetative growth to suppress weeds, tolerance for and resistance to major stresses such as drought, blast, African rice gall midge (ARGM), RYMV, nematodes and soil acidity has been identified. Growth duration (days from planting to maturity) of West African rice germplasm ranged from 80 to 140 days (Jones *et al.* 1993). Several accessions with very early duration of between 65 and 75 days to maturity were also identified, confirming earlier reports that some accessions of *O. glaberrima* evaluated during the dry season at Los Baños, Philippines matured within 50-60 days (Ng *et al.* 1991). Of 123 *O. glaberrima* entries screened in 1993 at WARDA, 14 were rated as resistant to leaf and neck blast (Jones *et al.* 1993). Forty-four accessions of

Table 19.5. The value of rice genetic resources: resistance to 13 disease and insect pests in *Oryza sativa* germplasm evaluated at IRRI.

	O. sativa accessions	
Biotic stress	**Number**	**% resistant**
Bacterial blight (*Xanthomonas oryzae* pv. Oryzae)	48203	11.2
Blast (*Pyricularia oryzae*)	36305	26.2
Sheath blight (*Thanatephorus cucumeris* (*Rhizoctonia solani*))	22754	9.2
Rice tungro disease	15795	3.5
Rice ragged stunt virus	13759	4.7
Brown planthopper biotype 1 (*Nilaparvata lugens*)	47268	1.6
Brown planthopper biotype 2	13652	1.5
Brown planthopper biotype 3	16643	1.9
Whitebacked planthopper (*Sogatella furcifera*)	56237	1.6
Green leafhopper (*Nephotettix* spp.)	57437	2.7
Rice whorl maggot (*Hydrellia philippina*)	22598	3.0
Zigzag leafhopper (*Recilia dorsalis*)	2732	10.1
Rice leaf folder (Cnaphalocrosis medinalis; Marasmia patnalis)	8005	0.6

Table 19.6. Rice germplasm accessions tested against major insect pests and diseases of rice by IITA scientists for the development of resistant varieties in Africa.

	Number of accessions/lines	
Insect	**Tested**	**Identified as resistant**
Stalk-eyed fly (*Diopsis longicornis*)	3767	35
Pink stem borer (*Sesamia calamistis*)	1369	8
White stem borer (*Maliarpha separatella*)	2415	5
Striped stem borer (*Chilo zacconius*)	600	10
Gall midge (*Orseola oryzivora*)	635	11
Whorl maggot (*Hydrellia prosternalis*)	430	2
Case worm (*Nymphula stagnalis*)	700	4
Angoumois grain moth (*Sitotroga cerealella*)	234	2
Rice yellow mottle virus	1531	95
Rice blast	1531	167

Sources: IITA 1975, 1976, 1977, 1982, 1983, 1984, 1985, 1986; Soto and Siddiqi 1976; Ng *et al.* 1980a, 1991; John *et al.* 1985; Alam and Masajo 1986; Abifarin 1991; Thottappilly and Rossel 1993; Paul *et al.* 1995.

Table 19.7. 'Hot spots' of useful resistance to diseases and pests in *Oryza sativa* germplasm.

Disease or pest	**Country**
Bacterial blight (*Xanthomonas oryzae* pv. *Oryzae*)	Bangladesh, Philippines
Blast (*Pyricularia oryzae*)	Vietnam, Lao PDR, Myanmar, Thailand
Sheath blight (*Thanatephorus cucumeris* (*Rhizoctonia solani*))	Malaysia, Sri Lanka, Vietnam
Rice tungro disease	Bangladesh
Brown planthopper (biotypes 1, 2 and 3) (*Nilaparvata lugens*)	Sri Lanka
Whitebacked planthopper (*Sogatella furcifera*)	Lao PDR
Green leafhopper (*Nephotettix* spp.)	Bangladesh

O. glaberrima inoculated with RYMV were highly resistant at IITA (Ng *et al.* 1980b). Jones *et al.* (1993) indicated that *O. glaberrima* could serve as a source of increased biomass, grain yield and improved grain quality.

Table 19.8.   Varietal classification of *Oryza sativa* L. based on isozyme polymorphism (adapted from Glaszmann 1986).

Isozyme group	Germplasm
I	Typical indica: Aman (Bangladesh, northeast India), Tjereh (Indonesia), Hsien (China)
II	Varieties from foothills of Himalayas, from Iran to Assam
III	Bhadoia, Aswina (two deepwater varieties from Bangladesh)
IV	Rayada varieties (Bangladesh)
V	Very diverse group (along Himalayas from Iran to Myanmar): basmati rices (Pakistan, India, Nepal); some special rices from Myanmar
VI	Japonica rices (Japan, Korea), keng (China), bulu (Indonesia), upland rices (Southeast Asia), high-altitude varieties (Himalayas)

The application of isozyme and molecular techniques now permits further insight into genetic diversity. Glaszmann (1986) developed a classification of *O. sativa* based on the allelic pattern of 21 isozyme loci.   Varieties of *O. sativa* were classified into six groups based on isozyme polymorphism that showed distinct geographic distribution (Table 19.8). The indica and japonica rices are placed in Groups I and VI, respectively, and this classification has proved extremely useful for the effective utilization of different germplasm in rice breeding because of the potential reproductive barriers. The javanica rices also fall within Group VI, and are often now referred to as tropical japonicas for this reason. They have become the basis of the new plant types being developed at IRRI and described by Khush (1993). The other isozyme groups II, III, IV and V include rice varieties with rather restricted distribution, such as the Rayada varieties of Bangladesh (Group IV) and the Basmati rices from Pakistan (Group V) renowned for their aroma. Analysis of nuclear and mitochondrial DNA has also added a new dimension to our understanding of the pattern of differentiation and diversity in rice (Second 1985; Second and Wang 1992).

The identification of duplicate accessions is a major concern for the IRG because the size of the collection is not necessarily a true reflection of genetic diversity of its germplasm.   Efforts are underway to identify germplasm based on passport information, morphological comparison and molecular markers.   In a collaborative research project between the University of Birmingham (UK) and IRRI, Virk *et al.* (1995a, 1995b) have demonstrated the utility of the PCR-based technique of analysis of RAPD. In one study, 'true' duplicates were included for comparative purposes, as well as suspected duplicate accessions and several that just represented a broad range of rice varieties and were not expected to be duplicates of any of the samples included in the study (Virk *et al.* 1995b). Not only did the study demonstrate that duplicate accessions could be identified, but it also raised the probability of identifying duplicate accessions with a given number of primers, and it indicated how many primers must be used to have confidence that duplicate accessions can be identified. In another study, these researchers showed that RAPD markers could be used to predict quantitative traits in the field (Virk *et al.* 1996).

Germplasm Exchange

Rice researchers throughout the world consider the IRG as a reliable source of rice germplasm. Since 1973, more than 740 000 samples of 10 g of seeds of each (for wild species just 10-20 seeds per packet) have been distributed to rice researchers free of charge, which includes more than 18% to collaborators outside IRRI. The requests are increasing every year, and during 1990 to 1994 alone, 157 363 samples of the cultivated and 9644 samples of wild relatives were supplied in response to requests from diverse sources. The major recipients were universities, research institutes and NARS. The germplasm is used most frequently by scientists at IRRI to identify sources of resistance to biotic and abiotic stresses. To send rice germplasm outside the Philippines requires an import permit from the requesting country and a Philippine phytosanitary certificate to accompany all shipments. Although it no longer has a mandate for rice improvement, IITA continues to distribute rice germplasm on request from its genebank.

Properties and Uses

Table 19.9 lists the caloric properties of rice for human use. Rice provides 20% of global human per capita energy and 15% of per capita protein (Juliano 1993). Unmilled (brown) rice of 17 587 cultivars in the IRRI germplasm collection averages 9.5% protein content, ranging from 4.3 to 18.2%. Rice also provides minerals, vitamins and fibre, although all constituents except carbohydrates are reduced by milling. Milling also removes roughly 80% of its thiamine. Environmental factors (soil fertility, wet or dry season, solar radiation and temperature during grain development) and crop management (added N fertilizer, plant spacing) affect rice protein content (Juliano and Bechtel 1985).

Where rice is the main item of the diet, it is frequently the basic ingredient of every meal and is normally prepared by boiling or steaming. In Asia, bean curd, fish, vegetables, meat and spices are added to rice. A small proportion of rice is consumed in the form of noodles, which serve as a bed for highly spiced specialities and as the bulk ingredient in soups.

Most rice is consumed in its polished state and when it constitutes a high proportion of food intake, dietary deficiencies may result (Juliano 1993). By contrast, parboiling rough rice before milling, a common practice in India and Bangladesh, allows a portion of the vitamins and minerals in the bran to permeate the endosperm and be retained in the polished rice (Juliano 1993). This treatment also lowers protein loss during milling and increases whole grain recovery.

Breeding

Landraces have become widely adapted to a range of agro-ecological conditions and to some pests and diseases. Initial rice breeding activities were limited to selection and purification of locally adapted landraces, which resulted in marginal gains of 10-15% increase in yield (Parthasarathy 1972). The first success in international rice breeding, which led to the release of Mahsuri and ADT27, was achieved at the Central Rice Research Institute, Cuttack, India, by crossing the tall tropical indica varieties with shorter japonicas from Japan and other regions of Eastern Asia. Rice breeding at IRRI began in 1962 following the acquisition of an array of rice germplasm from different sources. A major advance in rice breeding was the development of semidwarf varieties. These possessed high yield potential (10-11 t/ha), shorter crop duration from 150 days or longer to around 100 days, and greater yield stability through genetic resistance or tolerance for pests, diseases and problem soils. The first significant rice breeding achievement was the release of IR8 from a cross between the Chinese dwarf variety Dee-geo-woo-gen and the Indonesian variety Peta. Rice germplasm has been used to develop new plant types, reduce crop duration and incorporate resistance to biotic stresses (Chang 1985; Plucknett *et al.* 1987; Khush 1987, 1993; Chang and Li 1991; Jackson and Huggan 1993; Jackson 1995).

Breeding objectives vary from one rice ecosystem to another, but all emphasize high yield potential, good grain quality and yield stability (Khush 1993). IRRI breeders use pedigree breeding to develop germplasm with multiple resistance to diseases and insects important to Asian rice. Major genes control resistance to blast, tungro, grassy stunt, green leafhopper, brown planthopper and gall midge. Recurrent selection is the preferred method for quantitative trait loci (QTLs). In Latin America and the Caribbean (LAC), breeding aims to stabilize yields and reduce production costs by developing

Table 19.9. Rice consumption, caloric intake and percent of calories from rice, 1990.

Country	Milled rice consumption[†] (kg capita[-1] yr[-1])	Calories capita[-1] year[-1] Total[‡]	Rice	% calories from rice
Myanmar	190	2448	1893	77
Bangladesh	155	2100	1580	75
Indonesia	138	2631	1519	58
Thailand	128	2271	1258	55
Madagascar	104	2162	1091	50
Philippines	99	2452	995	41
People's Rep. of China	94	2706	959	35
India	66	2243	673	30
Japan	62	2926	699	24
Brazil	43	2723	448	16
Egypt	28	3318	300	9
Pakistan	19	2377	189	8
Nigeria	12	2147	130	6
South Africa	8	3158	86	3
Mexico	6	2986	64	2
USA	6	3680	69	2
Turkey	6	3262	57	2
CIS	6	3391	56	2
World	55	2712	574	21

Source: IRRI 1993.
[†] Amount available for human consumption.
[‡] Data include all food available for human consumption.

lines with higher and more stable resistance to major diseases and insects (particularly blast, RHBV, *Tagosodes oryzicolus* and leaf scald), grain discolouration, greater lodging resistance and better grain quality. A recurrent selection programme to increase the yield potential of lowland rice was initiated by CIAT-Palmira in 1993, using several genepools developed by CIRAD and CNPAF. Breeding for upland rice in LAC is concentrated in Brazil, Colombia and Mexico, and each programme deals with a different set of constraints. The Mexican group working on upland rice improvement for unfavourable environments is exploiting an Asian upland indica genetic base to identify drought tolerance and blast resistance. The semidwarf varieties have been adopted in 65% of the rice areas of the world and have doubled rice production from 256 million t in 1965 to 520 million t in 1990 (Khush 1993). The subsequent breeding efforts done by CIAT's Rice Program steadily improved the semidwarf prototype obtained from IRRI, particularly for early maturity, improved grain type and better pest resistance while maintaining the yield potential. Pests rapidly evolve to overcome resistances, however, and breeders constantly bring in new resistance sources just to maintain varietal performance. To increase yield potential further, IRRI scientists have developed a new plant type based on crosses between indica rices and the bulu rices of Indonesia – the so-called tropical japonicas. The discovery of cytoplasmic male sterility wild abortion type (CMS-WA) has permitted exploitation of hybrid vigour, which is reported to produce a 10-20% increase in yield (Virmani 1994).

The wild rices represent a reservoir of useful genes for resistance to diseases, insect pests and tolerance for abiotic stresses. Several useful traits from wild species (Table 19.10) have been transferred into elite breeding lines of rice through backcross breeding (Jena and Khush 1990). Four varieties (MTL98, MTL103, MTL105, MTL110) with resistance to the brown planthopper and whitebacked planthopper from *O. officinalis* (IRGC 100896) have been released in Vietnam. Two new CMS lines – IR66707 A and IR69700 A having cytoplasm from *O. rufipogon* and *O. glumaepatula*, respectively, and in the nuclear background of IR64 – also have been developed. The CMS source of these

Table 19.10. Transfer of useful characteristics from wild species into *Oryza sativa* using embryo rescue.

Species (genome)	Resistance to / tolerance for pests and diseases
O. brachyantha (FF)	Yellow stemborer, bacterial leaf blight (BB)
O. australiensis (EE)	Brown planthopper (BPH), BB
O. latifolia (CCDD)	BPH
O. minuta (BBCC)	BPH, blast, BB; sheath blight
O. officinalis (CC)	BPH, whitebacked planthopper, tungro
O. ridleyi (4x)	Yellow stemborer
O. nivara (AA)	BB, grassy stunt virus
O. longistaminata (A'A')	BB (*Xa-21*)
O. rufipogon (AA)	Tungro, acid sulphate, elongation ability

Source: D.S. Brar, pers. comm.

lines is different from WA cytoplasm, the most commonly used source in hybrid rice breeding (Virmani 1994).

Founded in 1975 as the International Rice Testing Program (IRTP), the International Network for Genetic Evaluation of Rice (INGER) today still provides an important mechanism for the safe exchange of elite germplasm.

Prospects

The collaborative efforts of the IARCs and scientists in many countries over more than three decades have contributed significantly toward the safe preservation of the rice genepool. Without this collaboration, much of the diversity of rice would have been lost altogether. Although much has been accomplished, the task has not yet been completed. Many countries in Asia are aiming to complete the task of collecting rice varieties before the end of the decade. In some countries, this means only collecting in a few remote areas that have not been surveyed in the past. In others, such as the Lao PDR, where collecting activities have been limited and few modern varieties have been introduced, much of the cultivation of rice is based on traditional varieties. During collecting in the second half of 1995, it was possible to identify many hundreds of different rice landraces being grown by farmers under rain-fed lowland and upland conditions.

As the economies of Asian countries in particular grow rapidly and as long as population growth remains unchecked, the demand for rice production from improved varieties will increase, thereby placing further pressure on traditional farming systems. On-farm conservation or farmers' management of diversity is being actively promoted as an alternative, and even better, strategy to *ex situ* conservation in genebanks. Beyond the generally accepted advantage of continued evolution of diversity, there is opportunity for seed exchange in dynamic systems, varietal improvement and the establishment of linkages between farmers' management of diversity and *ex situ* conservation (Bellon *et al.* 1997). Advocating these linkages is necessary, but developing a sustainable strategy is essential. There is a remarkable lack of research to back up the claims of the most ardent proponents of on-farm conservation. IRRI has initiated a research project to remedy this situation, and a multidisciplinary team led by a social anthropologist and population geneticist will evaluate the social and genetic consequences of different farmer management systems.

A question that is often asked concerns the value of rice genetic resources and what has been the return on investment for genetic conservation. Since genetic conservation does not aim to establish museum collections, the assumption has been that germplasm collections have value and genetic conservation has had an impact on rice improvement. Yale economist Robert Evenson and his colleague, Doug Gollin, made a study of the flows of rice germplasm and their impact on national rice production in a number of countries (Evenson and Gollin 1994). They showed that the use of landrace varieties had indeed increased over the past 10-15 years. Plucknett *et al.* (1987) also described IR36, in which no fewer than 15 landraces and one wild species figure in this variety's pedigree. The impact of particular alleles, such as that conferring grassy stunt virus from *O. nivara* (accession IRGC 101508), is easy to demonstrate. Unfortunately, such examples are few and far between.

We should not restrict our definition of use or value of germplasm to whether or not a particular accession has been used in breeding or appears in the pedigree of a released rice variety. Generating knowledge for rice science is equally important. The use of rice germplasm in research contributes to our knowledge about rice genetics, physiology, biochemistry, molecular biology and the reaction of this important crop to its many pests and diseases. All of these factors affect rice production.

The genebanks of the world contain much of the diversity of the rice genepool. This diversity has been used effectively by plant breeders to increase the productivity of rice varieties and thereby contribute to the well-being of growing populations in Asia and elsewhere in the world. It remains to be seen whether the application of biotechnology and molecular biology, which now permit the exploitation of novel and even alien genes to transfer important traits to rice varieties, will supersede the use of rice genetic resources. We do know that these tools are helping rice scientists to understand better the nature of genetic diversity in rice. In future they should help us to produce new varieties and to make the conservation of rice genetic resources more efficient.

Limitations

Irrigated rice production problems in Latin America and the Caribbean differ in many respects from those observed in Southeast Asia, South Asia and West Africa. Rice *hoja blanca* virus (RHBV) and its vector *Tagosodes oryzicolus* are present only in this region, whereas some constraints found in Asia, such as tungro and grassy stunt virus diseases and the brown planthopper, are not present. Similarly, yellow mottle virus found in Africa is not found in Latin America and the Caribbean. Some pests like blast are common to all rice-growing regions, but virulence diversity and frequencies are different. Direct seeding has always been a prevalent feature of rice in Latin America and the Caribbean, which together with often poor water control has resulted in serious weed problems that have led to the widespread use and abuse of herbicides in the region.

About one-fourth of the world's total riceland or approximately 40 million ha is rain-fed, contributing 18% of the global rice supply. Adverse climate, poor soils, a lack of suitable modern varieties and poverty keep farmers from being able to increase productivity. Technologies for the irrigated rice sector can also be applied in the favourable rain-fed lowland subecosystem.

Although upland rice constitutes a relatively small proportion of the total rice area, it is the dominant rice culture in Latin America and West Africa. In Asia, the area of the upland ecosystem is much larger than the area under rice, because rice is grown in rotation with many other crops (De Datta 1981). Upland rice soils range from erodible, badly leached alfisols in West Africa to fertile volcanic soils in some areas in Southeast Asia (Oldeman and Woodhead 1986). In Brazil, where upland rice is a major crop, soils have extremely low CEC values, high P fixation and high levels of exchangeable Al. Upland rice soils in most of Africa have low available water-holding capacity because of coarse texture, are often kaolinitic and have severe nutrient deficiencies and Al and Mg toxicities (Kang and Juo 1984).

Around 10 million ha of ricelands in South and Southeast Asia are subject to uncontrolled flooding. Although average yields are only about 1.5 t/ha, these areas support more than 100 million people. Farmers deal with problems of excess water in places ranging from stagnant >50 cm deep to the very deeply flooded (up to as much as 8 m) areas where floating rice is grown and the cropping systems vary depending upon the time, depth and duration of flooding (Catling 1992). Boro rices are grown in flood-prone ecosystems during the dry season in Bangladesh and India. Traditionally, boro rices were cultivated only in local land depressions that contained sufficient residual water in the soil for a crop during the dry season. With improved irrigation, mainly from tube wells, boro rice cultivation has now spread to other floodplains having low water percolation. Irrigated boro is fast replacing floating rices and has become a second rice crop in some young deltas. Such changes in rice cropping also have occurred in Vietnam after the construction of river canals.

Acknowledgement

Part of this chapter on rice production is adapted from the *IRRI Rice Almanac 1993-95*. The authors appreciate permission to use this information.

REFERENCES

Abifarin, A.O. 1991. Rice evaluation in Africa. Pp. 187-195 *in* Crop Genetic Resources of Africa. Vol. 1 (F. Attere, H. Zedan, N.Q. Ng and P. Perrino, eds.). IBPGR, UNEP, IITA, CNR, Rome.

Adair, C.R. , J.H. Martin, T.H. Johnston, C.N. Bollich, J.R. Gifford, N.E. Jodon, R.J. Smith, Jr., E.C. Tullis, B.D. Webb and J.N. Rutger. 1975. A summary of rice production investigations in the U. S. Department of Agriculture, 1898-1972. I-II. Rice J. 78 (3):26-29.

Alam, M.S. and T.M. Masajo. 1986. Resistance of rice varieties to gallmidge *Orseolia oryziora* H. & G. Abstract paper in Annual Plant Resistance to Insect Newsletter, Volume 12, June 1986, Purdue University, USA.

Bellon, M.R., J.L. Pham and M.T. Jackson. 1997. Rice farmers and genetic conservation. *In* Plant Conservation: the *In Situ* Approach (N. Maxted, B.V. Ford-Lloyd and J.G. Hawkes, eds.). Chapman & Hall, London. (in press).

Bettencourt, E. and J. Konopka. 1990. Directory of germplasm collections. 3. Cereals: *Avena, Hordeum, Millets, Oryza, Secale, Sorghum, Triticum, Zea* and Pseudocereals. IBPGR, Rome.

Bezançon, G., A. De Kochko and K. Goli. 1984. Cultivated and wild species of rice collected in Guinea. Plant Genet. Resour. Newsl. 57:43-46.

Catling, D. 1992. Rice in deep water. IRRI, Manila, Philippines and Macmillan, London and Basingstoke.

Chang, T.T. 1976a. Rice. Pp. 98-104 *in* Evolution of Crop Plants (N.W. Simmonds, ed.). Longman, London.

Chang, T.T. 1976b. The rice cultures. Phil. Trans. R. Soc. Lond. B. 275:143-157.

Chang, T.T. 1976c. The origin, evolution, cultivation, dissemination, and diversification of Asian and African rices. Euphytica 25:425-441.

Chang, T.T. 1985. Crop history and genetic conservation: rice—a case study. Iowa State J. Res. 59:425-456.

Chang, T.T. and E.A. Bardenas. 1965. The morphology and varietal characteristics of the rice plant. Technical Bull. 4. IRRI, Manila, Philippines.

Chang, T.T. and C.C. Li. 1991. Genetics and breeding. Pp. 23-101 *in* Rice: Production, Vol. I, 2nd edn. (B.S. Luh, ed.). Van Nostrand Reinhold, New York.

Chang, T.T. and D.A. Vaughan. 1991. Conservation and potentials of rice genetic resources. Pp. 531-552 *in* Biotechnology in Agriculture and Forestry, Vol. 14. Rice (Y.P.S. Bajaj, ed.). Springer-Verlag, Berlin, Heidelberg.

Charrier, A. and S. Hamon. 1991. Germplasm collection, conservation and utilization activities of the Office de la recherche scientifique et technique outre-mer (ORSTOM). Pp. 41-52 *in* Crop Genetic Resources of Africa, Vol. II (N.Q. Ng, P. Perrino, F. Attere and H. Zedan, eds.). Sayce Publishing, UK.

De Datta, S.K. 1981. Principles and Practices of Rice Production. John Wiley and Sons, New York.

De Kochko, A. 1985. Collecting rice in the Lake Alaotra region in Madagascar. Plant Genet. Resour. Newsl. 63:6-7.

Ellis, R.H., T.D. Hong and M.T. Jackson. 1993. Seed production environment, time of harvest, and the potential longevity of seeds of three cultivars of rice (*Oryza sativa* L.). Ann. Bot. 72:583-590.

Ellis, R.H., T.D. Hong and E.H. Roberts. 1985. Handbook of Seed Technology for Genebanks. Vol. I. IBPGR, Rome, Italy.

Evenson, R.E. and D. Gollin. 1994. Genetic resources, international organizations, and rice varietal improvement. Center Discussion Paper No. 713, Economic Growth Center, Yale University, New Haven, Connecticut, USA.

Ford-Lloyd, B. and M. Jackson. 1986. Plant Genetic Resources: an Introduction to their Conservation and Use. Edward Arnold, London.

Glaszmann, J.C. 1986. A varietal classification of Asian cultivated rice (*Oryza sativa* L.) based on isozyme polymorphism. Pp. 83-90 *in* Rice Genetics. IRRI, Manila, Philippines.

Grist, D.H. 1986. Rice. 6th edn. Tropical Agriculture Series. Longman, London and New York.

Huke, R.E. and E.H. Huke. 1990. Rice: then and now. IRRI, Manila, Philippines.

IITA. 1975. Annual Report for 1974. IITA, Ibadan, Nigeria.

IITA. 1976. Annual Report for 1975. IITA, Ibadan, Nigeria.

IITA. 1977. Annual Report for 1976. IITA, Ibadan, Nigeria.

IITA. 1982. Annual Report for 1981. IITA, Ibadan, Nigeria.

IITA. 1983. Annual Report for 1982. IITA, Ibadan, Nigeria.

IITA. 1984. Annual Report for 1983. IITA, Ibadan, Nigeria.

IITA. 1985. Annual Report for 1984. IITA, Ibadan, Nigeria.

IITA. 1986. Annual Report for 1985. IITA, Ibadan, Nigeria.

IRRI. 1993. IRRI rice almanac, 1993-1995. IRRI, Manila, Philippines.

IRRI. 1995. World rice statistics, 1993-1994. IRRI, Manila, Philippines.

Jackson, M.T. 1995. Protecting the heritage of rice biodiversity. GeoJournal 35:267-274.

Jackson, M.T. and R. Huggan. 1993. Sharing the diversity of rice to feed the world. Diversity 9 (3):22-25.

Jena, K.K. and G.S. Khush. 1990. Introgression of genes from *Oryza officinalis* Well ex Watt to cultivated rice, *O. sativa* L. Theor. Appl. Genet. 80 (6):737-745.

John, V.T., G. Thottappilly, Q. Ng, K. Alluri and J.W. Gibbons. 1985. Varietal reaction to rice yellow mottle virus disease. FAO Plant Prot. Bull. Vol. 33, No. 3.

Jones, M.P., E.A. Heinrichs, D.E. Johnson and C. Riches. 1993. Characterization and utilization of *Oryza glaberrima* in upland rice improvement. Pp. 3-12 *in* WARDA Annual Report 1993. Bouaké, Côte d'Ivoire.

Juliano, B.O. 1993. Rice in human nutrition. FAO Food and Nutrition Series, No. 26. FAO, Rome, Italy and IRRI, Manila, Philippines.

Juliano, B.O. and D.B. Bechtel. 1985. The rice grain and its gross composition. Pp. 17-57 *in* Rice: Chemistry and Technology, 2nd edn. (B.O. Juliano, ed.). American Association of Cereal Chemists, Inc., St. Paul, Minnesota, USA.

Kameswara Rao, N. and M.T. Jackson. 1996a. Seed longevity of seventeen rice cultivars and strategies for germplasm conservation in genebanks. Ann. Bot. 77:251-260.

Kameswara Rao, N. and M.T. Jackson. 1996b. Seed production environment and storage longevity of japonica rices (*Oryza sativa* L.). Seed Sci. Res. 6:17-21.

Kang, B.T. and A.S.R. Juo. 1984. Review of soil fertility management and cropping systems for wetland rice production in West Africa. Pp. 493-502 *in* An Overview of Upland Rice Research. IRRI, Manila, Philippines.

Khush, G.S. 1987. Rice breeding: past, present and future. J. Genet. 66 (3):195-216.

Khush, G.S. 1993. Breeding rice for sustainable agricultural systems. Pp. 189-199. *in* International Crop Science I (D.R. Buxton, R. Shibles, R.A. Forsberg, B.L. Blad, K.H. Asay, G.M. Paulsen and R.F. Wilson, eds.). Crop Science Society of America, Madison, USA.

Ng, N.Q., I.W. Buddenhagen and S. Sarkarung. 1980a. Screening of rice germplasm for rice yellow mottle virus resistance. *In* The role of genetics in food and health in the eighties. Abstract of the Genetic Societies of Nigeria, 4th Annual Conference, University of Benin, Benin City, 4-5 December, 1980.

Ng, N.Q., T.T. Chang, D.A. Vaughan and C. Zuno-Alto Veros. 1991. African rice diversity: conservation and prospects for crop improvement. Pp. 213-227 *in* Crop Genetic Resources of Africa, Vol. II (N.Q. Ng, P. Perrino, F. Attere and H. Zedan, eds.). Sayce Publishing, UK.

Ng, N.Q., A.T. Perez and S. Sarkarung. 1980b. Screening of rice germplasm for leaf blast resistance. *In* The role of genetics in food and health in the eighties. Abstract of the Genetic Societies of Nigeria, 4th Annual conference, University of Benin, Benin City, 4-5 December, 1980.

Oldeman, L.R. and T. Woodhead. 1986. Physical aspects of upland rice environments (an introductory statement). Pp. 3-5 *in* Progress in Upland Rice Research. IRRI, Manila, Philippines.

Parthasarathy, N. 1972. Rice breeding in tropical Asia up to 1960. Pp. 5-29 *in* Rice Breeding. IRRI, Manila, Philippines.

Paul, C.P., N.Q. Ng and T.A.O. Ladeinde. 1995. Diallel analysis of resistance to rice yellow mottle virus in African rice *Oryza glaberrima* Steud. J. Genet. & Breed. 49:217-222.

Plucknett, D.L., N.J.H. Smith, J.T. Williams and N. Murthi Anishetty. 1987. Gene Banks and the World's Food. Princeton University Press, Princeton, USA.

Second, G. 1985. Geographic Origins, Genetic Diversity and the Molecular Clock Hypothesis in the *Oryzeae*. NATO Adv. Stud. Inst. Ser. G5:41-56.

Second, G. and Wang. Z.Y. 1992. Mitochondrial DNA RFLP in genus *Oryza* and cultivated rice. Genet. Res. Crop Evol. 39:125-140.

Solheim, W.G. 1972. An earlier agricultural revolution. Sci. Am. 226 (4):34-41.

Soto, P.E. and Z. Siddiqi. 1976. Screening for resistance to African insect pests. Paper presented at the WARDA Varietal Improvement Seminar. Bouaké, Ivory Coast.

Swaminathan, M.S. 1982. Beyond IR36 - Rice research strategies for the 80s. Paper presented at the International Centers' Week, World Bank, November 10, 1982, Washington, D.C.

Thottappilly, G. and H.W. Rossel. 1993. Evaluation of resistance to rice yellow mottle virus in *Oryza* species. Indian J. Virol. 9(1):65-73 (January 1993).

Vaughan, D.A. 1989. The genus *Oryza* L.: current status of taxonomy. IRRI Res. Paper Ser. No. 138. IRRI, Manila, Philippines.

Vaughan, D.A. 1991. Choosing rice germplasm for evaluation. Euphytica 54:147-154.

Vaughan, D.A. 1994. The wild relatives of rice—a genetic resources handbook. IRRI, Manila, Philippines.

Vergara, B.S. 1991. Rice plant growth and development. Pp. 13-22 *in* Rice. Vol. I (2nd edn.) (B.S. Luh, ed.). Van Nostrand Reinhold, New York.

Virk, P.S., B.V. Ford-Lloyd, M.T. Jackson and H.J. Newbury. 1995a. The use of RAPD for the study of diversity within plant germplasm collections. Heredity 74:170-179.

Virk, P.S., B.V. Ford-Lloyd, M.T. Jackson, H. Pooni, T.P. Clemeno and H.J. Newbury. 1996. Predicting quantitative traits in rice using molecular markers and diverse germplasm. Heredity 76:296-304.

Virk, P.S., H.J. Newbury, M.T. Jackson and B.V. Ford-Lloyd. 1995b. The identification of duplicate accessions within a rice germplasm collection using RAPD analysis. Theor. Appl. Genet. 90:1049-1055.

Virmani, S.S. 1994. Heterosis and hybrid rice breeding. Monographs on Theoretical and Applied Genetics 22. Springer-Verlag Berlin and Heildelberg and IRRI, Manila, Philippines.

Yoshida, S. 1981. Fundamentals of Rice Crop Science. IRRI, Manila, Philippines.

Chapter 20

Sorghum[1]

J.W. Stenhouse, K.E. Prasada Rao, V. Gopal Reddy and S. Appa Rao

Cultivated sorghum, *Sorghum bicolor* (L.) Moench, is the fifth most important cereal in the world after rice, wheat, maize and barley. The countries with the greatest areas of sorghum production are India (12.9 million ha), Sudan (5.7 million ha), Nigeria and the USA (4.0 million ha each) and Niger (2.35 million ha). Other major producers include Mexico, Argentina, Colombia and Venezuela in the Americas; Burkina Faso, Ethiopia, Mali, Tanzania, Chad, Cameroon, Somalia and Mozambique in Africa; China, Pakistan and Yemen in Asia. In terms of production, the major sorghum-growing countries of the world are the USA (17.5 million t), India (12.4 million t), China (5.1 million t), Mexico (4.4 million t), Nigeria (4.0 million t) and Sudan (3.3 million t). In many other countries, production is much lower but sorghum is a significant part of the agricultural production and a very important food crop for millions of poor farmers. This is particularly true for rain-fed areas of Asia and Africa. In many areas, the stalks and foliage (used as fodder, fuel, thatching and fencing material) are valued as much as the grain.

BOTANY AND DISTRIBUTION

Sorghum bicolor (L.) Moench ($2n=20$) is synonomous with *Holcus bicolor* L., *Andropogon sorghum* (L.) Brot. and *Sorghum vulgare* Pers. Sorghum has been called great millet, guinea corn, milo, sorgo (English); *sorgho* (French); *sorgo* (Spanish, Portuguese); *jowar, cholam, jonna* (India); *kaoliang* (China) and *durra* (Sudan).

Sorghum bicolor is considered an extremely variable crop-weed complex. It comprises wild, weedy and cultivated annual forms which are fully interfertile. The cultivated forms fall in *S. bicolor* subsp. *bicolor* and are classified, in the most widely accepted system, into 5 basic and 10 intermediate races or groups of cultivars (Harlan and de Wet 1972; de Wet 1978) on the basis of spikelet morphology and panicle shape. Subrace names are being added to this system (Doggett and Prasada Rao 1995; Prasada Rao *et al.* 1989). The wild forms are classified into *S. bicolor* (L.) Moench subsp. *verticilliflorum* (Steud.) Piper (synonyms: *S. arundinaceum* (Desv.) Stafp, *S. bicolor* (L.) Moench subsp. *arundinaceum* (Desv.) de Wet & Harlan). The subspecies is further divided into four overlapping races, the most widely distributed and variable of which is verticilliflorum, found across the African savanna and introduced into tropical Australia, parts of India and the Americas. The weedy forms are classified into *S. bicolor* (L.) Moench subsp. *drummondii* (Steud.) de Wet (synonyms: *S. sudanense* (Piper) Stapf, *S. drummondii* (Steud.) Millsp. & Chase) which arose and probably continue to arise from crossing between cultivated grain sorghums and their close wild

---

[1] Submitted as Journal Article no. 1896 by ICRISAT.

relatives wherever in Africa they are sympatric. The hybrids have stabilized and occur as very persistent weeds in abandoned fields and field margins. A well-known forage grass, Sudan grass, belongs to this complex.

The primary genepool includes the *S. bicolor* complex described above and also *S. propinquum* (Kunth) Hitchc., a wild diploid complex found in Southeast Asia (Acheampong *et al.* 1984). The secondary genepool includes *S. halepense* (L.) Pers., a rhizomatous tetraploid fodder, thought to be an autotetraploid of *S. propinquum* (Acheampong *et al.* 1984) or the result of chromosome doubling (Doggett and Prasada Rao 1995). Commonly known as Johnson grass, this is a native of southern Eurasia east to India, but has now been introduced as a weed to all warm temperate regions of the world. The tertiary genepool includes all other sections/subgenera of *Sorghum*.

Origin, Domestication and Distribution

Sorghum is thought to have been domesticated between 5000 and 7000 years ago by selection from wild sorghum (Purseglove 1972), in the northeastern quarter of Africa, and this region shows the greatest variability of cultivated and wild sorghum (Doggett 1988). Other theories suggest that it may have had separate origins from wild species in western, eastern and eastern-central Africa (Snowden 1935; de Wet and Huckabay 1967). Remains of sorghum grains, 8000 years old, have been recovered from archaeological sites in southern Egypt (Wendorf *et al.* 1992). After spreading through Africa, sorghum reached India along shipping and trade routes through the Middle East at least 3000 years ago (Mann *et al.* 1983). It followed the Silk Route to China and to Southeast Asia by coastal shipping, and to the Americas and Australia from West Africa, North Africa, South Africa and India towards the end of the 19th century. The evolution and distribution of the races is illustrated in Figure 20.1.

Sorghum is now distributed in the drier areas of Africa, Asia, the Americas and Australia from sea level to 2200 m asl and up to 50°N in Russia and 40°S in Argentina. Modern cultivars predominate in the Americas, China and Australia, but probably occupy less than 10% of the cultivated area in Africa. In India, approximately half the sorghum area is sown to traditional landraces and the other half to modern varieties. This average figure masks considerable variation between the growing seasons and different regions of the country.

Reproductive Biology

Flowering begins when the peduncle completes elongation and typically lasts 7-9 days. Stigmas are receptive for 1-2 days before flowering and remain so for up to a week if not pollinated. They are pollinated from stamens of the same flower, from other flowers of the same panicle, or from other inflorescences. The range of cross-pollination is from 0 to 50% with an average of about 5%. The grain reaches its maximum dry weight 25-55 days after flowering, depending on environmental conditions. The moisture content at this stage usually varies between 25 and 35%. Grain is usually harvested when the moisture content drops below 15%, which may require an additional 10-20 days.

GERMPLASM CONSERVATION AND USE

The sorghum world collection was begun in the 1960s by the Rockefeller Foundation as part of the Indian Agricultural Research Programme (Murty *et al.* 1967; House 1985). This collection was transferred to ICRISAT in 1974. About 3000 accessions missing were obtained from duplicate materials held in the USA but there remains a gap of about 4000 accessions (Mengesha and Prasada Rao 1982). Many new sorghum accessions have been added; currently, 35 186 samples of sorghum from 90 countries are stored at ICRISAT Asia Center (IAC), Patancheru, India. The collection

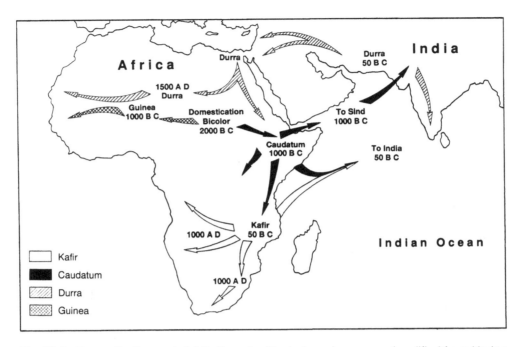

Fig. 20.1. Domestication and distribution of cultivated sorghum races (modified from Harlan and Stemler 1976).

comprises 29 474 landrace accessions, 418 accessions of 13 wild species, 935 unclassified samples and 4359 improved cultivars and breeding lines. Countries that have contributed more than 100 accessions are listed in Table 20.1, and the wild species represented are listed in Table 20.2.

The coverage achieved in the collection varies considerably from country to country. The diversity from many countries has been adequately sampled but others remain poorly represented and are high priority for future collecting. These include Angola, Chad, China, Côte d'Ivoire, Eritrea, Mozambique, Thailand, Turkey and Zaire. Several remain inaccessible for reasons of civil strife and political ideology. High priority for collecting of wild relatives of sorghum are Ethiopia, Indonesia, Myanmar and Philippines, and parts of West Africa.

Although sorghum is a partially outbreeding species, germplasm accessions are maintained and multiplied by selfing. During collecting and initial multiplication at IAC, diverse types have been separated such that each accession is a true-breeding single type. Individual panicles are covered with selfing bags as soon as they emerge from the boot prior to anthesis. They are kept covered for at least 3 weeks, after which the bags are removed to assist seed drying and maturity. Seed from at least 30 selfed plants is bulked to maintain an accession. Wild relatives are also maintained by selfing. All accessions are stored in an active collection in medium-term storage conditions (4°C, 20% RH). For each accession, 300-400 g of seed is harvested from post-rainy season multiplication plots, field-dried to 8% moisture content and stored in screw-capped aluminium containers. About 17% or 5954 accessions are also maintained at 18°C for long-term storage. From 3000 to 4000 seeds (125 g) are cleaned and dried to 5-6% moisture content by equilibration with air at 20°C and 15% RH for approximately 3 weeks. The dried grain is vacuum-sealed in aluminium foil bags and stored after confirming that initial viability is more than 95%. Viability is monitored by germination testing at 5-year intervals in medium-term storage and 10-year intervals in long-term

Table 20.1. Geographic distribution of races of the world collection of sorghum germplasm.

Country	B†	C	CB	D	DB	DC	G	GB	GC	GD	GK	K	KC	KD	KB	UC	Total
Benin	1	–	2	1	–	2	181	2	3	–	–	–	–	–	–	5	197
Botswana	3	17	11	50	3	18	13	2	9	1	3	54	8	24	2	–	218
Burkina Faso	7	4	10	23	5	7	414	5	68	2	–	–	–	–	–	1	546
Burundi	–	102	3	4	1	3	3	2	11	–	–	–	–	1	–	6	136
Cameroon	12	1304	34	360	20	274	202	37	218	2	1	–	–	–	–	16	2479
Central African Rep.	3	90	4	19	5	72	18	13	18	2	2	–	–	3	–	1	249
Chad	9	43	11	10	12	18	22	6	46	2	–	2	–	–	–	6	187
China	19	146	82	3	14	49	1	4	20	–	1	5	–	3	27	6	380
Ethiopia	107	246	59	656	627	169	7	3	151	4	–	4	–	5	–	32	2070
Ghana	5	13	5	4	1	2	83	3	27	2	–	1	–	–	–	–	146
India	337	204	123	3088	442	356	769	23	171	57	1	31	8	33	10	100	5753
Japan	17	35	10	7	2	11	17	5	2	–	2	–	–	–	–	–	108
Kenya	25	741	26	5	5	36	9	1	101	2	–	1	1	2	–	14	967
Lebanon	18	30	51	23	13	142	2	2	36	12	–	17	12	6	–	–	360
Lesotho	4	7	5	1	1	2	1	–	62	–	8	102	60	–	1	9	269
Malawi	5	19	13	5	1	8	256	18	59	13	–	–	1	–	–	5	403
Mali	13	25	11	61	24	40	446	8	13	1	–	–	1	–	–	23	666
Namibia	2	47	8	21	7	1	8	4	48	–	2	13	11	–	–	9	181
Niger	14	32	51	63	43	100	34	5	62	3	–	4	5	1	3	1	408
Nigeria	30	220	59	58	34	225	558	94	168	16	–	4	2	1	1	9	1484
Former USSR	57	44	23	85	63	37	1	2	12	6	–	–	–	–	–	14	351
Rwanda	2	217	–	44	1	19	–	–	2	1	–	–	–	–	–	4	290
Senegal	4	2	6	10	2	5	190	2	18	–	–	–	–	–	–	–	239
Sierra Leone	–	–	–	–	–	–	106	–	–	–	–	–	–	–	–	1	107
Somalia	–	11	–	421	1	3	–	–	2	–	–	–	–	–	–	7	445
South Africa	23	188	60	14	3	22	22	2	23	–	1	452	49	22	9	12	902
Sudan	78	983	121	229	47	307	42	10	504	27	–	23	39	2	11	19	2442
Swaziland	5	13	12	–	–	11	15	–	62	3	6	45	14	10	–	5	201
Tanzania	12	105	16	50	5	17	373	9	96	7	–	1	4	–	2	4	701
Togo	3	9	4	11	1	1	236	4	17	–	–	–	–	–	–	8	294
Uganda	33	928	34	10	12	38	27	3	200	2	2	20	6	11	3	8	1337
USA	316	278	151	125	49	259	66	21	159	12	65	235	107	66	41	10	1960
Yemen	8	115	30	307	215	1301	1	–	87	8	–	2	1	5	–	48	2129
Venezuela	2	35	10	7	2	26	2	–	11	2	–	62	6	4	4	–	173
Zambia	21	41	49	3	1	8	83	6	126	1	–	–	1	–	–	1	341
Zimbabwe	14	240	63	75	11	109	248	7	535	18	9	147	28	34	2	59	1599
Others	142	186	72	138	60	108	127	5	85	4	1	29	13	14	6	18	1008
Total	1351	6720	1229	5991	1733	3806	4583	309	3232	210	102	1254	377	246	122	461	31726

† B=bicolor, C=caudatum, D=durra, G=guinea, K=kafir, UC=unclassified.

Table 20.2. Wild relatives of sorghum assembled at ICRISAT Asia Center, to January 1996.

Species	Subspecies	Race	Subrace	No. of accessions
Genus *Sorghastrum*; Section *Sorghastrum rigidifolium*				
–	–	–	–	8
Genus *Sorghum*; Section *Para sorghum*				
versicolor	–	–	–	17
purpureosericeum	*deccanense*	–	–	5
purpureosericeum	*dimidiatum*	–	–	3
nitidum	–	–	–	3
australiense	–	–	–	3
Genus *Sorghum*; Section *Chaeto sorghum*				
macrospermum	–	–	–	1
Genus *Sorghum*; Section *Stipo sorghum*				
intrans	–	–	–	5
brevicallosum	–	–	–	1
stipodeum	–	–	–	9
plumosum	–	–	–	4
matarankense	–	–	–	3
Genus *Sorghum*; Section *Sorghum*				
halepense	–	Halepense	Halepense	15
halepense	–	Halepense	Johnson grass	5
halepense	–	Halepense	Almum	5
halepense	–	Miliaceum	–	4
halepense	–	Controversum	–	4
propinquum	–	–	–	3
bicolor	*drummondii*	–	–	155
bicolor	*verticilliflorum*	Verticilliflorum	–	97
bicolor	–	Arundinaceum	–	36
bicolor	–	Virgatum	–	16
bicolor	–	Aethiopicum	–	16
Total				**418**

Source: Prasada Rao *et al.* 1989.

storage. Any samples with viability of less than 85% are identified for rejuvenation. A number of wild species of sorghum that do not set seed readily are maintained as living plants in a field genebank.

The sorghum germplasm collections at IAC and at Fort Collins, Colorado, USA overlap considerably but they are poorly cross-referenced. ICRISAT has agreements for duplicate safety-conservation with the SADC Regional Genebank, Lusaka, Zambia and with the Genebank of Kenya, Nairobi, Kenya, but only a small number of accessions have been transferred for safety-duplication. More than 250 000 samples have been distributed to 99 countries over the past 20 years (Table 20.3).

The bicolor race is the most primitive cultivated sorghum. Other sorghum races were probably derived from crosses between bicolor and wild species. Approximately 4% of the accessions, collected mainly from Ethiopia and India, belong to bicolor. The most common intermediate race involving bicolor is durra-bicolor, which is found mainly in Ethiopia, India and Yemen.

Guinea, comprising 15%, is considered to be the oldest specialized race and is found mostly in West Africa, with a secondary centre in eastern Africa. This race includes the margaritiferum group from West Africa which have extremely hard, very small, oval grains. The conspicuum group with large drooping panicles and large flattened grain is found predominantly in Tanzania and Malawi; and the roxburghii group in southern Africa and in the hilly regions of central India (de Wet *et al.* 1972). The most common intermediate race involving guinea is guinea-caudatum, comprising 10% of the total collection, which has been particularly important in sorghum improvement.

Table 20.3. Number of samples of sorghum germplasm distributed by ICRISAT.

Year	Asia	Africa	America	Europe	Oceania	Total
1973	–	–	–	–	3	3
1994	8	–	354	–	–	362
1975	2 272	152	753	15	–	3 192
1976	2 425	1 802	195	100	–	4 522
1977	3 297	2 148	527	61	225	6 258
1978	2 268	482	75	27	–	2 852
1979	4 255	4 551	435	269	–	9 510
1980	2 307	303	481	207	397	3 695
1981	3 927	4 298	1 350	38	16	9 629
1982	6 094	3 380	4 847	142	–	14 463
1983	3 815	16 574	284	804	18	21 495
1984	2 455	7 602	5 903	748	83	16 791
1985	3 989	8 146	655	1384	18	14 192
1986	15 934	2 759	4 004	976	20	23 693
1987	13 710	4 582	524	300	–	19 116
1988	21 497	5 655	1 418	30	99	28 699
1989	11 754	842	3 971	237	1	16 805
1990	7 610	4 069	522	133	40	12 374
1991	7 473	2 580	58	422	–	10 533
1992	5 577	884	4 074	222	6	10 763
1993	7 864	6 773	7	26	–	14 670
1994	2 789	3 704	878	92	–	7 463
1995	2 056	690	187	20	5	2 958
Total	**133 376**	**81 976**	**31 502**	**6253**	**931**	**254 038**

Durra was probably the next sorghum race to evolve, as it is the only other race that reached India before the 18th century. This race is distributed in a belt across Africa in the dry zone immediately south of the Sahara between 10° and 15°N and is predominant in Ethiopia. It is also found on the fringes of the Kalahari desert in Botswana and Namibia and is the dominant race grown in the Nile valley in Sudan and Egypt. In Asia, it is widely distributed in India and Yemen. It is adapted to areas of low rainfall. Maldandi, the post-rainy season cultivar that is grown on over 6 million ha in India, is a durra. The milo sorghum that formed the basis of much of the early sorghum breeding programmes in the USA and was an important contributor to the discovery of cytoplasmic male-sterility systems in sorghum (Stephens and Holland 1954) also belongs to this race. Introgression between durra and *S. propinquum* probably gave rise to the kaoliang sorghums of China (Doggett and Prasada Rao 1995). Durra comprises approximately 20% of the collection. Intermediate races of durra comprise a further 19.5% of the collection, durra-caudatums forming the largest group.

The kafir race probably originated by introgression between cultivated sorghum and the wild species *S. verticilliflorum*, a relationship that has been confirmed by electrophoretic studies (Mann *et al.* 1983). Kafir reached West Africa or India in modern times (Doggett 1988) and has contributed to cytoplasmic male-sterility systems. Most commercially important male-sterile lines are derived from kafir germplasm. Kafir sorghum landraces tend to be insensitive to photoperiod (Appa Rao and Mushonga 1987), a relatively unusual trait. Kafirs make up approximately 4% of the germplasm collection.

The caudatum race is the most recent in origin and is the largest representative, comprising 21% of the world collection. The high-altitude sorghums of eastern Africa, many of which are used for brewing traditional opaque beers (Prasada Rao and Murty 1982; House 1985), belong predominantly to the caudatum race. Feteritas and Hegaris from Sudan, Dobbs from Uganda, Nkumba from Tanzania and Zangada from Ethiopia are some of the widely grown caudatum landraces. Intermediate races of caudatum are particularly numerous in the collection, making up approximately 29%. The so-called 'half caudatums' have contributed significantly to sorghum breeding programmes worldwide. It has been suggested these sorghums were developed in Sudan and introduced into the Gambella region of Ethiopia (Prasada Rao and Mengesha 1981). Durra-caudatum mixtures are also very numerous, particularly in Yemen and Nigeria.

The collection has been fully characterized for most of the botanical descriptors agreed and published by IBPGR and ICRISAT (IBPGR/ICRISAT 1993). Table 20.4 lists the important traits. Among the traits, substantial variation has been found for both the length and strength of the central rachis, primary and secondary branches, and combinations that confer distinct panicle shapes and densities (Fig. 20.2). The proportions of the collection showing different combinations of panicle shape and compactness are shown in Table 20.5 and seed colours in Table 20.6. The presence of a phenolic compound, flavan-4-ol, in the pericarp results in red pigmentation that has been associated with resistance to grain moulds (Jambunathan *et al.* 1990). The 100-grain mass of sorghum ranges from 0.29 g to more than 8 g (Table 20.7). Approximately 96% of accessions carry the red and black purple pigmentation; the remainder are tan (Table 20.8). Tan plant colour has been associated with resistance to leaf diseases and grain weathering (Frederiksen and Duncan 1982; Duncan *et al.* 1991). White midrib, found in 76% of accessions, is the most common, followed by dull green midrib, found in 23% (Table 20.8). Brown midrib appears to be often associated with low lignin content and high digestibility of the plant parts similar to the brown midrib trait in maize (Cherney *et al.* 1986). In southern Africa, farmers use dull green midrib to identify sweet stalk sorghums for chewing (Appa Rao *et al.* 1989). White midrib, in contrast, is associated with dry pithy stems. Variation in glume colour is similar to that in grain colour and to some extent the traits are governed by the same genes (Doggett 1988).

The accessions that have been identified as resistant or possessing some valuable trait are maintained at IAC as a separate collection of genetic stocks (Prasada Rao *et al.* 1989). The status of the genetic stocks collection is shown in Table 20.9. Over the years, ICRISAT has conducted large-scale screening of sorghum germplasm accessions for insect resistance under artificial infestation and enhanced infestation of natural populations, using screening techniques developed for the purpose (Sharma *et al.* 1992). Lists of resistant germplasm accessions have been published (Taneja and Leuschner 1985a, 1985b; Sharma *et al.* 1992, 1993). Over 25 000 accessions were screened by ICRISAT entomologists for resistance to shoot fly, *Atherigona soccata* Rondani, using a field-screening technique, and 40 sources of resistance have been identified (Sharma *et al.* 1992). Earlier, 42 resistant germplasm lines were identified (Taneja and Leuschner 1985a). The Indian sorghum programme has also screened over 10 000 accessions from the world collection for resistance to shoot fly (Rana *et al.* 1985). Most of the resistant lines are of Indian origin. Over 17 000 accessions also have been screened for the glossy trait that has been associated with shoot fly resistance (Maiti and Bidinger 1979) and almost 500 glossy lines identified (Maiti *et al.* 1984). Almost 20 000 germplasm accessions have been screened for resistance to stem borer, *Chilo partellus* Swinhoe, using a combination of testing under natural and artificial infestations, and 77 have been identified as resistant (Sharma *et al.* 1992).

The Indian sorghum breeding programme also has screened over 10 000 accessions and identified numerous resistance sources (Rana *et al.* 1985). Breeding for stem borer resistance has produced several improved sources, mainly using IS 2205, a durra accession from India, as the original source of resistance. Over 15 000 accessions have been screened for resistance to midge, *Contarinia sorghicola* Coq., under enhanced natural infestation and under no-choice conditions using headcages, and 27 have been identified as resistant (Sharma *et al.* 1993). Breeding for midge resistance at IAC has resulted in many high-yielding lines with high resistance levels (Sharma *et al.* 1992).

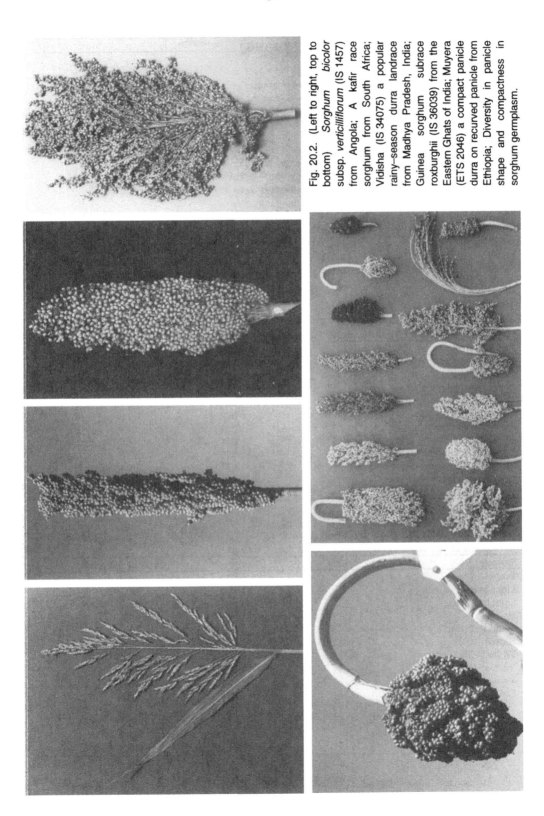

Fig. 20.2. (Left to right, top to bottom) *Sorghum bicolor* subsp. *verticilliflorum* (IS 1457) from Angola; A kafir race sorghum from South Africa; Vidisha (IS 34075) a popular rainy–season durra landrace from Madhya Pradesh, India; Guinea sorghum subrace roxburghii (IS 36039) from the Eastern Ghats of India; Muyera (ETS 2046) a compact panicle durra on recurved panicle from Ethiopia; Diversity in panicle shape and compactness in sorghum germplasm.

Table 20.4. Range of variation in selected characters in sorghum germplasm.

Character	Range of variation
Days to 50% flowering (d)	36 – 199
Plant height (cm)	55 – 655
Pigmentation	Tan – Pigmented
Midrib colour	White – Brown
Peduncle exsertion (cm)	0 – 55.0
Head length (cm)	2.5 – 71.0
Head width (cm)	1.0 – 29.0
Head compactness and shape	Very loose stiff branches – Compact oval
Glume colour	Straw – Black
Glume covering	Fully covered – Uncovered
Grain colour	White – Dark Brown
Grain size (mm)	1.0 – 7.5
100-seed mass (g)	0.29 – 8.56
Endosperm texture	Completely starchy – Completely corneous
Threshability	Freely threshable – Difficult to thresh
Lustre	Lustrous – Non-lustrous
Subcoat (testa)	Present – Absent

Source: Prasada Rao *et al.* 1989.

Table 20.5.   Classification by panicle shape and compactness of 31 752 sorghum germplasm accessions maintained at ICRISAT Asia Center, Patancheru, India.

Panicle shape	Number	Percentage
Compact elliptic	3 102	9.8
Compact oval	935	2.9
Loose, drooping branches	1 689	5.3
Loose, stiff branches	3 857	12.1
Semi-compact elliptic	9 622	30.3
Semi-compact oval	759	2.4
Semi-loose drooping branches	324	1.0
Semi-loose stiff branches	10 420	32.8
Very loose drooping branches	438	1.4
Very loose erect branches	173	0.5
Unclassified	433	1.4

Table 20.6.   Classification by seed colour of 31 752 sorghum germplasm accessions maintained at ICRISAT Asia Center, Patancheru, India.

Colour	Number	Percentage
Brown	2 750	8.7
Light brown	1 624	5.1
Light red	3 695	11.6
Red	841	2.6
Reddish brown	4 703	14.8
Straw	6 611	20.8
White	6 175	19.4
Yellow	1 919	6.0
Xenia	19	<0.1
Cream white	1 176	3.7
Grey	1 671	5.3
Purple	355	0.1
Black	1	<0.1
Unclassified	212	0.1

Table 20.7. Seed mass of sorghum germplasm accessions at ICRISAT Asia Center.

Seed mass (g)	Number of accessions
<1.00	175
1.01-2.00	3 428
2.01-3.00	11 015
3.01-4.00	8 493
4.00-5.00	3 346
5.01-6.00	998
6.01-7.00	234
7.01-8.00	38
>8.01	4

Table 20.8. Classification by plant pigmentation and midrib colour of 31 752 sorghum germplasm accessions maintained at ICRISAT Asia Center, India.

Character	Number	Percentage
Plant pigmentation		
Pigmented	30 483	96.0
Tan	1 097	3.5
Unclassified	172	0.5
Midrib colour		
White	24 070	75.8
Dull	7 242	22.8
Yellow	252	<0.1
Brown	13	<0.1
Unclassified	175	<0.1

Table 20.9. Genetic stocks collection maintained at ICRISAT Asia Center, Patancheru, India, as of June 1986.

Type of accession	No. of accessions
Promising lines for pest resistance	
Shoot fly (*Atherigona soccata*)	60
Stem borer (*Chilo partellus*)	70
Midge (*Contarinia sorghicola*)	14
Headbug (*Calocoris angustatus*)	6
Promising lines for disease resistance	
Grain mould	256
Anthracnose (*Colletotrichum graminicola*)	15
Rust (*Puccinea purpurea*)	31
Downy mildew (*Peronosclerospora sorghi*)	155
Striga low stimulant lines (lab screening)	645
Striga resistant lines (field screening)	24
Other characters	
Glossy lines	501
Pop sorghum lines	36
Sweet stalk sorghum lines	76
Scented sorghum lines	17
Twin-seeded lines	131
Large-glume lines	71
Bloomless sorghum lines	207
Broomcorn sorghum lines	52
Cytoplasmic A and B lines	240

The main source of resistance used in this breeding work was DJ 6514 (IS 18700), a durra line from Karnataka in India. Other sources of midge resistance have been used successfully in breeding midge-resistant cultivars in the USA and Australia. More than 15 000 accessions were screened for resistance to head bug, *Calocoris angustatus* Leth., between 1980 and 1990, under natural infestation and headcage conditions, and 38 accessions were identified as less susceptible (Sharma and Lopez 1992). More than 28 000 accessions have been screened for greenbug (White *et al.* 1994). Russian accessions possess valuable resistance to greenbug biotype I (Andrews *et al.* 1993).

Germplasm accessions have been screened for disease resistance using standardized screening techniques and results have been published by Mughogho *et al.* (1987). For grain moulds, 7132 accessions were screened in the field for resistance during the 1980-85 rainy seasons (Bandyopadhyay *et al.* 1988). Recently, a petri-plate screening method for grain mould resistance has been developed. Photosensitive germplasm that could not be screened in the field has been screened using this method (ICRISAT 1995) and new sources of resistance identified in completely new groups of material, particularly in the conspicuum group of guinea sorghums from Tanzania and Malawi and among conversion lines of zera zera sorghums (Prasada Rao *et al.* 1989).

More than 13 000 accessions were screened for resistance to downy mildew, *Peronosclerospora sorghi* (Weston and Uppal) C.G. Shaw, in field screening at Dharwad, Karnataka, between 1981 and 1988 and some 46 were found resistant (Y.D. Narayana, 1996, pers. comm.) and 2700 early flowering accessions have been screened in the greenhouse against the ICRISAT isolate of *P. sorghi* (Karunakar *et al.* 1994a). Out of 308 accessions tested further, 29 were found to be free of downy mildew and a further 8 showed very high levels of resistance (Karunakar *et al.* 1994b). Resistance to downy mildew was also identified in late-flowering (more than 80 days to flowering) accessions (Navi and Singh 1994). More than 13 000 germplasm accessions and advanced breeding lines were screened for resistance to anthracnose (*Colletotrichum graminicola* (Ces.) G.W. Wills) using a field-screening technique at Pantnagar in northern India between 1982 and 1991. Eleven lines with stable resistance were identified, of which six were germplasm accessions (Pande *et al.* 1994). More than 5000 accessions have been screened for resistance to rust, *Puccinia purpurea* Cooke, at IAC and 15 from various countries in Africa have been shown to have stable rust resistance over three seasons (Singh *et al.* 1994). Six of these lines, three of which were zera zera conversion lines from Ethiopia, showed very high levels of rust resistance and also showed moderate to high resistance to leaf blight. Earlier reports have listed other rust-resistant germplasm accessions from various locations (Mughogho *et al.* 1987). For sorghum ergot, *Sphacelia sorghi* McRae, numerous screening techniques have been tested. However, breeding lines and germplasm accessions reported as being resistant have proved susceptible in subsequent tests (Bandyopadhyay 1992). An effective screening technique has been developed recently (Tegegne *et al.* 1994) and has been applied to test limited numbers of germplasm accessions. Twelve accessions from among 246 with adaptation to high-altitude areas of eastern Africa were tested in Rwanda and proved resistant over 2 years of testing (Musabyimana *et al.* 1995). More than 14 000 germplasm accessions have been screened in the laboratory for low stimulant production, one of the factors determining resistance to *Striga* species (ICRISAT 1981). A total of 646 accessions that produced less than 10% of the stimulant level of control varieties were identified. A number of these lines have been further tested in field-screening trials and used in breeding resistant cultivars (Obilana and Ramaiah 1992).

More than 11 000 accessions were tested for protein and 9900 for lysine. Of 9000 accessions tested for sweet stalks, 96 were identified with high sugar content (Seetharama *et al.* 1987). Almost half (42) of the sweet-stalk lines identified belonged to the caudatum race, followed by durra (16). The accessions, originating from over 35 countries but predominantly from Ethiopia, India and Uganda, were evaluated for 26 characters at three locations in India and the results published in the form of a catalogue (Mathur *et al.* 1991). A similar evaluation was done at one location for 3943 accessions from India and the results published in a second catalogue (Mathur *et al.* 1992).

Properties and Uses

Sorghum grain is an important staple food in many parts of Africa and Asia and an important feed crop in the Americas and Australia. It is extensively used for producing beer and spirit drinks. Dried stems can be used as fuel, roofing or fencing materials. Sorghum is also grown for forage for direct feeding to ruminants or for preservation as hay or silage. Sweet-stemmed sorghums are used to produce sugar, syrup and alcohol.

In the simplest food preparations, whole sorghum grain can be boiled, roasted (usually at the dough stage) or popped like corn. Normally, however, the grain is ground or pounded to form a flour, often after dehulling to remove the pigmented pericarp. The flour is used to prepare porridge, bread or beer. The processes and types of food products vary considerably from region to region. Fermentation is often used to improve digestibility, particularly for high-tannin sorghum.

Beer production is important in Africa, where sorghum grain is germinated, dried and ground to form malt, and sorghum flour is used as a substrate for fermentation. In China, sorghum is extensively distilled to form a popular spirit drink and vinegar.

Sorghum grain is a major component of cattle, pig and chicken feeds in the USA, Central and South America, Australia and China, and is becoming important in chicken feed in India. The grain is usually processed by grinding, rolling, flaking or steaming to improve digestibility. Dried sorghum stems that remain standing in the field after harvest are often grazed or cut and fed to livestock. Sorghum is also grown for forage as a single- or multi-cut crop. Single-cut systems are usually rain-fed, with low levels of inputs, while multi-cut systems are often irrigated and use high levels of fertilizers. The forage produced can either be fed directly to animals or preserved as hay or silage.

The grain normally contains water (8-16%), protein (8-15%), fat (2-6%), carbohydrate (70-90%), fibre (1-3%) and ash (1-2%) (Hulse *et al.* 1980). The energy content ranges from 13.4 to 15.1 kJ/g dry matter. Tannin, which is found in some coloured sorghums, is important in determining nutritional value, as tannins bind proteins and reduce their digestibility. Much of the protein in sorghum is prolamine which has no nutritional value. The maximum available protein is usually 8-9%. Sorghum grain is deficient in lysine, methionine and threonine and is usually grossly deficient in calcium and phosphorus.

The composition of the green plant varies according to age and variety but it normally contains 78-86% water. Dried plant parts usually contain approximately 12% protein, 40-50% carbohydrate and 20-30% fibre. The aerial parts of most sorghum contain the cyanogenic glucoside dhurrin. The quantity depends on the variety, the plant part, age and environmental conditions. It is particularly concentrated in young leaves and tillers and in plants that are affected by drought. Dhurrin is hydrolyzed to hydrocyanic acid (HCN), which is highly toxic and can kill grazing animals. The HCN content usually declines with age, reaching non-toxic levels after approximately 45 days; it is destroyed by drying.

Alternative uses of sorghum grain and plant parts are becoming increasingly important as food uses decline. Grain can be used for various processed foods, for starch extraction or for alcohol production. Sugars can be extracted from plant stems to produce syrups for use in industrial food and drink preparations. Such sugars can also be used to produce alcohol. The fibrous stem materials can be used in the manufacture of fibre boards for construction purposes. Technologies for various alternative uses are well researched and developed but their translation into commercial production depends on the relative price of sorghum and other source materials.

Breeding

As with most crops, the earliest phase of sorghum improvement in Africa and India utilized pure line selection from among and within indigenous landraces, exploiting the variation that arose from mutation and natural hybridization between different plant types (Doggett 1988). A second phase involved deliberate hybridization between landraces. Both these phases used predominantly local germplasm that was readily available to achieve modest improvements in yield. Similar work in the USA produced a series of dwarf sorghum varieties suitable for combine harvesting (Duncan et al. 1991). Some of the important derivatives of landraces that were selected during this period are Maldandi in India, which still occupies millions of hectares of the post-rainy season sorghum belt; Naga White (IS 17632) in Ghana, notable for its exceptional seedling vigour; Segaolane (IS 18535) in Botswana; Framida (IS 8744), selected in South Africa from a variety from Chad, is notable for its excellent resistance to *Striga asiatica*; Serena (IS 18520) in Tanzania, subsequently shown to have some resistance to *Striga* and shoot fly, and to be resistant to bird damage because of its high-tannin brown grain.

The discovery of cytoplasmic male sterility in crosses between milo and kafir sorghum in the USA (Stephens and Holland 1954) transformed sorghum improvement by opening up the possibility of commercial production of F_1 hybrids. Immediate gains in productivity followed the introduction of hybrids in the USA in the 1950s and in India in the 1960s. These first hybrids were produced mainly on a male-sterile version of Combine Kafir 60, a dwarf line released in the USA in 1950 (Doggett 1988).

Lack of diversity among available male-sterile lines rapidly placed limits on the progress that could be made in breeding improved hybrids. In response, programmes were initiated to breed new and better male-sterile lines in the USA (Duncan et al. 1991) and in India (Doggett 1988). The Texas Agricultural Experiment Station (TAES)-United States Department of Agriculture (USDA) Sorghum Conversion Program also was initiated in response to the perceived lack of variability in sorghum suitable for use as hybrid parents. This programme aimed to backcross genes for short height and photoperiod insensitivity into selected sorghum germplasm lines (Stephens et al. 1967) and has now produced several hundred converted lines originating mainly from Africa with a much broader genetic base and a range of resistances to biotic and abiotic stresses (Duncan et al. 1991).

The effect of the TAES-USDA programme on sorghum improvement has been considerable. In the USA, two conversion products of germplasm lines from Ethiopia, SC110 and SC170, have been widely used in developing hybrid parent lines (Duncan et al. 1991). CS 3541, a line produced from IS 3541 in a similar conversion programme undertaken in India, has been released as CSV 4 and used as the male parent of hybrids CSH 5, CSH 6 and CSH 9, which are by far the most widely cultivated hybrids in India, accounting for 2-3 million ha annually. It also has been widely used as a parent in breeding new lines and varieties, by both the Indian national programme and ICRISAT. At ICRISAT, SC108-3 has been used as a parent in breeding elite lines and varieties and features extensively in the pedigrees of the elite materials. The same lines and their derivatives have been very widely used in other breeding programmes throughout the world.

These lines belong to the zera zera group of sorghum landraces that originate in Sudan and Ethiopia and their contribution to sorghum improvement in the 1970s and 1980s has been quite extraordinary. The Indian sorghum improvement programme was among the first to use them and the result has been that the majority of its breeding materials show the tan plant type and corneous grain characteristic of the zera zeras. The ICRISAT breeding programme was similarly influenced to the extent that serious concern over the narrow genetic base was raised. In recent years, there has been some effort to broaden the genetic base of ICRISAT breeding materials by introgression of diverse germplasm from other sources.

The conversion programme and germplasm evaluation exercises also have identified a number of sources of resistance for major yield-limiting stresses that have been incorporated into breeding programmes. This has resulted in some broadening of the genetic bases of the main breeding programmes. These programmes have been initiated in a large number of countries in the last 30 years. The size and scope vary from country to country, reflecting the different priorities accorded to sorghum. Generally, sorghum is a relatively low priority crop and receives less attention than it probably deserves. A variety of different breeding approaches is being used. Many programmes rely heavily on introduced varieties and hybrids, which they screen and select for local adaptation. However, many national breeding programmes are now hybridizing local and exotic germplasm and screening and selecting segregating populations for general adaptation. Most programmes in Africa concentrate exclusively on variety breeding as there are no facilities for hybrid seed production. In Kenya, Nigeria, South Africa, Sudan and Zimbabwe, however, active hybrid breeding and seed production programmes exist. In other regions, there is usually a balance between variety and hybrid breeding; only the strongest programmes are breeding their own male-sterile lines, however. Population breeding is less frequently used because of its generally longer-term horizon, relatively resource-intensive nature and the fact that many of its products are intermediate in nature. In the developed countries and a few developing ones, private-sector sorghum breeders contribute significantly to the supply of improved cultivars, especially hybrids.

Prospects

The demand for sorghum grain as food is likely to remain high in the traditional producing and consuming countries which have little opportunity to produce alternative crops. Growing populations will require increased production in these countries for the foreseeable future, something which will be difficult, given that the crop is generally grown on marginal lands with erratic rainfall. Efforts will be required to produced high-yielding, disease- and insect-resistant, improved varieties to replace traditional cultivars. The introduction of simple mechanization, better soil management and fertilizer application will be needed. Changes in policy to ensure adequate price incentives for sorghum production also will be required. The importance of feed uses of stover in maintaining sorghum in some cropping systems is likely to decline as mechanical traction replaces animal traction.

The demand in developed countries for sorghum as feed is likely to increase, but at a slower rate than in the past as consumption of animal products stabilizes. In developing countries, the use of sorghum as feed will probably increase in Asia and Latin America. The extent to which this happens will depend on its price relative to maize and on the demand for animal products, which will in turn depend on increases in income and changes in food consumption patterns. International trade in sorghum is likely to continue to be dominated by the USA, but there is also likely to be a place for other exporters who can supply good-quality grain at competitive prices.

Forage uses of sorghum in developed countries are likely to increase, but at a slower rate than in the past in line with future demand for animal products. In developing countries of Asia and Latin America, growth is likely to continue as improvements in income lead to increased demand for milk and other dairy products. The demand from this sector will be offset in part by decreases in food consumption as better-off consumers switch to preferred cereals.

There is likely to be little change in the use of sorghum for purposes other than food and animal feed. A possible exception is in the use of sorghum in commercial beer production, particularly in Africa, where there is scope to use increasing quantities for both opaque and lager-type beer.

Limitations

Sorghum competes with maize in many growing environments and also in many end uses. In more favourable environments, maize is usually higher yielding. For most purposes, maize is preferred by consumers or is easier and more efficient to process. The success of sorghum in any particular area or end use, therefore, depends on its price relative to maize; usually sorghum can only compete when it is cheaper than maize. This is normally the case only in very dry areas where maize production is impossible or extremely erratic.

Compared with other cereals, sorghum is usually less preferred as food and is consumed by the poor largely because it is cheap. As consumers become better off, they tend to switch to other cereals, even when more expensive. This is leading to reduction in per capita consumption of sorghum throughout the world, even in areas of traditional production and consumption, a trend that is likely to continue. The fact that sorghum processing for food is often more time- and labour-intensive than other crops exacerbates this trend.

Sorghum grows in and is adapted to a very wide range of environments. It is subject to a large number of yield-limiting diseases and pests, many of which occur in specific climatic zones. It also has a wide variety of end uses, many of which require particular grain qualities. It is, therefore, difficult to breed widely adapted improved varieties or hybrids of sorghum because specific combinations of resistances and quality traits are needed for different locations. Location-specific breeding programmes are required to produce suitable adapted cultivars and very often these are lacking or weak, particularly in the developing countries that need them most. Sorghum is grown predominantly in marginal environments that are subject to extreme variation in rainfall. Production, therefore, is subject to major fluctuations according to the weather; production is high when rainfall is good and low when rainfall is poor or badly distributed. This leads to similar fluctuations in price and changes in the availability and quality of sorghum in the market, which influence its adoption for uses which demand an even and reliable supply. Sorghum seed production also suffers from the erratic production. Demand for seed depends on timely arrival of adequate seed; when rains fail, demand for seed also fails. This makes it difficult and risky for seed producers to predict the demand for seed, leading to few producers and chronic shortages of planting materials in many seasons.

REFERENCES

Acheampong, E., N. Anishetty Murthi and J.T. Williams. 1984. A World Survey of Sorghum and Millets Germplasm. IPGRI, Rome, Italy.

Andrews, D.J., P.J. Bramel-Cox and G.E. Wilde. 1993. New sources of resistance to greenbug, biotype I, in sorghum. Crop Sci. 33:198-199.

Appa Rao, S. and J.N. Mushonga. 1987. A catalogue of passport and characterization data of sorghum, pearl millet and finger millet from Zimbabwe. IBPGR, Rome, Italy.

Appa Rao, S., L.R. House and S.C. Gupta. 1989. A review of sorghum, pearl millet and finger millet production in SADCC countries. Southern African Center for Cooperation in Agricultural Research (SACCAR) Gaborone, Botswana, and SADCC/ICRISAT, Bulawayo, Zimbabwe.

Bandyopadhyay, R. 1992. Sorghum Ergot. Pp. 235-244 *in* Sorghum and millets diseases: a second world review. (W.A.J. de Milliano, R.A. Frederiksen and G.D. Bengston, eds.). ICRISAT, Patancheru, India.

Bandyopadhyay, R., L.K. Mughogho and K.E. Prasada Rao. 1988. Sources of resistance to sorghum grain molds. Plant Dis. 72:504-508.

Cherney, J.H., K.J. Moore, J.J. Volenac and J.D. Axtell. 1986. Rate and extent of digestion of cell wall components of brown midrib sorghum species. Crop Sci. 26:1055-1059.

de Wet, J.M.J. 1978. Systematics and evolution of *Sorghum* Sect. *Sorghum* (Gramineae). Am. J. Bot. 65:477-484.

de Wet, J.M.J. and J.P. Huckabay. 1967. The origin of *Sorghum bicolor*. II. Distribution and domestication. Evolution 21:787-802.

de Wet, J.M.J., J.R. Harlan and B. Kurmarohita. 1972. Origin and evolution of guinea sorghums. East African Agric. and Forestry J. 38:114-119.

Doggett, H. 1988. Sorghum (2nd edn.). Longman, Harlow, UK.

Doggett, H. and K.E. Prasada Rao. 1995. Sorghum. Pp. 173-180 *in* Evolution of Crop Plants (J. Smartt and N.W. Simmonds, eds.). Longman, Harlow, UK.

Duncan, R.R., P.J. Bramel-Cox and F.R. Miller. 1991. Contributions of introduced sorghum germplasm to hybrid development in the USA. Pp. 69-102 *in* Use of Plant Introductions in Cultivar Development, Part I. CSSA Special Publication no. 17. Crop Science Society of America, Madison, USA.

Frederiksen, R.A. and R.R. Duncan. 1982. Sorghum diseases in North America. Pp. 85-88 *in* Sorghum and millets diseases: a second world review. (W.A.J. de Milliano, R.A. Frederiksen and G.D. Bengston, eds.). ICRISAT, Patancheru, India.

Harlan, J.R. and J.M.J. de Wet. 1972. A simple classification of cultivated sorghum. Crop Sci. 12:172-176.

Harlan, J.R. and A.B.L. Stemler. 1976. The races of sorghum in Africa. Pp. 465-478 *in* Origins of African Plant Domestication (J.R. Harlan, J.M.J. de Wet and A.B.L. Stemler, eds.). Mouton, The Hague, The Netherlands.

House, L.R. 1985. A Guide to Sorghum Breeding (2nd edn.). ICRISAT, Patancheru, India.

Hulse, J.H., E.M. Laing and O.E. Pearson. 1980. Sorghum and the Millets: Their Composition and Nutritive Value. Academic Press, London, UK.

IBPGR/ICRISAT. 1993. Descriptors for sorghum [*Sorghum bicolor* (L.) Moench]. IBPGR, Rome, Italy and ICRISAT, Patancheru, India.

ICRISAT. 1981. Annual Report 1979/80. ICRISAT, Patancheru, India.

ICRISAT. 1995. Sorghum grain molds. P. 63 *in* ICRISAT Asia Region Annual Report 1994. ICRISAT, Patancheru, India.

Jambunathan, R., M.S. Kherdekar and R. Bandyopadhyay. 1990. Flavan-4-ols concentration in mold-susceptible and mold-resistant sorghum at different stages of grain development. J. Agric. Food Chem. 38:545-548.

Karunakar, R.I., Y.D. Narayana, S. Pande, L.K. Mughogho and S.D. Singh. 1994a. Evaluation of early-flowering sorghum germplasm accessions for downy mildew resistance in the greenhouse. Int. Sorghum and Millets Newsl. 35:102-103.

Karunakar, R.I., Y.D. Narayana, S. Pande, L.K. Mughogho and S.D. Singh. 1994b. Evaluation of wild and weedy sorghums for downy mildew resistance. Int. Sorghum and Millets Newsl. 35:104-106.

Maiti, R.K. and F.R. Bidinger. 1979. A simple approach to the identification of shoot fly tolerance in sorghum. Ind. J. Plant Prot. 7:135-140.

Maiti, R.K., K.E. Prasada Rao, P.S. Raju and L.R. House. 1984. The glossy trait in sorghum: its characteristics and significance in crop improvement. Field Crops Res. 9:279-289.

Mann, J.A., K.T. Kimber and F.R. Miller. 1983. The origin and early cultivation of sorghums in Africa. Tex. Agric. Exp. Sta. Bull. 1454.

Mathur, P.N., K.E. Prasada Rao, I.P. Singh, R.C. Agrawal, M.H. Mengesha and R.S. Rana. 1992. Evaluation of Forage Sorghum Germplasm. Part 2. National Bureau of Plant Genetic Resources, Pusa Campus, New Delhi 110 012, India.

Mathur, P.N., K.E. Prasada Rao, T.A. Thomas, M.H. Mengesha, R.L. Sapra and R.S. Rana. 1991. Evaluation of forage sorghum germplasm. Part 1. National Bureau of Plant Genetic Resources, Pusa Campus, New Delhi, India.

Mengesha, M.H. and K.E. Prasada Rao. 1982. Current situation and future of sorghum germplasm. Pp. 323-333 *in* Sorghum in the Eighties: Proceedings of the International Symposium on Sorghum, 2-7 Nov 1981, ICRISAT Center, India, Vol. I. ICRISAT, Patancheru, India.

Mughogho, L.K., R. Bandyopadhyay and S. Pande. 1987. Sources of resistance to sorghum diseases in India. Pp. 405-427 *in* Proceedings of the Third Regional Workshop on Sorghum and Millets for Southern Africa, 6-10 Oct 1986, Lusaka, Zambia. SADCC/ICRISAT, Bulawayo, Zimbabwe.

Murty, B.R., V. Arunachalam and M.B.L. Saxena. 1967. Classification and catalogue of a world collection of sorghum. Ind. J. of Genet. & Pl. Breed. 27:1-194.

Musabyimana, T., C. Sehene and R. Bandyopadhyay. 1995. Ergot resistance in sorghum in relation to flowering, inoculation technique and disease development. Plant Pathol. 44:109-115.

Navi, S.S. and S.D. Singh. 1994. Identification of sources of resistance to downy mildew in late-flowering sorghum germplasm. Int. Sorghum and Millets Newsl. 35:104.

Obilana, A.T. and K.V. Ramaiah. 1992. *Striga* (Witchweeds) in Sorghum and Millet: Knowledge and Future Research Needs. Pp. 187-201 *in* Sorghum and Millets Diseases: a Second World Review (W.A.J. de Milliano, R.A. Frederiksen and G.D. Bengston, eds.). ICRISAT, Patancheru, India.

Pande, S., R.P. Thakur, R.I. Karunakar, R. Bandyopadhyay and B.V.S. Reddy. 1994. Development of screening methods and identification of stable resistance to anthracnose in sorghum. Field Crops Res. 38:157-166.

Prasada Rao, K.E. and M.H. Mengesha. 1981. A pointed collection of zera-zera sorghums from the Gambella area of Ethiopia. Genetic Resources Progress Report no. 33. ICRISAT, Patancheru, India.

Prasada Rao, K.E. and D.S. Murty. 1982. Sorghums for special uses. Pp 129-134 *in* Proceedings of the International Symposium on Sorghum Grain Quality, 28-31 Oct 1981, ICRISAT, Patancheru, India. ICRISAT, Patancheru, India.

Prasada Rao, K.E., M.H. Mengesha and V.G. Reddy. 1989. International use of sorghum germplasm collection. Pp. 49-67 *in* The Use of Plant Genetic Resources (A.H.D. Brown, O.H. Frankel, D.R. Marshall and J.T. Williams, eds.). Cambridge University Press, Cambridge, UK.

Purseglove, J.W. 1972. Tropical Crops: Monocotyledons. Longman, Harlow, UK.

Rana, B.S., B.U. Singh and N.G.P. Rao. 1985. Breeding for Shoot Fly and Stem Borer Resistance in Sorghum. Pp. 347-360 *in* Proceedings of the International Sorghum Entomology Workshop, 15-21 July 1984, Texas A & M University, College Station, Texas, USA. ICRISAT, Patancheru, India.

Seetharama, N., K.E. Prasada Rao, V. Subramanian and D.S. Murty. 1987. Screening for sweet stalk sorghums, and environmental effect of stalk sugar concentrations. Pp. 169-179 *in* Technology and Application for Alternate Uses of Sorghum (U.M. Ingle, D.N. Kulkarni and S.S. Thorat, eds.). Proceedings of the National Seminar, 2-3 Feb 1987, Parbhani, Maharashtra, India. Marathwada Agricultural University, Parbhani, Maharashtra, India.

Sharma, H.C. and V.F. Lopez. 1992. Genotypic resistance in sorghum to head bug, *Calocoris angustatus* Lethiery. Euphytica 58:193-200.

Sharma, H.C., B.L. Agrawal, P. Vidyasagar, C.V. Abraham and K.F. Nwanze. 1993. Identification and utilization of resistance to sorghum midge, *Contarinia sorghicola* (Coquillett), in India. Crop Prot. 12:343-350.

Sharma, H.C., S.L. Taneja, K. Leuschner and K.F. Nwanze. 1992. Techniques to screen sorghum for resistance to insect pests. Information Bulletin no. 32. ICRISAT, Patancheru, India.

Singh, S.D., P. Sathiah and K.E.P. Rao. 1994. Sources of rust resistance in purple-colored sorghum. Int. Sorghum and Millets Newsl. 35:100-101.

Snowden, J.D. 1935. The Cultivated Races of Sorghum. Adlard, London, UK.

Stephens, J.C. and R.F. Holland. 1954. Cytoplasmic male-sterility for hybrid sorghum seed production. Agron. J. 46:20-23.

Stephens, J.C., F.R. Miller and D.T. Rosenow. 1967. Conversion of alien sorghums to early combine genotypes. Crop Sci. 7:396.

Taneja, S.L. and K. Leuschner. 1985a. Resistance screening and mechanisms of resistance in sorghum to shoot fly. Pp. 115-129 *in* Proceedings of the International Sorghum Entomology Workshop, 15-21 July 1984, Texas A & M University, College Station, Texas, USA. ICRISAT, Patancheru, India.

Taneja, S.L. and K. Leuschner. 1985b. Methods of rearing, infestation, and evaluation for *Chilo partellus* resistance in sorghum. Pp. 175-188 *in* Proceedings of the International Sorghum Entomology Workshop, 15-21 July 1984, Texas A & M University, College Station, Texas, USA. ICRISAT, Patancheru, India.

Tegegne, G., R. Bandyopadhyay, T. Mulatu and Y. Kebede. 1994. Screening for ergot resistance in sorghum. Plant Dis. 78:873-876.

Wendorf, F., A.E. Close, R. Schild, K. Wasylikowa, R.A. Houslely, J.R. Harlan and H. Krolik. 1992. Saharan exploitation of plants 8000 years B.P. Nature 359:721-724.

White, G.A., R.A. Norris and M.F. Loftus. 1994. Sorghum germplasm exchanges 1989-93. Int. Sorghum and Millets Newsl. 35:85-88.

Chapter 21

Wheat

H.J. Dubin, R.A. Fischer, A. Mujeeb-Kazi,
R.J. Peña, K.D. Sayre, B. Skovmand and
J. Valkoun

Wheat is grown in almost all cropping environments of the world, except in the humid lowland tropics. Winter wheat under rain-fed conditions dominates in Europe, the USA, Ukraine and southern Russia, followed by spring-sown spring wheat in semi-arid conditions (Canada, Kazhakstan and Siberia) in the developed world. In the developing world, wheat is usually not a subsistence crop and its production is concentrated in several well-defined cropping systems: double-cropped with paddy rice in Asia (22 million ha) or with maize, cotton, soybean or berseem (15 million ha); with soybean in the Southern Cone (7 million ha). Only in the wetter parts of the developed world, such as the eastern USA, Europe and southern Russia, and in parts of the Southern Cone, is wheat grown in more complex rotational systems involving pulses, oilseeds, other cereals, alfalfa and pastures. An exception is southern Australia, where, despite low rainfall, complex rotations and ley farming are practised.

BOTANY AND DISTRIBUTION

Origin, Distribution and Diffusion

Wheat belongs to the genus *Triticum*, which originated about 10 000 years ago in what is now the Middle East. Polyploid *Triticum* arose when two diploid wild grasses crossed naturally to produce tetraploid wheat, which today includes cultivated durum wheat (*Triticum turgidum* L. var. group *durum* Desf. $2n=4x=28$). Tetraploid wheat later outcrossed to goat grass (*T. tauschii*, considered a troublesome weed in many wheat-growing areas) and gave rise to hexaploid bread wheat (*T. aestivum* L. em Thel. $2n=6x=42$).

Triticum turgidum durum, the main modern variety of tetraploid durum or macaroni wheat, is widely grown in drier areas of the world, such as those found in India, the Mediterranean Basin, the former Soviet Union, Argentina and the rain-fed areas of the North American Great Plains. The other tetraploid wheat, *T. timopheevi*, is only grown in some areas of the Transcaucasian regions. Cultivation of the diploid *T. monococcum* is restricted to some areas of the Middle East and Mediterranean regions.

Triticum aestivum is the most widely grown wheat in the world today. Generally called bread or common wheat, its flour is best suited for making bread. Bread wheat encompasses several thousand cultivars that are widely adapted over a range of environments and grown worldwide.

Placing the female parent first, durum and bread wheats are genomically represented as BBAA and BBAADD, respectively. The mode of chromosome pairing in triploid $2n=3x=21$ *T. turgidum* x *T. monococcum* and pentaploid $2n=5x=35$ *T. turgidum* x *T. aestivum* hybrids has revealed the diversity of ancestral species in the allopolyploid wheats. Chromosomal configurations at meiosis have unequivocally demonstrated that diploids, tetraploids and hexaploids have either one (AA), two (AABB) or three (AABBDD) sets of seven chromosomes each and that the A genome is common to durum and bread wheats. This led to the inference that bread wheat is an allohexaploid.

More recent indications are that *T. urartu* Tum., rather than *T. monococcum* L. subsp. *boeoticum* Boiss., a more widely spread diploid, is the source of the A genome in cultivated wheat. *Triticum urartu* inhabits areas in the Fertile Crescent from southern Lebanon to southern Syria, northward through southeastern Turkey into the Caucasus mountains and southeastward through northern Iraq into southwestern Iran. It was suggested as the A-genome donor by Johnson (1975) and more recently by Kerby and Kuspira (1987).

Wheat's B genome donor is thought to be similar to *Aegilops speltoides* Tausch., which grows wild throughout the Fertile Crescent. Another possible donor of the B genome is *Ae. searsii*, which is restricted to southern Palestine, but the issue is still somewhat unsettled. The source of the D genome is *T. tauschii*, which was readily accepted as the donor species based on analyses of synthesized and cultivated wheats that revealed certain genomic similarities.

The main centre of diversity and origin of *Triticum* is southwest Asia, around the Fertile Crescent hill sites, extending from the Mediterranean coast in the west around the Syrian Desert to the Tigris-Euphrates plain in the east. In that area, the diploid and polyploid *Triticum* species grow in mixed populations exhibiting a great diversity of morphological and ecological variation.

Classification
The Triticeae tribe is comprised of approximately 350 species, of which about 250 are perennials. Perennials not only include many important forage grasses but also serve as a vital genetic reservoir for improving annual Triticeae species, which include the major cereals – bread wheat, durum wheat, triticale (X *Triticosecale* Wittmack, a man-made cereal resulting from a cross between wheat and rye), barley (*Hordeum vulgare* L.) and rye (*Secale cereale*).

The genomic system for classifying the perennial Triticeae (Dewey 1984) recognizes ten genera with defined genomes or genome combinations. The ten genera, with their type species and genome compositions, are: *Agropyron* (*A. cristatum*; P), *Pseudoroegneria* (*P. strigosa*; S), *Psathyrostachys* (*Ps. lanuginosa*; N), *Critesion* (*C. jubatum*; H), *Thinopyrum* (*T. junceum*; J-E), *Elytrigia* (*E. repens*; SX), *Elymus* (*E. sibiricus*; SHY), *Leymus* (*L. arenarius*; JN), *Pascopyrum* (*Pa. smithii*; SHJN) and *Secale* (*S. montanum*; R).

Annual plants of the Triticeae are confined largely to the *Triticum* and *Aegilops* species, with the notable exception of certain species belonging to *Hordeum, Secale, Haynaldia, Eremopyrum, Heteranthelium, Taeniantherum* and *Henrardia*. The large number of generic and specific names of the interrelated *Triticum* and *Aegilops* groups has led to considerable confusion over the years. The multitude of names not only expresses the opinions of various taxonomists, but also represents the diversity of the species themselves. To clear up the confusion, Kimber and Feldman (1987) compiled a synonym list of the most commonly used names among the *Triticum/Aegilops* groups.

GERMPLASM CONSERVATION AND USE

CIMMYT's wheat genebank contains almost 122 000 accessions representing more than 50 years of breeding, collecting and acquisition. The collection includes different types of Triticeae genetic resources, as shown in Table 21.1. Currently (1996), the entire collection is maintained as an active collection. However, hexaploid wheat and triticale germplasm recently have been stored in both base and active collections. Durum wheats and wild relatives are maintained as active collections, in accordance with the ICARDA-CIMMYT agreement. Barley (*Hordeum vulgare*) germplasm is stored at CIMMYT as a working collection for the ICARDA-CIMMYT Barley Program.

Multiplication and regeneration are accomplished in a screenhouse at CIMMYT headquarters in El Batan, Mexico. This facility expedites the production of quality seed for medium- and long-term storage. It also minimizes accidental mechanical mixing and other handling errors by allowing sowing to be programmed year-round rather than strictly by the annual crop season. Thus, a manageable number of accessions can be multiplied through sequential plantings.

CIMMYT's Seed Health Unit inspects all introductions before planting, and any samples presenting a potential quarantine risk are destroyed. If no problems are encountered, introductions are released for planting in the greenhouse or designated introduction fields. All introductions are inspected periodically during the crop cycle, and fungicides and insecticides are routinely applied.

The key to most wheat genetic resources work in the future is the development of a database, or an interconnected system of databases, with the capacity to manage and integrate all wheat information, including passport, characterization and evaluation data. In the early 1990s, CIMMYT's Wheat Program established just such a strategy for integrating and managing all data pertaining to germplasm regardless of where they were generated. The goal was to facilitate the unambiguous identification of wheat genetic resources and remove barriers to handling and accessing information. As a result, the International Wheat Information System (IWIS), a system that seamlessly joins conservation, utilization and exchange of genetic material, came into being. The system is fast, user-friendly and is available on an annually updated CD-ROM.

IWIS has two major components: the Wheat Pedigree Management System, which assigns and maintains unique wheat identifiers and genealogies, and the Wheat Data Management System, which manages performance information and data on known genes.

The power of IWIS has been demonstrated in several ways. For example, it has been used to trace genealogies of modern cultivars to their parental landraces or to lines of unknown pedigrees. The system has also revealed that the number of parental landraces in CIMMYT's bread wheats has increased markedly over four decades, from six ancestors in Yaqui 50 to 68 ancestors in Weaver. Using information generated by IWIS, cytoplasmic diversity in CIMMYT wheats was found to be restricted. Other analyses using the system have demonstrated that landraces from certain regions of wheat's centre of origin do not, or only rarely, appear in the pedigrees of modern wheats. Such genealogical analyses, which could have broad implications for genetic resources utilization, are evidence of the system's great utility and many potential applications.

Byerlee and Moya (1993) have indicated that 40 million ha in the developing world are sown to wheat cultivars originating directly from CIMMYT crosses or from national agricultural research systems (NARS) crosses using a CIMMYT parent. In industrialized countries, at least 20 to 25 million ha are sown to cultivars of CIMMYT ancestry. These cultivars are the direct result of active seed exchange between NARS and CIMMYT over the last three decades, which has made it

Table 21.1. Numbers of accessions in the CIMMYT Wheat Germplasm Bank by species as of February 1995.

Crop	No. of accessions
Bread wheat (*Triticum aestivum* L. em Thell.)	71 171
Durum wheat (*T. turgidum* L.)	15 940
Triticale (X *Triticosecale* Wittmack)	15 200
Barley (*Hordeum vulgare* L.)	9 084
Rye (*Secale cereale* L.)	202
Primitive wheats (*T. monococcum* L.,	7 245
T. dicoccon Schrank)	
Wild relatives (*Triticum* spp.)	4 549
Total	**121 944**

possible to introduce important traits into widely adapted germplasm. There is no indication that seed exchange will become less important in the future; on the contrary, it probably will be of greater importance. IWIS will assist in the exchange of seed and associated information, increase efficiency and minimize unnecessary duplication of evaluations.

Finally, IWIS makes it possible to estimate the degree of relatedness among wheats, allows breeders to increase genetic diversity by selecting materials of divergent parentage for crosses, thereby reducing wheat's vulnerability to diseases and climatic changes, and automatically updates family trees as additional ancestry is discovered.

About 640 000 accessions of *Triticum* spp., *Aegilops* spp. and X *Triticosecale* can be found in collections around the world (Table 21.2). The degree of duplication in these collections is difficult to ascertain without some type of global wheat genetic resources database. Given this situation, the level of priority that should be placed on collecting more materials is uncertain, except where there is a real threat of genetic erosion to native species in specific areas. Accessions in collections around the world may or may not be preserved properly, and some may not even be catalogued. It may thus be more cost-effective to place such collections into secure storage than to collect more materials in the field.

Several accessions of diploid wild relatives that have the A, B or D genomes are potential candidates for use in interspecific crosses. Accessions with other ploidy levels and partial genomic similarity are also good candidates. CIMMYT maintains working collections of these wild grass accessions.

The different genepools within the annual and perennial grasses of the Triticeae tribe also provide tremendous genetic variability for improving wheat (Dewey 1984). However, in contrast with the annual *Triticum/Aegilops* spp., the perennial genera we use in our intergeneric crosses are genomically quite diverse and rather difficult to cross with wheat. Hence, accomplishing beneficial alien transfers through intergeneric hybridization is quite time consuming.

Evaluation

Two evaluation approaches are used by the genetic resources unit. One is demand-driven, meaning that evaluations are conducted for specific traits or characteristics where the breeding programmes lack variation. Demand-driven evaluations may be requested by CIMMYT and national programme breeders. The other approach is systematic evaluation of germplasm groups that the Bank considers potential sources of variation for use in breeding programmes. The systematic evaluation of groups of germplasm as potential sources of variation for use in breeding programmes is done by selecting a specific set of underutilized germplasm and evaluating it for all possible characters.

Table 21.2. Number of accessions available in collections around the world.

Type of wheat	No. of accessions
Hexaploid	266 589
Tetraploid	78 726
Diploid	11 314
Unspecified *Triticum*	252 530
Aegilops spp.	17 748
Triticale	23 659
Total	640 603

Source: Information collated from IBPGR (1990).

Pre-breeding

Genetic resources available for wheat improvement include perennial and annual grasses belonging to the tribe Triticeae. Their enormous genetic inheritance can be used for improving wheat thanks to the patterns of relationships within the tribe and to the possibility of making wide crosses. Wide crosses of wheat have received considerable attention because of the crop's global importance and the availability of a wide range of wild species; also, wide crossing techniques and subsequent genetic manipulations are well established.

Wheat wide crossing at CIMMYT was established during the late 1970s. At that time, the main goals were intergeneric hybridization and subsequent introgression of agronomically important traits into modern high-yielding strains, and *Aegilops, Hordeum, Secale* and *Thinopyrum* spp. received the most attention.   Major achievements to date include the registration of Karnal bunt and *Helminthosporium sativum* resistant stocks and the release in Mexico and Pakistan of Karnal bunt-resistant and salt-tolerant cultivars, respectively (Mujeeb-Kazi and Hettel 1995).

During the late 1980s, the wide crosses section expanded to include interspecific hybridization and started exploiting the variability locked in the three genome donors of modern wheat. As a result, more than 500 synthetic wheats having exotic A, B or D genomes were produced and are being maintained at CIMMYT. These synthetics are proving to be extremely valuable sources of resistance to various biotic and abiotic stresses, as well as of yield-related traits (Mujeeb-Kazi and Hettel 1995).

Properties and Uses

Today wheat is grown as an autumn/early winter-sown cool season crop in latitudes from 45° to 55° to as close to the equator as latitude 12° at low altitudes (<1500 m asl).   Further from the equator, winters are too cold to permit overwintering but commercial spring-sown spring wheat takes over, being found up to latitude 62°N in Europe. Spring-type wheat also occurs from latitude 20° right to the equator at high altitudes (>1500 m) in cool tropical highlands, where it is usually grown during the wet summer season.

Daylength sensitivity (wheat is a long-day plant) and temperature-conditioned vernalization requirements, both genetically determined traits, are principal determinants of adaptation. A major phenological separation exists between spring-habit wheats, with no vernalization response and marked daylength sensitivity, and winter-habit wheats, which almost always show a strong or obligate requirement for vernalizing temperatures to induce timely floral initiation, as well as daylength sensitivity. There are also intermediate or facultative habit wheats. Traditionally, spring wheats were sown at high latitudes in the spring, and winter or facultative wheats at middle latitudes in the autumn. As wheat growing has shifted to lower latitudes and with hybridization and selection, these distinctions have become less clear. In particular, autumn-sown spring wheats with reduced daylength sensitivity have come to dominate at lower latitudes (less than 35°S and N).

Wheat is relatively tolerant to drought, requiring as little as 200-250 mm average evapotranspiration for commercial production. It is moderately tolerant to frost, except during the period from stem elongation through anthesis. Wheat is also quite tolerant of heat, at least under irrigated cultivation in locations with low atmospheric humidity (e.g. in the Sudan). It is not well adapted to strongly acid or saline soils, but genetic differences exist, especially for tolerance to acid soils with aluminium toxicity problems. The widely varied environments where wheat is grown, however, are solid testimonial to its genetic diversity for adaptation. It is interesting to speculate that the hexaploid nature of bread wheat may explain a significant part of this broad adaptation.

Rain-fed wheat production is found from where average growing season rainfall is close to nil (growth on stored residual summer moisture in moisture-retentive soils of low latitude environments like central India) to up to 1000 mm. On less favoured soils and without fallow, the minimum average growing season rainfall is about 200 mm, meaning an average annual rainfall of 250 mm in winter rainfall mid-latitude environments like North Africa. One year of fallow to conserve moisture if soils are retentive improves yield, greatly reduces risk in such environments, and is essential for commercial production in more continental locations (e.g. Great Plains of the USA, prairies of Canada, steppes of Siberia, Central Anatolian Plateau) if annual precipitation falls below about 350 mm. In the last 30 years, supplemental and full irrigation have permitted the expansion of wheat into many arid and semi-arid environments.

Cereal grains, although low in protein content and deficient in protein nutritional quality (low levels of the essential amino acid lysine), constitute the major source of energy and nutrients in the world (Roderick and Fox 1987). Wheat accounts for the largest proportion of global cereal production (USDA 1995) and, chiefly in the form of bread, is the principal food item in most developing countries, for it provides the population with more energy and nutrients than any other single food source (Pomeranz 1987). The chemical composition and nutritional value of wheat and wheat-based products have been extensively reviewed elsewhere (Roderick and Fox 1987; Ranhotra 1991).

The baking industry's principal reason for preferring wheat over other cereals as an essential raw material is its unique ability to form an insoluble and viscoelastic protein complex, known as gluten, that confers viscoelasticity to flour doughs. Wheat varieties may vary significantly in gluten quality. Although genetically controlled, gluten quality can be greatly influenced by grain productivity factors such as soil fertility and crop management practices (grain yield is negatively correlated to protein content), and by environmental conditions such as frost, heat and rainfall during grain development (Sander *et al.* 1987). For these reasons, wheat quality improvement has been a task for breeders in NARS and international agricultural research centres (IARCs) around the world.

Selecting for specific genes is accompanied by a decrease in genetic diversity. This is demonstrated by the regional predominance of a few glutenin subunits controlled by genes situated on the long arm of wheat group 1 chromosomes, most likely as a function of selection pressure (Morgunov *et al.* 1993; Peña 1995). Fortunately, NARS and CGIAR centres in particular are taking action to maintain and enhance biodiversity while breeding for quality. Genes for increased protein content from *T. dicoccoides* and for novel, quality-desirable glutenins from *T. tauschii* can be transferred into durum wheat and/or bread wheat (Grama *et al.* 1984; Lagudah and Halloran 1988; Khan *et al.* 1989; William *et al.* 1993; Peña *et al.* 1995). They are being exploited at CIMMYT and ICARDA to improve the nutritional and/or industrial qualities of wheat (Nachit 1992; William *et al.* 1993; R.J. Peña, unpublished data). Finally, landraces and other alien diploid (*T. monococcum, T. boeoticum, T. urartu*) and tetraploid (*T. dicoccon*) species are being examined with the aim of finding additional novel genes that could improve wheat's end-use qualities and enhance its genetic endowment.

Aiming to provide NARS with high-yielding germplasm possessing acceptable quality attributes, CIMMYT and ICARDA have included quality improvement as part of their breeding activities. Traditionally, quality improvement has involved crossing a high-quality wheat parent with a parent possessing other desirable attributes; only offspring carrying the desired traits are selected and advanced throughout the segregating stages. Although somewhat slow and expensive, this breeding approach has succeeded in generating many wheat varieties of acceptable quality around the world.

Breeding

In their efforts to meet the increasing worldwide demand for food, plant breeders everywhere are finding very little germplasm of cultivated crops having the desired traits with which to make needed improvements in those crops. In conventional bread wheat improvement, breeders have normally made crosses between varieties. Such crosses have few constraints, and all associations of parental traits and segregation are invariably based on genetic recombination. Fortunately, useful genetic resources (i.e. useful in crop improvement) are being found among uncultivated plants in the wild. The challenge is to be able to exploit and incorporate this 'new' germplasm routinely into existing food crops.

Most efforts to transfer alien germplasm from wild plants into cultivated crops have involved the *Triticum* grass species, with the greatest emphasis being placed on improving bread wheat. Introgressing alien variability into bread wheat involves working in two distinct areas: long-term intergeneric and short-term interspecific hybridization. This separation is essentially based on wheat/alien genomic similarity and level of genetic recombination. Because it is short term, interspecific hybridization is the favoured approach for genetic introgression.

In interspecific crosses, the most useful materials are the numerous alien accessions of Triticeae species having genomes similar to the A, B or D genomes of bread wheat. Crosses with these materials allow relatively easy alien gene transfers, are compatible with normal field research and set the stage for simultaneous introgression of several genes.

Interspecific hybridization

Triticum tauschii, which has the D genome, could be very useful for transferring unique multiple diversity to wheat. At CIMMYT we are using *T. tauschii* accessions in the following ways:

- Producing synthetic hexaploids by crossing durum cultivars with *T. tauschii* accessions and using synthetic hexaploid wheats for crosses onto wheat (Fig. 21.1). This permits exchanges across all three genomes.
- Crossing elite, but susceptible, bread wheat cultivars with resistant *T. tauschii* accessions and backcrossing the ABDD F_1 hybrids with the elite bread wheat cultivar used in the initial cross. The progenies will reflect D genome exchanges only.
- Extracting the AABB genomes from commercial bread wheat cultivars and then developing hexaploids through crosses with desired *T. tauschii* accessions. This partitions the D genome very precisely.

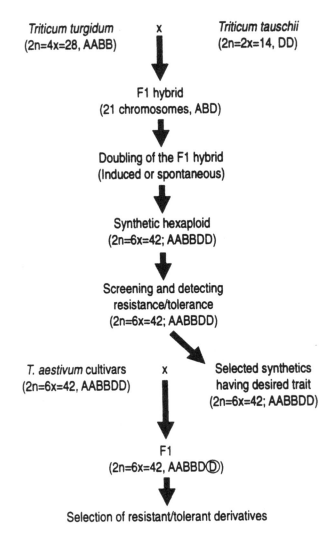

Fig. 21.1. Schematic showing the production of synthetic hexaploids derived from crossing *Triticum turgidum* x *T. tauschii* and their utilization.

Intergeneric hybridization

A few new hybrids have been obtained since Wang (1989) reviewed intergeneric crosses involving perennial Triticeae. All genomes of the perennial Triticeae have been combined, either singly or in combination, with the A, B and D bread wheat genomes (Wang 1989; Mujeeb-Kazi *et al.* 1994). To date, there are at least 89 different combinations involving the three wheat genomes and the eight genomes of the perennial species. Of these, 52 involve hexaploid wheats, 30 involve tetraploid wheats and 7 involve diploid primitive wheats.

The wide array of genetic variability residing in the above Triticeae relatives supplies a superb arsenal of new defences against biotic and abiotic stresses in cereals. However, the use of this variation has its constraints, since genomic homoeology does not offer a satisfactory level of chromosomal association in the F_1 hybrids to promote alien gene transfers. Use of the *ph1* locus may provide a way to overcome the recombination constraint. Other genetic manipulations progressively revolve around production of alien disomic chromosome additions or substitutions, which could lead to translocations or subtle genetic exchanges through cytogenetic and novel manipulative procedures. Although the transfer process is slow, the potential benefits of incorporating these diverse genetic resources into wheat are extremely high.

Outputs from intergeneric programmes are long term, and even a 10-year (two cycles per year) span could be considered a short-term involvement because of the research complexity associated with inducing alien introgression, identifying expression and achieving stability. Novel genes have unique potential for contributing to stress resistance in wheat. The release in Pakistan of two wheat varieties (Pasban 90 and Rohtas 90) derived from wheat/*Th. distichum* advanced lines and the registration of five *Helminthosporium sativum* resistant lines from *Th. curvifolium* advanced lines (Mujeeb-Kazi *et al.* 1996) are two successful examples that encourage continuing the pursuit of this long-term procedure.

Prospects

Variability is needed to further increase wheat's yield potential; provide new sources of disease and pest resistance and maintain the yield levels achieved so far; develop germplasm adapted to more marginal environments, and to improve quality.

Most wheat breeding programmes dedicate a major portion of their efforts to protecting gains in yield potential by incorporating new and better genes or combinations of genes for disease and pest resistance. Collections of adapted and unadapted wheats have been rich sources of resistance to various diseases, and their greatest underlying value is as a reservoir of undetected resistance genes (Williams 1989). For most wheat diseases there is a need to identify more genes for resistance of the hypersensitive type to achieve combinations of genes that confer resistances similar to stem rust resistance, which so far has been very effective. The wild relatives of wheat will most likely be major contributors to this type of resistance.

There is also a need to identify the quantitative type of resistance (partial resistance), characterized by durability and a reduced rate of epidemic build-up (Parlevliet 1988). This type of resistance may be very important in diseases such as yellow rust, where race-specific resistance has not been very long lasting. The most likely sources of quantitative resistance are landraces and obsolete cultivars that have been grown extensively over many years in areas where particular diseases are endemic.

Fungal diseases are the predominant causes of losses of yield and quality in wheat, with nematodes, viruses, insects and bacteria of much less importance. Globally the most important of the fungal diseases are the rusts. All three rusts of wheat are potentially dangerous and research efforts aimed at combating them have developed various control strategies, of which genetic resistance is the most cost-effective and environmentally sound measure. For example, durable resistance to stem rust (*Puccinia graminis* Pers. f.sp. *tritici* Eriks. & Henn.) has kept this traditional pathogen of wheat at bay for the last 40 years. Durable resistance to leaf rust (*P. trecondita* Rob. ex Desm. f.sp. *tritici*), presently the most significant rust, is currently being incorporated into the wheats. Resistance to yellow or stripe rust (*P. tstriiformis* West. f.sp. *tritici*), important in the wetter, cooler areas, is still inadequate; however, it appears that good resistance will be incorporated in the near future.

Further details on diseases and others of less widespread importance can be obtained from Wiese (1987), Heyne (1987) and Roelfs *et al.* (1992).

Overall, it is worth noting that growing conditions have generally become more favourable for wheat diseases and pests with time, particularly as irrigated wheat cropping has spread, N fertilizer levels have increased and cropping intensity has risen. The developed world has tended to respond to this situation by increasing the use of fungicides and pesticides, whereas the developing world, often being unable to afford these chemicals, has had to rely largely on host plant resistance for control of losses. More recently, in view of heightened environmental concerns and diminishing profitability of wheat, the developed world is also turning in this direction. Nevertheless, a huge job remains to incorporate adequate durable resistance to all significant diseases and pests for each wheat environment, and scope also exists for better non-chemical management strategies. Molecular biology is expected to have a significant impact on the former endeavour.

The introduction of wheat cropping into marginal areas will present many abiotic stress challenges. Mineral ion deficiencies and toxicities, drought, wind, salinity and temperature extremes are some of the factors that will limit wheat production in these environments. Primitive wheats and wild relatives that originated in such environments can be expected to provide genes for tolerance to these abiotic stresses. The genetic variation that man has exploited in wheat, along with the agronomic modification of the crop environment, has permitted the wheat crop to be pushed into almost all cropping environments of the world, with the major exception of the humid lowland tropics. Here, diseases and high mean temperatures will probably permanently exclude wheat growing, and rice and maize will remain the main cereals.

At the dry and cold limits of wheat adaptation one tends to find that wheat is replaced by barley. Wheat lacks the vigour and frost resistance that give barley an advantage in such marginal environments. Wheat also is not as tolerant of salinity as barley. Introduction and breeding of traits from wheat progenitors and landraces may permit some gains in adaptation of wheat in this respect; certainly breeding for salt tolerance is receiving attention and some progress has been made. Better resistance to waterlogging, lodging and shattering would reduce the not insignificant and recurring losses these problems cause in most environments. Certain soil micro-element deficiencies (e.g. zinc, copper, boron) and toxicities (e.g. aluminium, manganese, boron) limit wheat production in an increasing number of environments (as soils become depleted and/or soils inexorably acidify), yet useful genetic variation for tolerance of these problems exists. This needs to be exploited to lessen soil amelioration costs.

Genetic improvement of wheat is likely to remain a major source of productivity gains. The crop's yield potential has risen about 1% annually over the past 30 years. This trend is expected to continue but may require greater breeding resources, especially as recent gains in efficiency that resulted from computerization and mechanization of breeding begin to dwindle. Innovations in molecular biology appear unlikely to effect an impact on yield progress at any time. Thus it is likely that more resources will need to be invested in other potentially useful improvement strategies such as exploiting heterosis in wheat through the use of F_1 hybrids and developing input-saving wheats that compete better against weeds or extract greater amounts of available soil nutrients (e.g. phosphorus and zinc). Despite considerable progress in the last 100 years, huge scope still exists for strengthening and making more durable the resistance of wheat to diseases, viruses and insects. This appears to be the area in which molecular biology will make its first impact on wheat breeding. Molecular biology is also likely to aid conventional breeding in changing the quality of wheat grain by developing it for novel industrial uses and improving its nutritional structure in ways that would clearly benefit consumers (increasing its content of available iron, zinc, vitamin A and certain amino acids).

The last 30 years have witnessed an unprecedented level of international wheat germplasm exchange and the development of a greater degree of genetic relatedness among successful cultivars globally; the concept of broad adaptation has thus been well vindicated. However, this is seen by some as increasing genetic vulnerability to pathogens, although such vulnerability depends more on similarities in resistance genes, which may actually be more diverse now than before. This notwithstanding, various new factors (including the growing strength of national breeding programmes in the developing world and the advent of breeders' rights) should result in increased diversity among cultivars and perhaps lead to the exploitation of hitherto-overlooked specific adaptation in wheat. This would be especially important if climate change accelerates. Just as increasing nitrogen supply and improving weed control have been almost universal driving factors of wheat cultivation in the last 50 years, higher atmospheric concentrations of CO_2 and warmer temperatures could significantly influence breeding objectives in the next.

Evans (1993) argued convincingly that agronomic advances and improvements in yield, efficiency or sustainability also create breeding opportunities; this was the case with the improved nitrogen-use efficiency and reduced herbicide use mentioned above. Looking ahead, the global move toward reduced tillage and increased crop residue retention – so important for sustaining the soil resource – will require wheats to possess new traits such as resistance to diseases encouraged by crop residues or adaptation to less than ideal seedbed conditions. Although some scope probably exists for reducing the need for fertilizer applications through breeding for nutrient-use efficiency (e.g. phosphorus and zinc), it may be that the current interest in breeding for low nitrogen input will be overtaken by agronomic solutions to this yield constraint (optimal timing of fertilizer application, rotations with legumes). There will, however, inevitably be other moves toward more sustainable wheat cropping systems and these will undoubtedly place new demands on wheat improvement.

REFERENCES

Byerlee, D. and P. Moya. 1993. Impacts of international wheat breeding research in the developing world, 1966-90. CIMMYT, Mexico, D.F.

Dewey, D.R. 1984. The genomic system of classification as a guide to intergeneric hybridization with the perennial Triticeae. Pp. 209-279 *in* Gene Manipulation in Plant Improvement (J.P. Gustafson, ed.). Plenum Press, New York.

Evans, L.T. 1993. Crop Evolution, Adaptation and Yield. Cambridge University Press, Cambridge, UK.

Gale, M.D. and S. Youssefian. 1986. Dwarfing genes in wheat. *In* Progress in Plant Breeding (G.E. Russell, ed.). Butterworths, London, UK.

Grama, A., Z.K. Gerechter-Amitai, A. Blum and G.L. Rubenthaler. 1984. Breeding bread wheat cultivars for high protein content by transfer of protein genes from *Triticum dicoccoides*. Pp. 145-153 *in* Cereal Grain Protein Improvement: Proceedings of the Final Research Coordinated Meeting of the FAO/IEAE/GSF/SIDA. International Atomic Energy Agency, Vienna.

Heyne, E.G. (ed.). 1987. Wheat and Wheat Improvement. Second edition. Agronomy Series No. 13. American Society of Agronomy, Madison, WI, USA.

IBPGR. 1990. Directory of crop germplasm collections. 3. Cereals: *Avena, Hordeum*, Millets, *Oryza, Secale*, Sorghum, *Triticum, Zea* and Pseudocereals (E. Bettencourt and J. Konopka, eds.), pp. 155-212. International Board for Plant Genetic Resources, Rome.

Johnson, B.L. 1975. Identification of the apparent B-genome donor of wheat. Can. J. Genet. Cytol. 17:21-39.

Kerby, K. and J. Kuspira. 1987. The phylogeny of the polyploid wheats *Triticum aestivum* (bread wheat) and *Triticum turgidum* (macaroni wheat). Genome 29:722-737.

Khan, K., R. Frohberg, T. Olson and L. Huckle. 1989. Inheritance of gluten protein components of high-protein hard red spring wheat lines derived from *Triticum turgidum* var. *dicoccoides*. Cereal Chem. 66:397-401.

Kimber, G. and M. Feldman. 1987. Wild wheat. An introduction. Special report 353. College of Agriculture, Univ. of Missouri, Columbia, USA.

Lagudah, G.S. and G.M. Halloran. 1988. Phylogenetic relationships of *Triticum tauschii*, the D genome donor to hexaploid wheat. 1. Variation in HMW subunits of glutenin and gliadins. Theor. Appl. Genet. 75:851-856.

Morgunov, A.I., R.J. Peña, J. Crossa and S. Rajaram. 1993. Worldwide distribution of *Glu-1* alleles in bread wheat. J. Genet. Breed. 47:53-60.

Mujeeb-Kazi, A. and G.P. Hettel, eds. 1995. Utilizing Wild Grass Biodiversity in Wheat Improvement: 15 Years of Wide Cross Research at CIMMYT. CIMMYT Research Report No. 2. CIMMYT, Mexico, D.F.

Mujeeb-Kazi, A., V. Rosas and S. Roldan. 1994. Conservation of the genetic variation of *Triticum tauschii* (*Aegilops squarrosa*) in synthetic hexaploid wheats (*T. turgidum* x *T. tauschii*; 2n=6x=42, AABBDD) and its potential utilization for wheat improvement. Genet. Resour. Cr. Evol. (in press).

Mujeeb-Kazi, A., R.L. Villareal, L.A. Gilchrist and S. Rajaram. 1996. Registration of five wheat germplasm lines resistant to helminthosporium leaf blight. Crop Sci. 36:216-217.

Nachit, M.M. 1992. Durum wheat breeding for Mediterranean drylands of North Africa and West Asia. Pp. 14-27 *in* Durum Wheats: Challenges and Opportunities (S. Rajaram, E.E. Saari and G.P. Hettel, eds.). Wheat Special Report No. 9. CIMMYT, Mexico, D.F.

Parlevliet, J.E. 1988. Strategies for the utilization of partial resistance for control of cereal rusts. *In* Breeding Strategies for Resistance to the Rusts of Wheat (N.W. Simmonds and S. Rajaram, eds.). CIMMYT, Mexico, D.F.

Peña, R.J. 1995. Quality improvement of wheat and triticale. Pp. 100-111 *in* Wheat Breeding at CIMMYT: Commemorating 50 years of research in Mexico for global wheat improvement (S. Rajaram and G.P. Hettel, eds.). Wheat Special Report No. 29. CIMMYT, Mexico, D.F.

Peña, R.J., J. Zarco-Hernandez and A. Mujeeb-Kazi. 1995. Glutenin subunit compositions and bread-making quality characteristics of synthetic hexaploid wheats derived from *Triticum turgidum* X *Triticum tauschii* (coss.) Schmal crosses. J. Cereal Sci. 21:15-23.

Pomeranz, Y. (ed.). 1987. Modern Cereal Science and Technology. VCH Publishers, Inc., New York, NY.

Ranhotra, G.S. 1991. Nutritional quality of cereals and cereal-based foods. Pp. 845-861 *in* Handbook of Cereal Science and Technology (K.J. Lorenz and K. Kulp, eds.). Marcel Dekker, Inc., New York, Basel, Hong Kong.

Roderick, C.E. H. and Fox. 1987. Nutritional value of cereal grains. Pp. 1-10 *in* Nutritional Quality of Cereal Grains: Genetic and Agronomic Improvement. Agronomy Monograph No. 20. ASA-CSSA-SSSA, Madison, WI.

Roelfs, A.P., R.P. Singh and E.E. Saari. 1992. Rust Diseases of Wheat: Concepts and Methods of Disease Management. CIMMYT, Mexico, D.F.

Sander, D.H., W.H. Allaway and R.A. Olson. 1987. Modification of nutritional quality by environment and production practices. Pp. 43-82 *in* Nutritional Quality of Cereal Grains: Genetic and Agronomic Improvement. Agronomy Monograph No. 20. ASA-CSSA-SSSA, Madison, WI.

USDA. 1995. Production Highlights for 1994/95. FAS, USDA.

Wang, R.R.-C. 1989. Intergeneric hybrids involving perennial Triticeae (Invited review). Genet. (Life Sci. Adv.) 8:57-64.

Wiese, M.V. 1987. Compendium of Wheat Diseases. 2nd ed. American Phytopathological Society, St. Paul, MN.

William, M.D.H.M., R.J. Peña and A. Mujeeb-Kazi. 1993. Seed protein and isozyme variations in *Triticum tauschii* (*Aegilops squarrosa*). Theor. Appl. Genet. 87:257-263.

Williams, P.H. 1989. Screening for resistance to diseases. *In* The Use of Plant Genetic Resources (A.H.D. Brown *et al.*, eds.). Cambridge University Press, Cambridge, UK.

Chapter 22

Forages

B.L. Maass, J. Hanson, L.D. Robertson, P.C. Kerridge and A.M. Abd El Moneim

Forages are usually associated with grasslands, but they also occur and are used widely on roadsides, in fallow areas, and as crop residues. In temperate areas livestock is grazed on domesticated forages. In the tropics livestock use native grasslands (Africa, Australia), naturalized grasslands (tropical America, Asia), or grasslands improved with selected wild species (tropical America). The worldwide exchange of forage germplasm has involved relatively few accessions of some widely adapted species, and the need exists to increase the diversity of these cultivated forages.

BOTANY AND DISTRIBUTION

Forages comprise different species, mostly from the Gramineae and Leguminosae families. Forages that have been domesticated and cultivated since historical times include lucerne (*Medicago sativa* L.), berseem (*Trifolium alexandrinum* L.) and red clover (*Trifolium pratense* L.) (Mannetje *et al.* 1980). Others have been cultivated more recently, such as white clover (*Trifolium repens* L.) and subterranean clover (*Trifolium subterraneum* L. *sensu lato*), ryegrasses (*Lolium*), *Festuca* and *Phalaris*. Wild species have been distributed and used widely: 40 million ha of *Brachiaria decumbens* Stapf in South America (Miles *et al.* 1996) and about 1 million ha of *Stylosanthes* species in Australia.

A great number of families, genera and species can be considered as forage and it is not practical to give botanical details of every species, including synonyms. Vernacular names of forage species have been documented (Mejía 1984; Skerman *et al.* 1988; Skerman and Riveros 1989; Wiersema *et al.* 1990; Mannetje and Jones 1992; Barnes *et al.* 1995; Brako *et al.* 1995). Because most forage species are wild or partially domesticated, their centres of diversity correspond with centres of origin. The ecophysiological characteristics of forages are, to a large extent, reflected in the environmental conditions under which they were collected. In this chapter, we will describe major genera from the legumes and grasses, used as native or cultivated forages (for taxonomic classification see Tables 22.1 and 22.2).

Tropical grasses from Africa have been spread widely and became naturalized so that they now have a pan-tropical distribution: guinea grass (*Panicum maximum* Jacq.), para grass (*Brachiaria mutica* (Forssk.) Stapf), molasses grass (*Melinis minutiflora* P. Beauv.), jaragua (*Hyparrhenia rufa* (Nees) Stapf), kikuyu grass (*Pennisetum clandestinum* Hochst. ex Chiov.) and pangola grass (*Digitaria eriantha* Steudel, syn. *D. decumbens* Stent) (Parsons 1972; Mannetje and Jones 1992). Some herbaceous legumes are becoming naturalized in tropical regions other than their geographic origin (Thomas 1994). Tropical kudzu (*Pueraria phaseoloides* (Roxb.) Benth.), of Asian origin, is widespread in the humid tropics of South America and *Centrosema pubescens* Benth. and *C. plumieri* (Turp. ex Pers.) Benth., both of American origin, have become naturalized in Southeast Asia and parts of West

Table 22.1. Systematic position of important forage grasses.

Subfamily	Tribe	Tropical and subtropical genera	Mediterranean and temperate genera
Chloridoideae	Eragrostideae	*Eragrostis* von Wolf	
	Chlorideae	*Chloris* Swartz	
		Cynodon Richard	
Panicoideae	Paniceae	*Axonopus* P. Beauvois	
		Brachiaria (Trin.) Grisebach	
		Cenchrus L.	
		Digitaria Haller	
		Melinis P. Beauvois	
		Panicum L.	
		Paspalum L.	
		Pennisetum Richard	
		Setaria P. Beauvois	
		Urochloa P. Beauvois	
	Andropogoneae	*Andropogon* L.	
		Dichanthium Willemet	
		Hemarthria R. Brown	
		Hyparrhenia Fournier	
		Sorghum Moench	
		Tripsacum L.	
		Zea L.	
Pooideae (Festucoideae)	Stipeae		*Stipa* L.
	Poeae		*Bromus* L.
			Dactylis L.
			Festuca L.
			Lolium L.
			Poa L.
	Aveneae		*Avena* L.
	Phalarideae		*Phalaris* L.
	Agrostideae		*Calamagrostis* Adanson
			Phleum L.
	Triticeae (Hordeae)		*Agropyron* Gaertner
			Hordeum L.
			Secale L.
			Triticum L.

Africa. Shrub legumes of Mesoamerican origin, such as *Leucaena leucocephala* (Lamk.) de Wit and *Gliricidia sepium* (Jacq.) Kunth ex Walp., were transported around the world in colonial times and became naturalized in particular niches (Shelton and Brewbaker 1994; Simons and Stewart 1994).

GERMPLASM CONSERVATION AND USE

Australian institutions were leaders in assembling and using exotic forage genetic resources (Davidson and Davidson 1993). Subsequently, the IARCs – CIAT, ICARDA and ILRI (created from the merger of ILCA (International Livestock Centre for Africa), Ethiopia, and ILRAD (International Livestock Research Institute on Animal Diseases), Kenya, in 1995)

Table 22.2. Systematic position of important forage legumes.

Subfamily	Tribe[†]	Tropical and subtropical genera	Mediterranean and temperate genera
Caesalpinoideae	Cassieae	*Chamaecrista* Moench	
Faboideae	Robinieae	*Gliricidia* Kunth	
		Sesbania Scop.	
	Desmodieae	*Alysicarpus* Desv.	*Lespedeza* Mich.
		Desmodium Desv.	
	Phaseoleae	*Cajanus* DC.	
		Calopogonium Desv.	
		Centrosema (DC.) Benth.	
		Lablab Adans.	
		Macroptilium (Benth.) Urban	
		Macrotyloma (Wight & Arn.) Verdc.	
		Neonotonia Lackey	
		Pueraria DC.	
		Vigna Savi	
	Aeschynom-eneae	*Aeschynomene* L.	
		Arachis L.	
		Stylosanthes Swartz	
	Galegeae		*Astragalus* L.
	Hedysareae		*Hedysarum* L.
			Onobrychis Mill.
	Loteae		*Lotus* L.
			Tetragonolobus Scop.
	Coronilleae		*Ornithopus* L.
	Vicieae		*Lathyrus* L.
			Pisum L.
			Vicia L.
	Trifolieae		*Medicago* L.
			Melilotus L.
			Trifolium L.
			Trigonella L.
	Crotalarieae	*Lotononis* (DC.) Eckl. & Zeyh.	
	Genisteae		*Lupinus* L.
Mimosoideae	Mimoseae	*Desmanthus* Willd.	
		Leucaena Benth.	
	Ingeae	*Calliandra* Benth.	

[†] Order of tribes according to Polhill and Raven (1981).

– with forages in their mandate have assembled large *ex situ* collections (Table 22.3) in collaboration with NARIs and IPGRI. CSIRO and NARIs undertook systematic expeditions in tropical America from the 1960s until the mid-1980s to search for more diverse germplasm (Burt and Williams 1975). As a result, the novel species *Stylosanthes scabra* Vogel and *S. hamata* (L.) Taubert were selected and used widely for pasture improvement in Australia (Cameron *et al.* 1993b) and other tropical regions. In contrast to the Mediterranean and tropical forage germplasm collections, temperate collections are more numerous and hold larger numbers of accessions, mostly of a very small number of genera, showing thus a higher degree of specialization (Schultze-Kraft *et al.* 1993b).

Table 22.3. Holdings of the germplasm banks at CIAT, ICARDA, and ILRI (August 1995).

| | Family | | | Other | |
	Gramineae	Leguminosae[†]	Other browse	herbaceous‡	Total
ICARDA					
genera (no.)	–	23	–	–	23
species (no.)	–	339	–	–	339
accessions (no.)	–	26198	–	–	26198
CIAT					
genera (no.)	47	89	2	–	138
species (no.)	162	592	2	–	756
accessions (no.)	2081	18751	2	–	20834
ILRI					
genera (no.)	78	127	67	36	308
species (no.)	297	706	114	97	1214
accessions (no.)	2716	8849	163	236	11964
Total access.[§]	4114	51069	163	236	55582

[†] Herbaceous and browse legumes.
[‡] More than half of the accessions belong to Chenopodiaceae (*Atriplex* L. and *Chenopodium* L.) and Amaranthaceae (*Amaranthus* L.).
[§] Estimated number of distinct accessions.

Table 22.4. Major donors (more than 100 samples) to the CIAT forage collection.

Country	Institute	Accessions (no.)	Main genera (>100 accessions)
Argentina	INTA	367	Various legumes
Australia	CSIRO-ATFGRC, Brisbane	591	*Stylosanthes*
	QDPI, Walkamin	179	Various legumes
Belize	IDRC	300	Various legumes
Brazil	EMBRAPA	863	*Stylosanthes, Centrosema*
	EPAMIG	326	*Stylosanthes*
	IRI	146	Various legumes
France	ORSTOM, Côte d'Ivoire	291	*Panicum*
Indonesia	IBPGR, Bogor	847	*Desmodium*
USA	University of Florida	153	Various legumes

Forage genetic resources have been reviewed from different aspects: globally (McIvor and Bray 1983); by region, such as Africa (Dzowela 1988; Le Houérou 1991) and the Andes of South America (Paladines and Delgadillo 1990); by environment, such as semi-arid temperate (Smith and Robertson 1991); or by genus or plant type within state-of-the-art reviews of legumes – *Stylosanthes* (Stace and Edye 1984; Leeuw et al. 1994), *Centrosema* (Schultze-Kraft and Clements 1990); forage – *Arachis* (Kerridge and Hardy 1994); grasses – *Andropogon gayanus* Kunth. (Toledo et al. 1990) and *Brachiaria* (Miles et al. 1996); and forage tree and shrub legumes – *G. sepium* (Withington et al. 1987), *Sesbania* (Macklin and Evans 1990; Kategile and Adoutan 1993), *Erythrina* (Westley and Powell 1993), *Leucaena* (Shelton et al. 1995), *Cratylia* (Pizarro and Coradin 1996) and shrub legumes in general (Gutteridge and Shelton 1994).

Collections of CIAT, ILRI and ICARDA

Tropical forage germplasm is held in the genebanks at CIAT and ILRI. At CIAT, the forage germplasm collection (Table 22.3) was assembled to find forage plants that are well adapted to the acid, low-fertility soils that prevail in the humid and subhumid tropics, and where disease and pest pressures are high and rainfall patterns are seasonal (Schultze-Kraft and Giacometti 1979; Schultze-Kraft 1985). The collection was initiated in 1971, with the first prospective collecting trips in 1974 (Schultze-Kraft and Alvarez 1984). About three-quarters of the almost 21 000 accessions maintained in the CIAT germplasm bank were collected by CIAT scientists in collaboration with NARIs. Important collection trips were made in conjunction with EMBRAPA/CENARGEN in Brazil (Coradin and Schultze-Kraft 1990), FONAIAP in Venezuela (Flores and Schultze-Kraft 1994) and NARIs in Thailand, Malaysia, Indonesia and tropical China (Schultze-Kraft *et al.* 1989a). The remainder of the collection has been acquired as donations from other forage germplasm collections, such as that of EMBRAPA/CENARGEN; IBPGR, Bogor; CSIRO-ATFGRC, Brisbane, Australia; INTA, Argentina; and the Forage Legume and Pasture Research Programme of IDRC, Canada in Belize (Table 22.4).

The ILRI genebank was established in 1982 and reached 75% of its present size of about 12 000 accessions by 1988 (Hanson and Lazier 1989). A diverse collection was assembled (Table 22.3) in order to identify appropriate forage germplasm acceptable to farmers and adapted to the wide range of environments and farming systems in sub-Saharan Africa. About half of the accessions were collected from African countries with participation of ILCA scientists, the other half being acquired mainly from CIAT, CSIRO and ICARDA (Table 22.5). Major partner institutions involved in collecting were CATIE in Costa Rica, CIAT, IBPGR (now IPGRI), and ICRAF, in addition to sub-Saharan NARIs.

The ICARDA pasture and forage collection was established in 1979 with the purpose to conserve and promote the utilization of native germplasm of the West Asia and North Africa (WANA) region for pasture and forage production in the region and elsewhere. It is the largest and most specialized of the forage germplasm collections in the CGIAR system, in terms of number of accessions, species and genera (Table 22.3). Acquisition policy for the collection maintained at ICARDA was guided by the recommendations of a working group (IBPGR 1985), based on paucity of existing collections, known areas of genetic erosion and the potential for improvement. During the last 10 years, the forage legume germplasm collection doubled from 13 440 accessions in 1985 (IBPGR 1985) to more than 26 000 accessions in 1995. Germplasm of the ICARDA collection of pasture and forages has been widely used in the WANA region and also in southern Europe, the USA and Australia.

Both collecting and donation have contributed in almost equal proportion to the ICARDA germplasm bank. The Germplasm Institute at Bari, Italy, and the United States Department of Agriculture (USDA), USA, donated the most accessions (Table 22.6). In addition, other important donations were received from SADAF, the John Innes Institute and the University of Southampton, UK, and IPK, Gatersleben, Germany. Collecting was concentrated heavily in the WANA region. The most important collaborators in collecting missions were NARIs from Syria, Turkey, Morocco, Jordan, Tunisia and Algeria.

Conservation

The forage germplasm held in the genebanks of the CGIAR centres is mainly stored as seed. Seed storage provides a long-term and economical means of storage for those species which can be dried and which behave as orthodox seeds. CIAT, ILRI and ICARDA store seeds of their base collections under long-term conditions at low seed moisture content (5-8%) and temperatures from –18 to –22°C in line with international standards for use in base collections (FAO/IPGRI 1994). Storage conditions for the active collections vary between Centres. ILRI stores seed dried to 5% moisture content in laminated foil packets at 8°C, CIAT stores seed in closed plastic jars at 35% RH and 5-8°C, while ICARDA stores seeds in cloth bags at 25% RH and 2°C. The storage conditions of these active collections are designed to ensure maintenance of viability from 10-20 years; they are also used for distribution of germplasm.

Seeds of the forage legumes *Calopogonium mucunoides* Desv., *Centrosema pubescens*, *Desmodium ovalifolium* Wall. (now *D. heterocarpon* (L.) DC. subsp. *ovalifolium* (Prain) Ohashi), *Pueraria phaseoloides* and *Stylosanthes guianensis* (Aubl.) Sw. showed little loss of germination when stored for 16 years at low temperatures (Bass 1984). Akinola *et al.*

Table 22.5. Major donors (more than 100 samples) to the ILRI forage collection.

Country	Institute	Accessions (no.)	Main genera (>100 accessions)
Australia	CSIRO	860	*Stylosanthes, Trifolium*
	DSIR	162	Various legumes
Belize	Ministerio de Agricultura	297	Various legumes, fodder trees
Colombia	CIAT	1354	*Stylosanthes, Phaseolous, Centrosema, Zornia*
Great Britain	Royal Botanic Gardens, Kew	232	Various grasses, legumes, others
Italy	IBPGR, Rome	1055	Various legumes, fodder trees
Mexico	CIMMYT	387	*Triticum*
Syria	ICARDA	788	*Vicia, Lathyrus, Pisum, Avena*
USA	University of Florida	115	Various legumes

Table 22.6. Major donors (more than 100 samples) to the ICARDA forage collection.

Country	Institute	Accessions (no.)	Main genera (>100 accessions)
Australia	SADAF, Adelaide	1247	*Medicago*
	WADA, Perth	184	*Medicago*
Germany	IPK, Gatersleben	773	*Vicia, Lathyrus*
Great Britain	IBPGR Seed Handling Unit, Kew	116	*Medicago*
	John Innes Institute, Norwitch	1076	*Pisum*
	University of Southampton	1034	*Vicia, Lathyrus*
Hungary	Research Centre for Agrobotany, Tapioszele	161	*Pisum*
Iran	National Program (SPII, Karaj)	102	*Medicago*
Italy	Germplasm Institute, Bari	3698	*Pisum, Vicia*
Lebanon	ALAD, Beirut	285	*Medicago*
Morocco	INRA, Rabat	248	*Scorpiurus*
Russia	VIR, Leningrad	99	Various legumes
USA	USDA	2247	*Medicago, Onobrychis, Trifolium*

(1991) studied dormancy and germination in *Desmodium velutinum* (Willd.)DC.and found that as seed dormancy declined, germination increased for up to 3 years of storage at room temperature. Similar studies in *Desmodium incanum* DC. showed about 80% hard seeds after harvest with some dormancy remaining after storage at room temperature for 1 year (Keya and Eijnatten 1975). Recent research has shown a correlation between the degree of hard-seededness, measured through water imbibition, and physiological quality of seed samples in *C. mucunoides*, suggesting that hard seeds may retain viability for a longer period during storage (Souza and Marcos-Filho 1993).

Some leguminous trees and shrubs have larger seeds, which may be more difficult to store for long periods without loss of viability. For example, ye-eb (*Cordeauxia edulis* Hemsl., Caesalpinieae) has protein-rich seeds which are short-lived and lose viability after only a few months (NAS 1979). Experiments on storage of seeds of *Panicum maximum* have shown that seeds begin to lose viability after 2-3 years of storage at room temperature (Harty *et al.* 1983). Seeds of *B. decumbens* maintained viability during storage at 10°C and low moisture content for 4.5 years (Whiteman and Mendra 1982). Loch *et al.* (1988) reported that seeds of *A. gayanus, Bothriochloa insculpta* Hochst. ex A. Rich. and *B. pertusa* (L.) A. Camus showed little damage from threshing because the caryopses are tightly held in the chaffy seed units. Studies by Whiteman and Mendra (1982) showed that two dormancy mechanisms operated in *B. decumbens*. Primary dormancy was physiological while long-term dormancy was caused by a mechanical restriction of the seed coat, which declined over a period of 1 year. Dormancy in *P. maximum* is probably also physiological and disappears after 50-300 days of storage (Harty *et al.* 1983) with more rapid loss at 22°C than at 10°C (Smith 1979). Other species have a very short dormancy period of up to 3 months. Butler (1985) has indicated that in *Cenchrus ciliaris* the dormancy mechanism is in the caryopsis rather than the spikelet.

The most common alternative to seed storage is to store the materials as living collections in field genebanks. This is used for the collections of *A. gayanus*, *Brachiaria*, *P. maximum* and *Leucaena* at CIAT and the collections of *Brachiaria*, *Cynodon* and *Pennisetum purpureum* Schumach. held by ILRI. However, maintenance in field genebanks is not secure since the plants are at risk from pests and diseases. It is also costly in terms of management and requires large areas of land to accommodate sufficient numbers of individuals to represent the genetic diversity in the population, especially in large fodder tree species. Research on the physiology of seed production would make a major impact in defining conditions for production of high-quality seed and thus reduce the need for maintaining field collections.

A simple technique was developed for field collection of vegetative material of *Digitaria eriantha* subsp. *pentzii* (Stent) Kok and *Cynodon dactylon* (L.) Pers. using sterilized axillary buds cultured on sterile media with fungicides and antibiotics (Ruredzo 1991). Research on methods of culture of axillary buds and use of low temperatures to slow growth in culture were tested on *D. eriantha* subsp. *pentzii*, *Cynodon aethiopicus* Clayton & Harlan and *C. dactylon* (Ruredzo and Hanson 1990). Cultures of some accessions of all species were maintained at 15°C for 18 months without subculturing and could be recovered into normal growth conditions (Hanson and Ruredzo 1992). Plantlets of *C. dactylon* that had been stored for 5 years without subculturing also have been recovered and showed no signs of abnormality or retarded growth in normal growth culture conditions (J. Hanson, unpublished data). CIAT and ILCA developed a technique for *in vitro* culture of axillary buds in *Brachiaria* spp. that helped to transfer a large portion of the *Brachiaria* germplasm collected by CIAT and ILCA in Africa to Colombia by uncontaminated *in vitro* culture, and subsequent germplasm distribution to screening sites in Brazil, Peru and Costa Rica in the same form (Keller-Grein *et al.* 1996).

In situ conservation in national parks and game reserves will assist in protecting wild species. Wildlife reserves are widespread in East and southern Africa, especially in Kenya, Tanzania, Zambia, Zimbabwe, Malawi, Botswana, Swaziland, Lesotho, South Africa and Namibia. These reserves are specifically for wildlife, but there are also protected areas for vegetation studies in the natural rangeland where the forages are conserved as part of the ecosystem. A field study of vegetation and its pastoral use at 30 sites along a north-south transect between 150 and 600 mm rainfall isohyets was carried out over 10 years in the Gourma region of Mali (Hiernaux 1994). In a few cases plants with forage potential have been collected from national parks, for example, *Chamaecytisus* on the Canary islands (Fancisco-Ortega *et al.* 1990). *In situ* conservation of vegetation types ensures that plants that serve as forages will be preserved. Species inventories of forages in these protected areas give some indication about interspecific biodiversity. However, more information is needed about the intraspecific diversity and its spatial distribution, and how well this is represented in *ex situ* conserved germplasm. Research on genetic diversity assisted by the use of geographic information systems (GIS) could contribute considerably to the field of conservation of genetic resources.

Tropical Forages

Legumes make up the major part in the collections held by CIAT and ILRI. Most of the legume accessions (65%) have been collected from the primary centre of diversity in tropical America, 20% originated from Southeast Asia and 15% from Africa. On the other hand, about 90% of the grasses held in the two collections originated from Africa. At CIAT a collection of over 4000 strains of *Rhizobium sensu lato* is maintained as a working collection and added to when a specific need has been identified. Around 100 requests for forage legume inoculants and ampoules are serviced per annum. An updated fifth version of the *Rhizobium* collection catalogue for forage legumes was completed and published in 1993 with the further characterization of some 2056 strains (Franco *et al.* 1993). ILRI also maintains a small collection of *Rhizobium* for use and distribution with germplasm. Both CIAT and ILRI keep passport data in computerized databases, which are periodically published in germplasm catalogues (Schultze-Kraft *et al.* 1987; Schultze-Kraft 1990, 1991a, 1991b; Kidest Shenkoru *et al.* 1991a, 1991b; Belalcázar and Schultze-Kraft 1994). Information generated in characterization and performance is attached to every accession, and thus is available through the respective databases. In 1992, CIAT transferred its database to the relational database management system ORACLE. The information at ILRI is held in dBASE software. CIAT and ILRI collaborate with the CGIAR System-wide Information Network on Genetic Resources (SINGER). Passport data on all CGIAR collections will be available in the Internet through the SINGER system. Large parts of the forage germplasm at CIAT were screened for disease and pest susceptibility and results compiled by species (Lenné 1990; Lenné and Trutmann 1994). From 1980 to 1994, CIAT distributed more than 100 genera to 55 countries. The result has been the release of many accessions as commercial cultivars (CIAT 1993).

Germplasm of some regions or countries is not represented in the collections at CIAT and ILRI, and there is potential for further collecting. For example, Ferrufino *et al.* (1991) observed rich leguminous diversity in the humid and subhumid tropics of Bolivia and Hacker *et al.* (1996a) have recognized the Chaco region of Paraguay as a potential source of new pasture legumes for the subtropics; neither country has provided germplasm to the CGIAR forage collections. Other areas are threatened by genetic erosion because of rapidly changing land-use patterns, such as the Cerrados in Brazil (Klink *et al.* 1993). Large areas in Southeast Asia have not been collected for forage genetic resources, where genetic erosion is likely to have occurred because of very intensive agricultural production systems (Schultze-Kraft *et al.* 1989a). The first collection of forage genetic resources in Vietnam, supported by IBPGR, was only carried out in 1992 (Schultze-Kraft *et al.* 1993a).

Tropical Legumes

The collection of tropical legumes held by CIAT and ILRI is very diverse in terms of number of genera, species and accessions maintained (Tables 22.3 and 22.7). Tropical America is the primary centre of diversity of tropical legumes (Williams 1983).

Grof *et al.* (1989) evaluated a collection of 215 accessions of *Aeschynomene* from 15 species for adaptation to poorly drained savannas in tropical South America. Bishop *et al.* (1988) characterized a large collection of diverse *Aeschynomene* germplasm in Australia. Three cultivars selected out of the natural variation of *Aeschynomene* species have been released in Australia (Oram 1990). Another large collection of *Aeschynomene* is held at the Fort Pierce Research and Education Center of the University of Florida, USA. Its agronomic importance in the USA, where cv. Glenn has been released, was briefly documented by Kretschmer and Pitman (1995).

In a recent review, Valls *et al.* (1994) estimated wild *Arachis* species kept in 10 major *ex situ* germplasm collections to be about 1000 accessions; however, not all of them have forage potential. Nevertheless, wild species of *Arachis* that show promise as a forage or cover crop have created much interest among researchers all over the tropical world (Kerridge and Hardy 1994). Background information on the genetic resources of important species, such as pinto peanut (*A. pintoi* Krap. & Greg.) and rhizoma peanut (*A. glabrata* Benth.), is given in several chapters of the book on the biology and agronomy of forage *Arachis* (Kerridge and Hardy 1994). A germplasm catalogue of wild *Arachis* species is in preparation (H.T. Stalker 1994, pers. comm.), which will take into account the new taxonomic treatment by Krapovickas and Gregory (1994). CIAT currently maintains about 100 accessions (Table 22.7), of which morphological characterization and isozyme fingerprinting by PAGE (polyacrylamide gel electrophoresis) revealed large intraspecific

Table 22.7. Germplasm (number of species and accessions) of tropical and temperate herbaceous forage legumes held in *ex situ* collections.

Genus	CIAT		ILRI		ICARDA		Total[†]
	spp.	access.	spp.	access.	spp.	access.	acces.
Tropical legumes							
Aeschynomene	32	999	11	129	–	–	1075
Alysicarpus	9	274	10	241	–	–	463
Arachis	19	98	5	8	–	–	105
Calopogonium	4	536	3	41	–	–	554
Centrosema	31	2410	12	325	–	–	2522
Chamaecrista	16	05	5	100	–	–	381
Crotalaria	24	283	34	226	–	–	508
Desmodium	52	2940	28	165	–	–	3026
Galactia	12	570	2	4	–	–	573
Indigofera	16	37	32	255	–	–	483
Lablab	1	22	1	176	–	–	200
Lotononis	2	3	3	7	–	–	10
Macroptilium	10	610	10	71	–	–	669
Macrotyloma	6	51	5	32	–	–	75
Neonotonia	1	68	1	254	–	–	293
Phaseolus	3	34	7	276	–	–	310
Pueraria	4	260	2	7	–	–	266
Rhynchosia	14	448	13	139	–	–	562
Stylosanthes	25	3587	15	1140	–	–	4148
Teramnus	4	386	4	65	–	–	434
Vigna	33	747	20	418	–	–	746
Zornia	17	1030	10	274	–	–	1108
Other tropical	112	1486	72	421	–	–	1500[†]
Temperate legumes							
Astragalus	1	1	3	5	21	885	890
Hippocrepis	–	–	–	–	4	292	292
Lathyrus	–	–	8	163	42	1619	1635
Medicago	3	5	19	240	39	8456	8570
Onobrychis	–	–	17	28	40	801	804
Pisum	–	–	3	126	2	3553	3557
Scorpiurus	–	–	–	–	2	445	445
Trifolium	8	29	49	1494	60	3401	4889
Trigonella	–	–	2	2	22	595	597
Vicia	1	1	20	251	62	5350	5449
Other temperate[‡]	–	–	40	158	45	813	965
Total	**460**	**17420**	**466**	**7241**	**339**	**26210**	**48104**[†]

[†] Estimated number of different accessions.
[‡] Includes the genera *Anthyllis, Biserrula, Coronilla, Hedysarum, Hymenocarpus, Lotus, Lupinus, Melilotus, Ononis, Ornithopus, Psoralea, Securigera, Tetragonolobus.*

variation (CIAT 1993; Maass *et al.* 1993; Valls *et al.* 1994; Maass and Ocampo 1995). A total of eight commercial cultivars of forage *Arachis* species have been released in Australia and the Americas (Valls *et al.* 1994).

A large collection of 215 *Calopogonium* accessions was characterized and agronomically evaluated at Planaltina, Brazil (Pizarro *et al.* 1996). Wide variation of agronomic characters was also found in a core collection, assembled on the basis of geographic representation, evaluated at Carimagua and Villavicencio, Colombia (CIAT 1993).

Centrosema genetic resources have been revised by Schultze-Kraft *et al.* (1990), who estimated the world germplasm held in five major *ex situ* collections to be of more than 3000 accessions. Large gaps in collections from the geographical areas of Paraguay and Bolivia have been identified (Schultze-Kraft *et al.* 1990). A world germplasm catalogue of *Centrosema* was published by Schultze-Kraft *et al.* (1989b). *Centrosema* species have been extensively evaluated at CIAT for acid soil adaptation in the savannas of Colombia and Brazil, and in the humid tropics of Peru. Results have been compiled and published by Schultze-Kraft and Clements (1990) and more recently by Keller-Grein (1993a, 1993b) and Keller-Grein and Passoni (1993). Accessions of *Centrosema macrocarpum* Benth. do not show much morphological variation, but they differ in their adaptation to acid soils (Schultze-Kraft 1986). Lack of persistence of *C. macrocarpum* under grazing may be because of its limited seed production. *Centrosema brasilianum* (L.) Benth., on the other hand, is morphologically very diverse (Schultze-Kraft and Belalcázar 1988), which may be partly because of its high rate of outcrossing (Maass and Torres 1992). Some breeding has been carried out in *Centrosema* species in Australia and Colombia (Miles *et al.* 1990), which led to the release of *C. pascuorum* cv. Cavalcade in Australia. Other cultivars released were selected from germplasm accessions, such as *C. acutifolium* Benth. cv. Vichada in Colombia and *C. schiedeanum* (Schlecht.) Clements & Williams *nom. nud.* cv. Belalto and *C. pascuorum* cv. Bundey in Australia (Kretschmer and Pitman 1995; Oram 1990). At ILRI, the agronomic performance of different *Centrosema* species was evaluated in the subhumid environment of Kaduna, Nigeria (Peters *et al.* 1994a, 1994b).

The genus *Desmodium* is very diverse in the number of species, several of which have been used successfully as forages, in limited climatic zones (Imrie *et al.* 1983). Germplasm of the agronomically important *D. heterocarpon* (L.) DC. subsp. *ovalifolium* (Prain) Ohashi (syn. *D. ovalifolium* Wall.) was characterized by Schultze-Kraft and Benavides (1988). In Brazil, cv. Itabela has been released for pasture improvement (CIAT 1993). Schultze-Kraft and Benavides (1988) established a core collection on the basis of geographic origin and some important agronomic attributes, part of which is presently being studied for genotype by environment interaction in Colombia (Schmidt *et al.* 1997). The potential agronomic importance of *Desmodium* has driven research, and led to the release of several commercial cultivars of different *Desmodium* species in Australia, Brazil and the USA (Kretschmer and Pitman 1995).

A core collection of *Macroptilium*, assembled on the basis of geographic representation, was characterized and agronomically evaluated at Planaltina, Brazil (CIAT 1995). Research in Australia led to the release of the commercial cultivars Siratro, Murray, Maldonado (Cameron 1991; Oram 1990) and the rust-resistant Aztec. Pengelly and Eagles (1993) concluded from their characterization and preliminary evaluation of about 150 germplasm accessions from 10 different species of *Macroptilium* that further collection of *M. affine* (new combination, basionym *Phaseolus affinis* Piper, according to B.C. Pengelly 1996, pers. comm.) and Brazilian forms of *M. panduratum* (Mart. ex Benth.) Maréchal & Baudet should be considered a priority. They also characterized diversity in *M. gracile* germplasm (Pengelly and Eagles 1995). The Fort Pierce Research and Education Center of the University of Florida, USA also holds a large collection of *Macroptilium*. The agronomic importance of *Macroptilium* in the USA was highlighted by Kretschmer and Pitman (1995).

Portions of the collection of *Pueraria phaseoloides* have been evaluated at several sites in the savannas of Colombia and Brazil, and the humid tropics of Colombia and Peru (CIAT 1988, 1993). The agronomic attribute showing the most variation within the species was seasonal dry matter yield (CIAT 1993). Selected accessions are presently being evaluated in a wider environmental range, with special emphasis on selecting early seed producing materials (CIAT, unpublished data).

Genetic resources of *Stylosanthes* have been reviewed by Schultze-Kraft *et al.* (1984b) and recently by Hanson and Heering (1994). Several large species collections have been morphologically and agronomically characterized: *S. fruticosa* (Retz.) Mohlenbrock by Hakiza *et al.* (1988a), *S. macrocephala* by Schultze-Kraft *et al.* (1984a), *S. scabra* by Maass (1989) and *S. viscosa* (L.) Sw. by Keller-Grein and Schultze-Kraft (1992). The morpho-agronomic variation encountered in *S. scabra* was clustered and the groups formed helped in determining a core collection for further studies (Maass and Schultze-Kraft 1993). The botanical variety *S. guianensis* var. *pauciflora* Brandão *et al.* was detected to be the most resistant to anthracnose disease. Costa and Schultze-Kraft (1993) have described the biogeography of this variety together with that of *S. capitata* Vogel. Fingerprinting of accessions by isozyme and seed protein electrophoresis has been carried out at CIAT, which helped to identify regions of greatest diversity for *S. capitata* and *S. guianensis* (B.L. Maass, S.I. Marulanda and C.H. Ocampo, unpublished). Biochemical characterization will also help to detect collection gaps. Breeding projects in *Stylosanthes* aim at anthracnose resistance in *S. scabra* in Australia (Cameron *et al.* 1993a) and combining anthracnose resistance with high seed production in *S. guianensis* at CIAT (CIAT 1995). Besides several cultivars released in Australia (Oram 1990), screening and selection has resulted in the release of commercial materials in South America: *S. capitata* cv. Capica in Colombia, *S. guianensis* cv. Pucallpa in Peru and cvs. Bandeirante and Mineirão, and *S. macrocephala* cv. Pioneiro in Brazil; and *S. guianensis* cv. Bihuadou in China (CIAT 1993).

A large *Vigna* collection has been assembled from many *Vigna* species, almost a third of which originate from Colombia. However, there are small numbers of most species, except *V. adenantha* (G.F. Meyer) Maréchal *et al.*, a species with considerable tolerance of waterlogging (Kretschmer and Pitman 1995), which is represented by more than 100 accessions (18%). Hacker *et al.* (1996b) characterized germplasm of the genus *Vigna* in Australia with regard to potential as forage.

The genus *Zornia* was considered very promising for pasture improvement in the late 1970s and early 1980s (Thomas and Grof 1986). A group of 161 accessions of four *Zornia* species was characterized by ILCA (Hakiza *et al.* 1988b). The entire collection of *Zornia* held at CIAT was donated to CSIRO-ATFGRC, where a morpho-agronomic characterization is presently being conducted (B.C. Pengelly 1996, pers. comm.).

Alysicarpus and *Neonotonia* both have value as forage and are indigenous to Africa. Gramshaw *et al.* (1987) classified an *Alysicarpus* germplasm collection in Australia, using morphological and agronomic attributes, and indicated potential for agronomic use. *Neonotonia wightii* (Wight & Arn.) Lackey is sown for pasture in association with grasses, for example in Australia where four cultivars have been developed (Oram 1990), Brazil and Bolivia in subhumid areas. It is considered to have potential for drier areas in Africa. There has been some agronomic evaluation of *Neonotonia* at ILRI (Larbi *et al.* 1992). In a germplasm characterization and preliminary evaluation for acid soil adaptation in the Colombian Llanos, reasonable morphological and agronomic variation was found in large collections of *Chamaecrista rotundifolia* (Pers.) Greene and *Galactia striata* (Jacq.) Urban (CIAT 1993). Larbi *et al.* (1993b) have conducted agronomic evaluations of *C. rotundifolia* accessions under grazing in southern Ethiopia. Screening of more comprehensive germplasm of *C. rotundifolia* (Strickland *et al.* 1985) resulted in the selection and release of cv. Wynn in Australia (Oram 1990). Harding *et al.* (1989) evaluated and classified a diverse collection of *Rhynchosia* in Australia, using morphological and agronomic attributes.

There are considerable gaps in germplasm held in CGIAR collections for forage legume genera and species, such as *Lotononis bainesii* Baker (10 accessions), *Macrotyloma* species (75) and *Lablab* (200). These genera seem to merit further acquisition although reasonable collections are held, for instance, at CSIRO-ATFGRC with *Lablab* (151 accessions), *Lotononis* (105) and *Macrotyloma* (160) (B.C. Pengelly 1995, pers. comm.). It is necessary to assess the forage potential and identify the niches for these genera on the basis of the characterization and evaluation of comprehensive collections, as performed by Blumenthal and Staples (1993) for *Macrotyloma*, before new collecting is undertaken, unless some regions of origin are under immediate threat of genetic erosion.

Tropical Browse

Three species of the tribe Mimosoideae have shown potential as forage and have been widely distributed outside their native habitat in Mexico and Central America: *Leucaena leucocephala, Calliandra calothyrsus* Meissn. and *Desmanthus virgatus* (L.) Willd., the latter occurring naturally throughout tropical America (Mannetje and Jones 1992; Gutteridge and Shelton 1994; Shelton *et al.* 1995). The Mesoamerican *Gliricidia sepium* of the tribe Papilionoideae has been cultivated for millenia (Simons and Stewart 1994). Research at ILRI and CIAT has shown that multipurpose tree species (MPTS) are very diverse in occurrence and quantity of secondary plant compounds (Heering 1995a; Lascano *et al.* 1995; Raaflaub and Lascano 1995).

Both ILRI and CIAT have assembled diverse germplasm of about 3000 accessions of browse (Table 22.8), partly as donations, partly acquired in collecting missions, such as *Gliricidia* (Sumberg 1985) and *Sesbania* species by ILRI and ICRAF (Ndungu and Boland 1992). *Acacia* species studied include *A. cyanophylla* Lindley, *A. nilotica* Del., *A. seyal* Del., *A. sieberiana* DC. and *A. tortilis* (Forsk.) Hayne and *Faidherbia albida* (Reed and Soller 1987; Reed *et al.* 1990; Tanner *et al.* 1990). A morphological characterization of 17 *Cajanus cajan* accessions (J. Belalcázar and B.L. Maass, unpublished data) did not reveal much variation in the germplasm maintained at CIAT. Almost three-quarters of the germplasm held at ILRI (Table 22.8) belongs to one species, *Chamaecytisus proliferus* (L. fil.) Link *sensu lato* (tagasaste), and was acquired from the Canary Islands, Spain. A comprehensive germplasm of wild tagasaste was collected and morphologically described by Francisco-Ortega *et al.* (1990); seed of that collection is conserved in the Centro de Conservación de Recursos Fitogenéticos, Madrid, Spain. Systematic germplasm screening of *Desmanthus* resulted in the release of three accessions of *D. virgatus* (cvs. Marc, Bayamo and Uman) in Australia for use on heavy clay soils in subhumid areas (Cook *et al.* 1993).

CIAT has assembled a considerable collection of *Flemingia* species (Table 22.8), in particular *Flemingia macrophylla* (Willd.) Merr., a species well adapted to acid, low-fertility soils. Part of the *Flemingia* collection has been morphologically characterized and evaluated for its adaptation to acid, low-fertility soils in the savannas and the humid tropics of Colombia (CIAT 1993; Argel and Maass 1995). The most important limitation for its use as a forage is its high tannin content that consequently leads to low intake and digestibility; there is limited variation in tannin content among accessions (Lascano *et al.* 1993, 1995).

ILRI assembled a collection of *Gliricidia sepium* in collaboration with CATIE in Costa Rica (Table 22.8), and characterized some agronomically important attributes (Sumberg 1985). It has concentrated on the evaluation of *G. sepium* for use in alley farming in West Africa (Reynolds and Jabbar 1994). A range of *Gliricidia* accessions were evaluated and assessed for both productivity and palatability (Larbi *et al.* 1993a). These studies indicated that the most productive agronomically were not the most palatable to livestock. Biological and agronomic aspects of *G. sepium* have been reviewed by Simons and Stewart (1994), who participated in collecting and evaluation of the large germplasm collection held at the Oxford Forestry Institute (OFI), UK.

Genetic resources of seven major collections of *Leucaena* have been reviewed by Hughes *et al.* (1995), who pointed out that most collections are derived from three large collections, of the University of Hawaii (UH), CSIRO and OFI. It has been well characterized for morphology, agronomic properties and breeding behaviour (Sorensson 1995). Although *L. leucocephala* is best represented, there are numerous other species kept in the collections. *Leucaena* breeding projects have targeted low mimosine content (Bray 1995), acid soil tolerance (Hutton and Chen 1993) or psyllid resistance (Shelton and Brewbaker 1994). In 1994, a network for research into *Leucaena* was established which aims to evaluate germplasm accessions as well as breed lines for adaptation, forage value and psyllid resistance (Shelton *et al.* 1995).

Table 22.8. Germplasm (number of species and accessions) of tropical shrubs and trees with forage potential held in *ex situ* collections.

Genus	CIAT species	CIAT access.	ILRI species	ILRI access.	Total accessions
Shrub legumes					
Acacia	8	23	69	180	201
Albizia	2	7	10	23	30
Cajanus	1	54	1	156	209
Calliandra	3	21	2	6	27
Chamaecytisus	1	1	2	211	212
Codariocalyx	2	37	1	27	38
Cratylia	2	12	1	1	12
Desmanthus	7	95	7	112	201
Erythrina	9	63	10	38	100
Flemingia	6	147	1	6	151
Gliricidia	1	9	1	88	96
Leucaena	11	203	20	148	316
Prosopis	1	11	4	11	22
Sesbania	12	38	25	414	435
Other	74	610	86	187	750a
Total shrub legumes	132	1331	238	1608	2800[†]
Other browse	2	2	114	163	165
Total	**134**	**1333**	**354**	**1771**	**2965**[†]

[†] Estimated number of different accessions.

The *Sesbania* collection was characterized on morphological and agronomic characters at ILRI (Heering 1995a, 1995b). The results of multilocational germplasm evaluation in East and southern Africa were compiled by Kategile and Adoutan (1993). Other research has been carried out on genetics and reproductive behaviour of different *Sesbania* species (Heering 1994; Heering and Hanson 1993). Research into the variation of *Desmodium velutinum* (Willd.) DC. showed this species to be of good forage value and moderately well adapted to acid soils (Cárdenas 1990; Argel and Maass 1995). *Codariocalyx gyroides* (Roxb. ex Link) Hassk., a shrub legume for the humid tropics, showed reasonable variation in tannin content among accessions (Maass *et al.* 1996). In a recent workshop on *Cratylia* in Brazil, future research needs were identified (Pizarro and Coradin 1996).

Significant collections of several multipurpose tree species are held by institutions, such as OFI (Barnes and Burley 1990) or the Nitrogen Fixing Tree Association (NFTA)/UH (Chamberlain 1993). On the other hand, there may be a very urgent need for collecting or protection of regions for *in situ* conservation purposes because large areas with indigenous shrub and tree species, whose potential has not yet been assessed, are being turned into crop land, for example in the Brazilian Cerrados (Klink *et al.* 1993; Schultze-Kraft *et al.* 1993b). Several international workshops are being conducted to assess the current knowledge on the potential of different MPTS species and to outline research priorities for improving the utility of them, such as those organized by NFTA and collaborating institutions on *Albizia* and *Paraserianthes* in the Philippines in 1994, *Calliandra* in Indonesia in 1996, *Dalbergia* (Westley and Roshetko 1995), *Erythrina* (Westley and Powell 1993) and *G. sepium* (Withington *et al.* 1987), and by other institutions such as *Prosopis* by the National Academy of Sciences in Washington, USA in 1996.

Tropical Grasses

The major *ex situ* collections of tropical grasses in the CGIAR system are located at ILRI and CIAT (Table 22.9), a large proportion of which has been assembled in joint collecting

Table 22.9. Germplasm (number of species and accessions) of tropical and temperate forage grasses held in *ex situ* collections.

Genus	CIAT species	CIAT access.	ILRI species	ILRI access.	Total[†] accessions
Tropical grasses					
Andropogon	3	99	5	42	134
Brachiaria	26	684	23	547	843
Cenchrus	2	54	6	112	139
Chloris	4	55	9	105	128
Cynodon	3	17	4	100	103
Digitaria	8	29	13	50	70
Eragrostis	7	55	16	58	105
Hyparrhenia	11	60	11	35	75
Panicum	10	598	16	175	736
Paspalum	14	119	9	50	146
Pennisetum	9	55	19	183	206
Setaria	5	46	8	54	74
Other tropical	62	210	118	430	580
Temperate grasses					
Avena	–	–	2	121	121
Triticum	–	–	1	441	441
Other temperate[‡]	–	–	37	213	213
Total	164	2081	297	2716	4114

[†] Estimated number of different accessions.
[‡] Incudes the genera *Agropyron, Agrostis, Arrhenaterum, Bromus, Dactylis, Elymus, Festuca, Hordeum, Lolium, Phalaris, Phleum, Poa.*

trips of both institutions to African countries during 1984/1985 (Mengistu 1985a, 1985b, 1986; Keller-Grein *et al.* 1996). For *Andropogon*, CIAT and ILRI hold a relatively small germplasm collection of less than 150 accessions. According to Keller-Grein and Schultze-Kraft (1990) the world germplasm collection probably does not contain many more accessions. Miles and Grof (1990) compiled information about the variation and heritability of agronomically important characters, such as plant height, number of flowering culms, dry matter yield, leaf:stem ratio, leaf width and regrowth after cutting. The only breeding project was carried out in the 1980s at CIAT, with the purpose to generate leafy, high-yielding genotypes with good forage quality (Miles and Grof 1990). In the Brazilian Cerrados alone, pastures sown to *A. gayanus* were projected at more than 1.5 million ha in 1993 (R.R. Sáez and R.P de Andrade 1990, unpublished report).

Keller-Grein *et al.* (1996) reviewed *Brachiaria* genetic resources worldwide; about 1000 distinct accessions of 35 different species are held in seven major collections, of which both the CIAT and ILRI collections together account for about 85%. They also characterized *Brachiaria* germplasm morphologically and by isozymes, and identified gaps in the *ex situ* collection (Keller-Grein *et al.* 1996). The mode of reproduction of a large proportion of the collection has been determined (Valle and Savidan 1996) to access sexuality for genetic recombination in this essentially apomictic genus. Multilocational agronomic evaluation has been carried out using a large portion of the germplasm maintained at CIAT and regional experiences have been compiled in the review edited by Miles *et al.* (1996). A germplasm catalogue of *Brachiaria* and other forages from Africa is in preparation (G. Keller-Grein and B.L. Maass unpublished data). Current breeding projects at CIAT, Colombia and CNPGC of EMBRAPA, Campo Grande, Brazil aim to create apomictic genotypes combining acid-soil adaptation with spittlebug resistance and high forage value (Miles and Valle 1996).

The collection of *P. maximum* originated mainly from a donation by ORSTOM to CIAT. It had been collected from 1964 to 1969 in Kenya, Tanzania and Côte d'Ivoire. *Panicum maximum* is a tall bunch grass that reproduces predominantly by apomixis. Sexual populations of *Panicum* were located near Korogwe in Tanzania (Savidan *et al.* 1984). Comprehensive morphological, agronomic and cytological characterization have been carried out in Côte d'Ivoire (Pernès 1975; Savidan *et al.* 1984) and in Brazil (Costa *et al.* 1989), and the assembled germplasm has been extensively used for selection and breeding (Savidan *et al.* 1989; Nakagawa 1990; Sato *et al.* 1992; Jank 1995). While EMBRAPA/ CNPGC evaluated germplasm agronomically with the aim to select material superior to the naturalized, common ecotype for pasture improvement in the Brazilian Cerrados (Jank *et al.* 1993), CIAT researchers gave more attention to the selection of ecotypes and breeding new lines, adapted to acid, low-fertility soils in the Colombian Llanos (Hutton 1989; Thomas and Lapointe 1989). A breeding project was initiated in Brazil (Savidan *et al.* 1989; Jank 1995). Several cultivars have been released in Brazil by NARIs and in Japan, where they resulted both from selection out of natural variation and from a breeding programme (Nakagawa 1990; Sato *et al.* 1992). A catalogue of *Panicum* germplasm, which provides information on passport and evaluation data, has been published (Jank *et al.* 1996).

Toledo *et al.* (1989) described the variation encountered in productivity and shade tolerance of *Paspalum* in the humid tropics. A group of 86 new *Paspalum* accessions donated by EMBRAPA/CENARGEN to CIAT originated from recent collections by EMBRAPA/CENARGEN in Brazil, where it also has been characterized morphologically and agronomically (CIAT 1993; Valls 1993).

Characterization of *Pennisetum purpureum* Schumach. (napier or elephant grass) has been carried out on morphological characters and using molecular techniques, which are also being used to identify duplicates (ILRI unpublished data). The CNPGL of EMBRAPA at Coronel Pacheco, MG, Brazil, holds a substantial working collection of *Pennisetum*, research on which has resulted in new cultivars in Brazil (Carvalho *et al.* 1994). Breeding projects for napier grass exist in the USA, where dwarf types, such as Mott elephant grass, are considered highly desirable. Researchers in both Florida and Georgia are attempting to develop seed-propagated hybrids of pearl millet, *Pennisetum americanum* (L.) Leeke x elephant grass (Ocumpaugh and Sollenberger 1995).

The *Cenchrus* collection has been morphologically characterized (ILRI, unpublished data). CSIRO-ATFGRC also holds a large *Cenchrus* collection, part of which was classified, using morphological and agronomic attributes (Eagles *et al.* 1992; Pengelly *et al.* 1992). Bermuda grass (*Cynodon dactylon* Pers.) has been bred since the 1950s in the USA by Burton and collaborators whose research resulted in the release of several cultivars that now occupy some 10 million ha worldwide (Burton and Hanna 1995). *Chloris gayana* Kunth (rhodes grass) is widely adapted and widely grown in sub-Saharan Africa and also very productive. There was an important breeding programme at Kitale, Kenya, on *Chloris* (Boonman 1993). In Africa and Australia, rhodes grass has been one of the important cultivated grass species, in contrast to the USA, where it is only grown in Texas (Kretschmer and Pitman 1995).

Other tropical grasses of recognized importance, such as *Digitaria* and *Setaria*, are under-represented in the CGIAR collections and have not received much research attention. The forage potential of others, such as *Axonopus*, the *Bothriochloa-Dichanthium* complex, *Hemarthria* and *Urochloa* has been recognized; however, they are represented by less than 50 accessions. On the other hand, *ex situ* germplasm of these genera is available from important forage grass collections, such as those of CSIRO-ATFGRC (B.C. Pengelly 1995, pers. comm.) and the USDA.

Temperate and Tropical Highland Legumes

The four genera *Medicago*, *Vicia*, *Pisum* and *Trifolium* (in order of collection size) account for almost 80% of the large collection of Mediterranean and temperate forage and pasture legumes held in the ICARDA germplasm bank (Table 22.7). ILRI, on the other hand, only keeps a relatively small collection of temperate legumes that originated from the East African highlands, in particular, *Trifolium* species (Table 22.7); other temperate genera with large holdings, such as *Vicia, Medicago, Lathyrus* and *Pisum*, are mainly active duplicates of material obtained from ICARDA.

Selection within germplasm accessions and hybridization have been started at ICARDA for some *Vicia* and *Lathyrus* species. This work has led to the release of several cultivars of *Vicia sativa* L. subsp. *sativa, V. villosa* Roth. subsp. *dasycarpa* (Ten.) Cavill., *Lathyrus ochrus* and *L. sativus*.

There has been a systematic screening of the forage germplasm collection for resistances to major biotic stresses. In *Vicia* and *Lathyrus* species, several resistance sources for powdery mildew (induced by *Erysiphe pisi* Syn.), botrytis blight (induced by *Botrytis cinerea* Pers. ex Fr.) and ascochyta blight (induced by *Ascochyta pisi* Lib.) have been found (Robertson *et al.* 1996). Similarly, field and laboratory screening techniques were developed for root-knot nematode (*Meloidogyne artiellia* Frankland) and several accessions with resistance were found (Abd El Moneim and Bellar 1993). High intra- and interspecific variation in reaction to the parasitic weed *Orobanche crenata* Forsk. was found in *Vicia* species (Robertson *et al.* 1996). Screening for cold tolerance has started with vetch and chickling at ICARDA (Robertson *et al.* 1996).

At ICARDA, special attention has been given to evaluating *Lathyrus* species for low neurotoxin content (a free amino acid, ODAP[1]) in the germplasm collection (Aletor *et al.* 1994). There appears to be good potential for breeding for low ODAP content (Aletor *et al.* 1994; Robertson *et al.* 1996).

Sainfoin (*Onybrychis sativa* Lam.) is cultivated in some areas with cold winters in the WANA countries, and the genus contains several very palatable species (Nemati 1991). The *Trifolium* characterization carried out at ILRI described variation in flowering time, dry matter and seed yield of the annual *Trifolium* species from Ethiopia (Kahurananga and Tsehay 1991).

It is well known that many legumes are not adequately covered and further collecting is urgently needed (Reid *et al.* 1989; Schultze-Kraft *et al.* 1993b); in particular, in those regions that show drastic signs of genetic erosion, for example the rangelands of Beluchistan in Pakistan (ICARDA 1993). Arora and Bhag Mal (1991) suggested more collecting of the diverse temperate legumes found in the western Himalayas.

At ICARDA, a collection of rhizobia is maintained for pasture and forage legumes. This includes over 1000 strains of *Rhizobium meliloti, R. trifolii* and *R. leguminosarum* which have been collected from the WANA region. All cultures have been characterized for specificity and symbiotic properties. Passport and evaluation data for the rhizobia collection are available within the germplasm database. At present, approximately 20% of this collection is kept lyophilized and is maintained in the long-term cold store of the Genetic Resources Unit with the remainder maintained by the Microbiology Laboratory as cultures on host media in vials.

ICARDA keeps both passport and available characterization data in a computerized database, and has recently revised the *Lathyrus* and most important *Vicia* species. This germplasm database is maintained in the overall genebank database, which is managed on an xBase-based system. These data are provided to interested users in test file or dBASE format on request. All information generated on characterization and performance is attached to every accession. The production of several germplasm catalogues is in progress. A germplasm catalogue for temperate and tropical highland forages maintained at ILRI was produced by Kidest Shenkoru *et al.* (1991c).

---

[1] A free amino acid, known as 8-N-Oxalyl-L-",8-Diaminopropionic Acid, which may cause lathyrism, if *L. sativus* is excessively consumed by humans or domestic animals.

Temperate and Tropical Highland Grasses

Only small numbers of temperate grasses were held at ILRI; ICARDA has donated its collection of 775 accessions to ILRI (Table 22.9), and other collections are available elsewhere (Bettencourt *et al.* 1992). Documentation, research and utilization are well documented and coordinated by the European Forage Working group (IBPGR 1989, 1993) and others such as Tyler *et al.* (1992). Large regions of the world have not been adequately sampled, such as the high Andes that are rich in species of *Calamagrostis*, *Festuca* and *Poa*, and in Peru alone there are 30-40 different species of each of these genera (Tovar 1993). Intraspecific diversity has not yet been determined. Similarly, the Himalayan region offers a large resource (e.g. Sah and Ram 1989; Koul 1992) that may contribute to the improvement of temperate grasses.

Prospects

As more land is coming under intensive cultivation, much of the natural diversity of species will be lost. Using the existing species in the CGIAR and other genetic resources collections, geneticists can compare diversity in that of conserved species with that *in situ*. There is also the opportunity to focus on ecosystems with great genetic diversity and devise strategies for *in situ* conservation in these areas. Thus the forage genetic resources held in the CGIAR centres can be used not only for improvement for increased productivity but to develop strategies to ensure conservation of the biodiversity for future generations.

Some degraded areas require massive re-seeding of rangelands. It will not be possible or even desirable to always re-seed with the original vegetation, especially in rangelands used for livestock where species should be able to be used by the animals. But there should be an effort to seed with a mixture of species of grasses, legumes and shrubs, that is, to develop mixed vegetation systems that are in harmony with the mosaic of soils, temperature and water availability found across a landscape.

Other factors that are likely to require management of biodiversity in the future are related to climate changes that will lead to changes in distribution of vegetation and the need to have species available that are better adapted to the changing climate. Demand for use of forage genetic resources for increasing livestock production and for contributing to more sustainable agricultural systems is expected to increase, especially in the tropics and subtropics (Schultze-Kraft *et al.* 1993b). There will be an increased demand for forages for non-livestock uses in soil conservation and for recreational areas. The role of national and regional organizations in the conservation of genetic resources needs to be developed. In order to realize the full potential of the germplasm, functional collaborative networks exist for the Mediterranean and temperate European regions. Additional permanent working groups or networks should be formed for those regions not yet covered by those groups in order to share acquisition, conservation, evaluation, documentation, research and utilization of the germplasm within a global framework (Davies 1984; Reid 1993; Schultze-Kraft *et al.* 1993b).

Limitations

The demand for livestock products is increasing with population increase and a general increase in living standards in most countries. For example, even in the densely farmed and populated area of Southeast Asia, it has been estimated that the demand for forage would double by the year 2000 (Remenyi and McWilliam 1986). Nevertheless, it is essential that development of forage systems be considered in relation to the sustainable use of vegetation, land and water resources.

Development of forage systems also needs to be carried out within the context of the production systems in which forages are used. The role that the CGIAR forage genetic resource centres can play is through some form of intervention with other partners in these production systems. There is a need to determine where intervention is feasible and what other interacting factors need to be taken into account. The nature of the intervention will generally be through introduction or re-introduction of species and accessions with forage potential in production systems, where there is a need for livestock feed. This implies that more priority will be devoted to intensive production systems where it is feasible to consider material as well as management inputs rather than to extensive systems where only management inputs are economically feasible. Examples of such intensive production systems are found in the crop-livestock systems in Africa; crop-livestock systems in the Middle East, upland farming systems in South East Asia and intensive livestock and crop-livestock systems in South America.

Technology available for identifying areas for acquisition has progressed considerably since the early studies of Hartley (1949) with the advent of powerful geographical information systems (GIS) that allow areas of genetic diversity to be mapped and related to environmental parameters. There has been slower progress in identifying or developing high disease and insect resistance in otherwise desirable species than in defining climate and soil adaptation. This is partly because diseases and pests only become evident after large areas of a species are planted, such as the spittlebug insect in *Brachiaria* (Miles *et al.* 1996). Larger resources are required to mount a plant improvement programme of recombination and selection where there is initially a lack of knowledge of the mechanism of host-plant resistance and no knowledge of sources of resistance other than to screen wild species for adaptation. Resources for breeding will only be used for forages of high commercial value, for example, lucerne (*Medicago sativa*), where aphid resistance was rapidly incorporated into commercial varieties in the USA (Lehman *et al.* 1985) and Australia; *Stylosanthes guianensis*, where recurrent selection has resulted in apparently durable resistance to anthracnose in the Colombian Llanos (CIAT 1995) and *Brachiaria*, where recombination and selection for spittlebug resistance is underway (Miles and Valle 1996). New technology that allows molecular characterization of the host and the pathogen and the precise location of resistance genes will facilitate future programmes of identifying and incorporating these genes.

Low forage quality in many species and the decrease in quality with maturity is a major limitation to increased productivity of forages. In general, forage quality is lower in tropical grasses and legumes than in temperate species (Minson 1980). There are examples of success: breeding for improved forage quality in *Cynodon dactylon* (Burton and Hanna 1995) and selection for low phytoestrogen content in *Trifolium subterraneum* L. (Mannetje *et al.* 1980).

In most forage and livestock production systems, absence of good management is a major limitation to the efficient use of forages for increased animal production. This is due to a variety of factors, technical, economic and social. The main technical practices that optimize nutritive value are the selection of appropriate species or accessions, the supply of deficient soil nutrients and controlled defoliation, and grazing to achieve high nutritive value and favourable conditions for regrowth. Availability of credit may limit the purchase of fertilizers and new cultivars, but in many cases poor management can be attributed to a lack of knowledge which can be overcome by education.

The reality is that many farmers do not manage forage with the same care that crops are managed. This is well illustrated by the use of the same word for 'forage' and 'weed' in many languages. Animals may be held as a source of wealth or capital, they tend to be given first consideration over the management of a pasture, and it is difficult to rapidly adjust herd size to availability of feed.

Current production systems are not sustainable in the long term. This applies as much to crop systems as to livestock systems. Sustainable production requires that inputs of technology, management and policy work toward a common agenda for the welfare of the farmer, the ecosystem, country and region. Inputs to increased productivity should also take issues of equity and the environment into account.

Improved forage systems can contribute to increased equity, and improve or minimize damage to the soil and water resource (Thomas *et al.* 1995a, 1995b; Boddey *et al.* 1996). Conversely, poorly managed forage systems may contribute to decreased productivity through overgrazing (Talamucci and Chaulet 1989), use of poorly adapted forage species and inappropriate location of livestock systems (Hecht and Cockburn 1989). In many instances these negative effects will only be overcome with a change in and administration of new land-use policies.

Judicious use of forages can contribute to more sustainable production systems (Humphreys 1994). The inclusion of legumes in forage systems provides a high quality forage for animals and contributes to soil health through increased microbial and faunal activity which in turn enhances nutrient cycling and improves physical structure. This has been demonstrated for grazed pastures (Thomas *et al.* 1995b), in fodder banks (Mohamed-Salem 1994) and in the use of legumes as soil covers (Cruz *et al.* 1994).

It is important that new forage species be evaluated for their nutritive value in addition to their agronomic value. An important limitation is to establish reliable relations between a chemical test and feeding value. A particular problem arises in the case of factors that effect palatability and digestibility, such as tannins. Additional difficulties arise when forages or feeds are fed in combination with each other as in many cases, the effect is interactive rather than additive (Leng 1990a, 1990b).

Local seed production and distribution systems need to be developed (Ferguson 1994). Most of the work on forage seed production has come out of tropical Australia. However, both CIAT and ILRI established seed production units to ensure seed supply of promising forages to national partners (Ferguson and Sauma 1993). Recently ILRI and ICARDA held a research planning workshop on smallholder seed production to assess the current situation and assist in the development of future research in this area (Hanson 1994).

The greatest potential for adoption of new forages is in crop-livestock systems, where agricultural production is being intensified. There is a need to incorporate forages in farming systems in order to maintain soil fertility, to support and supplement livestock to fully utilize roughage not suitable for humans or non-ruminant livestock, to feed animals used for draft, or to generate cash flow for farmers. It is also in these systems that farmers are most likely to apply the management needed for efficient utilization of improved forage species.

Within the production system context, there is a need to develop forage systems that are more in harmony with the concept of a sustainable environment. Thus, there is a need to move away from monocultures to the use of diverse species. The use of one apomictic cultivar of *Brachiaria decumbens* over 40 million ha in South America, where it is susceptible to the spittlebug insect (Miles *et al.* 1996), presents a similar danger to the large losses in production of *Leucaena leucocephala* in Southeast Asia due to the psyllid insect (Shelton and Brewbaker 1994). There should also be an attempt to develop systems that use a mixture of trees, shrubs and herbaceous plants that occur naturally in many climax or induced savannas.

Within a land-use system context, it needs to be recognized that there are areas that should not be developed for livestock and hence promoted for forage development. Population pressure and government policies of 'assisting' farmers on marginal lands is resulting in heavier grazing pressure on these lands (Talamucci and Chaulet 1989). However, different outcomes might be anticipated in different areas with different population and food pressures. In the developed world, there is now a strong argument to contain animal production to the more fertile lands and allow the marginal lands to be used for conservation and recreation purposes. In the developing world, there is more scope for increased animal production in intensive crop-livestock systems than on extensive grazing lands.

REFERENCES

Abd El Moneim, A.M. and M. Bellar. 1993. Response of forage vetches and forage peas to root-knot nematode (*Meloidogyne artiellia*) and cyst nematode (*Heterodera ciceri*). Nematol. Medit. 21:67-70.

Akinola, J.O., R.A. Afolayan and S.A.S. Olorunju. 1991. Effects of storage, testa colour and scarification method on seed germination of *Desmodium velutinum* (Willd.) DC. Seed Sci. Technol. 19:159-166.

Aletor, V.A., A.M. Abd El Moneim and A.V. Goodchild. 1994. Evaluation of the seeds of selected lines of three *Lathyrus* spp. for BOAA, tannins, trypsin inhibitor activity and certain *in vitro* characteristics. J. Sci. Food Agric. 65:143-151.

Argel, P.J. and B.L. Maass. 1995. Evaluación y adaptación de leguminosas arbustivas en suelos ácidos infertiles de América tropical. Pp. 215-227 *in* Nitrogen Fixing Trees for Acid Soils (D.O. Evans and L.T. Szott, eds.). Nitrogen Fixing Tree Research Reports (Special issue). Winrock International and NFTA, Morrilton, Arkansas, USA.

Arora, R.K. and Bhag Mal. 1991. Genetic resources of forage and food legumes for temperate and cold arid regions of India. Pp. 65-78 *in* Legume Genetic Resources for Semi-arid Temperate Environments (A. Smith and L.D. Robertson, eds.). ICARDA, Aleppo, Syria.

Barnes, R.D. and J. Burley. 1990. Tropical forest genetics at the Oxford Forestry Institute: exploration, evaluation, utilization and conservation of genetic resources. For. Ecol. and Manage. 35:159-169.

Barnes, R.F., D.A. Miller and C.J. Nelson (eds.). 1995. Forages: An introduction to grassland agriculture (5th edn.). Iowa State University Press, Ames Iowa, USA.

Bass, L.N. 1984. Storage of seeds of tropical legumes. Seed Sci. Technol. 12:395-402.

Belalcázar, J. and R. Schultze-Kraft (comps.). 1994. La colección de forrajeras tropicales mantenida en CIAT. 4. Catálogo de germoplasma de Colombia. Documento de Trabajo No. 137. CIAT, Cali, Colombia.

Bettencourt, E., Th. Hazecamp and M.C. Perry (eds.). 1992. Directory of Germplasm Collections. 7. Forages - Legumes, Grasses, Browse Plants and Others. IBPGR, Rome, Italy.

Bishop, H.G., B.C. Pengelly and D.H. Ludke. 1988. Classification and description of a collection of the legume genus *Aeschynomene*. Trop. Grasslands 22:160-175.

Blumenthal, M.J. and I.B. Staples. 1993. Origin, evaluation and use of *Macrotyloma* as forage - a review. Trop. Grasslands 27:16-29.

Boddey, R.M., I.M. Rao and R.J. Thomas. 1996. Nutrient cycling and environmental impact of *Brachiaria* pastures. Pp. 72-86 *in* Brachiaria: Biology, Agronomy, and Improvement (J.W. Miles, B.L. Maass and C.B. do Valle, eds.). CIAT and EMBRAPA/CNPGC, Cali, Colombia.

Boonman, J.G. 1993. East Africa's grasses and fodders: their ecology and husbandry. Kluwer Academic Publishers, Dordrecht, Netherlands.

Brako, L., A.Y. Rossman and D.F. Farr (eds.). 1995. Scientific and common names of 7000 vascular plants in the United States. APS Press, St. Paul, Minn., USA.

Bray, R.A. 1995. Possibilities for developing low mimosine *Leucaena*. Pp. 119-124 *in* Leucaena - Opportunities and Limitations (H.M. Shelton, C.M. Piggin and J.L. Brewbaker, eds.). Proceedings of a Workshop, 24-29 January 1994, Bogor, Indonesia. ACIAR Proceedings No. 57. Canberra, Australia.

Burt, R.L. and W.T. Williams. 1975. Plant introduction and the *Stylosanthes* story. Aust. Meat Res. Committee 25:1-23.

Burton, G.W. and W.W. Hanna. 1995. Bermudagrass. Pp. 421-429 *in* Forages: An Introduction to Grassland Agriculture (R.F. Barnes, D.A. Miller and C.J. Nelson, eds., 5th edn.). Iowa State University Press, Ames Iowa, USA.

Butler, J.E. 1985. Germination of Buffel grass (*Cenchrus ciliaris*). Seed Sci. Technol. 13:583-591.

Cameron, A.G. 1991. *Macroptilium longipedunculatum* (Bentham) Urban (Llanos macro) cv. Maldonado. Trop. Grasslands 25(4):375-376.

Cameron, D.F., R.A. Boland, S. Chakraborty, B. Jamieson and J.A.G. Irwin. 1993a. Recurrent selection for partial resistance to anthracnose disease in shrubby stylo (*Stylosanthes scabra*). Pp. 2137-2138 *in* Proc. XVII Inter. Grassl. Congr., Vol. 3, New Zealand and Australia.

Cameron, D.F., C.P. Miller, L.A. Edye and J.W. Miles. 1993b. Advances in research and development with stylosanthes and other tropical pasture legumes. Pp. 796-801 *in* Grasslands for Our World (M.J. Baker, ed.). SIR Publishing, Wellington, New Zealand.

Cárdenas, J.E. 1990. Evaluación preliminar de una colección de la leguminosa forrajera tropical *Desmodium velutinum* (Willd.) DC. BSc Thesis, Universidad Nacional de Colombia, Palmira, Colombia.

Carvalho, M.M., M.J. Alvim, D.F. Xavier and L. de A. Carvalho (eds.). 1994. Capim-elefante: produçäno e utilizaçäno. EMBRAPA-CNPGL, Coronel Pacheco, MG, Brazil.

Chamberlain, J. 1993. The Nitrogen Fixing Tree Association research strategy relative to the CGIAR. Pp. 102-109 *in* Proc. of the Internat. Consultation on the Development of the ICRAF Multipurpose Tree Germplasm Resource Centre, 2-5 June 1992, Nairobi, Kenya.

CIAT. 1988. Annual report 1988 Tropical Pastures. Working Document No. 58. CIAT, Cali, Colombia.]

CIAT. 1993. Biennial report 1992-1993 Tropical Forages. Working Document No. 166. CIAT, Cali, Colombia.

CIAT. 1995. Biennial report 1994-1995 Tropical Forages. Working Document No. 152. CIAT, Cali, Colombia.

Cook, B.G., T.W.G. Graham, R.L. Clem, T.J. Hall and M.F. Quirk. 1993. Evaluation and development of *Desmanthus virgatus* on medium- to heavy-textured soils in Queensland. Pp. 2148-2149 *in* Proc. XVII Inter. Grassl. Congr., Vol. 3, New Zealand and Australia.

Coradin, L. and R. Schultze-Kraft. 1990. Germplasm collection of tropical pasture legumes in Brazil. Tropic. Agric. (Trinidad) 67(2):98-100.

Costa, J.C.G., Y.H. Savidan, L. Jank and L.H.R. Castro. 1989. Morphological studies as a tool for the evaluation of wide tropical forage grass germplasms. Pp. 277-278 *in* Proc. XVI Inter. Grassl. Congr., Nice, France.

Costa, N.M. de S. and R. Schultze-Kraft, R. 1993. Biogeografía de *Stylosanthes capitata* Vog. y de *Stylosanthes guianensis* Sw. var. *pauciflora*. Pasturas Trop. 15(1):10-15.

Cruz, R. de la, S. Suárez and J.E. Ferguson. 1994. The contribution of Arachis pintoi as a ground cover in some farming systems of tropical America. Pp. 102-108 *in* Biology and Agronomy of Forage *Arachis* (P.C. Kerridge and B. Hardy, eds.). CIAT, Cali, Colombia.

Davidson, B.R. and H.F. Davidson. 1993. Legumes: The Australian Experience. The Botany, Ecology and Agriculture of Indigenous and Immigrant Legumes. Reasearch Studies Press, Somerset, UK.

Davies, W.E. 1984. A Plan of Action for Forage Genetic Resources. IBPGR, Rome, Italy.

Dzowela, B.H. (ed.). 1988. African forage plant genetic resources, evaluation of forage germplasm and extensive livestock production systems. Proceedings of the third PANESA workshop, Addis Ababa, Ethiopia. ILCA, Addis Ababa, Ethiopia.

Eagles, D.A., J.B. Hacker and B.C. Pengelly. 1992. Morphological and agronomic attributes of a collection of buffel grass (*Cenchrus ciliaris*) and related species. Genet. Resour. Commun. 14:1-15.

FAO/IPGRI. 1994. Genebank standards. FAO and IPGRI, Rome, Italy.

Ferguson, J.E. (ed.). 1994. Semilla de especies forrajeras tropicales: conceptos, caso y enfoque de la investigación y la producción. Memorias de la octava reunión del Comité Asesor de la Red Internacional de Evaluación de Pastos Tropicales (RIEPT), Villavicencio, Colombia, noviembre 1992. CIAT, Cali, Colombia.

Ferguson, J.E. and G. Sauma. 1993. Towards more forage seeds for small farmers in Latin America. Pp. 666-671 *in* Grasslands for Our World (M.J. Baker, ed.). SIR Publishing, Wellington, New Zealand.

Ferrufino, A., A. Vallejos and S. Beck. 1991. Recolección de leguminosas forrajeras en el trópico húmedo de Cochabamba y Santa Cruz, Bolivia. Pasturas Trop. 13(2):46-48.

Flores, A.J. and R. Schultze-Kraft. 1994. Recolección de recursos genéticos de leguminosas forrajeras tropicales en Venezuela. Agronomía Trop. 44(3):357-371.

Francisco-Ortega, J., M.T. Jackson, A. Santos-Guerra and M. Fernandez-Galvan. 1990. Genetic resources of the fodder legumes tagasaste and escobón (*Chamaecytisus proliferus* (L. fil.) Link *sensu lato*) in the Canary Islands. FAO/IBPGR Plant Genet. Resour. Newsl. 81/82:27-32.

Franco, M.A., G.I. Ocampo, E. Melo and R. Thomas (comps.). 1993. Catalogue of *Rhizobium* strains for tropical forage legumes (5th edn.). Working Document No. 14. CIAT, Cali, Colombia.

Gramshaw, D., B.C. Pengelly, F.W. Muller, the late W.A.T. Harding and R.J. Williams. 1987. Classification of a collection of the legume *Alysicarpus* using morphological and preliminary agronomic attributes. Aust. J. Agric. Res. 38:355-372.

Grof, B., D. Thomas, R.P. de Andrade, J.L.F. Zoby and M.A. de Souza. 1989. Perennial legumes and grass-legume associations adapted to poorly drained savannas in tropical South America. Pp. 189-190 *in* Proc. XVI Inter. Grassl. Congr., Nice, France.

Gutteridge, R.C. and H.M. Shelton (eds.). 1994. Forage Tree Legumes in Tropical Agriculture. CAB International, Wallingford, UK.

Hacker, J.B., A. Glatzle and R. Vanni. 1996a. Paraguay - a potential source of new pasture legumes for the subtropics. Trop. Grasslands 30:273-281.

Hacker, J.B., R. J. Williams and B.C. Pengelly. 1996b. A characterisation of the genus *Vigna* with regard to potential as forage. Genet. Resour. Commun. 22:1-9.

Hakiza, J.J., J.R. Lazier and A.R. Sayers. 1988a. Characterization and evaluation of forage legumes in Ethiopia: Preliminary examination of variation between accessions of Stylosanthes fruticosa (Retz.) Alston. Pp. 174-191 *in* Pasture Network for Eastern and Southern Africa (PANESA). African Forage Plant Genetic Resources, Evaluation of Forage Germplasm and Extensive Livestock Production Systems. ILCA, Addis Ababa, Ethiopia.

Hakiza, J.J., J.R. Lazier and A.R. Sayers. 1988b. Characterization and preliminary evaluation of accesion of *Zornia* species from the ILCA collection. Pp. 149-173 *in* Pasture Network for Eastern and Southern Africa (PANESA). African Forage Plant Genetic Resources, Evaluation of Forage Germplasm and Extensive Livestock Production Systems. ILCA, Addis Ababa, Ethiopia.

Hanson, J. (ed.) 1994. Seed production by smallholder farmers. Proceedings of the ILCA/ICARDA Research Planning Workshop held at ILCA, Addis Ababa, Ethiopia, 13-15 June 1994. ILCA, Addis Ababa, Ethiopia.

Hanson, J. and J.H. Heering. 1994. Genetic resources of *Stylosanthes* species. Pp. 55-61 *in Stylosanthes* as a Forage and Fallow Crop (P.N. de Leeuw, M.A. Mohamed-Saleem and A.M. Nyamu, eds.). Proceedings of the regional workshop on the use of Stylosanthes in West Africa held in Kaduna, Nigeria, 26-31 October 1992. ILCA, Addis Ababa, Ethiopia.

Hanson, J. and R.J. Lazier. 1989. Forage germplasm at the International Livestock Centre for Africa (ILCA): an essential resource for evaluation and selection. Pp. 265-266 *in* Proc. XVI Inter. Grassl. Congr., Nice, France.

Hanson, J. and T.J. Ruredzo. 1992. *In vitro* culture techniques for forage genetic resources. Pp. 149-151 *in* Biotechnology and Crop Improvement in Asia (J.P. Moss, ed.). ICRISAT, Patancheru, India.

Harding, W.A.T. the late, B.C. Pengelly, D.G. Cameron, L. Pedley and R.J. Williams. 1989. Classification of a diverse collection of *Rhynchosia* and some allied species. Genet. Resour. Commun. 13:1-30.

Hartley, W. 1949. Plant collecting expedition to subtropical South America, 1947/8. A report, with notes on climate, vegetation and principal economic plants, with an inventory of the collections. CSIRO Div. Plant Ind., Div. Rep. No. 7.

Harty, R.L., J.M. Hopkinson, B.H. English and J. Alder. 1983. Germination, dormancy and longevity in stored seeds of *Panicum maximum*. Seed Sci. Technol. 11:341-351.

Hecht, S. and A. Cockburn. 1989. The Fate of the Forest: Developers, Destroyers and Defenders of the Amazon. Penguin Books, London, UK.

Heering, J.H. 1994. The reproductive biology of three perennial *Sesbania* species (Leguminosae). Euphytica 74:143-148.

Heering, J.H. 1995a. Botanical and agronomic evaluation of a collection of *Sesbania* sesban and related perennial species. PhD dissertation, Landbouw Universiteit Wageningen, Netherlands.

Heering, J.H. 1995b. The effect of cutting height and frequency on the forage, wood and seed production of six *Sesbania* sesban accessions. Agroforestry Syst. (Netherlands) 30(3):341-350.

Heering, J.H. and J. Hanson. 1993. Karyotype analysis and interspecific hybridisation in three perennial *Sesbania* species (Leguminosae). Euphytica 71:21-28.

Hiernaux, P. 1994. The crisis of Sahelian pastoralism: ecological or economic. ILCA, Addis Ababa, Ethiopia.

Hughes, C.E., C.T. Sorensson, R. Bray and J.L. Brewbaker. 1995. *Leucaena* germplasm collections, genetic conservation and seed increase. Pp. 66-74 *in Leucaena* - Opportunities and Limitations (H.M. Shelton, C.M. Piggin and J.L. Brewbaker, eds.). Proceedings of a Workshop, 24-29 January 1994, Bogor, Indonesia. ACIAR Proceedings No. 57. Canberra, Australia.

Humphreys, L.R. 1994. Tropical Forages: their Role in Sustainable Agriculture. Longman Scientific and Technical, Essex, UK.

Hutton, E.M. 1989. Breeding acid-soil tolerant lines of the tropical grass *Panicum maximum* Jacq. Pp. 355-356 *in* Proc. XVI Inter. Grassl. Congr., Nice, France.

Hutton, E.M. and C.P. Chen. 1993. Meeting the challenge of adapting *Leucaena leucocephala* to acid oxisols and ultisols of South America and south-east Asia. Pp. 2124-2125 *in* Proc. XVII Inter. Grassl. Congr., Vol. 3, New Zealand and Australia.

IBPGR. 1985. Forages for Mediterranean and adjacent arid/semi-arid areas. Report of a Working Group held at Limassol, Cyprus, 24-26 April 1985. IBPGR, Rome, Italy.

IBPGR. 1989. Report of a working group on forages (third meeting) held in Monpellier, France, 9-12 January 1989. European Cooperative Programme for the Conservation and Exchange of Crop Genetic Resources/International Board for Plant Genetic Resources (ECP/GR/IBPGR), Rome, Italy.

IBPGR. 1993. Report of a working group on forages (fourth meeting) held in Budapest, Hungary, 28-30 October 1991. European Cooperative Programme for Crop Genetic Resources Networks/International Board for Plant Genetic Resources (ECP/GR/IBPGR), Rome, Italy.

ICARDA. 1993. Genetic resources unit - annual report for 1993. ICARDA, Aleppo, Syria.

Imrie, B.C., R.M. Jones and P.C. Kerridge. 1983. Desmodium. Pp. 97-140 *in* The Role of *Centrosema, Desmodium*, and *Stylosanthes* in Improving Tropical Pastures (R.L. Burt, P.P. Rotar, J.L. Walker and M.W. Silvey, eds.). Westview Tropical Agriculture Series, No. 6. Boulder, Colorado, USA.

Jank, L. 1995. Melhoramento e seleção de variedades de *Panicum maximum*. Pp. 21-58 *in* Anais do 121 Simpósio sobre Manejo da Pastagem - Tema: O Capim Coloniao (A.M. Peixoto, J.C. de Moura and V. P. de Faria, eds.). Fundação de Estudos Agrários Luiz de Queiroz (FEALQ), Piracicaba, SP, Brazil.

Jank, L., S. Calixto, J.C.G. Costa and J.B.E. Curvo. 1996. Catalog of the characterization and evaluation of the *Panicum maximum* germplasm: morphological description and agronomical performance. EMBRAPA-CNPGC, Documentos. EMBRAPA/CNPGC, Campo Grande, MS, Brazil. (In press).

Jank, L., J.C.G. Costa, Y.H. Savidan and C.B. do Valle. 1993. New *Panicum maximum* cultivars for diverse ecosystems in Brazil. Pp. 509-511 *in* Proc. XVII Inter. Grassl. Congr., Vol. 1, New Zealand and Australia.

Kahurananga, J. and A. Tsehay. 1991. Variation in flowering time, dry matter and seed yield among annual *Trifolium* species, Ethiopia. Trop. Grasslands 25:20-25.

Kategile, J.A. and S.B. Adoutan (eds.). 1993. Collaborative research on *Sesbania* in East and Southern Africa. African Feed Research Network (AFRNET). ILCA, Nairobi, Kenya.

Keller-Grein, G. 1993a. Agronomic evaluation of *Centrosema* germplasm on an acid ultisol in the humid tropics of Peru. I. *Centrosema macrocarpum* Benth. Pp. 2141-2143 *in* Proc. XVII Inter. Grassl. Congr., Vol. 3, New Zealand and Australia.

Keller-Grein, G. 1993b. Agronomic evaluation of *Centrosema* germplasm on an acid ultisol in the humid tropics of Peru. II. *Centrosema tetragonolobum* Schultze-Kraft & Williams. Pp. 2143-2144 *in* Proc. XVII Inter. Grassl. Congr., Vol. 3, New Zealand and Australia.

Keller-Grein, G. and F. Passoni. 1993. Agronomic evaluation of *Centrosema* germplasm on an acid ultisol in the humid tropics of Peru. III. Selected *Centrosema pubescens* Benth. accessions. Pp. 2145-2146 *in* Proc. XVII Inter. Grassl. Congr., Vol. 3, New Zealand and Australia.

Keller-Grein, G. and R. Schultze-Kraft, 1992. Preliminary agronomic evaluation of a *Stylosanthes viscosa* Sw. collection. Genet. Resour. Commun. 15:1-35.

Keller-Grein, G. and R. Schultze-Kraft. 1990. Botanical description and natural distribution of *Andropogon gayanus*. Pp. 1-18 *in* Andropogon gayanus Kunth: a Feed Grass for Tropical Acid Soils (J.M. Toledo, R. Vera, C. Lascano and J.M. Lenné, eds.). CIAT, Cali, Colombia.

Keller-Grein, G., B.L. Maass and J. Hanson. 1996. Natural variation in *Brachiaria* and existing germplasm collections. Pp. 16-42 *in* Brachiaria: Biology, Agronomy, and Improvement (J.W. Miles, B.L. Maass and C.B. do Valle, eds.). CIAT and EMBRAPA/CNPGC, Cali, Colombia.

Kerridge, P.C. and B. Hardy (eds.). 1994. Biology and Agronomy of Forage Arachis. CIAT, Cali, Colombia.

Keya, N.C.O. and C.L.M. van Eijnatten. 1975. Studies on oversowing of natural grasslands in Kenya 1. The effects of seed threshing, scarification and storage on the germination of *Desmodium uncinatum* (Jacq.) DC. E. Afr. Agric. For. J. 40:261-263.

Kidest Shenkoru, J. Hanson and T. Metz. 1991a. ILCA Forage Germplasm Catalogue 1991. 1. Multipurpose Trees and Large Shrubs. ILCA, Addis Ababa, Ethiopia.

Kidest Shenkoru, J. Hanson and T. Metz. 1991b. ILCA Forage Germplasm Catalogue 1991. 2. Tropical Lowland Forages. ILCA, Addis Ababa, Ethiopia.

Kidest Shenkoru, J. Hanson and T. Metz. 1991c. ILCA Forage Germplasm Catalogue 1991. 3. Temperate and Tropical Highland Forages. ILCA, Addis Ababa, Ethiopia.

Klink, C.A., A.G. Moreira and O.T. Solbrig. 1993. Ecological impact of agricultural development in the Brazilian Cerrados. Pp. 259-282 *in* The World's Savannas (M.D. Young and O.T. Solbrig, eds.). UNESCO, Man and the Biosphere Series, Vol. 12. UNESCO/Parthenon Publishing Group, Paris, France.

Koul, K.K. 1992. Collecting and evaluating the fodder grasses in the Kashmir Himalayas. FAO/IBPGR Plant Genet. Resour. Newsl. 88/89:60-63.

Krapovickas, A. and W.C. Gregory. 1994. Taxonomáa del género *Arachis* (Leguminosae). Bonplandia 8(1-4):1-186.

Kretschmer, A.E. Jr. and W.D. Pitman. 1995. Other tropical and subtropical forages. Pp. 283-304 *in* Forages: An Introduction to Grassland Agriculture (R.F. Barnes, D.A. Miller and C.J. Nelson, eds., 5th edn.). Iowa State University Press, Ames Iowa, USA.

Larbi, A., J. Ochang, J. Hanson and J. Lazier. 1992. Agronomic evaluation of *Neonotonia wightii*, *Stylosanthes scabra* and S. *guianensis* accessions in Ethiopia. Trop. Grasslands 26:115-119.

Larbi, A., I.I. Osakwe and J.W. Lambourne. 1993a. Variation in relative palatability to sheep among *Gliricidia sepium* provenances. Agroforestry Syst. 22:221-224.

Larbi, A., D. Thomas and J. Ochang. 1993b. Agronomic evaluations of Chamaecrista rotundifolia accessions under grazing in southern Ethiopia. E. Afr. Agric. For. J. 59:75-78.

Lascano, C.E., B.L. Maass and G. Keller-Grein. 1995. Forage quality of shrub legumes evaluated in acid soils. Pp. 228-236 *in* Nitrogen Fixing Trees for Acid Soils (D.O. Evans and L.T. Szott, eds.). Nitrogen Fixing Tree Research Reports (Special issue). Winrock International and NFTA, Morrilton, Arkansas, USA.

Lascano, C.E., B.L. Maass and R.J. Thomas. 1993. Multipurpose trees and shrubs at CIAT. Pp. 157-163 *in* Proc. of the Internat. Consultation on the Development of the ICRAF Multipurpose Tree Germplasm Resource Centre, 2-5 June 1992, Nairobi, Kenya.

Le Houéreou, H.N. 1991. Forage diversity in Africa: An overview of the genetic resources. Pp. 99-117 *in* Crop Genetic Resources of Africa, Vol. 1 (F. Attere, H. Zedan, N.Q. Ng and P. Perrino, eds.). IBPGR, IITA, and UNEP, Nairobi, Kenya.

Leeuw, P.N. de, M.A. Mohamed-Saleem and A.M. Nyamu (eds.). 1994. *Stylosanthes* as a forage and fallow crop. *In* Proceedings of the regional workshop on the use of *Stylosanthes* in West Africa held in Kaduna, Nigeria, 26-31 October 1992. ILCA, Addis Ababa, Ethiopia.

Lehman, W.F., W.L. Marble and M.W. Nelson. 1985. Key factors in the rapid development and seed increase of alfalfa with insect resistance. Pp. 232-233 *in* Proc. XV Inter. Grassl. Congr., Kyoto, Japan.

Leng, R.A. 1990a. Factors affecting the utilization of 'poor-quality' forages by ruminants particularly under tropical conditions. Nutr. Res. Rev. (UK) 3:277-303.

Leng, R.A. 1990b. Nutrition of ruminants at pasture in the tropics: implications for selection criteria. Pp. 298-309 *in* Proc. 4th World Congress on Genetics Applied to Livestock Production, Edinburgh, UK.

Lenné, J.M. 1990. A World List of Fungal Diseases of Tropical Pasture Species. CAB International, Wallingford, UK.

Lenné, J.M. and P. Trutmann (eds.). 1994. Diseases of Tropical Pasture Plants. CAB International, Wallingford, UK.

Loch, D.S., W.A.T. Harding and G.L. Harvey. 1988. Cone threshing of chaffy grass seeds to improve handling characteristics. Queensland J. Agric. Anim. Sci. 45:205-212.

Maass, B.L. 1989. Die tropische Weideleguminose *Stylosanthes scabra* Vog. - Variabilität, Leistungsstand und Möglichkeiten züchterischer Verbesserung. Sonderheft 97. Landbauforschung Völkenrode, Braunschweig, Germany.

Maass, B.L. and C.H. Ocampo. 1995. Isozyme polymorphism provides fingerprints for germplasm of *Arachis glabrata* Bentham. Genet. Resour. Crop Evol. 42:77-82.

Maass, B.L. and R. Schultze-Kraft. 1993. Characterisation and preliminary evaluation of a large germplasm collection of the tropical forage legume *Stylosanthes scabra* Vog. Pp. 2151-2153 *in* Proc. XVII Inter. Grassl. Congr., Vol. 3, New Zealand and Australia.

Maass, B.L. and A.M. Torres. 1992. Outcrossing in the tropical forage legume *Centrosema brasilianum* (L.) Benth. Pp. 465-466 *in* Abstr. XIII EUCARPIA Congress, Angers, France.

Maass, B.L., C.E. Lascano and E.A. Cárdenas. 1996. La leguminosa arbustiva *Codariocalyx gyroides*. 2. Valor nutritivo y aceptabilidad en el piedemonte amazónico, Caquetá, Colombia. Pasturas Tropicales 18(3) (in press).

Maass, B.L., A.M. Torres and C.H. Ocampo. 1993. Morphological and isozyme characterisation of *Arachis pintoi* Krap. et Greg. *nom. nud.* germplasm. Euphytica 70:43-52.

Macklin, B. and D.O. Evans. 1990. Perennial *Sesbania* species in agroforestry systems. Special Publication 90-01. Nitrogen Fixing Tree Association (NFTA), Waimanalo, Hawaii, USA.

Mannetje, L. 't and R.M. Jones (eds.). 1992. Plant resources of South-East Asia. No. 4 Forages. Pudoc Scientific Publishers, Wageningen, Netherlands.

Mannetje, L. 't, K.F. O'Connor and R.L. Burt. 1980. The use and adaptation of pasture and fodder legumes. Pp. 537-551 *in* Advances in Legume Science (R.J. Summerfield and A.H. Bunting, eds.). Royal Botanic Gardens, Kew, UK.

McIvor, J.G. and R.A. Bray (eds.). 1983. Genetic Resources of Forage Plants. CSIRO, Melbourne, Australia.

Mejía M., M. 1984. Scientific and Common Names of Tropical Forage Species. CIAT, Cali, Colombia.

Mengistu, S. 1985a. ILCA-CIAT forage germplasm collection in Kenya. PGRC/E-ILCA Germplasm Newsl. 8:23-29.

Mengistu, S. 1985b. Survey and collection of forage germplasm in Tanzania 1985. PGRC/E-ILCA Germplasm Newsl. 10:16-27.

Mengistu, S. 1986. Survey and collection of forage germplasm in Rwanda, Burundi and eastern Zaire. PGRC/E-ILCA Germplasm Newsl. 13:3-18.

Miles, J.W. and B. Grof. 1990. Genetics and plant breeding of *Andropogon gayanus*. Pp. 19-35 *in Andropogon gayanus* Kunth: a Feed Grass for Tropical Acid Soils (J.M. Toledo, R. Vera, C. Lascano and J.M. Lenné, eds.). CIAT, Cali, Colombia.

Miles, J.W. and C.B. do Valle. 1996. Manipulation of apomixis in *Brachiaria* breeding. Pp. 164-177 *in Brachiaria*: Biology, Agronomy, and Improvement (J.W. Miles, B.L. Maass and C.B. do Valle, eds.). CIAT and EMBRAPA/CNPGC, Cali, Colombia.

Miles, J.W., R.J. Clements, B. Grof and A. Serpa. 1990. Genetics and breeding of *Centrosema*. Pp. 245-270 *in Centrosema*: Biology, Agronomy, and Utilization (R. Schultze-Kraft and R.J. Clements, eds.). CIAT, Cali, Colombia.

Miles, J.W., B.L. Maass and C.B. do Valle (eds.). 1996. *Brachiaria*: Biology, Agronomy, and Improvement. CIAT and EMBRAPA/CNPGC, Cali, Colombia.

Minson, D.J. 1980. Nutritional differences between tropical and temperate pastures. Pp. 143-157 *in* Grazing Animals (F.H.W. Morley, ed.). Elsevier Scientific Publishing Company, Amsterdam, Netherlands.

Mohamed-Salem, M.A. 1994. *Stylosanthes* for pasture development: An overview of ILCA's experience in Nigeria. Pp. 17-23 *in Stylosanthes* as a Fallow and Forage Crop (P.N. de Leeuw, M.A. Mohamed-Saleem and A.M. Nyamu, eds.). ILCA, Addis Ababa, Ethiopia.

Nakagawa, H. 1990. Embryo sac analysis and crossing procedure for breeding apomictic guineagrass (*Panicum maximum* Jacq.). JARQ (Japanese Agricultural Research Quarterly) 24(2):163-168.

NAS (National Academy of Sciences). 1979. Tropical legumes: resources for the future. NAS, Washington, USA.

Ndungu, J.N. and D.J. Boland. 1992. *Sesbania* seed collections in Southern Africa. Agroforestry Syst. (Netherlands) 27(2):129-143.

Nemati, N. 1991. Genetic resources of cool season pasture, forage and food legumes for semi-arid temperate environments of Iran. Pp. 89-103 *in* Legume Genetic Resources for Semi-arid Temperate Environments (A. Smith and L.D. Robertson, eds.). ICARDA, Aleppo, Syria.

Ocumpaugh, W.R. and L.E. Sollenberger. 1995. Other grasses for the humid South. Pp. 441-449 *in* Forages: An Introduction to Grassland Agriculture, 5th edn. (R.F. Barnes, D.A. Miller and C.J. Nelson, eds.). Iowa State University Press, Ames Iowa, USA.

Oram, R.N. (comp.). 1990. Register of Australian Herbage Plant Cultivars. Commonwealth Scientific and Industrial Research Organisation (CSIRO), Canberra, A.C.T., Australia.

Paladines, M., O. and J. Delgadillo A. (eds.) 1990. Seminario pastizales andinos: Importancia, producción y mejoramiento. Red de Pastizales Andinos (REPAAN), Cochabamba, Bolivia.

Parsons, J.J. 1972. Spread of African grasses to the American tropics. J. Range Manage. 25:12-17.

Pengelly, B.C. and D.A. Eagles. 1993. Diversity and forage potential of some *Macroptilium* species. Genet. Resour. Commun. 20:1-14.

Pengelly, B.C. and D.A. Eagles. 1995. Geographic distribution and diversity in a collection of the tropical legume *Macroptilium gracile* (Poeppig ex Bentham) Urban. Aust. J. Agric. Res. 46:569-580.

Pengelly, B.C., J.B. Hacker and D.A. Eagles. 1992. The classification of a collection of buffel grass and related species. Trop. Grasslands 26:1-6.

Pernès, J. 1975. Organisation évolutive d'un groupe agamique: la section des Maximae du genre *Panicum* (Graminées). Thèse, Univ. Paris. Mémoires ORSTOM, 75.

Peters, M., S.A. Tarawali and J. Alkämper. 1994a. Evaluation of tropical pasture legumes for fodder banks in subhumid Nigeria. 1. Accessions of *Centrosema brasilianum*, *C. pascuorum*, *Chamaecrista rotundifolia* and *Stylosanthes hamata*. Trop. Grasslands 28:65-73.

Peters, M., S.A. Tarawali and J. Alkämper. 1994b. Evaluation of tropical pasture legumes for fodder banks in subhumid Nigeria. 2. Accessions of *Aeschynomene hystrix*, *Centrosema acutifolium*, *C. pascuorum*, *Stylosanthes guianensis* and *S. hamata*. Trop. Grasslands 28:74-79.

Pizarro, E.A. and L. Coradin (eds.). 1996. Potencial del género *Cratylia* como leguminosa forrajera. Memorias del taller de trabajo sobre *Cratylia* realizado el 19 y 20 de julio de 1995, Brasilia, DF, Brazil. CIAT, Cali, Colombia.

Pizarro, E.A., A.K.B. Ramos and M.A. Carvalho. 1996. Potencial forrajero y producción de semillas de accesiones de *Calopogonium mucunoides* preseleccionadas en el Cerrado brasileño. Pasturas Trop. 18(2):9-13.

Polhill, R.M. and P.H. Raven (eds.). 1981. Advances in Legume Systematics. Part 1. Royal Botanic Gardens, Kew, UK.

Raaflaub, M. and C.E. Lascano. 1995. The effect of wilting and drying on intake rate and acceptability by sheep of the shrub legume *Cratylia argentea*. Trop. Grasslands 29:97-101.

Reed, J.D and H. Soller. 1987. Phenolics and nitrogen utilization in sheep fed browse. Pp. 47-48 *in* Herbivore Nutrition Research (M. Rose, ed.). Australian Society of Animal Production, Occasional Publication. Armidale, A.C.T., Australia.

Reed, J.D., H. Soller and A. Woodward. 1990. Fodder tree and straw diets for sheep: intake, growth, digestibility and the effects of phenolics on nitrogen utilization. Anim. Feed Sci. Technol. 30:39-50.

Reid, R. 1993. Establishing and sharing collections of a valuable global resource. Pp. 77-82 *in* Grasslands for Our World (M.J. Baker, ed.). SIR Publishing, Wellington, New Zealand.

Reid, R., J. Konopka, P.M. Perret and L. Guarino. 1989. Forage germplasm resources for the Mediterranean and adjacent arid/semi-arid areas. Pp. 289-290 *in* Proc. XVI Inter. Grassl. Congr., Nice, France.

Remenyi, J.V. and J.R. McWilliam. 1986. Ruminant production trends in South-eastern Asia and the South Pacific, and the need for forages. Pp. 1-6 *in* Forages in Southeast Asian and South Pacific agriculture (G.J. Blair, D.A. Ivory and T.R. Evans, eds.). ACIAR Proceedings No. 12. ACIAR, Canberra, Australia.

Reynolds, L. and M. Jabbar. 1994. The role of alley farming in African livestock production. Outlook Agric. Oxon 23(2):105-113.

Robertson, L.D., K.B. Singh, W. Erskine and A.M. Abd El Moneim. 1996. Useful genetic diversity in germplasm collections of food and forage legumes from West Asia and North Africa. Genet. Resour. and Crop Evol. 43:447-460.

Ruredzo, T.J. 1991. A minimum facility method for *in vitro* collection of *Digitaria eriantha* ssp. *pentzii* and *Cynodon dactylon*. Trop. Grasslands 25:56-63.

Ruredzo, T.J. and J. Hanson. 1990. Practical applications of *in vitro* techniques to forage germplasm. Pp. 578-591 *in* Utilization of Research Results on Forage and Agricultural By-product Materials as Animal Feed Resources in Africa (B.H. Dzowela, A.N. Said, Asrat Wendem-Agenehu and J.A. Kategile, eds.). Panesa/ARNAB (Pasture Network for Eastern and Southern Africa/African Research Network for Agricultural By-products). Proceedings of the first joint workshop held in Lilongwe, Malawi, 5-9 December 1988. ILCA, Addis Ababa, Ethiopia.

Sah, V.K. and J. Ram. 1989. Biomass dynamics, species diversity and net primary production in a temperate grassland of central Himalaya, India. Tropical Ecol. 30:294-302.

Sato, H., N. Shimizu, H. Nakagawa and H. Matsuoka. 1992. A new registered cultivar "Natsuyutaka" of guineagrass, *Panicum maximum* Jacq. JARQ (Japanese Agricultural Research Quarterly) 25(4):259-266.

Savidan, Y.H., D. Combes and J. Pernès. 1984. *Panicum maximum*. Pp. 6-42 *in* Gestion des ressources génétiques des plantes (J. Pernès, ed.). Vol. 1: Monographies. Agence de Cooperation Culturelle et Technique, Paris, France.

Savidan, Y.H., L. Jank, J.C.G. Costa and C.B. do Valle. 1989. Breeding *Panicum maximum* in Brazil. 1. Genetic resources, modes of reproduction and breeding procedures. Euphytica 41:107-112.

Schmidt, A., C.E. Lascano, B.L. Maass and R. Schultze-Kraft. 1997. An approach to define G x E interaction in a core collection of *Desmodium ovalifolium*. *In* Proc. XVIII Inter. Grassl. Congr., 8-19 June 1997, Winnipeg, Manitoba, and Saskatoon, Saskatchewan, Canada. (In press).

Schultze-Kraft, R. 1985. Development of an international collection of tropical forage germplasm for acid soils. Pp. 109-111 *in* Proc. XV Inter. Grassl. Congr., Kyoto, Japan.

Schultze-Kraft, R. 1986. Natural distribution and germplasm collection of the tropical pasture legume *Centrosema macrocarpum* Benth. Angew. Bot. 60(5-6):407-419.

Schultze-Kraft, R. (comp.). 1990. The CIAT Collection of Tropical Forages. 1. Catalog of Germplasm from Southeast Asia. Working Document No. 76. CIAT, Cali, Colombia.

Schultze-Kraft, R. (comp.). 1991a. La colección de forrajeras tropicales del CIAT. 2. Catálogo de germoplasma de Venezuela. Documento de Trabajo No. 85. CIAT, Cali, Colombia.

Schultze-Kraft, R. (comp.). 1991b. La colección de forrajeras tropicales del CIAT. 3. Catálogo de germoplasma de Centroamérica, México y el Caribe. Documento de Trabajo No. 90. CIAT, Cali, Colombia.

Schultze-Kraft, R. and G. Alvarez. 1984. CIAT tropical forage collection: A status report. Plant Genet. Resour. Newsl. 57:15-18.

Schultze-Kraft, R. and J. Belalcázar. 1988. Germplasm collection and preliminary evaluation of the pasture legume *Centrosema brasilianum* (L.) Benth. Tropic. Agric. (Trinidad) 65(2):137-144.

Schultze-Kraft, R. and G. Benavides. 1988. Germplasm collection and preliminary evaluation of *Desmodium ovalifolium* Wall. Genet. Resour. Commun. 12:1-20.

Schultze-Kraft, R. and R.J. Clements (eds.). 1990. *Centrosema*: Biology, Agronomy, and Utilization. CIAT, Cali, Colombia.

Schultze-Kraft, R. and D.C. Giacometti. 1979. Genetic resources for the acid, infertile savannas of tropical America. Pp. 55-64 *in* Pasture Production in Acid Soils of the Tropics (P.A. Sánchez and L.E. Tergas, eds.). CIAT, Cali, Colombia.

Schultze-Kraft, R., J.A. Arenas, M.A. Franco, J. Belalcázar and J. Ortiz. 1987. Catálogo de germoplasma de especies forrajeras tropicales. Tomos 1-3. CIAT, Cali, Colombia.

Schultze-Kraft, R., N.M.S. Costa and A. Flores. 1984a. *Stylosanthes macrocephala* M.B. Ferr. et S. Costa - collection and preliminary evaluation of a new pasture legume. Tropic. Agric. (Trinidad) 61(3):230-240.

Schultze-Kraft, R., Ha Dinh Tuan and Nguyen Phung Ha. 1993a. Collection of native forage legume germplasm in Vietnam. Pp. 233-234 *in* Proc. XVII Inter. Grassl. Congr., Vol. 1, New Zealand and Australia.

Schultze-Kraft, R., S. Pattanavibul, A. Gani, C. He and C.C. Wong. 1989a. Collection of native germplasm resources of tropical forage legumes in Southeast Asia. Pp. 271-272 *in* Proc. XVI Inter. Grassl. Congr., Nice, France.

Schultze-Kraft, R., R. Reid, R.J. Williams and L. Coradin. 1984b. The existing *Stylosanthes* collection. Pp. 125-146 *in* The Biology and Agronomy of *Stylosanthes* (H.M. Stace and L.A. Edye, eds.). Academic Press, North Ryde, N.S.W., Australia.

Schultze-Kraft, R., R.J. Williams and L. Coradin. 1990. Biogeography of *Centrosema*. Pp. 29-76 *in Centrosema*: Biology, Agronomy, and Utilization (R. Schultze-Kraft and R.J. Clements, eds.). CIAT, Cali, Colombia.

Schultze-Kraft, R., R.J. Williams, L. Coradin, J.R. Lazier and A.E. Kretschmer Jr. 1989b. 1989 World Catalog of *Centrosema* Germplasm. CIAT and IBPGR, Cali, Colombia.

Schultze-Kraft, R., W.M. Williams and J.M. Keoghan. 1993b. Searching for new germplasm for the year 2000 and beyond. Pp. 70-76 *in* Grasslands for Our World (M.J. Baker, ed.). SIR Publishing, Wellington, New Zealand.

Shelton, H.M. and J.L. Brewbaker. 1994. *Leucaena leucocephala* - the most widely used forage tree legume. Pp. 15-29 *in* Forage Tree Legumes in Tropical Agriculture (R.C. Gutteridge and H.M. Shelton, eds.). CAB International, Wallingford, UK.

Shelton, H.M., C.M. Piggin and J.L. Brewbaker (eds.). 1995. *Leucaena* - Opportunities and Limitations (H.M. Shelton, C.M. Piggin and J.L. Brewbaker, eds.). Proceedings of a Workshop, 24-29 January 1994, Bogor, Indonesia. ACIAR Proceedings No. 57. Canberra, Australia.

Simons, A.J. and J.L. Stewart. 1994. *Gliricidia sepium* - a multipurpose forage tree legume. Pp. 30-48 *in* Forage Tree Legumes in Tropical Agriculture (R.C. Gutteridge and H.M. Shelton, eds.). CAB International, Wallingford, UK.

Skerman, P.J. and F. Riveros. 1989. Tropical Grasses. FAO Plant Production and Protection Series No. 23. Food and Agriculture Organization of the United Nations, Rome, Italy.

Skerman, P.J., D.G. Cameron and F. Riveros. 1988. Tropical Forage Legumes (2nd edn.). FAO Plant Production and Protection Series No. 2. Food and Agriculture Organization of the United Nations, Rome, Italy.

Smith, A. and L.D. Robertson (eds.). 1991. Legume Genetic Resources for Semi-arid Temperate Environments (A. Smith and L.D. Robertson, eds.). ICARDA, Aleppo, Syria.

Smith, R.L. 1979. Seed dormancy in *Panicum maximum* Jacq. Trop. Agric. (Trinidad) 56:233-239.

Sorensson, C.T. 1995. Potential for improvement of *Leucaena* through interspecific hybridisation. Pp. 47-53 *in* Leucaena - Opportunities and Limitations (H.M. Shelton, C.M. Piggin and J.L. Brewbaker, eds.). Proceedings of a Workshop, 24-29 January 1994, Bogor, Indonesia. ACIAR Proceedings No. 57. Canberra, Australia.

Souza, F.H.D. and J. Marcos-Filho. 1993. Physiological characteristics associated with water imbibition by *Calopogonium mucunoides* Desv. seeds. Seed Sci. Technol. 21:561-572.

Stace, H.M. and L.A. Edye (eds.). 1984. The biology and Agronomy of *Stylosanthes*. Academic Press, North Ryde, N.S.W., Australia.

Strickland, R.W., R.G. Greenfield, G.P.M. Wilson and G.L. Harvey. 1985. Morphological and agronomic attributes of *Cassia rotundifolia* Pers., *C. pilosa* L., and *C. trichopoda* Benth. potential forage legumes for northern Australia. Aust. J. Exp. Agric. 25:100-108.

Sumberg, J.E. 1985. Collection and initial evaluation of *Gliricidia sepium* from Costa Rica. Agroforestry Syst. 3:357-361.

Talamucci, P. and C. Chaulet. 1989. Influence of constraints on the evolution of forage resources: The Mediterranean basin. Pp. 1731-1740 *in* Proc. XVI Inter. Grassl. Congr., Nice, France.

Tanner, J.C., J.D. Reed and E. Owen. 1990. The nutritive value of fruits (pods and seeds) from four *Acacia* spp. compared with extracted noug (*Guizotia abyssinica*) meal as supplements to maize stover for Ethiopian highland sheep. Anim. Prod. 51:127-133.

Thomas, D. 1994. Tropical pastures and the importance of plant diseases. Pp. 1-17 *in* Diseases of Tropical Pasture Plants (J.M. Lenné and P. Trutmann, eds.). CAB International, Wallingford, UK.

Thomas, D. and B. Grof. 1986. Some pasture species for the tropical savannas of South America. 2. Species of *Centrosema*, *Desmodium*, and *Zornia*. Herbage Abstr. 56(11):511-525.

Thomas, D. and S.L. Lapointe. 1989. Testing new accessions of guinea grass (*Panicum maximum*) for acid soils and resistance to spittlebug (*Aeneolamia reducta*). Trop. Grasslands 23:232-239.

Thomas, R.J., J.K. Ladha and M.B. Peoples. 1995a. Role of legumes in providing N for sustainable tropical pasture systems. Plant and Soil 174:103-118.

Thomas, R.J., C.E. Lascano and J.M. Powell. 1995b. The benefits of forge legumes for livestock production and nutrient cycling in pasture and agropastoral systems of acid-soil savannahs of Latin America. Pp. 277-291 *in* Livestock and Sustainable Nutrient Cycling in Mixed Farming Systems of Sub-Saharan Africa (S. Fernández Rivera, T.O. Williams and C. Renard, eds.). Vol. 2, Technical papers, proceedings of an international symposium held in Addis Ababa, Ethiopia, 22-26 November 1993. ILCA, Addis Ababa, Ethiopia.

Toledo, J.M., A. Arias and R. Schultze-Kraft. 1989. Productivity and shade tolerance of *Axonopus* spp., *Paspalum* spp. and *Stenotaphrum secundatum* in the humid tropics. Pp. 221-222 *in* Proc. XVI Inter. Grassl. Congr., Nice, France.

Toledo, J.M., R. Vera, C. Lascano and J.M. Lenné (eds.). 1990. *Andropogon gayanus* Kunth: a grass for tropical acid soils.

Tovar, O. 1993. Las gramíneas (Poaceae) del Perd. Ruizia 13:1-480.

Tyler, B.F., K.H. Chorlton and I.D. Thomas. 1992. Activities in forage grass genetic resources at the Welsh Plant Breeding Station, Aberystwyth. FAO/IBPGR Plant Genet. Resour. Newsl. 88/89:37-42.

Valle, C.B. do and Y.H. Savidan. 1996. Genetics, cytogenetics and reproductive biology of *Brachiaria*. Pp. 147-163 *in* Brachiaria: Biology, Agronomy, and Improvement (J.W. Miles, B.L. Maass and C.B. do Valle, eds.). CIAT and EMBRAPA/CNPGC, Cali, Colombia.

Valls, J.F.M. 1993. Origem do germoplasma de *Paspalum* disponivel no Brasil para a area tropical. Pp. 69-80 *in* Red Internacional de Evaluación de Pastos Tropicales, RIEPT, 1a. Reunión Sabanas (E.A. Pizarro, ed.). Documento de Trabajo No. 117. CIAT, Cali, Colombia.

Valls, J.F.M., B.L. Maass and C.R. Lopes. 1994. Genetic resources of wild *Arachis* and genetic diversity. Pp. 28-42 *in* Biology and Agronomy of Forage *Arachis* (P.C. Kerridge and B. Hardy, eds.). CIAT, Cali, Colombia.

Westley, S.B. and M.H. Powell (eds.). 1993. *Erythina* in the new and old worlds. Nitrogen Fixing Tree Research Reports (Special Issue). Nitrogen Fixing Tree Association (NFTA), Waimanolo, Hawaii, USA.

Westley, S.B. and J.M. Roshetko (eds.). 1995. *Dalbergia*: proceedings of an international workshop. Institute of Forestry, Nepal, 31 May-4 June 1993. Nitrogen Fixing Tree Association (NFTA), Morrilton, Arkansas, USA.

Whiteman, P.C. and K. Mendra. 1982. Effects of storage and seed treatments on germination of *Brachiaria decumbens*. Seed Sci. Technol. 10:233-242.

Wiersema, J.H., J.H. Kirkbride Jr. and Ch.R. Gunn. 1990. Legume (Fabaceae) nomenclature in the USDA germplasm system. Technical Bulletin No. 1757. USDA, Springfield, VA, USA.

Williams, R.J. 1983. Tropical legumes. Pp. 17-37 in Genetic Resources of Forage Plants (J.G. McIvor and R.A. Bray, eds.). CSIRO, Melbourne, Australia.

Withington, D., N. Glover and J.L. Brewbaker (eds.). 1987. *Gliricidia sepium* (Jacq.) Walp.: management and improvement. Proceedings of a Workshop held at CATIE, Turrialba, Costa Rica, 21-27 June 1987. Special Publication 87-01. Nitrogen Fixing Tree Association (NFTA), Waimanolo, Hawaii, USA.

Index

Note: Page references in **bold** indicate major chapter headings.

Printed in the United States
By Bookmasters

This book is dedicated to all who have caught or want to catch the vision of ambient computing.

Table of Contents